Lecture Notes in Computer Science   9381

Commenced Publication in 1973
Founding and Former Series Editors:
Gerhard Goos, Juris Hartmanis, and Jan van Leeuwen

More information about this series at http://www.springer.com/series/7409

Paul Johannesson · Mong Li Lee
Stephen W. Liddle · Andreas L. Opdahl
Óscar Pastor López (Eds.)

Conceptual Modeling

34th International Conference, ER 2015
Stockholm, Sweden, October 19–22, 2015
Proceedings

 Springer

Editors

Paul Johannesson
Stockholm University
Kista
Sweden

Mong Li Lee
National University of Singapore
Singapore
Singapore

Stephen W. Liddle
Brigham Young University
Provo, UT
USA

Andreas L. Opdahl
University of Bergen
Bergen
Norway

Óscar Pastor López
Universidad Politécnica de Valencia
Valencia
Spain

ISSN 0302-9743   ISSN 1611-3349   (electronic)
Lecture Notes in Computer Science
ISBN 978-3-319-25263-6   ISBN 978-3-319-25264-3   (eBook)
DOI 10.1007/978-3-319-25264-3

Library of Congress Control Number: 2015950889

LNCS Sublibrary: SL3 – Information Systems and Applications, incl. Internet/Web, and HCI

Springer Cham Heidelberg New York Dordrecht London

Printed on acid-free paper

Springer International Publishing AG Switzerland is part of Springer Science+Business Media
(www.springer.com)

Preface

This volume contains a collection of research papers that comprise the technical program of the 34th International Conference on Conceptual Modeling (ER 2015), held in Stockholm, Sweden, during October 19–22, 2015. Also known as the "Entity Relationship" or "ER" conference, this conference series began in 1979 as a forum for researchers from around the world to gather and discuss the nascent field of conceptual modeling. During the past four decades, the conference has been held at an interesting variety of locations, rotating in successive years between Europe, Asia, and the Americas, attracting a diverse international community of scholars.

Chen's seminal work on the entity relationship (ER) model coincided with the emergence of conceptual modeling as a distinct field. Conceptual modeling is an activity that involves capturing aspects of the real world and representing them in the form of a model that can be used for communication; conceptual models typically are used in the development of computer-based information systems. The technical program for ER 2015 included papers addressing a number of current and emerging topics in conceptual modeling.

In response to the call for papers, we received 144 abstracts and 131 full papers. The Program Committee provided at least three reviews for each paper, and on the basis of these reviews we selected 26 full papers (an acceptance rate of 19.85 %) and 19 short papers (a combined acceptance rate of 34.35 %). These papers group into several topical areas, including business process and goal models, ontology-based models and ontology patterns, constraints, normalization, interoperability and integration, collaborative modeling, variability and uncertainty modeling, modeling and visualization of user-generated content, schema discovery and evolution, process and text mining, domain-based modeling, data models and semantics, and applications of conceptual modeling.

We express gratitude to all who helped make ER 2015 a success. It required the significant efforts of many people to make this conference possible. We thank the 91 Program Committee members along with the numerous external referees who reviewed and discussed the submitted manuscripts. In addition, 18 members of the Program Committee served on the senior review board to help oversee the review process. We especially thank the authors who took the time to carefully write up the results of their research and submit papers for consideration. It is abundantly clear that a great deal of high-quality and interesting work continues to be done by researchers from many corners of the globe.

August 2015

Paul Johannesson
Mong Li Lee
Stephen W. Liddle
Andreas L. Opdahl
Óscar Pastor López

Conference Organization

General Co-chairs

Paul Johannesson   Stockholm University, Sweden
Andreas L. Opdahl   University of Bergen, Norway

Technical Program Co-chairs

Mong Li Lee   National University of Singapore, Singapore
Stephen W. Liddle   Brigham Young University, USA
Oscar Pastor   Universidad Politécnica de Valencia, Spain

Organization Co-chairs

Maria Bergholtz   Stockholm University, Sweden
Jelena Zdravkovic   Stockholm University, Sweden
Eric-Oluf Svee   Stockholm University, Sweden

Workshop Co-chairs

Manfred Jeusfeld   Tilburg University, The Netherlands
Kamlakar Karlapalem   CDE, IIIT Hyderabad, India

Tutorial Co-chairs

Eva Söderström   University of Skövde, Sweden
Sandeep Purao   Bentley University, USA

Panel Co-chairs

Bernhard Thalheim   University of Kiel, Germany
Andreas L. Opdahl   University of Bergen, Norway

Demonstration Chair

Giancarlo Guizzardi   Federal University of Espirito Santo, Brazil

PhD Colloquium Chairs

Michael Rosemann   Queensland University of Technology, Australia
Lois Delcambre   Portland State University, USA

Industry and Student Chairs

Vandana Kabilan   Accenture, Sweden
Parisa Aasi   Stockholm University, Sweden

Treasurer and Registration Chair

Martin Henkel   Stockholm University, Sweden

Publicity Co-chairs

Ilia Bider   Stockholm University, Sweden
Erik Perjons   Stockholm University, Sweden

Sponsorships and Exhibits Chair

Gustaf Juell-Skielse   Stockholm University, Sweden

Technical Chair

Iyad Zikra   Stockholm University, Sweden

Steering Committee Liaison

Heinrich C. Mayr   Alpen-Adria-Universität Klagenfurt, Austria

Program Committee

Jacky Akoka   CNAM and TEM, France
Gove Allen   Brigham Young University, USA
Yuan An   Drexel University, USA
Joao Araujo   Universidade Nova de Lisboa, Portugal
Zhifeng Bao   University of Singapore, Singapore
Sourav S. Bhowmick   Nanyang Technological University, Singapore
Sandro Bimonte   IRSTEA, France
Shawn Bowers   Gonzaga University, USA
Stephane Bressan   National University of Singapore, Singapore
Stefano Ceri   Politecnico di Milano, Italy
Roger Chiang   University of Cincinnati, USA
Dickson Chiu   University of Hong Kong, SAR China
Byron Choi   Hong Kong Baptist University, SAR China
Fabiano Dalpiaz   Universiteit Utrecht, The Netherlands
Karen Davis   University of Cincinnati, USA
Valeria De Antonellis   University of Brescia, Italy
Sergio De Cesaro   Brunel University, USA

Lois Delcambre	Portland State University, USA
Oscar Diaz	University of the Basque Country, Spain
Gill Dobbie	University of Auckland, New Zealand
Johann Eder	Alpen Adria Universität Klagenfurt, Austria
Sergio España	PROS Research Centre, Spain
Xavier Franch	Universitat Politècnica de Catalunya, Spain
Avigdor Gal	Technion-Israel Institute of Technology, Israel
Sepideh Ghanavati	Luxembourg Institute of Science and Technology, Luxembourg
Aditya Ghose	University of Wollongong, Australia
Paolo Giorgini	University of Trento, Italy
Georg Grossmann	University of South Australia, Australia
Giancarlo Guizzardi	Federal University of Espirito Santo, Brazil
Renata Guizzardi	Federal University of Espirito Santo, Brazil
Arantza Illarramendi	Basque Country University, Spain
Matthias Jarke	RWTH Aachen University, Germany
Manfred Jeusfeld	Tilburg University, The Netherlands
Ivan Jureta	University of Namur, Belgium
Dimitris Karagiannis	University of Vienna, Austria
Kamlakar Karlapalem	CDE, IIIT Hyderabad, India
David Kensche	RWTH Aachen University, Germany
Alberto Laender	Federal University of Minas Gerais, Brazil
Dik Lun Lee	Hong Kong University of Science and Technology, China
Julio Cesar Leite	PUC-Rio, Brazil
Guoliang Li	Tsinghua University, China
Tok Wang Ling	National University of Singapore, Singapore
Sebastian Link	The University of Auckland, New Zealand
Fred Lochovsky	The Hong Kong University of Science and Technology, China
Pericles Loucopoulos	Harokopio University of Athens, Greece
Hui Ma	Victoria University of Wellington, New Zealand
Wolfgang Maaß	Saarland University, Germany
Heinrich C. Mayr	Alpen-Adria-Universität Klagenfurt, Austria
Jan Mendling	Wirtschaftsuniversität Wien, Austria
Haralambos Mouratidis	University of Brighton, UK
John Mylopoulos	University of Toronto, Canada
Wilfred Ng	HKUST, SAR China
Antoni Olivé	Universitat Politècnica de Catalunya, Spain
José Palazzo M. De Oliveira	Universidade Federal do Rio Grande do Sul, Brazil
Jeffrey Parsons	Memorial University of Newfoundland, Canada
Chris Partridge	Brunel University, BORO Solutions, USA
Zhiyong Peng	Wuhan University, China
Barbara Pernici	Politecnico di Milano, Italy
Geert Poels	Ghent University, Belgium
Henderik Proper	Public Research Centre Henri Tudor, Luxembourg

Christoph Quix	RWTH Aachen University, Germany
Jolita Ralyté	University of Geneva, Switzerland
Sudha Ram	University of Arizona, USA
Iris Reinhartz-Berger	University of Haifa, Israel
Stefano Rizzi	University of Bologna, Italy
Colette Rolland	Université Paris 1 Panthéon—Sorbonne, France
Antonio Ruiz-Cortés	University of Seville, Spain
Bernhard Rumpe	RWTH Aachen University, Germany
Mehrdad Sabetzadeh	University of Luxembourg, Luxembourg
Motoshi Saeki	Tokyo Institute of Technology, Japan
Camille Salinesi	Université Paris 1 Panthéon—Sorbonne, France
Peretz Shoval	Ben-Gurion University, Israel
Fernando Silva Parreiras	FUMEC University, Brazil
Pnina Soffer	University of Haifa, Israel
Il-Yeol Song	Drexel University, USA
Veda Storey	Georgia State University, USA
Arnon Sturm	Ben-Gurion University of the Negev, Israel
David Taniar	Monash University, Australia
Ernest Teniente	Universitat Politècnica de Catalunya, Spain
James Terwilliger	Microsoft Corporation, USA
Bernhard Thalheim	University of Kiel, Germany
Aibo Tian	The University of Texas at Austin, USA
Juan Trujillo	University of Alicante, Spain
Panos Vassiliadis	University of Ioannina, Greece
Gerd Wagner	Brandenburg University of Technology at Cottbus, Germany
Barbara Weber	University of Innsbruck, Austria
Roel Wieringa	University of Twente, The Netherlands
Carson Woo	University of British Columbia, Canada
Huayu Wu	Agency for Science, Technology and Research, Singapore
Eric Yu	University of Toronto, Canada
Esteban Zimányi	Université Libre de Bruxelles, Belgium

Additional Reviewers

Adhinugraha, Kiki	Cabanillas, Cristina	Fernandez, Pablo
Anthonysamy, Pauline	Cappiello, Cinzia	Gallinucci, Enrico
Argyropoulos, Nikos	Costal, Dolors	García, José María
Baldacci, Lorenzo	de La Vara, Jose Luis	Golan, Sapir
Berges, Idoia	Dhiman, Ruchi	Graziani, Simone
Bhamborae, Mayur	Di Ciccio, Claudio	Greifenberg, Timo
Bianchini, Devis	Fan, Zhe	Hartmann, Sven
Bork, Dominik	Fang, Qiong	Heim, Robert
Burattin, Andrea	Feltus, Christophe	Jenkins, Jeffrey

Journaux, Ludovic
Juric, Damir
Koehler, Henning
Köpke, Julius
Leopold, Henrik
Li, Yuchen
Lopez, Lidia
Lucassen, Garm
Ma, Qin
Malinova, Monika

Manousis, Petros
Melchiori, Michele
Memari, Mozhgan
Mena, Eduardo
Niño, Mikel
Pittke, Fabian
Plataniotis, Georgios
Quer, Carme
Roth, Alexander
Ruiz, Marcela

Sawczuk Da Silva,
    Alexandre
Schulze, Christoph
Svirsky, Jonathan
Tantouris, Nikolaos
Trinidad, Pablo
Valverde, Francisco
von Wenckstern, Michael
Walch, Michael
Woodfield, Scott

Sponsoring Institution

Department of Computer and Systems Sciences (DSV), Stockholm University, Sweden

Keynote Abstract

Conceptual Modelling in the Digital Age

Michael Rosemann

Queensland University of Technology
Science and Engineering Faculty, Information Systems School
2 George Street, Brisbane Qld 4000, Australia
m.rosemann@qut.edu.au

Abstract. The digital transformation of the society has disruptive implications on all sectors, and conceptual modelling will be no exception. This keynote covers some of the emerging impacts digital opportunities will have on the way conceptual modelling is conducted and the type of conceptual models needed. First, the move from an age of automation to an age of digitisation shifts the attention from corporations to people leading to a higher demand for digital identities, "personal conceptual models" and a requirement to capture birth-to-death value chains. Second, the pressure to reduce innovation latency will require improvement patterns and predictive capabilities demanding "proactive conceptual models" which will serve as recommender systems for their users. Third, a shift towards "real-time conceptual modelling" will increase the relevance of mining and context-aware solutions leading to a higher alignment between conceptual models and the reality they are depicting. The presentation ends with some further predictions on the nature and research opportunities of "conceptual modelling 3.0".

Contents

Ontology Patterns

Constraints

Normalization

Interoperability and Integration

Collaborative Modeling

Variability and Uncertainty Modeling

Modeling and Visualization of User Generated Content

Schema Discovery and Evolution

Process and Text Mining

Applications and Domain-based Modeling

Data Models and Semantics

Keynotes

Why Philosophize; Why not Just Model?

Brian Henderson-Sellers[✉]

School of Software, University of Technology Sydney, Broadway, Australia
brian.henderson-sellers@uts.edu.au

Abstract. Conceptual modelling relies on the availability of good quality modelling languages (MLs). However, many of these MLs have not been created in any well-organized and consistent manner leading to identified flaws and ambiguities. These result, in part, from the lack of an ontological commitment, the neglect of language use and speech act theory and possibly incoherent philosophical underpinnings. These various disciplines are examined in the context of their potential integration into the creation of the next generation of modelling languages.

Keywords: Modelling languages · Conceptual modelling · Foundational ontologies · Language use · Set theory

1   Context

Conceptual modelling requires a good quality modelling language in order to explore and document the models produced. It is arguable whether conceptual modelling should focus on creating an ontological representation of the real world or, rather, focus on a representation of the real world as scoped in the requirements statement elicited by the requirements engineering from the various stakeholders – including of course the end users. Such a focus, probably the most dominant in information systems development – perhaps more correctly called an epistemology-first approach – only includes those aspects (properties or characteristics) of the concepts included in the model (for example, as derived from the requirements specification) and may offer greater possibility of attainment than an ontological approach that is aimed at representing the 'truth' of the real world [1]. In current modelling languages, like the UML, these two views – the ontological versus the epistemological – are often confounded. Indeed, the two tend to emerge together (e.g. see discussion in [2]) so the question may be which has dominance in the modeller's way of thinking: ontology-focussed first or epistemology-focussed first.

In contrast to this recognized dichotomy of ontological versus epistemological focus, many modelling languages, whilst claiming general applicability across the whole of the software development lifecycle, are actually much more focussed on documenting designs that can be coded (manually or automatically via MDE). As an exemplar of a so-called General Purpose Modelling Language (GPML), we take the Unified Modeling Language (UML [3]). Its heritage clearly is from design and programming with many programming idioms in it. Despite this bias, it has been argued to be useful for conceptual modelling [4] although many in the requirements engineering community would disagree.

© Springer International Publishing Switzerland 2015
P. Johannesson et al. (Eds.): ER 2015, LNCS 9381, pp. 3–17, 2015.
DOI: 10.1007/978-3-319-25264-3_1

We therefore conclude that the ontological commitment of a GPML like UML is ambiguous or, at best, unstated and unrecognized [5]. In the project overviewed here, we therefore seek to identify the underpinning ontological approach combined with an assessment of the potential contribution of language use (see early results in [6]), mathematics [7] and philosophy [8–10].

In the rest of the paper, we outline briefly basic notion of concepts (Sect. 2) and foundational ontologies (Sect. 3). The contribution of language use and speech act theories leads to modifications to the selected foundational ontology to support social or non-materialistic concepts in Sect. 4. The paper concludes in Sect. 5 with a brief discussion including potential benefits of such an integration as proposed here; whilst Sect. 6 offers our conclusions.

2 Conceptual Modelling Starts with Concepts

By its very name, one could deduce that conceptual modelling uses concepts. However, the notion of concepts from general language use and from philosophy is not clear-cut (e.g. [11]). Three main 'definitions' are discussed in detail in [9] in our current context and summarized here.

2.1 Concepts in Philosophy

The term 'concept' in philosophy has several different adherent schools. Firstly, we could consider concepts as mental constructs – probably the approach visualized in much of the software engineering literature, which relates concepts and conceptual modelling to Ullmann's triangle [12]. Here, concepts may be thought of as either (1) representations – for instance, the concept DOG is a general representation of a dog: its state and behaviour as well as its identity. Alternatively, (2) concepts may be regarded as abilities (still a mental conceptualization). In this case, the concept DOG is seen as an ability to tell dogs from non-dogs. Alternatively, there is a school of thought that regards concepts as being independent of the human mind. This leads to (3) concepts-as-meanings. In this context, concepts are meanings, or more technically, they are Fregean senses [13]. The concept DOG is thus the meaning of the general term 'dog'.
[For detailed discussion of these, we refer the reader to [9].]

Finally, we note that conceptual models represent things – possibly from one of two perspectives (see also discussion of epistemic divergence in [14]:

(1) things-as-they-are. We call these ontic models. The content of an ontic model is independent of an observer i.e. this is the domain of basic ontologies like that of Bunge [15, 16]. For example, every person has a date of birth (dob). This should not be nullable since permitting an attribute value to be null states that it is a possibility that such an attribute value does not exist. This is the ontic modelling approach.
(2) things-as-we-know-them. These are so-called epistemic models. These models capture knowledge about reality and are therefore subjective. Using the same example, if the software developer chooses to allow the dob to be unknown i.e. they choose to ignore information that *does* exist, then this gives an example of an epistemic way of modelling.

Clearly there are modelling situations where an ontic approach is needed and others where an epistemic approach is more useful. A GPML like UML cannot (or at least does not currently) capture these nuances for representing missing data: null for ontologically absent or unknown for epistemically absent data. In practice, most models are epistemic [17].

2.2 Concepts in Ontology Engineering

Initially the realm of philosophers, ontology, as a component part of philosophy, has become of interest to information systems developers and software engineers e.g. [1, 18–22]. This adoption into software engineering, essentially by non-philosophers/non-ontologists, has introduced some misdirection for later researchers. For instance, Smith [1] argues strongly that the notion of 'concept' has been much misused, even "exerting a damaging influence on the progress of ontology" [11]; whilst other ontologists suggest that the widespread adoption of the definition of ontology by Gruber [23] that ontology is "a specification of a conceptualization" has led the field astray. In addition, the common neglect of socially-constructed concepts, representing largely intangible entities[1], has focussed unnecessarily (from an IS development viewpoint) on materiality i.e. entities made of stuff (substance or material). Finally, there is a general confounding in the UML world of whether UML classes represent concepts in an information system or tangible objects in the real world (as noted above). Both need to be represented but clarification is required (see also discussion in [6]). It is a common mistake in modelling, especially with UML, to mistake a model entity for a real world entity and vice versa [24].

Whilst ontological engineering tends to phrase the underpinning approach in terms of types and individuals (of those types), philosophical treatises link particulars (the name given to individuals that exist in reality), to universals (which signify what corresponding instances have in common) (e.g. [11]). The corresponding representations in the software domain are shown in Fig. 1.

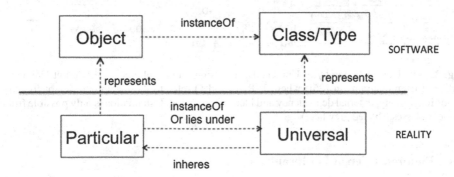

Fig. 1. Linking OO terminology to philosophical terminology

---

[1] Regarding social entities as non-materialistic comes from the language use community. In contrast, in both the endurant component of UFO and in the perdurantist BORO foundational ontologies, all such entities are considered to be materialistic.

3   Foundational Ontologies

Foundational or upper level ontologies provide something that could be considered, loosely, as a parallel to the metamodel that defines a GPML like UML [25]. Such metamodels tend to mix representations of the real world and representations of model characteristics, whilst a foundational ontology only addresses the former. One example of a foundational ontology that is targeted specifically at conceptual modelling is the Unified Foundational Ontology (UFO) [26], an overview of which is given in Fig. 2 for the Endurant part of the ontology – the most dominant part and later called UFO-A.

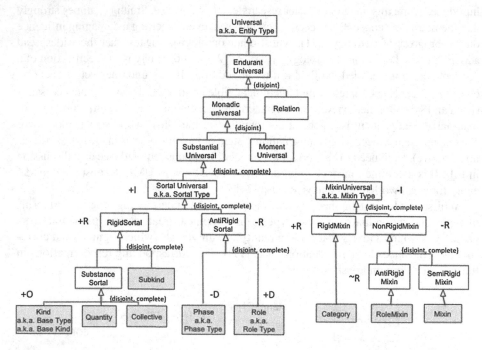

Fig. 2. The UFO-A hierarchy of Universals, focussing on the subtypes of Endurant Universal (adapted from diagrams in [26]) where I, R, D and O refer to the metaproperties of identity condition, rigidity, external dependency and identity provision. Instantiation is only possible from the leaf classes (shaded grey here)

3.1   Endurants Versus Perdurants

There are two kinds of persistent entities: endurants and perdurants. An endurant is an entity that is wholly present whenever it is present. Typically it is a material object (or class representation thereof). Examples are things like Person, Chair, Train, Postage Stamp. A perdurant on the other hand is an entity with a spatio-temporal extent such that at any time-slice the entity may 'look different' even though it can be regarded as having a holistic identity and presence. In other words, a perdurant evolves over time e.g. [27]. Typical examples include events such as a birthday party or a cricket match.

Although Fig. 2 depicts only EndurantUniversal and its subtypes, there is a second strand of UFO, called UFO-B (e.g. [28]), that is concerned with PerdurantUniversal (there are also upper level models describing the corresponding structures for particulars – both endurant and perdurant). Perdurants are also studied in detail by e.g. [29] in a recent survey that builds on earlier philosophical treatises – a philosophical approach often named '4D ontology'.

Furthermore, the temporality of roles and base kinds is distinctly different. Viewed from a philosophical stance, in an endurantist approach, as sometimes tied to set theory, the members of a set are essentially fixed. This means that when an individual leaves a set a new set must be constructed to model this change. On the other hand, when adopting a perdurantist philosophy (for example, BORO: [30]), temporality is depicted explicitly i.e. all particulars are modelled as changing over time. Furthermore, an identity criterion becomes a non-issue because it is defined simply by the spatio-temporal extent of the entity in question. In set theory terms, the members of the extension are thus immutable, making the perdurantist approach more amenable than an endurantist approach to a set theoretic representation.

In modelling domains with no infrastructure for representing change over time, one particular contentious kind of entity is the role or roletype; this is intended to capture where a particular temporarily acquires one or more characteristics and may later shed them. This represents some kind of dynamic and non-rigid behaviour such that endurantist modelling can have different semantics than modelling roles using a fully perdurantist approach [31]. In a perdurantist model, as noted above, temporality is explicitly depicted (e.g. [29]) such that one need not ask (as one would in an endurantist setting) what happens when a person loses a leg in an accident since, in a perdurantist setting, change is to be expected. Identity is no longer a difficult issue (as discussed for instance in [10]) but is defined simply by the spatio-temporal extent of the entity in question.

A second concept that is highly relevant and susceptible to a 4D analysis is that of mereology – whole-part relationships. Sider [29] discusses this topic in detail in his 4D treatise since this essentially perdurantist approach models not only parts but also wholes as temporal-spatial extents. In other words, parts have a temporality that is hard to capture in a 3D or endurantist approach (as discussed for instance in [32–34]) without resorting to complex structures including time-stamping etc.

3.2   Importance of Identity for Foundational Ontologies

As noted above and in Fig. 2, the identity of an object is used as a discriminant in constructing the UFO and similar foundational ontologies. It is important because, in UFO at least, only base kinds can supply identity although all other 'leaf' classes in Fig. 2 need to acquire it somehow – they are labelled as having the ability to 'carry identity'. Furthermore, when UFO is understood in terms of it being a UML class diagram, it should be noted that a leaf class like RoleType or PhaseType can only acquire identity by subsumption from one of its *direct* ancestors (i.e. NOT from a base kind) or by an association relationship to a base kind e.g. Fig. 3 and [6, 35]. This leads to the need to identify a "criterion of identity" (often extremely difficult in practice). However, the meaning of 'identity' from a software modelling perspective is fraught with ambiguity e.g. [10]. Identity could be regarded in terms of (i) a set of properties possessed by an entity, (ii) the substantial-extensional model; (iii) an enduring property of an entity; (iv) use of a rigid designator [36, 37]. These various approaches are most relevant to substantial entities – indeed the substantially-focussed UFO reflects each of these in various aspects of its exposition. Furthermore, language use studies, which lead to a focus on non-material (institutional) entities, also stress the need to supply (and carry) identity.

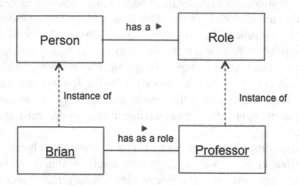

Fig. 3. Representing the role-base type relationship with an association not a generalization

3.3   Ontology-Based Conceptual Modelling

As noted above, foundational ontologies (e.g. BORO [30] and UFO [26]) have been developed to support conceptual modelling and to act as a reference for the design of new conceptual modelling languages. Before that can be carried out successfully, observations on foundational ontologies, and UFO in particular, are in order. UFO-A is based on an endurantist model – this may or may not be a good assumption given the need to model a sociomaterial world – alternatives being proposed recently by e.g. [38].

Rigidity is a key idea in foundational ontologies like UFO, especially when defining the important BaseKind. This is found necessary in UFO because of the need for a base kind that is not only a sortal (can be counted) but also is unchanged in all possible worlds. Rigidity is used as a key determinant in the UFO to divide Sortals into Rigid and Anti-Rigid Sortals. However, we suggest that a rigid designator should not be seen as being tied to a substantial thing but as a language construct used to identify an object because of its (Fregean) sense in different spatial-temporal and social contexts e.g. Mark Twain or Samuel Clemens.

Using this endurantist, sortal-based ontology leads to a paradox regarding how a roletype (and phasetype) can acquire its necessary identity. Whilst preventing a base kind from an inheriting from a role type, Guizzardi and Wagner [39] introduce a textual overwrite that permits a role type to inherit from a base kind (Fig. 4). Although done textually, this is in effect an introduction of an association relationship labelled 'subtypeOf' between RoleType and BaseType (a.k.a. BaseKind) – see Fig. 5 and [35].

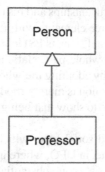

Fig. 4. A typical generalization relationship between a role type and a base type in UFO

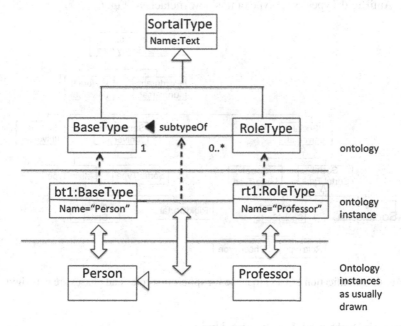

Fig. 5. The necessary introduction of a Sub-TypeOf relationship linking RoleType to BaseType in the UFO (modified from [35])

de Cesare et al. [31] examine the different consequences of adopting a perdurantist ontology (BORO) compared to an endurantist one (UFO). The choices in these two lead to quite a different account of what entities exist. Using an example of John as a student who, during his studentship, takes on a year-long role as the President of the Student Union (PSU), in UFO there is just John (who is identical to John qua Student and John qua PSU). In other words, the instance of Person and Student is the same object – John – across all possible worlds in which John exists. Hence, another mechanism is needed to precisely disentangle John, his studentships and his presidencies in this world and across possible worlds because these things all share the same identity and are the same thing.

In contrast, in BORO John, his studentships and his presidencies are all distinct objects, even across worlds, i.e. there are three elements with their own identity (spatio-temporal extents): John, John's Student state and John as PSU. Moreover, the explanatory power of BORO is enhanced by the temporal whole-part relationships that underpin its mereology. Roles of roles are modelled simply by adopting the whole-part pattern of the foundational ontology. In UFO, since PSUOccupation is merely modelled as a subtype of Student, then the representation lacks in being able to show that being a PSU is the state of a student and not a student.

In UFO Roletype is an antirigid sortal type, which is a subtype of SubstantialUniversal. Hence roles are made of matter in UFO; where in reality, roles are argued by many to be social i.e. that they are immaterial (non-substantial) created by agreed social acceptance. This led to a recommendation [40] to add a SocialUniversal entity type to UFO and locating AntiRigid Types as subtype of this new metaclass (Fig. 6).

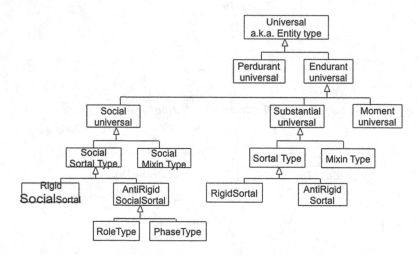

Fig. 6. Modification of UFO to make it explicit that roles and phases are anti-rigid

4   Speech Acts and Language Use

From speech act theory we must first bring an object "into existence" (i.e. into the conversation) before we can talk about classification and property values. For example, we might instantiate a particular by a statement like "Fido is a dog". This is an existential proposition that instantiates a substantial object using the identifier 'Fido' in the Dog class. Having effectively instantiated Fido in this manner, we can now ascribe moment universals to the substantial object (Table 1 and Fig. 7).

Table 1. Speech act introducing Dog (after [6])

Dog's name	Dog's height (cm)	Dog's colour
Fido	50	Brown

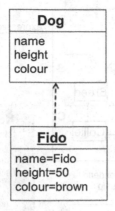

Fig. 7. UML class diagram (M1 level) representation of Table 1 (after [6])

However, having established Fido as a dog, when we encounter a further speech act that states "Fido is a Collie", we have a problem since Fido is an instance of the Dog concept and cannot therefore be (also) a member of the Collie class – since in IS modelling one cannot be an instance of two classes simultaneously – in reality one can of course have multiple categorization but one can conjecture that this usually depicts categorization at different points in time and/or for different purposes. So, rather than the depiction in Table 2, we must have instead (Table 3) a depiction in which the breed is not a supertype but an instance of a moment universal (Fig. 8).

Table 2. Speech act purporting to introduce Collie (after [6])

Collie
Collie's name
Fido

Table 3. Speech act adding a dog's breed (after [6])

Dog			
Dog's name	Dog's height (cm)	Dog's colour	Dog's breed
Fido	50	Brown	Collie

The introduction of language use and speech act theory into the conceptual modelling and modelling language definition space offers solution to the ontological/linguistic paradox [41] as well as suggesting the replacement of the OMG level M2 metamodel by "modelling language" – which may or may not be realized by a metamodel (Fig. 9).

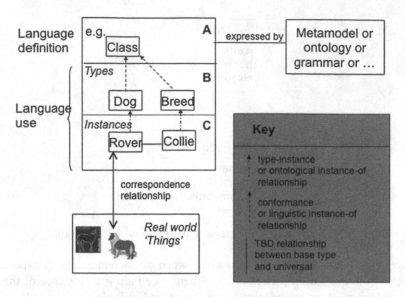

Fig. 8. Complete multilevel framework based on language use – to replace the strict metamodelling architecture of the OMG when modelling in information systems development and software engineering (after [7]) (with kind permission of Springer Science + Business Media)

5   Discussion and Perceived Benefits

A modelling language can be thought of as a collection of useful semantic representations. Even with a poor quality modelling language, expert modellers are able to produce high quality analyses and designs by selecting an appropriate representation. However, in contrast, less experienced modellers can have the quality of their work enhanced by having more and better semantically based notations made available to them in the modelling language (although too much 'choice' may have an adverse effect). For example, current metamodels for modelling languages (e.g. UML V2) ignore ontologically specified classes that are subtypes of Classifier. This means that the most granular visualization of an entity is that of a class. Whilst roles, phases etc. can of course be depicted with a stereotype (assuming that the stereotype is clearly and precisely defined before use), elevating these subtypes to 'first class citizens', standardized across all application domains, could provide an additional tool for modellers in describing both the real world and the software design – permitting representation of entities that are NOT time independent. In addition, as noted earlier, contemporary MLs often conflate representation of objects and concepts (classes) within the information system with representations of things in the real world (that the software classes are mapped to) e.g. [24].

Current modelling languages seldom state either their ontological commitment or their philosophical basis. Although natural language is used to add constraints to the language, and indeed a software modelling language IS-A language, the whole area of language use, speech acts and even linguistics is totally ignored in most cases. Language use introduces the notion that meaning is not fixed in all contexts but rather is in fact

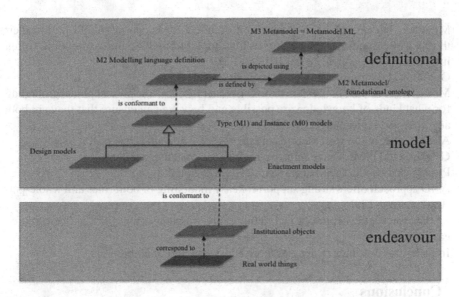

Fig. 9. Proposed revision of the OMG multi-level architecture, placing the modelling language and not the metamodel at the 'M2' level in the multilevel stack (now restricted to three levels with a linkage to a second multilevel stack – here shown as having only two levels) (after [35])

context-dependent. For example, while in a general information system, the class Dog would be a base kind and the class Collie an instance of MomentUniversal, in other contexts, such as a dog show, Collie could be the main base kind in the system [6]. However, if an anti-rigid type is thus elevated, the resultant information system becomes highly constrained to that particular problem space i.e. reuse and extensibility became difficult if not impossible. In another problem space, that of university administration, an information system that deals with students may well be tempted to take Student as an instance of Base Kind. That will work initially but when a decision is made to extend the student information system to also cater for, say, alumni or employment records post-graduation, then a new system will need to be constructed since the chosen onto-logical basis is no longer appropriate.

The replacement of 'analysis' and 'design' by 'modelling' in the early days of object technology (valid at the time) now reappears since the requirements engineering community and the conceptual modelling community – and indeed the CIM of the MDE approach – are much more aligned with 'analysis' than with 'design' whilst, at the same time, a GPML like UML is actually a design language with restricted applicability to 'analysis'. The question thus arises as to whether we (the modelling community) should aim for one all-purpose ML ([42] and others suggest this to be a vain hope) or a suite of languages spanning analysis and design (as in ConML [43, 44]) – although in the latter case we also need a mapping between each pair of languages recommended for use across the full lifecycle.

As noted in [5], some of the perceived benefits of an integrative or holistic approach to conceptual modelling languages include

- Modellers becoming aware of how philosophical ideas can or should inform their thinking by providing a context for its logic and illuminating the assumptions (often implicit) that are adopted e.g. [2]
- Ontological commitment is made explicit e.g. [1]
- Modellers are aware of their epistemology that provides "the philosophical grounding for what kinds of knowledge are possible and how we can ensure that they are both adequate and legitimate" [2]
- Alignment is made with theories of language use and speech acts e.g. [6]
- GPMLs and DSMLs have improved semantics (and are standards)
- Process and product modelling seamlessly integrated
- A 4D approach offers strong support for easy modelling of both roles and whole-part structures e.g. [29]
- Social issues and soft issues incl. temporality, vagueness e.g. [45] can be captured easily
- Industry use becomes easier and more consistent and unambiguous

6 Conclusions

Philosophical thinking can complement and augment the modelling experience on many (but not necessarily all) occasions – resolving ambiguities and ensuring model coherence. It is perhaps self-evident that the use of good models can significantly help in producing good quality software and that production is helped by the availability of high quality modelling languages (particularly for novice modellers). However, we have demonstrated that current ML approaches have many flaws e.g. [35] and incompatibilities with knowledge outside of SE (philosophy, language use, mathematics). In addition, there has been little philosophical influence on conceptual modelling to date e.g. [8].

The research synopsized here suggests strongly that modellers, and particularly model language developers and tool builders, need to be aware of the contributions that could be made by foundational ontologies, languages use and speech act theory as well as appreciating philosophical constraints and opportunities. In particular, understanding of the endurantist versus perdurantist viewpoint can lead to different models e.g. [31]. Similarly, recognition of whether data in an information system needs an epistemic or an ontic description, especially with regard to missing or unavailable data, needs attention that is currently not well supported in GPMLs like UML.

Role modelling provides an interesting and illuminating comparison for both the endurantist/perdurantist dichotomy as well as for an investigation of how a role, when modelled as an anti-rigid sortal, can acquire its identity – leading to a proposed relocation of RoleType in a foundational ontology like the UFO from a subtype of SubstantialUniversal to being a subtype of SocialUniversal (as proposed in Fig. 6). Notwithstanding, Eriksson et al. [6] also identified situations in which a role can in fact be equated to an instance of a BaseKind (from a foundational ontology) – although such a modelled universe of discourse may be limited being too specialized and specific to the current information systems development and precluding simple extension of the software to other application domains.

Furthermore, modern GPMLs take little cognizance of language use principles by which an entity needs to be 'introduced' into the modelling conversation – current UML modellers tend to assume this introduction has been done a priori. Indeed, replacing the M2 level of the OMG architecture (the so-called metamodel level) by a 'language definition' (which may be reified by a grammar, a metamodel or a foundational ontology for example), as shown in Fig. 9, offers a simpler framework with its three clearly defined 'layers'.

However, it should be noted that integrating philosophical thoughts into conceptual modelling languages provides a challenge to conceptual modellers since the discipline of philosophy is organized somewhat differently to that commonly encountered in information technology. Authority cannot be built upon such that there is no single philosophy that can be considered to be best or "correct" for conceptual modelling (an original but, it turns out, misdirected aim of this research). Instead, we recommend the utilization of philosophical insights as a resource to engender clarification and removal of ambiguities.

Successful integration of these differing viewpoints should encourage the development of higher quality in the next generation of modelling languages. The final challenge is thus to all conceptual modellers: Can we, as a community, create and utilize a new language architecture that simultaneously satisfies UML modellers, metamodellers, language use constraints, foundational ontologies and underpinning mathematical theories and is philosophically sound as well as being implementable in software support tools?

Acknowledgements. I wish to thank, in alphabetical order, Pär Ågerfalk, Sergio de Cesare, Owen Eriksson, Cesar Gonzalez-Perez, Chris Partridge and Greg Walkerden who have helped me co-author many of the papers cited herein as well as providing useful and constructive comments on an earlier draft of this keynote paper. I also acknowledge many discussions with Giancarlo Guizzardi over the past several years.

References

1. Smith, B.: Ontology. In: Floridi, L. (ed.) Blackwell Guide to the Philosophy of Computing and Information, pp. 155–166. Blackwell, Oxford (2003)
2. Crotty, M.: The Foundations of Social Research. Allen & Unwin, St Leonards (1998)
3. OMG: OMG Unified Modeling LanguageTM (OMG UML), Superstructure, Version 2.4.1, formal/2011-08-06, p. 748 (2011)
4. Falkovych, K., Sabou, M., Stuckenschmidt, H.: UML for the semantic web: transformation-based approaches. In: Omelayenko, B., Klein, M. (eds.) Knowledge Transformation for the Semantic Web, pp. 92–106. IOS Press, Amsterdam (2003)
5. Henderson-Sellers, B., Gonzalez-Perez, C., Eriksson, O., Ågerfalk, P.J., Walkerden, G.: Software modelling languages: a wish list. In: Proceedings of 2015 IEEE/ACM 7th International Workshop on Modeling in Software Engineering, pp. 72–77, IEEE, Los Alamitos (2015)
6. Eriksson, O., Henderson-Sellers, B., Ågerfalk, P.J.: Ontological and linguistic metamodelling revisited – a language use approach. Inf. Soft. Technol. **55**(12), 2099–2124 (2013)

7. Henderson-Sellers, B.: On the Mathematics of Modelling, Metamodelling, Ontologies and Modelling Languages, p. 106. Springer, Heidelberg (2012). Springer Briefs in Computer Science

8. Henderson-Sellers, B., Gonzalez-Perez, C., Walkerden, G.: An application of philosophy in software modelling and future information systems development. In: Franch, X., Soffer, P. (eds.) CAiSE Workshops 2013. LNBIP, vol. 148, pp. 329–340. Springer, Heidelberg (2013)

9. Partridge, C., Gonzalez-Perez, C., Henderson-Sellers, B.: Are conceptual models concept models? In: Ng, W., Storey, V.C., Trujillo, J.C. (eds.) ER 2013. LNCS, vol. 8217, pp. 96–105. Springer, Heidelberg (2013)

10. Henderson-Sellers, B., Eriksson, O., Ågerfalk, P.J.: On the need for identity in ontology-based conceptual modelling. In: Saeki, M., Köhler, H. (eds.): Conceptual Modelling 2015, Proceedings of the 11th Asia-Pacific Conference on Conceptual Modelling (APCCM 2015), Sydney, Australia. CRPIT 165, pp. 9–20 (27–30 January 2015)

11. Smith, B.: Beyond concepts: ontology as reality representation. In: Varzi, A., Vieu, L. (eds.): Proceedings of the Third International Conference on Formal Ontology in Information Systems (FOIS), pp. 73–84. IOS Press (2004)

12. Ullmann, S.: Semantics: An Introduction to the Science of Meaning. Basil Blackwell, Oxford (1972)

13. Frege, G.: Über Sinn und Bedeutung. Zeitschrift für Philosophie und philosophische Kritik **100**, 25–50 (1892)

14. Lycett, M., Partridge, C.: The challenge of epistemic divergence in IS development. Comm. ACM **52**(6), 127–131 (2009)

15. Bunge, M.: Treatise on basic philosophy. Ontology I: The Furniture of the World, vol. 3. Reidel, Boston (1977)

16. Bunge, M.: Treatise on basic philosophy. Ontology II: The Furniture of the World, vol. 4. Reidel, Boston (1979)

17. Partridge, C.: Private communication, email dated 1 June 2015 (2015)

18. Wand, Y., Weber, R.: On the ontological expressiveness of information systems analysis and design grammars. J. Inf. Syst. **3**, 217–237 (1993)

19. Wand, Y., Weber, R.: On the deep structure of information systems. Inf. Syst. J. **5**, 203–223 (1995)

20. Opdahl, A., Henderson-Sellers, B.: Ontological evaluation of the uml using the Bunge-Wand-Weber model. Softw. Syst. Model. **1**(1), 43–67 (2002)

21. Opdahl, A., Henderson-Sellers, B.: A template for defining enterprise modelling constructs. J. Database Manag. **15**(2), 39–73 (2004)

22. Hesse, W.: From conceptual models to ontologies - a software engineering approach. In: Presentation at Dagstuhl Seminar on Conceptual Modelling (2008)

23. Gruber, T.R.: A translation approach to portable ontology specifications. Knowl. Acquisition **5**, 199–220 (1993)

24. Unhelkar, B.: Process Quality Assurance for UML-based Projects, p. 394. Pearson Education (Addison-Wesley), Boston (2003)

25. Henderson-Sellers, B.: Bridging metamodels and ontologies in software engineering. J. Syst. Softw. **84**(2), 301–313 (2011)

26. Guizzardi, G.: Ontological foundations for structural conceptual models. CTIT PhD Thesis Series, no. 05–74, Enschede, The Netherlands (2005)

27. Bittner, T., Smith, B.: Normalizing medical ontologies using basic formal ontology. In: Kooperative Versorgung, Vernetzte Forschung, Ubiquitäre Information (Proceedings of GMDS Innsbruck, 26-30 September 2004), Niebüll: Videel OHG, pp. 199–201 (2004)

28. Rosa, D.E., Carbonera, J.L., Torres, G.M., Abel, M.: Using events from UFO-B in an ontology collaborative construction environment. CEUR-WSX **938**, 278–283 (2012)
29. Sider, T.: Four Dimensionalism: An Ontology of Persistence and Time. Oxford Univ. Press, Oxford (2002)
30. Partridge, C.: Business Objects: Re-Engineering for Re-Use. Butterworth-Heinemann, Oxford (1996)
31. de Cesare, S., Henderson-Sellers, B., Lycett, M., Partridge, C.: Improving model quality through foundational ontologies: two contrasting approaches to the representation of roles. In: Jeusfeld, M., Karlapalem, K. (eds.) Advances in Conceptual Modelling – ER 2015 Workshops, Springer (2015)
32. Keet, M., Artale, A.: Representing and reasoning over a taxonomy of part–whole relations. Appl. Ontol. **3**, 91–110 (2008)
33. Guizzardi, G.: Modal aspects of object types and part-whole relations and the de re/de dicto distinction. In: Krogstie, J., Opdahl, A.L., Sindre, G. (eds.) CAiSE 2007 and WES 2007. LNCS, vol. 4495, pp. 5–20. Springer, Heidelberg (2007)
34. Guizzardi, G.: Ontological foundations for conceptual part-whole relations: the case of collectives and their parts. In: Mouratidis, H., Rolland, C. (eds.) CAiSE 2011. LNCS, vol. 6741, pp. 138–153. Springer, Heidelberg (2011)
35. Henderson-Sellers, B., Eriksson, O., Gonzalez-Perez, C., Ågerfalk, P.J.: Ptolemaic metamodelling? The need for a paradigm shift. In: Garcia Diaz, V., Cueva Lovelle, J.M., Pelayo García-Bustelo, B.C., Sanjuán Martínez, O. (eds.) Progressions and Innovations in Model-Driven Software Engineering, pp. 90–146. IGI Global, Hershey (2013)
36. Kripke, S.: Identity and necessity. In: Munitz, M.K. (ed.) Identity and Individuation. New York Univ. Press, New York (1971)
37. Kripke, S.: Naming and Necessity. Harvard Univ. Press, Cambridge (1980)
38. Bergholtz, M., Eriksson, O., Johannesson, P.: Towards a sociomaterial ontology. In: Franch, X., Soffer, P. (eds.) CAiSE Workshops 2013. LNBIP, vol. 148, pp. 341–348. Springer, Heidelberg (2013)
39. Guizzardi, G., Wagner, G.: Towards ontological foundations for agent modeling concepts using the Unified Foundational Ontology (UFO). In: Bresciani, P., Giorgini, P., Henderson-Sellers, B., Low, G., Winikoff, M. (eds.) AOIS. LNCS, vol. 3508, pp. 110–124. Springer, Heidelberg (2005)
40. Henderson-Sellers, B., Gonzalez-Perez, C., Eriksson, O.: Improving the unified foundational ontology (UFO) architecture in its support of non-materialistic entities (2015, submitted for publication)
41. Atkinson, C., Kühne, T.: Model-driven development: a metamodelling foundation. IEEE Softw. **20**, 36–41 (2003)
42. Giraldo, F.D.: A framework for evaluating the ontological quality of languages in MDE environments. Paper presented in PhD Symposium, CAISE 2013 (2013)
43. Incipit: ConML Technical Specification. ConML 1.1 (2011). http://www.conml.org/Resources_TechSpec.aspx
44. Gonzalez-Perez, C.: A Conceptual Modelling Language for the Humanities and Social Sciences. In: Sixth International Conference on Research Challenges in Information Science (RCIS 2012), pp. 396–401. IEEE Computer Society. Valencia, Spain (2012)
45. Noonan, H.: Identity. Stanford Encyclopedia of Philosophy (2009). http://plato.stanford.edu/entries/identity/

Methodologies for Semi-automated Conceptual Data Modeling from Requirements

Il-Yeol Song[1(✉)], Yongjun Zhu[1], Hyithaek Ceong[2], and Ornsiri Thonggoom[3]

[1] College of Computing and Informatics, Drexel University, Philadelphia, 19104, USA
{song,zhu}@drexel.edu
[2] Department of Multimedia, Chonnam National University, Yeosu 550-749, Korea
htceong@chonnam.ac.kr
[3] Department of Textile Science and Technology, Thammasat University,
Bangkok 10200, Thailand
ornsirithonggoom@gmail.com

Abstract. Conceptual modeling is the foundation of system analysis and design methodologies. It is challenging because it requires a clear understanding of an application domain and the ability to translate the requirement specification into a conceptual data model. Semi-automated conceptual data modeling is a process of using an intelligent tool to aid the modeler for the purpose of building a quality conceptual data model. In this paper, we first present six categories of methodologies that can be used for developing conceptual data models. We then describe the characteristics of each category, compare these characteristics, and present related work of each category. We finally suggest a framework for semi-automatically generating conceptual data models from requirements and suggest challenging research topics.

1 Introduction

Conceptual modeling is the foundation of analysis and design methodologies for the development of information systems. For many years, researchers have proposed various conceptual data modeling formalisms such as Entity-Relationship (ER) model [13], variations of the ER model such as IDEF1X, Oracle CASE notation, and Information Engineering, Natural/Nijssen Language Information Analysis Method (NIAM), Extended Entity Relationship (EER) models, Object Role Modeling (ORM), object-oriented (OO) modeling, Unified Modeling Language (UML), EXPRESS, RM-ODP, and others. Thalheim [64] estimates that the total number of proposed variations of the ER model is over 80. Comparisons of these modeling formalisms are shown in studies by Song et al. [58], Kim and March [37], and Neill and Laplante [50].

Among the different conceptual modeling formalisms, ER models and UML models are the most widely used in practice [40]. The ER model originally proposed by Chen [13] has been widely used in systems analysis and conceptual modeling. The ER approach is easy to understand, models real-world concepts, and readily translates an ER diagram into a database schema [21]. Many extensions or variations of the ER model in the form of EER (Extended ER or Enhanced ER) model have been proposed and utilized in different

P. Johannesson et al. (Eds.): ER 2015, LNCS 9381, pp. 18–31, 2015.
DOI: 10.1007/978-3-319-25264-3_2

applications [28]. The UML class diagram is another widely used conceptual modeling approach, especially in software engineering.

Natural language (NL) is the most common tool for people to describe things and communicate. Studies [40, 50] show that nearly 90 % of all the requirements in industrial practices are written in NL. There are at least three limitations in translating a requirement in NL into a conceptual model [59]: (1) NL is ambiguous and an effective and accurate analysis is difficult; therefore, techniques and rules for modeling are required. (2) The same semantics can be represented in multiple ways; therefore, ways of handling these context-dependent semantic variations are necessary. (3) Concepts that are not explicitly expressed in a requirement specification but still important in the application are often difficult to discover and model. Expertise in domain knowledge to discover the hidden entities that were not explicitly stated is therefore needed.

Thus, this paper focuses on how to effectively translate a requirement specification into a conceptual model, wherein we review literature to find out the state-of-art in semi-automated conceptual data modeling.

The rest of this paper is organized as follows: Sect. 2 describes difficulties in creating conceptual data models; Sect. 3 discusses six categories of semi-automated conceptual data modeling methodologies; Sect. 4 proposes the next steps for semi-automated conceptual modeling as well as a framework for semi-automated conceptual modeling; and Sect. 5 concludes the paper.

2   The Difficulties in Creating Conceptual Data Models

The difficulties in creating conceptual data models have been discussed in a previous study [7]. In spite of its importance, it is shown that conceptual data modeling is not done well [56]. Through a literature review, we identified the factors that contribute to the difficulties of creating conceptual data models.

2.1   Combinatorial Complexity in Possible Relationships

A previous study [67] shows that novice designers experience more difficulties in modeling relationships than entities. Batra [7] reports that novice designers not only have difficulties in modeling unary and ternary relationships, but also have difficulties in modeling all kinds of meaningful binary relationships. As the number of entities increases, the number of possible relationships increases at a combinatorial rate. The primary challenge in modeling relationships is how to select a minimum set that captures the semantics effectively. For the meaningful identification of relationships, at least the following criteria should be met: (1) all semantics in the application should be modeled, (2) any relationship constructs should not be redundant, (3) and the degree of relationships should be minimal.

2.2   Scattered Modeling Rules

There is no single complete set of heuristics/rules that can be used in developing quality data models. In general, heuristics/rules are often useful, but sometimes they may lead to cognitive errors called biases [53] if they are not applied appropriately. Furthermore, there is always a trade-off in applying heuristics: applying more heuristics requires less human intervention while more complexities arise. Context-dependent conflicts among heuristics could be one kind of complexity.

2.3 Semantic Mismatch

Translating a requirement specification literally into a database structure could cause literal translation errors [7]. For instance, a sentence stating that "an order records a sale of products to customers" may incur an erroneous relationship between customer and product entities. The example shows that not all real-world relationships map into meaningful database relationships in the data model; some real-world relationships are derivable at the database level.

2.4 Inexperience/Incomplete Knowledge of Designers

Novice designers have limited knowledge and skills, while expert designers often draw their knowledge from past experiences. Even expert designers might fail to create a quality conceptual model due to their lack of domain knowledge, unless he/she has a clear perception on the requirement specification [36]. Expertise in domain knowledge to identify the hidden entities is therefore needed. Important issues are how novice designers can be efficiently trained and how domain knowledge can be effectively transferred to the designers.

2.5 Multiple Solutions

In conceptual design, there is no single best answer. Moody and Shanks [48] also state that one of the common problems encountered in design is a large number of alternative designs can be created for a particular problem. Therefore, they propose a six-element framework to evaluate the quality of conceptual data models [47]. Their framework is composed of the following six factors: completeness, simplicity, flexibility, understandability, integration, and implementability. Later this framework is increased into eight factors by including correctness and integrity by empirically validating the framework [49].

3 Methodologies for Semi-automated Conceptual Data Modeling

The field of conceptual data modeling has spawned numerous techniques for the identification of entities and their relationships. Most of these techniques heavily rely on manual processes and experiences of designers. Currently, there are several commercial graphical CASE tools used for automatically converting a conceptual data model into a logical model and then generate a DDL definition in SQL for physical implementation. However, there is still no commercial tool that directly translates a NL requirement specification into a conceptual data model. At present, a fully automated conceptual modeling approach seems extremely challenging due to the inherent ambiguities in NL, the context-dependent nature of modeling, and the incompleteness of domain knowledge in tools. It is desirable to develop a semi-automated intelligent tool that can alleviate those problems to assist modelers.

This section presents a broad scope of methodologies that can be used for semi-automatically developing conceptual data models. We classify them into six categories: linguistic-based, pattern-based, case-based, ontology-based, metamodeling-based, and multi-techniques-based.

3.1 The Linguistic-Based Approach

Natural language processing (NLP) techniques and linguistic theories are used for designing conceptual models. Chen [12] proposes eleven rules for translating English sentence structures into ER constructs. Since then, several studies [33, 51] have tried to refine and extend the approach. However, these rules are still neither fully complete nor accurate. Although entities can be identified by nouns in a requirements specification, not all the nouns are entities because nouns not only refer to entities but also refer to attributes, or other throw-away concepts which need not be modeled. An entity can also be identified from a verb phrase and implicit requirements [59]. Hartmann and Link [33] modified Chen's eleven rules for translation from English sentence structures into EER constructs in which they re-organize and extend those rules using twelve heuristics. However, even these heuristics are not complete. These rules cannot overcome the inherent ambiguities of NLs. In addition, these kinds of rules cannot be universally applied to multiple domains. Therefore, this approach can only serve as a basis for a manual or a semi-automated process of transforming an English specification into an ER diagram.

In order to overcome inherent ambiguities in NL requirements, some studies put constraints on the input by restricting either the vocabulary or sentence structures [3]. With these restrictions, simple linguistic processing such as tagging and chunking can achieve reasonably good results. However, the use of controlled languages has some limitations. They cannot be easily applied to existing requirement documents. Furthermore, they are not natural and place extra burdens on requirements writers. Several formal specification languages such as Z, Object-Z, VDM, B, and OCL have also been proposed for formal model-based specifications. They are very expressive but require formal knowledge of the language to write a correct specification. Moreover, these languages have been designed for some specific applications, and their usages for different purposes may become awkward and difficult. Other researchers propose dialogue tools that help elicit the NL requirement specification [9, 36]. The main disadvantage of these tools is that they require more laborious human interventions, and thus it is difficult to use them for large-scale projects. In [54], a requirement specification is rewritten based on a constrained grammar. This approach, however, could introduce semantic loss due to imposed constraints and requires the participation of experts who know the grammar.

Classification theory and entity categories have been applied to conceptual data modeling [59]. Entity categories are characterized by the properties shared by their members and a widely-used technique in identifying entities or classes. In addition, entity categories can be used to discover missing or hidden entities or classes that are not explicitly stated in the requirement but can be discovered by applying domain knowledge to the entity categories [59].

A trend in this technique is the collaborative use of huge linguistic dictionaries [23, 45] and common sense ontologies [39]. Linguistic dictionaries provide not only semantic links among concepts such as synonym, antonym, hypernym (is-a), and meronym/holonym (part-of), but also syntactical and morphological information. A variety of relationship types is discussed in [62]. WordNet [44] is a popular linguistic dictionary used for concept identification during conceptual modeling because it is readily available and extends to other languages such as European languages, Spanish, Chinese, and so forth. However, the main drawback of WordNet is that it does not cover many relationships. Therefore, WordNet++, the extension of WordNet, which contains special types of relationships that are not available in WordNet, is proposed in [19].

The domain independence is the strength of this approach. However, the strength of this technique is also its weakness because the tools or systems proposed have no domain knowledge incorporated in them. This technique does not provide an optimal solution to many sophisticated requirement specifications because of the inherent limitations of NL.

3.2   The Pattern-Based Approach

The important role of patterns in design is recognized in Alexander's book [4] on architecture and urban planning. It suggests that designers should produce and use patterns rather than solve problems from the first principle. Now patterns have been well established as a technique for reusing solutions of recurrent problems in the software development process. Pattern reuse provides many benefits such as higher productivity, software quality improvement, and a reduction of time and cost in software development. Design patterns have been proven very useful in speeding up the design process through reuse and in improving the overall quality of systems. Integrating patterns into conceptual design is challenging. The recognition of patterns in the context of conceptual data modeling is based on works by Coad et al. [15], Hay [34], and Fowler [26]. They created a library of proven modeling structures and provided some examples of adapting generic models to suit particular requirements.

Several authors have proposed various kinds of patterns [15, 26, 34]. Blaha [8] proposes several types of data modeling patterns: universal antipatterns are the patterns that we should avoid for all applications; archetypes are the common modeling patterns occurring across different applications; and canonical patterns are corresponding to metamodels of modeling formalisms. However, their utility in automated data modeling remains to be seen.

Empirical research shows that experts reuse patterns while novices do not [11].

The process in pattern reuse can be divided into three tasks: retrieval, adaptation, and integration [5]. Retrieval involves choosing patterns that may be relevant to a particular problem. After a pattern is chosen, it needs to be adapted or instantiated to fit the specific problem. Finally, it needs to be integrated with other patterns to form a complete model in the form of a conceptual data model. The advantage of reusable patterns aims to reuse not only schema components but also relationships between objects. However, building a repository of patterns involves the explication of human developers' knowledge, which is a major obstacle in facilitating the reuse of knowledge. To develop a pattern repository, designers should have very clear knowledge about the specific domain and identify the boundaries of what objects to include and what degree they should be abstracted. It takes a lot of time and effort to create a pattern repository. Currently, most of the proposed reusable pattern repositories used for conceptual data modeling are developed based on a manual approach that is time-consuming and skill-intensive for expert designers. Furthermore, most of the proposed tools in this technique use analysis patterns that require manual matching.

One solution to reduce the effort and time of human experts comes from extracting the pattern artifacts from existing designs [31]. If this could be done for various application domains, then it would assist in creating practically reusable pattern artifacts.

3.3   The Case-Based Approach

Case-based reasoning is a technology that utilizes a repository of cases for decision-making. The basic idea is, given the description of a new problem, to retrieve a similar problem from the case repository and adapt the retrieval to obtain a solution.

Very few have used case-based reasoning where cases of conceptual models are stored, indexed, and used for future design. We can find only three KBSs that use this technique, which are CSBR [63], DES-DS [52], and CABSYDD [14]. A comparison between these KBSs can be found in [14].

This technique involves storing conceptual models of a large number of applications and providing a keyword mechanism that enables users to search for a conceptual model that is a candidate solution for a problem statement. The technique takes advantage of reusing the previous design. The limitation of this technique is that if any adjustment is required in the conceptual model, it has to resort to the generic data modeling approach. Moreover, adjustments on the conceptual model are always required in order to cope with changes in requirement specification. The major disadvantage of this technique is that developing the conceptual model libraries and indexing mechanism are very expensive.

3.4   The Ontology-Based Approach

Ontologies have been proposed as an important way to represent real-world knowledge and, at some level, to support interoperability [57]. An ontology can represent in the form of a taxonomy, a thesaurus, a domain model, or a logical theory. Some papers point out some similarities and differences between ontologies and conceptual data models [20, 25]. According to Fonseca [25], two criteria that differentiate ontologies from conceptual data models are (1) the objective of modeling and (2) objects to model. Embley [21] suggests that ontologies are the key for solving the semantic problems of information systems.

In the conceptual modeling field, many researchers employ ontologies for evaluating, improving or developing conceptual modeling formalisms. Storey [61] proposes an ontology to classify the verb phrases of relationships based on research in linguistics and semantic data models. Evermann and Wand [22] examine the mapping between ontological elements and UML elements and propose guidelines on how to use UML elements to model real-world systems in particular. Purao and Storey [55] propose a multilayered ontology for classifying relationships by using data abstractions and by separating domain-dependent and domain-independent aspects of the relationship constructs.

The major advantage of using ontologies for conceptual modeling is the reusability of a knowledge repository. This can be developed into two levels: domain ontology and large scale or upper level ontology. A domain ontology [29] represents concepts, relationships between concepts, and inference rules for a particular domain. Several tools for creating and querying domain ontologies are available such as Protégé, OWL, SPARQL, etc. A detailed comparison of each tool is shown in [17]. Instead of developing individual ontologies, there has been interest in creating an upper level or large scale ontology. An upper level ontology [29] represents general concepts that are the same across all domains and always consist of a hierarchy of entities and rules that describe those general entities which do not belong to a specific problem domain. Examples of upper level ontologies are Cyc, ResearchCyc, BFO (Basic Formal Ontology), DOLCE, SUMO, geneontology, etc. For a review and comparison of upper level ontologies see [42]. The potential usefulness of upper level ontologies lies in the fact that they are domain independent. A major problem with existing upper level ontologies is the lack of good user interface and a good API. For example, Cyc is not an ontology of word sense like WordNet. As a result, there is no comprehensive mapping of Cyc concepts into words of NL [16]. Without the support of an adequate tool, it is difficult to work with them. Obviously, domain ontologies are more usable than large scale ontologies [29].

An ontology can be used as the source of domain knowledge, and designers can use the ontology to get initial domain knowledge. Corcho and his colleagues [17] suggest that a strategy for developing domain ontologies would be to reuse large scale or upper level ontologies. The same upper level ontology can be used for developing many domain ontologies, which share the same skeleton. Extensions of the skeleton should be made at a low level by adding domain-specific subconcepts.

3.5   The Metamodeling Approach

A metamodel describes the syntax of the models with a collection of modeling concepts and constraints. Metamodeling is an abstraction-based modeling technique; it is a process of creating models using the domain knowledge starting from the metamodel. One of the most active research areas utilizing metamodeling is model-driven development [6]. Metamodeling allows the user to use permitted structures and assign semantics of the domain to the structures [60]. Fill and Karagiannis conceptualize the components of modeling methods using the ADOxx metamodeling platform [24]. A detailed discussion of research challenges in metamodeling is presented in [60].

3.6 Multi-Techniques-Based Approaches

From our survey, most tools or systems for conceptual design require users' involvement during the process. No single technique works best all the time because each technique has some limitations. Ideally, various techniques should be integrated together during a modeling process. For example, Song et al. [59] have proposed a TCM (Taxonomic Class Modeling) methodology used for OO analysis in business applications. This method integrates several class modeling techniques under one framework. The framework integrates the noun analysis method, class categories, English structures, check lists, and modeling rules. Thonggoom et al. [65, 66] present two knowledge-based modeling tools that integrate the pattern-based technique, WordNet, and other heuristics-based modeling techniques. These tools show how domain knowledge stored in instance patterns can be reused together with other modeling techniques and improve the effectiveness of data modeling.

4   A Proposal for Semi-automated Conceptual Modeling

In this section, we propose a 3-step approach for semi-automated conceptual data modeling from requirement specification. The method includes the process of (1) analysis and refinement of the requirement specification, (2) generation of a conceptual model from the requirement, and (3) verification and validation of the generated model. Figure 1 shows our framework for semi-automated conceptual data modeling.

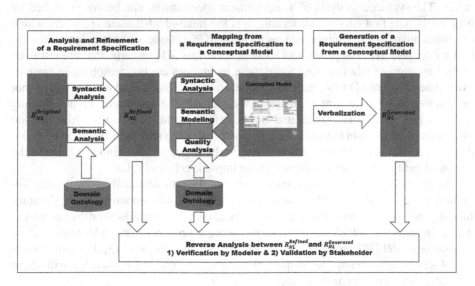

Fig. 1. A framework for semi-automated conceptual modeling

4.1   Analysis and Refinement of a Requirement Specification

In the first step, the syntax and semantics of the requirement specification in NL is analyzed and refined. The requirement specification in NL must be syntactically complete. Most of previous studies simply assume that requirements are complete and free from syntactic errors. Hence, it is necessary to check the syntax of the requirement specification before it is fed into an automated tool. An incomplete requirement specification is treated to have syntactic errors, and the modeler should be informed so that the errors could be corrected.

In order to check semantic quality of the specification, a domain ontology related to the specification is most effective [30]. In order to build the domain knowledge and associate it to the requirement specification, the methodology proposed in [31] can be used. A systematic analysis and refinement of a requirement specification using a sophisticated NLP tool that utilizes a domain-specific ontology is necessary. However, the checking of the quality of a specification, in terms of syntactic and semantic completeness and accuracy, needs to be studied more.

4.2   Mapping from a Requirement Specification to a Conceptual Model

In the second step, a conceptual model is generated from the refined requirement specification. The conceptual model should be generated by integrating syntactic analysis, semantic modeling with domain knowledge, and quality analysis. Ideally, multiple methodologies discussed in Sect. 3 should be combined and used.

The syntax and semantics of a requirement specification should be analyzed separately. The syntactic analysis of a requirement specification can be accomplished by NLP tools. The first easy-to-use technique is the noun-identification approach, which evaluates nouns in the requirement specification to find classes or entities in the conceptual model [18, 51]. The patterns of specific sentences [41] or verbs [54] are retrieved to find appropriate relationships. Since these techniques are not complete, accurate or even generic, we need to use other techniques to identify additional entities that cannot be discovered by pure syntax analysis. We can use the taxonomic class modeling methods [59] to identify entities that can be inferred from verbs and entities that are important to the domain but are not explicitly stated in the specification. Song et al. use rule-based methods to identify entities from verbs and apply domain knowledge to the notion of entity categories to discover those implicit entities [59].

Based on the results of the syntactic analysis, the semantic analysis of the requirement specification should be followed. A popular way to add semantics during the automated generation of a model is to exploit a lexical database such as WordNet as well as a domain ontology. Additional semantic modeling approaches that can be employed are Discourse model [32], Interlingua [38], ROM [69], and SemNet [43]. For more meaningful semantic modeling, techniques of utilizing the domain knowledge with these semantic modeling approaches should be studied.

Next, the ways of enforcing constraints of a domain or a modeling language and the quality of final conceptual models should be considered. A simple way to effectively enforce constraints is to employ a metamodel of the target conceptual model. Constraint types specified in the metamodel help the modeler identify and define domain-specific constraints. An automated quality analysis of conceptual models is challenging. Aguilera et al. [2] present a method that defines quality issues of a given model and detects quality issues of the model. Another approach is to define domain-specific quality issues by developing a UML profile [27]. An automated mapping methodology that incorporates the domain-specific quality issues needs much more investigation. To realize the automation of the aforementioned processes, a series of steps such as syntax analysis, semantic modeling, domain knowledge, and constraints and quality enforcement must be systematically integrated.

4.3 Reverse Analysis Between the Refined Requirement Specification and the Generated Requirement Specification

Evaluation methods are required to examine whether the generated conceptual data model represents the necessary and sufficient conditions of the semantics of the requirement specification. Ideally, both the data modeler and stakeholders (or domain experts) should participate in the evaluation process. The data modeler can verify the syntax and semantics of the model against the requirement specification as he/she is well versed with the employed data model. Stakeholders can validate the semantics of the generated model as they know more about the domain than the data modeler does. In order to help the stakeholders validate the model against the requirement specification, the generated conceptual model should be translated into a description in NL that could be easily understood by the stakeholders. The semantics of generated description can be evaluated against the original requirement specification and the domain. A verbalization method, used to produce a verbalization of a UML conceptual schema [1], may be used for this purpose.

In previous studies, the conceptual models are summarized in terms of recall and precision based on the experts' analyses [35]. Comparative analyses between the conceptual model in NL and the requirement specification are also necessary. For this purpose, both the comparison aspect of semantics at the sentence level and the development of quantitative measurements are also required. Semantic recall and precision [10] could be employed for that purpose.

4.4 Needs for Rational Empirical Studies

For effective comparisons of various conceptual modeling techniques and tools, it is necessary to have consistent, reliable, and replicable testbed and benchmark tests. Our survey shows that there have been no such environments in the conceptual modeling community. A lack of these standard testing environments limits comparative studies.

These testing environments should include multiple specific application areas with their requirement specifications; domain knowledge of the areas; automated conceptual modeling methodologies and tools to create conceptual models; methods of translating the generated conceptual model into diverse natural languages; and quantitative measures for comparing the results of the testing. In order to create these environments, contributions by crowdsourcing would be desirable.

5 Conclusions

This paper presented six categories of methodologies that can be used for semi-automated conceptual data modeling. These categories include linguistic-based, pattern-based, case-based, ontology-based, metamodeling-based, and multi-techniques-based conceptual modeling. Based on this survey, we proposed a three-step framework for semi-automated conceptual data modeling. It consists of an analysis and refinement of the requirement specification, a mapping of the requirement specification to a conceptual model, and a reverse analysis between the generated requirement specification from the generated conceptual model and the input requirement specification. We proposed the need for testbeds and benchmark testing environments for conceptual modeling research. We also discussed promising research topics in each step. Automated conceptual data modeling is far from satisfactory and there are many challenging issues ahead. Recent advances in cognitive computing [46] and machine learning [68] techniques could be introduced in automated conceptual modeling.

References

1. Aguilera, D., García-Ranea, R., Gómez, C., Olivé, A.: An eclipse plugin for validating names in UML conceptual schemas. In: De Troyer, O., Medeiros, C.B., Billen, R., Hallot, P., Simitsis, A., Van Mingroot, H. (eds.) ER 2011. LNCS, vol. 6999, pp. 323–327. Springer, Heidelberg (2011)
2. Aguilera, D., Gómez, C., Olivé, A.: A method for the definition and treatment of conceptual schema quality issues. In: Atzeni, P., Cheung, D., Ram, S. (eds.) ER 2012. LNCS, vol. 7532, pp. 501–514. Springer, Heidelberg (2012)
3. Ambriola, V., Gervasi, V.: On the systematic analysis of natural language requirements with circe. Autom. Softw. Eng. **13**, 107–167 (2006)
4. Alexander, C.: The Timeless Way of Building. Oxford University Press, New York (1979)
5. Anthony, S., Mellarkod, V.: Data modeling patterns: a method and evaluation. In: Proceedings of the Fifteenth Americas Conference on Information Systems, San Francisco, California (2009)
6. Atkinson, C., Kuhne, T.: Model-driven development: a metamodeling foundation. IEEE Softw. **20**(5), 36–41 (2003)
7. Batra, D.: Cognitive complexity in data modeling: causes and recommendations. Requirements Eng. **12**(4), 231–244 (2007)
8. Blaha, M.: Patterns of Data Modeling. CRC Press, Boca Raton (2010)
9. Buchholz, E., Cyriaks, H., Dsterhft, A., Mehlan, H., Thalheim, B.: Applying a natural language dialogue tool for designing databases. In: Proceedings of the first International Workshop on Applications of Natural Language to Databases (NLDB 1995) (1995)
10. Burton-Jones, A., Meso, P.: How good are these UML diagrams? An empirical test of the Wand and Weber good decomposition model. In: ICIS 2002 Proceedings, 10 (2002)
11. Chaiyasut, P., Shanks, G.: Conceptual data modeling process: a study of novice and expert data modelers. In: Proceedings of the 1st International Conference on Object-Role Modeling, Australia, University of Queensland (1994)
12. Chen, P.: English sentence structure and entity-relationship diagram. Inf. Sci. **1**(1), 127–149 (1983)

13. Chen, P.: The entity-relationship model: toward a unified view of data. ACM Trans. Database Syst. **1**(1), 9–36 (1976)
14. Choobineh, J., Lo, A.: CABSYDD: case-based system for database design. J. Manag. Inf. Syst. **21**(3), 242–253 (2004)
15. Coad, P., North, D., Mayfield, M.: Object Models – Strategies, Pattern, and Applications. Yourdon Press, Englewood Cliffs (1995)
16. Conesa, J., Storey, V.C., Sugumaran, V.: Experiences using the ResearchCyc upper level ontology. In: Kedad, Z., Lammari, N., Métais, E., Meziane, F., Rezgui, Y. (eds.) NLDB 2007. LNCS, vol. 4592, pp. 143–155. Springer, Heidelberg (2007)
17. Corcho, O., Fernandez-Lopez, M., Gomez-Perez, A.: Methodologies, tools and languages for building ontologies: where is their meeting point? Data Knowl. Eng. **46**, 41–64 (2003)
18. Deeptimahanti, D.K, Sanyal, R.: Semi-automatic generation of UML models from natural language requirements. In: Proceedings of the 4th India Software Engineering Conference (ISEC 2011), pp. 165–174 (2011)
19. Dehne, F., Steuten, A., van de Riet, R.P.: WordNet++: a lexicon for the color-X method. Data Knowl. Eng. **38**(1), 3–29 (2001)
20. El-Ghalayini, H., Odeh, M., McClatchey, R.: Engineering conceptual data models from domain ontologies: a critical evaluation. Int. J. Inf. Technol. Web Eng. **2**(1), 57–70 (2006)
21. Embley, D.: Toward semantic understanding: an approach based on information extraction ontologies. In: Proceedings of the 15th Australian Database Conference, Denedin, New Zealand, pp. 3–12 (2004)
22. Evermann, J., Wand, Y.: Towards ontologically-based semantics for UML constructs. In: Kunii, H.S., Jajodia, S., Sølvberg, A. (eds.) ER 2001. LNCS, vol. 2224, pp. 354–367. Springer, Heidelberg (2001)
23. Fellbaum, C.: WordNet: An Electronic Lexical Database. MIT Press, Cambridge (1998)
24. Fill, H.-G., Karagiannis, D.: On the conceptualization of modelling methods using the ADOxx meta modeling platform. Enterp. Model. Inf. Syst. Archit. **8**(1), 4–25 (2013)
25. Fonseca, F., Martin, J.: Learning the differences between ontologies and conceptual schemas through ontology-driven information systems. JAIS – J. Assoc. Inf. Syst. Spec. Issue Ontol. Context IS **8**(2), 129–142 (2007)
26. Fowler, M.: Analysis Patterns: Reusable Object Models. Addison Wesley, Menlo Park (1997)
27. Gailly F., Poels, G.: Conceptual modeling using domain ontologies: improving the domain-specific quality of conceptual schemas. In: Proceedings of the 10th Workshop on Domain-Specific Modeling, pp. 18–24 (2010)
28. Gogolla, M., Hohenstein, U.: Towards a semantic view of an extended entity-relationship model. ACM Trans. Database Syst. **16**(3), 369–416 (1991)
29. Conesa, J., Storey, V., Sugumaran, V.: Usability of upper level ontologies: the case of ResearchCyc. Data Knowl. Eng. **69**(4) (2010)
30. Gnesi, S., Fabbrini, F., Fusani, M., Trentanni, G.: An automatic tool for the analysis of natural language requirements. informe técnico, CNR Information Science and Technology Institute, pp. 53–62 (2004)
31. Han, T., Purao, S., Storey, V.: Generating large-scale repositories of reusable artifacts for conceptual design of information systems. Decis. Support Syst. **45**, 665–680 (2008)
32. Harmain, M., Gaizauskas, R.: CM-builder: a natural language-based CASE tool for OO analysis. J. Autom. Softw. Eng. **10**(2), 157–181 (2003)
33. Hartmann, S., Link, S.: English sentence structures and EER modeling. In: Proceedings of the 4th Asia-Pacific Conference on Conceptual Modeling (2007)
34. Hay, D.C.: Data Model Patterns: Conventions of Thought. Dorset House Publishing, New York (1996)

35. Jarvenpaa, S.L., Machesky, J.J.: Data analysis and learning: an experimental study of data modeling tools. Int. J. Man Mach. Stud. **31**(4), 367–391 (1989)
36. Kim, N., Lee, S., Moon, S.: Formalized entity extraction methodology for changeable business requirements. J. Inf. Sci. Eng. **24**, 649–671 (2008)
37. Kim, Y., March, S.: Comparing data modeling formalisms. Commun. ACM **38**(6), 103–115 (1995)
38. Kop, C., Fliedl, G. Mayr, H.: From natural language requirements to a conceptual model. In: Proceeding of the First International Workshop on Evolution Support for Model-Based Development and Testing (EMDT2010), pp. 67–73 (2010)
39. Lenat, D.B.: CYC: a large-scale investment in knowledge infrastructure. Commun. ACM **38**(11), 33–38 (1995)
40. Luisa, M., Mariangela, F., Pierluigi, N.I.: Market research for requirements analysis using linguistic tools. Requirements Eng. **9**(1), 40–56 (2004)
41. Mala, A., Uma, V.: Automatic construction of object oriented design models [UML diagrams] from natural language requirements specification. In: Proceedings of the 9th Pacific Rim international conference on Artificial intelligence (PRICAI 2006), pp. 1155–1159 (2006)
42. Mascardi, V., Cordì, V., Rosso, P.: A comparison of upper ontologies. Technical report DISI-TR-06-2 (2007)
43. Mich, L., Garigliano, R.: The NL-OOPS project: object oriented modeling using the natural language processing system LOLITA. In: Proceedings of the 4th International Conference on the Applications of Natural Language to Information Systems (NLDB 1999) (1999)
44. Miller, G.A.: WordNet: a lexical database for English. Commun. ACM **38**(11), 39–41 (1995)
45. Miyoshi, H., Sugiyama, K., Kobayashi, M., Ogino, T.: An overview of the EDR electronic dictionary and the current status of its utilization. In: Proceedings of the 16th International Conference on Computational Linguistics (1996)
46. Modha, D.S., Ananthanarayanan, R., Esser, S.K., Ndirango, A., Sherbondy, A.J., Singh, R.: Cognitive computing. Commun. ACM **54**(8), 62–71 (2011)
47. Moody, D.L.: Metrics for evaluating the quality of entity relationship models. In: Ling, T.W., Ram, S., Lee, M.L. (eds.) ER 1998. LNCS, vol. 1507, pp. 211–225. Springer, Heidelberg (1998)
48. Moody, D.L., Shanks, G.G.: What makes a good data model? Evaluating the quality of entity relationship models. In: Loucopoulos, P. (ed.) ER 1994. LNCS, vol. 881, pp. 94–111. Springer, Heidelberg (1994)
49. Moody, D.L., Shanks, G.G.: Improving the quality of data models: empirical validation of a quality management framework. Inf. Syst. **28**(6), 619–650 (2003)
50. Neill, C., Laplante, P.: Requirement engineering: the state of the practice. IEEE Softw. **20**(6), 40–45 (2003)
51. Omar, N,. Hanna, P., Mc Kevitt, P.: Heuristics-based entity-relationship modelling through natural language processing. In: Proceedings of the Fifteenth Irish Conference on Artificial Intelligence and Cognitive Science (AICS-04), pp. 302–313 (2004)
52. Paek, Y.K., Seo, J., Kim, G.C.: An expert system with case-based reasoning for database schema design. Decis. Support Syst. **18**(1), 83–95 (1996)
53. Parson, J., Saunders, C.: Cognitive heuristics in software engineering: applying and extending anchoring and adjustment to artifact reuse. IEEE Trans. Softw. Eng. **30**(12), 873–888 (2004)
54. Popescu, D., Rugaber, S., Medvidovic, N., Berry, D.M.: Reducing ambiguities in requirements specifications via automatically created object-oriented models. In: Martell, C. (ed.) Monterey Workshop 2007. LNCS, vol. 5320, pp. 103–124. Springer, Heidelberg (2008)
55. Purao, S., Storey, V.C.: A multi-layered ontology for comparing relationship semantics in conceptual models of databases. J. Appl. Ontol. **1**(1), 117–139 (2005)

56. Simsion, G.: Data Modeling Theory and Practice. Technique Publications, LLC, New York (2007)
57. Soares, A., Fonseca, F.: Ontology-Driven Information Systems at Development Time. IJCSS – J. Comput. Syst. Signals **8**(2) (2007)
58. Song, I.-Y., Evans, M., Park, E.: A comparative analysis of entity-relationship diagrams. J. Comput. Softw. Eng. **3**(4), 427–459 (1995)
59. Song, I.-Y., Yano, K., Trujillo, J., Lujan-Mora, S.: A taxonomic class modeling methodology for object-oriented analysis. In: Krostige, T.H.J., Siau, K. (eds.) Information Modeling Methods and Methodologies. Advanced Topics in Databases Series, pp. 216–240. Idea Group Publishing, Hershey (2004)
60. Sprinkle, J., Rumpe, B., Vangheluwe, H., Karsai, G.: Metamodeling - state of the art and research challenges. In: Giese, H., Karsai, G., Lee, E., Rumpe, B., Schätz, B. (eds.) MBEERTS. LNCS, vol. 6100, pp. 57–76. Springer, Heidelberg (2008)
61. Storey, V.C.: Classifying and comparing relationships in conceptual modeling. IEEE Trans. Knowl. Data Eng. **17**(11), 1–13 (2005)
62. Storey, V.C.: Understanding semantic relationships. VLDB J. **2**, 455–488 (1993)
63. Storey, V.C., Chiang, R., Goldstein, R., Dey, D., Sundaresan, S.: Database design with common sense business reasoning and learning. ACM Trans. Database Syst. **22**(4), 471–512 (1997)
64. Thalheim, B.: Entity-Relationship Modeling: Foundations of Database Technology. Springer, Berlin (2000)
65. Thonggoom, O., Song, I.-Y., An, Y.: EIPW: a knowledge-based database modeling tool. In: Salinesi, C., Pastor, O. (eds.) CAiSE Workshops 2011. LNBIP, vol. 83, pp. 119–133. Springer, Heidelberg (2011)
66. Thonggoom, O., Song, I.-Y., An, Y.: Semi-automatic conceptual data modeling using entity and relationship instance repositories. In: Jeusfeld, M., Delcambre, L., Ling, T.-W. (eds.) ER 2011. LNCS, vol. 6998, pp. 219–232. Springer, Heidelberg (2011)
67. Topi, H., Ramesh, V.: Human factors research on data modeling: a review of prior research, an extended framework and future research directions. J. Database Manag. **13**, 3–15 (2002)
68. Witten, I.H., Frank, E.: Data Mining: Practical Machine Learning Tools and Techniques. Morgan Kaufmann, Los Altos (2005)
69. Zeng, Y.: Recursive object model (ROM)-modeling of linguistic information in engineering design. J. Comput. Ind. **59**, 612–625 (2008)

Business Process and Goal Models

Business Process and Conference

Aligning Business Goals and Risks
in OSS Adoption

Dolors Costal[1], Lidia López[1(⊠)], Mirko Morandini[2], Alberto Siena[2],
Maria Carmela Annosi[3], Daniel Gross[2], Lucía Méndez[1],
Xavier Franch[1], and Angelo Susi[2]

[1] Universitat Politècnica de Catalunya (UPC),
c/Jordi Girona, 1-3, 08034 Barcelona, Spain
{dolors, llopez, franch}@essi.upc.edu,
emendez@lsi.upc.edu
[2] Fondazione Bruno Kessler (FBK), 38123 Trento, Italy
{morandini, siena, dgross, susi}@fbk.eu
[3] Ericsson Telecomunicazioni S.p.A., Italy (TEI), 84016 Pagani, Italy
mariacarmela.annosi@ericsson.com

Abstract. Increasing adoption of Open Source Software (OSS) requires a
change in the organizational culture and reshaping IT decision-makers mindset.
Adopting OSS software components introduces some risks that can affect the
adopter organization's business goals, therefore they need to be considered. To
assess these risks, it is required to understand the socio-technical structures that
interrelate the stakeholders in the OSS ecosystem, and how these structures may
propagate the potential risks to them. In this paper, we study the connection
between OSS adoption risks and OSS adopter organizations' business goals. We
propose a model-based approach and analysis framework that combines two
existing frameworks: the *i** framework to model and reason about business
goals, and the RiskML notation to represent and analyse OSS adoption risks.
We illustrate our approach with data drawn from an industrial partner organization in a joint EU project.

Keywords: Risk analysis · Open source software · *i** framework · i-star

1 Introduction

Open Source Software (OSS) has become a driver for business in various sectors,
namely the primary and secondary IT sector. Estimates exist that in 2016, a 95 % of all
commercial software packages will include OSS components [1].

OSS adoption impacts in fact far beyond technology, because it requires a change
in the organizational culture and reshaping IT decision-makers mindset. Hence, the
way in which organizations adopt OSS affects and shapes their businesses. At the same

This work is a result of the RISCOSS project, funded by the EC 7th Framework Programme
FP7/2007-2013, agreement number 318249.

© Springer International Publishing Switzerland 2015
P. Johannesson et al. (Eds.): ER 2015, LNCS 9381, pp. 35–49, 2015.
DOI: 10.1007/978-3-319-25264-3_3

time, OSS software components introduce various risks that may not be visible at the time of the adoption, but can manifest in later development and maintenance phases, causing unexpected failures. These risks may have an impact on the business goals of the adopter organization.

In this paper, we study the connection between risks and the business goals of the OSS adopter organizations, and how risks propagating through OSS ecosystem structures may compromise stakeholders' strategic goals. The present work builds upon the results of two previous papers:

(1) In [2] we proposed goal-oriented models using the i^* approach [3] to model the different existing OSS adoption strategies. The models describe the consequences of adopting one such strategy or another: which are the business goals that are supported, which are the resources that emerge, which are the dependencies that exist between the different actors of the OSS ecosystem, etc.
(2) In [4] we presented a framework for risk modelling and risk evaluation, which is tailored to assess OSS adoption risks. The framework is comprised by a risk modelling language (RiskML) and a quantitative reasoning algorithm that analyses risk models.

The present work proposes to align RiskML models and i^* models to analyse the propagation of the risk impact towards the business goals of the OSS adopter and the rest of actors of an OSS ecosystem. It is guided by two main research questions:

- **RQ1: What is the conceptual relationship between OSS adoption risks and the adopter organization business goals?**
- **RQ2: How do OSS adoption risks affect the adopter organization business goals?**

RQ1 explores how the risk and business goal-oriented modelling approaches can be integrated into a single modelling framework. This will be done by formulating an integrated metamodel that will serve as a basis for the design of risk-aware OSS ecosystems (RQ1.1), and then by offering means to examine the relationship between risks and business goals at the instance level (RQ1.2). RQ2 explores how existing risk and business model analysis techniques can be combined to propagate the results of risk analysis to business goals, considering the relationships that may exist among actors which collaborate in OSS ecosystems.

This research is part of an ongoing European FP7 project (RISCOSS, www.riscoss. eu), which aims to support organizations in understanding, managing and mitigating risks during OSS adoption [5]. The preliminary validation of our research results was performed at one of the industrial partners (Ericsson Italy at Pagani, TEI), where the approach helped illustrating how risks during the adoption and maintenance of OSS components may impact on its business goals.

As research method we adopted a design science approach following the engineering cycle described in [6]. This fits well with the research aim to create new metamodel artefacts, while also acquiring new knowledge. Figure 1 illustrates the cycle, which includes problem investigation, solution design and solution validation.

The rest of the paper is organized as follows. Section 2 briefly provides additional background which is illustrated through a running example drawn from Ericsson's

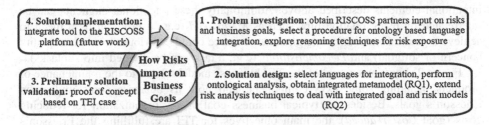

4. Solution implementation: integrate tool to the RISCOSS platform (future work)

1 . Problem investigation: obtain RISCOSS partners input on risks and business goals, select a procedure for ontology based language integration, explore reasoning techniques for risk exposure

How Risks impact on Business Goals

3. Preliminary solution validation: proof of concept based on TEI case

2. Solution design: select languages for integration, perform ontological analysis, obtain integrated metamodel (RQ1), extend risk analysis techniques to deal with integrated goal and risk models (RQ2)

Fig. 1. Steps of the engineering cycle following [6].

business and development environment. Sections 3–5 present the integrated risk and business modelling framework, the steps taken to align both risk and business models and illustrates the proposed analysis engine. Section 6 presents related work, while Sect. 7 concludes and points to future work.

2 Background

In this section we present the two modelling frameworks upon which we build our proposal, namely the *i** framework and the RiskML modelling language. In order to illustrate their concepts, we will use a running example from the RISCOSS project.

Running Example. TEI is part of Ericsson, one of the world's leading telecommunication corporations. Ericsson produces hardware (telecommunications infrastructure and devices) as well as the software to run it. The company's mission is to empower people, business and society at large, guided by a vision of a sustainable networked society. One of TEI's roles within the Ericsson ecosystem is to provide OSS alternatives to support efficient third party products handling. However, adopting OSS components also exposes TEI to risk, because OSS comes typically without legal contracts that guarantee the adopter over time about software functionalities and qualities, so the company may suffer towards its partners and customer for lacks in the adopted software. It must therefore undertake adequate actions to analyse, assess and possibly mitigate potential risks.

2.1 Business Goal Models: *i**

In OSS ecosystems, actors pursue their goals while interacting with other actors. The *i** framework [3] was formulated for representing, modelling and reasoning about socio-technical systems. Its modelling language (the *i** language) is composed by a set of graphic constructs which can be used in two diagrams. Firstly, the Strategic Dependency (SD) diagram, including the organizational *Actors*. Actors have *Dependencies*, one actor (*Depender*) depends on other (*Dependee*) for the achievement of some intention (*Dependum*). The main intentional elements are: *Resource, Task, Goal* and *Softgoal*. Softgoals represent goals with no clear criteria for their satisfaction.

Secondly, the Strategic Rationale (SR) diagram represents the internal actors' rationale. The rationality of each actor is represented using the same types of intentional elements described above. Additionally these intentional elements can be interrelated by using relationships such as *Means-end* (e.g., a task can be a mean to achieve a goal), *Contributions* (e.g., some resource could contribute to reach a quality concern or softgoal) and *Decompositions* (e.g., a task can be divided into subtasks).

Figure 2 shows an excerpt of the TEI business model related to the maintenance of products including some OSS component, and how its business goals impact on Ericsson's goals. Besides the typical business goal of any organization for reducing costs (goal *Cost reduced*), the main objectives for TEI are fulfilling the Ericsson's customers' requirements (softgoal *Product requirements achieved*) using a *Maintainable code* in order to secure the quality (*Quality of code*). For TEI it is crucial to use *Mature technology* and *Secure code*. When TEI decides to use an OSS component, there are three possibilities for maintaining this code: they can assume the activity (*Provide in-house maintenance*), rely to the community behind the OSS component (*Rely on the OSS community for maintenance*) or rely to a third party organization (*Contract 3PP organization for maintenance*). For this portion of the TEI's business model, the impacted Ericsson business goals are *Time-to-market reduced* and *Reputation kept*. Ericsson expects from TEI that the *Development time is reduced*, *Responsiveness* and *Reliable products* for achieving its business goals. Notice that the model only includes the third party organization (*3PP OSS Provider*) in order to illustrate that the maintenance is outsourced; for the sake of brevity, the description of this relation is not exhaustive and not all dependencies between both organizations are included in this model.

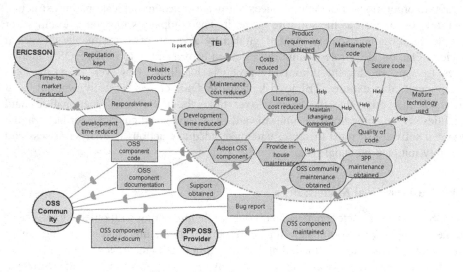

Fig. 2. TEI business goal model

2.2   Risk Models: RiskML

OSS ecosystems rotate around the production and use of OSS software components. RiskML is a modelling language introduced in [4] to capture knowledge related to risks of software components and to support automated analysis. RiskML uses the concepts of *Event* - a change in states of affairs, which may harm goals [7], with a certain *likelihood*, and *significance; Goal* - anything, which is of interest for a stakeholder to obtain or to maintain; and *Situation* - states of affairs, under which risks are possible [8, 9]. Additionally, *Indicators* represent one (simple indicator) or more (composite indicator) gathered measures about a certain property of a software component [8]. By means of transformation functions, indicators inform about the evidence of being in certain situations. The *Indicate* relation represents such transformation, propagating the value of an indicator to the evidence that a situation is satisfied. *Expose, Protect, Increase* and *Reduce* relations raise a target event's likelihood, lower it, raise an event's significance or lower it, respectively. The *Impact* relation represents the negative effect of a certain event on the satisfaction of a given goal. The higher the impact, the higher is the severity of the negative impact. The *exposure* to a certain risk is defined as a combination of the risky event's likelihood, its significance, and the severity of its impact to goals. For details on RiskML see [4].

Figure 3 shows an excerpt of a risk model related to OSS code maintainability (indicators in relationship to situations are not displayed for space reasons, as well as information like likelihood and significance). The model is based on (1) interviews with managers, (2) a literature study on OSS risks [10] and (3) various metrics for code quality, such as complexity metrics, lines of code, and test coverage; such metrics have been shown in [11] to serve as measures of the maintainability of the source code. These measures do not offer absolute measures of maintainability, but assist in identifying where code exhibits properties that are known to decrease (or increase) maintainability, showing correlation e.g. with the time and skills needed to maintain the code. For example, code that is more complex takes more time to be understood by the analysts; this can lead to delays in bug fixing and code maintenance and evolution activities.

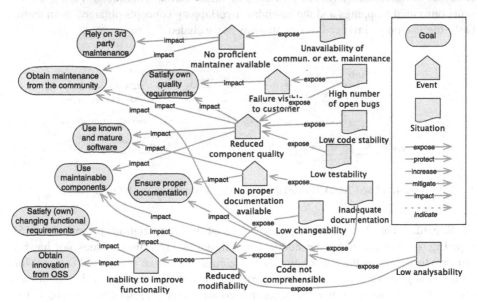

Fig. 3. RiskML risk model for OSS code maintainability

Following the definition of *risk* as a combination of event likelihood and the severity of its impact, the risk model comprises risk events, situations as a means to capture risk indicators, and the goals that are directly impacted by the risk events. For example, the central goal of this risk model is *Use maintainable component* (regarding the quality and modifiability of the code and the availability of documentation), there are high level goals like *Satisfying own quality and functional requirements* throughout the time of maintenance, and lower level goals such as to continuously *Obtain innovation from the OSS* community, to have access to a *proper documentation*, or to rely on a proficient *3rd party for maintenance*.

3 An Integrated Model for Risks and Goals

3.1 Analysis of Overlapping Concepts

Integrating the product models of two methods requires identifying concepts that have the same semantics, and to merge them afterwards. Following the approach proposed in [12], we have opted for using ontological analysis to identify these concepts, mapping the concepts of the two methods to the concepts of a reference ontology. Among possible options (e.g., BWW, Chisholm's, DOLCE, etc.), we have chosen the UFO ontology [13, 14]. UFO is a foundational ontology that has been used to analyse, redesign and integrate language models in a large number of domains.

The starting points are the mappings between both modelling languages concepts (*i** and RiskML) and those of UFO. The concepts which are mapped onto the same or related UFO concepts are considered candidate overlapping concepts for an integrated *i**-RiskML model. Finally, we analyse the overlap and decide whether to map the concepts unconditionally (i.e. in all cases) or under certain conditions. Figure 4 presents our initial mappings and the candidate overlapping concepts obtained from them. Only the mappings involved in our analysis are included.

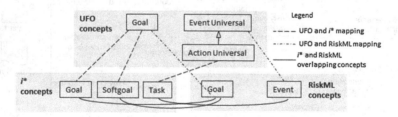

Fig. 4. *i** and RiskML overlapping concepts according to UFO mappings

Regarding the *i** and UFO mapping, we adopt the interpretations presented in [15] (see Fig. 4). We use the same notion for softgoal as [15] since it does not have a uniform treatment, as the paper points out. We consider that an *i* Softgoal* is a goal for which it is possible that two rational agents differ in their beliefs to which situations satisfy it. Conversely, for an *i* Goal* (or hardgoal), the set of situations that satisfy it is necessarily shared by all agents.

Next, we provide the mapping between RiskML and UFO[1]. A *Goal* in RiskML is defined as anything which is of interest for a stakeholder to obtain or maintain. As such, it is satisfied if the corresponding state of affairs is achieved. Goals in UFO are related to sets of intended states of affairs of an agent. UFO contemplates a relation between Situations and Goals such that one or more Situations may satisfy a Goal. In other words, a Goal is a proposition and a particular state of affairs can be the truthmaker of that proposition. Consequently, the RiskML *Goal* concept can be interpreted as a Goal in UFO. *Events* in RiskML model changes in states of affairs. From the UFO ontology, we have that an Event (instance of Event Universal) is a perduring entity, i.e., entities that occur in time, accumulating their temporal parts. Events are triggered by certain Situations in reality (termed their pre-situations) and they change the world by producing a different post-situation. Consequently, the RiskML *Event* can be interpreted as an UFO Event Universal.

Next we describe the *i**-RiskML candidate overlapping concepts identified after analysing the mappings (see Fig. 4). The RiskML *Goal* overlaps both the *i** *Goal* (also called hardgoal) and the *i** *Softgoal*. These concepts map unconditionally, meaning that any *i** *Goal* and any *i** *Softgoal* maps into a RiskML *Goal* and also any RiskML *Goal* maps into either a *i** *Goal* or a *i** *Softgoal*, since they all map into the same UFO concept. The RiskML *Event* overlaps the *i** *Task*. These two concepts map only under certain conditions because: (1) they map into UFO Event Universal and UFO Action Universal, respectively and, (2) according to UFO, only Events deliberately performed by Agents in order to fulfil their Intentions are Actions. Therefore, all *i** *Tasks* map into RiskML *Events* but not all RiskML *Events* map into *i** *Tasks*.

3.2 Analysis of the *Impact* Relation

The *Impact* relation in RiskML relates an *Event* E and a *Goal* G such that the occurrence of E has an effect on the satisfaction of G (e.g. the event post-situation does not satisfy the goal). Since RiskML *Goals* map into *i** *Goals* and *i** *Softgoals*, it follows that goals and softgoals are the only *i** elements that can be the target of an *Impact* relation. However, we argue that *i** *Tasks* and *i** *Resources* could also be involved in impact relations for shortcutting purposes.

As mentioned above, *i** *Tasks* are interpreted as UFO Action Universals. According to UFO, Actions are intentional Events, i.e., events with the specific purpose of satisfying some goals. Therefore, any Action Universal (*Task*) implies the existence of an underlying Goal, explicit or not (i.e., it is a hidden goal). If the hidden Goal of a *Task* is impacted by an *Event*, as a shortcut, we allow to use *Task* as target of the *Impact* relation to avoid making the goal explicit and thus to cause excessive model growth).

The case of *i** *Resources* is quite similar. Intuitively, for any *i** *Resource* we may assume the existence of the underlying Goal on getting the resource available to an agent, explicit or not in the *i** model. Again, for shortcutting purposes, we propose to specify *Resource* as the target of the *Impact* relation.

---

[1] UFO concepts appear underlined in the text whereas RiskML and *i** ones appear in italics.

3.3   Metamodel Integration: The *i**-RiskML Metamodel

We start from the *i** metamodel presented in [16] and the RiskML metamodel in [4]. To integrate them, we take into account the overlapping concepts (Sect. 3.1) and the shortcuts for the *Impact* relation (Sect. 3.2). For each set of overlapping concepts and assuming that each one has a corresponding metaclass in the initial metamodels, we need to decide whether to keep all the corresponding metaclasses of both metamodels, or just the metaclasses from one metamodel. To make such decisions, we slightly adapt the heuristics proposed in [12]. In cases where the concepts are totally equivalent, the simplest solution is to keep only the metaclasses from one metamodel (it must be decided which one); the others are removed. Clearly, the associations in which the removed metaclasses participated need to be reconnected accordingly. In the cases where the mapping of concepts is qualified with a condition specifying under which

Table 1. *i** and RiskML metamodel integration

RiskML	*i**	*i**- RiskML	
Concept	**Concept**	**Metaclasses kept**	**Rationale**
Event	Task	Both (Task subclass of Event)	Task → Event, not (Event → Task)
Goal	Goal, Softgoal	Goal, Softgoal (from *i**)	Equivalent (*i** more fine-grained representation for goals)
Assoc.		**Metaassociations reconnected**	**Rationale**
Impact		relates Event to IntentionalElement (superclass Goal, Softgoal, Task, Resource)	Goal and Softgoal are equivalent to Goal, Task and Resource for shortcutting

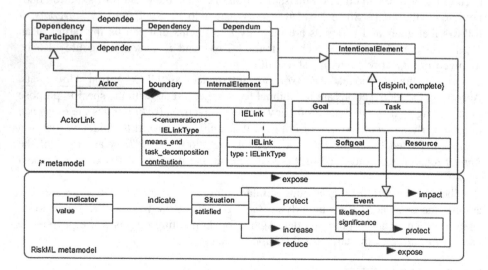

Fig. 5. *i** - RiskML integrated metamodel

circumstances they can be mapped, all metaclasses should be kept and relations should be defined between them. Table 1 summarises the application of these heuristics to our case and Fig. 5 shows the resulting metamodel.

4 Aligning Models for an Organization

In this section, we consider the alignment of a business model and a risk model in a concrete organization. We use the TEI case presented in Sect. 2.

It is worth to mention that the concrete form of the alignment may depend on the business case in which it is done. In this section, we assume that: the risk model is part of a catalogue of reusable models and as such, it cannot be modified. The business model is produced independently of the risk model (either because it already existed, or the business modeller was not aware of the risk model, or even it could have been a conscious decision to avoid bias in the business model). We think that this scenario can be quite usual when analysing the impact of risks in business goals.

4.1 Alignment Method for the *Impact* Relation

According to the results of Sect. 3, we analyse the possible mapping of every impacted goal in the risk model with some intentional element in the business model. In other words, the alignment problem may be stated as:

Given a business model B which contains a set I of intentional elements, and given a risk model R which contains a set G of goals, we want to combine them to produce a new model M in which the goals in G are semantically connected to the intentional elements in I according to the ontological framework defined in Sect. 3.

We define M as initially including B's actors with their corresponding SR diagram as in B. We analyse next the effect of each goal g in the risk model on the initial model M. Let's call M' the model resulting of this step.

Alignment case 1. There is an intentional element x in B such that g can be considered semantically equivalent to x. In this case, the model M' will keep x and will include the impacts from events in R to g but changing x by g.

Alignment case 2. There is an intentional element x in B which subsumes g. In this case, the model M' includes both related through the appropriate model construct (e.g., means-end if x is a goal, or contribution link if it is a softgoal). The impacts from events in R to g are also included in M'.

Alignment case 3. There is no x in B satisfying cases 1 or 2. This means that there is no obvious impact from the risk to any business goal. In this case, it is necessary to further interact with the business analyst with two possible outcomes:

a. g is in fact pointing out a business goal which has been neglected in the initial version of the business model B. The business model needs to be updated and g finally falls into the cases 1 or 2.

b. g is indicating a risk that is not important for the company. In this case, g is not added in M', as well as any other element in R (e.g. situation) related only to g.

4.2   Alignment Application and Results

The alignment between a general risk model (Fig. 3) and an organization business goal model (Fig. 2), follows the iterative process previously defined. The risk model gives awareness on specific issues, which need to be analysed to decide if they may or may not be needed to be addressed in a specific organization, e.g. because a goal is not important in the particular context. The addressed goals are then put in relation to the organization's goals, also joining semantically similar goals from both models.

Applying the guidelines presented above, Table 2 includes the alignment between the models presented in Sect. 2, a business model B (Fig. 2) and a risk model R (Fig. 3). In this concrete example, goals from g1 to g4 are equivalent (*Alignment case 1*). In this case, the elements from B are kept. For g1 and g2, the intentional elements in B are tasks, so the underlying goals to these tasks in B are equivalent to the goals in R. Goals g5 and g6 are subsumed by one element of B (*Alignment case 2*), therefore the elements from R are included in B as goals. Finally, for missing goals (*Alignment case 3*), g7 is included in the TEI business model (as a softgoal) and g8 is discarded.

Table 2. Alignment of Business and Risk models

	g in R	x in B	New link
	Alignment case 1 (equivalent)		
g1	Obtain maintenance from the community	Rely on the OSS community for maintenance	
g2	Rely on 3rd party maintenance	Contract 3PP organization for maintenance	
g3	Use known and mature software	Mature technology used	
g4	Use maintainable components	Maintainable code	
	Alignment case 2 (subsumes)		
g5	Satisfy (own) changing functional requirements	Product requirements achieved	contribution link
g6	Satisfy own quality requirements	Product requirements achieved	contribution link
	Alignment case 3.a (new business goals)		
g7	Ensure proper documentation	Ensure proper documentation (new)	task-decomposition (Adopt OSS component) contribution link (Maintainable code)
	Alignment rule 3.b (discarded)		
g8	Obtain innovation from OSS		

Figure 6 shows the part of M in the TEI running example which includes the alignments above. Concretely, it includes the SR diagram for the actor TEI, including the new elements and the impacts from the Risk Events included in R.

5   Risk Analysis

The alignments described in the previous sections allow us to perform a model-based analysis of OSS ecosystems, linking the metrics of OSS projects to their impact on business goals. Risk analysis in RiskML is a reasoning technique, described in [4],

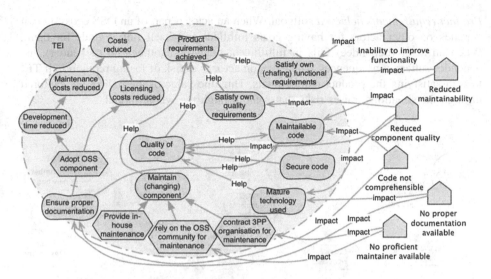

Fig. 6. TEI SR diagram connected to RiskML risk events

which uses forward quantitative inference algorithms to evaluate risk exposure. The algorithm starts from the gathered indicators about OSS projects, and applies inference rules to derive exposure to risky events: firstly, indicator values are mapped onto the satisfaction evidence of situations; afterwards, situation satisfaction raises or lowers the occurrence likelihood of events (*expose* and *protect* relations) or their significance (*increase* and *reduce* relations). The impact of risk events on the software ecosystems is captured by goal analysis, which is a technique for reasoning on *i** models, described in [17]. In a nutshell, it relies on the idea that actors want their goals to be achieved, so well-engineered goal models should ensure goal satisfiability. In goal analysis, intentional elements hold a satisfiability evidence and a deniability evidence, representing the evidence that the intentional element can be achieved or not achieved. Satisfiability and deniability evidence can hold at the same time for the same intentional element, thus representing the existence of contradictory information. Both value types are propagated across an *i** goal model: and-or decompositions propagate satisfiability evidence from operational goals to tactical and strategic goals, while contribution link represent partial or total, positive or negative effect of a source goal to a different one.

Risks affect negatively goals, because they can reduce their chance to be achieved. The *impact* relation represents this negative effect: the more the source event is likely and significant, the more there is evidence that the impacted goal is not achievable. This is depicted in the integrated model in Fig. 7, where the RiskML model concerning maintenance quality risks is plugged into the goal model of an adopter. E.g., the *Reduced component quality* risk event impacts on the *Quality of code* softgoal. If the event is exposed (likely and significant), there is evidence that the softgoal is denied. Once a goal has been impacted, this impact can be propagated across the goal model. The denial evidence of the *Quality of code* softgoal propagates through the model to other goals, having a negative effect on the *Maintainable code* softgoal and on the

Product requirements achieved softgoal. When an actor is part of an OSS ecosystem, it depends on other actors for having goals fulfilled, and itself fulfils goals for them. When an actor (dependee) fails in fulfilling a goal for another one (depender), the depender suffers consequences. The Ericsson actor is at risk of losing reputation if TEI fails in satisfying the product requirements. Thus the value of the indicators, captured through the situations, raises risks that span through the whole model.

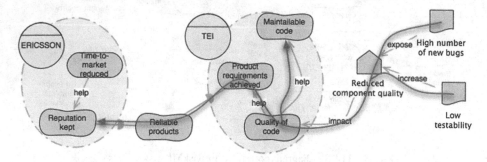

Fig. 7. Example flow of risk propagation in a OSS ecosystem

6 Related Work

Several works dealt with the modelling of risks in the context of organizations through goal-oriented languages. The Goal-Risk framework by Asnar et al. [7] uses *i** to capture, analyse and assess risk at an early stage of the requirements engineering process. They distinguish an asset layer to model business goals, an event layer to model risk events, and a treatment layer to model mitigation actions. We build on top of their approach and extend it in two directions: first, we support complex indicators, connecting risk models to data sources; and, second, we extend the goal analysis support, adding the capability to analyse how risks are propagated across a set of actors. Additionally, the choice of using two different languages - risk and goal modelling - instead of an integrated one, is of help in models reuse, allowing us to study how the same risks may impact on different strategies or different ecosystems.

The KAOS methodology [18] deals with risk management by complementing goal modelling with obstacle analysis that consists in identifying the adverse conditions that may prevent a goal to be achieved [19]. KAOS has a formal representation for goal models, and takes into account partial or probabilistic values for goal satisfaction, allowing quantitative reasoning on the models but does not integrate concrete measures and indicators. Sabetzadeh et al. [20] presents an approach, based on KAOS, which uses statistical reasoning to analyse the problem of introducing a new technology in an organization. Assessing the risk related to the possible non-compliance of the new technology to the fixed standards or internal regulations. CORAS [21] is a model-based approach for security risk analysis and assessment, comprised by a risk modelling language, a process for security analysis, and a tool for reporting risk analysis results. It limits to a defensive risk analysis to protect company assets, and does not rely on a particular reasoning technique. Finally, Grandry et al. [22] integrates an enterprise architecture model and an information system security risk management model by mapping concepts of two metamodels from both domains. A main difference to ours proposal is that it focuses on the management of an enterprise more than on the analysis of a larger organization where the presence of multiple interacting actors and the strategic dependencies among them is of crucial importance. Moreover, our approach also proposes a set of guidelines to align the different models at the level of the model instances.

7 Conclusions and Future Work

In this paper we have presented an ongoing work to analyse OSS adoption from a risk management perspective. The work is motivated by the need of industry actors, represented here by Ericsson, to understand the impact of OSS adoption risks on their business goals, and how that impact can spread over a whole ecosystem. The approach is quite general, though, and could be applied to other kinds of risks, providing the adequate business and risks models. We have chosen two modelling languages that are able to represent the business environment and the underlying risks, and support goal and risk analysis, respectively. To explore the interaction of risks and goals in the OSS ecosystem, we have developed a formal alignment of concepts, and shown how this reflects on the analysis, applying it to the case of OSS adoption.

The main contributions of this work are: (1) the integrated metamodel including goals and risks related concepts (RQ1.1), using the foundational ontology UFO in order to provide an ontological matching between the concepts *Goal* and *Event* from RiskML and *Goal*, *Softgoal* and *Task* of *i**, concluding that a *Risk* can *Impact* on any type of *i** Intentional Elements (*Goal, Softgoal, Task* and *Resource*); (2) a methodology to plug a risk model to a goal model. (RQ1.2); (3) the propagation techniques in order to show how risk exposure can reflect in an evidence of goal denial, which intuitively means that higher risk causes higher possibility that the goal will not be achieved (RQ2). The Ericsson case has been used as preliminary validation. Besides the validation of the formal framework, this case has shown the adequacy of the proposal in an industrial setting. Although we may expect that models may grow in more complete cases, the business and risk models themselves are not expected to grow proportionally, supporting then scalability of the approach. Of course, further validation of this statement is required.

The conceptual alignment described in this paper allowed us to map a formalism for reasoning on risk exposure onto another well-suited to reason on goal satisfaction. While this is good for integration purposes, having a finer-grained analysis technique would help in providing more specific results. Future work goes along several directions. First, we are interested in assigning importance degree to goals, in order to classify the risks impact on the basis of their severity. Also, this will allow us to develop reasoning techniques to select mitigation strategies to reduce risk exposure. Second, it will be important to explore the risk–goal relation in the other way, understanding how OSS adoption can impact the measures and modify the risk exposure. Third, we also need to work further in the alignment between risk and goal models, so that the process that has been depicted in Sect. 4 becomes more prescriptive (e.g. risks impacting on dependums). Lastly, further validation of the approach is needed in order to evaluate its practical implications such as the effort required to align models.

References

1. Driver, M.: Hype cycle for open-source software. Technical report, Gartner (2013)
2. López, L., Costal, D., Ayala, C.P., Franch, X., Glott, R., Haaland, K.: Modelling and applying OSS adoption strategies. In: Yu, E., Dobbie, G., Jarke, M., Purao, S. (eds.) ER 2014. LNCS, vol. 8824, pp. 349–362. Springer, Heidelberg (2014)
3. Yu, E.: Modelling Strategic Relationships for Process Reengineering. Ph.D. thesis, University of Toronto, Toronto, Ontario, Canada (1995)
4. Siena, A., Morandini, M., Susi, A.: Modelling risks in open source software component selection. In: Yu, E., Dobbie, G., Jarke, M., Purao, S. (eds.) ER 2014. LNCS, vol. 8824, pp. 335–348. Springer, Heidelberg (2014)
5. Franch, X. et al.: Managing risk in open source software adoption. In: ICSOFT, pp. 258–264 (2013)
6. Wieringa, R.: Design Science Methodology for Information Systems and Software Engineering. Springer, Berlin (2014)
7. Asnar, Y., Giorgini, P., Mylopoulos, J.: Goal-driven risk assessment in requirements engineering. Requirements Eng. J. **16**(2), 101–116 (2011)
8. Barone, D., Jiang, L., Amyot, D., Mylopoulos, J.: Reasoning with key performance indicators. In: Johannesson, P., Krogstie, J., Opdahl, A.L. (eds.) PoEM 2011. LNBIP, vol. 92, pp. 82–96. Springer, Heidelberg (2011)
9. Siena, A., Jureta, I., Ingolfo, S., Susi, A., Perini, A., Mylopoulos, J.: Capturing variability of law with *Nómos* 2. In: Atzeni, P., Cheung, D., Ram, S. (eds.) ER 2012 Main Conference 2012. LNCS, vol. 7532, pp. 383–396. Springer, Heidelberg (2012)
10. Morandini, M., Siena, A., Susi, A.: Risk awareness in open source component selection. In: Abramowicz, W., Kokkinaki, A. (eds.) BIS 2014. LNBIP, vol. 176, pp. 241–252. Springer, Heidelberg (2014)
11. Heitlager, I., Kuipers, T., Visser, J.: A practical model for measuring maintainability. In: QUATIC, pp. 30–39 (2007)
12. Ruiz, M., Costal, D., España, S., Franch, X., Pastor, Ó.: Integrating the goal and business process perspectives in information system analysis. In: Jarke, M., Mylopoulos, J., Quix, C., Rolland, C., Manolopoulos, Y., Mouratidis, H., Horkoff, J. (eds.) CAiSE 2014. LNCS, vol. 8484, pp. 332–346. Springer, Heidelberg (2014)
13. Guizzardi, G.: Ontological Foundations for Structural Conceptual Models. Ph.D. thesis, University of Twente, The Netherlands (2005)
14. Santos Jr., P.S., Almeida, J.P.A., Guizzardi, G.: An ontology-based semantic foundation for ARIS EPCs. In: SAC, pp. 124–130 (2010)
15. Guizzardi, R.S.S, Franch, X., Guizzardi, G.: Applying a foundational ontology to analyze means-end links in the *i** framework. In: RCIS, pp. 1–11 (2012)
16. López, L., Franch, X., Marco, J.: Making explicit some implicit *i** language decisions. In: Jeusfeld, M., Delcambre, L., Ling, T.-W. (eds.) ER 2011. LNCS, vol. 6998, pp. 62–77. Springer, Heidelberg (2011)

17. Giorgini, P., Mylopoulos, J., Nicchiarelli, E., Sebastiani, R.: Reasoning with goal models. In: Spaccapietra, S., March, S.T., Kambayashi, Y. (eds.) ER 2002. LNCS, vol. 2503, pp. 167–181. Springer, Heidelberg (2002)
18. van Lamsweerde, A., Letier, E.: Handling obstacles in goal-oriented requirements engineering. IEEE Trans. Software Eng. **26**(10), 978–1005 (2000)
19. Cailliau, A., van Lamsweerde, A.: Assessing requirements-related risks through probabilistic goals and obstacles. Requirements Eng. J. **18**(2), 129–146 (2013)
20. Sabetzadeh, M, Falessi, D., Briand, L.C, Di Alesio, S., McGeorge, D., Åhjem, V., Borg, J.: Combining goal models, expert elicitation, and probabilistic simulation for qualification of new technology. In: HASE, pp. 63–72 (2011)
21. Lund, M.S., Solhaug, B., Stølen, K.: Model-Driven Risk Analysis - The CORAS Approach. Springer, Berlin (2011)
22. Grandry, E., Feltus, C., Dubois, E.: Conceptual integration of enterprise architecture management and security risk management. In: EDOC Workshops, pp. 114–123 (2013)

Pragmatic Requirements for Adaptive Systems: A Goal-Driven Modeling and Analysis Approach

Felipe Pontes Guimaraes[1,4]([✉]), Genaina Nunes Rodrigues[2],
Daniel Macedo Batista[1], and Raian Ali[3]

[1] Universidade de São Paulo, São Paulo, Brazil
{felipepg,batista}@ime.usp.br
[2] Universidade de Brasília, Brasília, Brazil
genaina@cic.unb.br
[3] Bournemouth University, Poole, UK
rali@bournemouth.ac.uk
[4] Instituto de Ensino Superior de Brasília - IESB, Brasília, Brazil

Abstract. Goal-models (GM) have been used in adaptive systems engineering for their ability to capture the different ways to fulfill the requirements. Contextual GM (CGM) extend these models with the notion of context and context-dependent applicability of goals. In this paper, we observe that the interpretation of a goal achievement is itself context-dependent. Thus, we introduce the notion of Pragmatic Goals which have a dynamic satisfaction criteria. However, the specification of context-dependent goals' applicability as well as their interpretations make it hard for stakeholders to decide whether the model is achievable for all possible context combinations. Thus we also developed and evaluated an algorithm to decide on the Pragmatic CGM's achievability. We performed several experiments to evaluate our algorithm regarding correctness and performance and concluded that it can be used for deciding at runtime the tasks to execute under a given context to achieve a quality constraint as well as for pinpointing context sets in which the model is intrinsically unachievable.

Keywords: Context dependency · Goal-models · Requirements engineering

1 Introduction

Goal-Models (GM) are well established requirements engineering tools to depict and break-down systems using socio-technical concepts [16]. In other words, it provides the goals for which the system should be designed and the various possible ways to reach those goals.

The variability of goal achievement strategies is the baseline for an actor to adapt by deciding which alternative to adopt as a response to certain triggers or adaptation drivers, e.g. faults, errors, availability of computational resources and newly available services and packages. The dynamic environment in which

© Springer International Publishing Switzerland 2015
P. Johannesson et al. (Eds.): ER 2015, LNCS 9381, pp. 50–64, 2015.
DOI: 10.1007/978-3-319-25264-3_4

the system operates, *i.e.* its context, could also be an adaptation driver. The Contextual Goal Model (CGM) [1] extends the traditional goal model [5,6,15] with the notion of context. Context may be an activator of goals, a precondition on the applicability of certain alternatives to reach a goal and a factor to consider when evaluating the quality provided by each of these alternatives.

However, we advocate another effect of context on CGMs and requirements in general. The interpretation of a goal achievement is itself context dependent. This means that, in certain contexts, the mere achievement of the sub-goals in a goal model does not imply that the parent goal has been achieved. As an example, consider an ambulance dispatch. The goal of arriving at the patient's location in timely fashion would be seen as achieved when this takes 15 min and he/she suffers from dizziness. However, the same goal would not be achieved if the patient suffered from a heart condition. The pragmatism, i.e. dynamic interpretation, is not about the quality but the boolean decision whether a goal is achieved.

In this paper, we introduce the concept of *pragmatic goals* to grasp and model the idea that a goal's interpretation varies according to context. We define the achievability of pragmatic goals as being the capability of fulfilling a goal as interpreted within the context of operation. We also develop and implement an algorithm to compute the execution plan which is likely to achieve a pragmatic goal in a certain context.

In order to validate our approach, we compared the performance and reliability of the answers generated by volunteers and our algorithm [9]. Results showed that volunteers took up to 17 min to provide answers with 26.81 % reliability without considering our algorithmic approach. This brought to our attention the need for an algorithmic approach. We then evaluated the applicability of our modeling and reasoning algorithm by applying it on a case study of a Mobile Personal Emergency Response System. Finally, we performed a scalability analysis to show the usability of our algorithm in pinpointing context sets in which the CGM as a whole may become unachievable, as well as the possibility of using it to support runtime adaptation by laying out an execution plan which is likely to achieve the necessary constraints. Results show that our algorithm is able to lay out an execution plan in less than a few milliseconds in average. Even in the worst case scenario, for a model with 10000 CGM nodes and 20 contexts, it was able to find a suitable plan in less than 1.2 s.

The paper is organized as follows. Section 2 presents the CGM concept on which our model is based. Section 3 presents the pragmatic goals and pragmatic goal achievability concepts. Section 4 presents the proposed model and automated reasoning to decide the pragmatic achievability. Section 5 demonstrates the applicability of the modeling and analysis approach. Section 6 presents related work and Sect. 7 concludes the paper and outlines our future work.

2   The Contextual Goal-Model

Contextual Goal Model, proposed in [1], explicitly captures the relation between the goal model and dynamic environment of a system. It considers context as

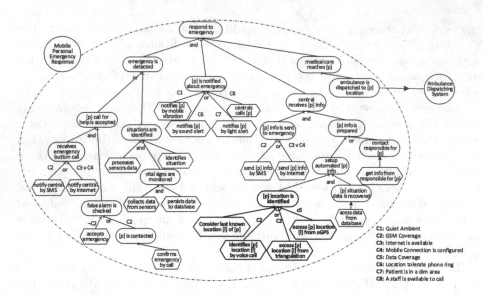

Fig. 1. A CGM for responding to emergencies in an assisted living environment (adapted from [10])

an adaptation driver when deciding the goals to activate and the alternatives - subgoal, task or delegation - to adopt and reach the activated goals. Context can also have an effect on the quality of those alternatives and this is captured through the notion of contextual contribution to softgoals.

Context is defined as the reification of the system's environment, *i.e.*, the surrounding in which it operates [8]. For goal models, context is defined as a partial state of the world relevant to an actor's goals [1]. An actor is an entity that has goals and can decide autonomously how to achieve them. A context may be the time of a day, a weather condition, patient's chronic cardiac problem, etc.

The CGM presented in Fig. 1 depicts the goals to be achieved by a Mobile Personal Emergency Response System which is meant to respond to emergencies in an assisted living environment. The root goal is "respond to emergency", which is performed by the actor **Mobile Personal Emergency Response**. The root goal is divided into 4 subgoals: "emergency is detected", "[p] is notified about emergency", "central receives [p] info" and "medical care reaches [p]" ([p] stands for "patient"). Such goals are then further decomposed, within the boundary of an actor, to finally reach executable tasks or delegations to other actors. A task is a process performed by the actor and a delegation is the act of passing a goal on to another actor that can perform it.

It is important to highlight that not all the subgoals, delegations and/or tasks are always applicable. Some of them depend on certain contexts whether they hold.

3   Conceptualizing Pragmatism in Requirements

Traditionally in a CGM, achieving one (OR-Decomposition) or all (AND-Decomposition) of the subgoals is seen as a satisfactory precondition for achieving the parent goal. We argue that the achievement of some goals would need to be seen from a more pragmatic point-of-view and not as a straightforward implication of the achievement of other goals or the execution of certain tasks. The decision whether a goal is achieved could be context-dependent. Thus we need a more flexible definition of goals to accommodate their contextual interpretation and achievement measures.

The representation of the quality of achievement of a goal as a softgoal is different from the pragmatism of the goal achievement. The pragmatic nature of a goal is not a matter of achieving it with higher or lower quality, but achieving it at all. Also, it has to do with the context at the time of execution and not with the model itself, making the quality requirements more strict or even relaxing them when some contexts apply.

Take the example of Fig. 1: in general, the ambulance may take up to ten minutes to arrive. For a patient with a minor discomfort it can take its time and arrive about ten minutes later without suffering any penalty. On the other hand, if the patient is having a heart attack, one cannot say the goal was achieved. In these situations, the delivered level of quality may not be a separate part from the boolean answer of whether a goal is achieved or not.

A pragmatic goal describes the means to achieve it and also the interpretation of that achievement. This interpretation, which depicts the goal's pragmatic fulfillment criteria, can be expressed as a set of quality constraints (QCs). Unlike softgoals, which are a special type of goal with no clearcut satisfaction criteria [12], these QCs are mandatory and crisp, therefore quantifiable, constraints needed for the fulfillment of a goal and an inherent part of its definition. For instance, take goal "[p] location is identified" from Fig. 1 ($\mathbf{G}_{loc}$ for short): it could be defined as "in order to reach $\mathbf{G}_{loc}$, the location must be identified within an error radius of maximum 500 m and in less than 2 min". Again, this would not suffice, as a radius of 500 m and 2 min might be an over-relaxed condition for patients under critical conditions.

This brings into light another aspect to be taken into account for the pragmatic requirements: the fact that the interpretation for the achievement of a goal is itself context-dependent. We consider that there is a default condition for achieving a goal. However, for specific contexts, we could relax or further strengthen the condition which interprets whether a goal is achieved. We propose that the contextual QCs on the achievement of a goal should be captured together with the other effects of context in the CGM. One advantage of capturing the pragmatic goals within the CGM is to enable reasoning on the possibility of achieving a goal under the current context and QCs. We differentiate these interpretations in the sense that a relaxation condition is not mandatory but a requirement that further strengthen the QC must necessarily be considered.

In the previous example, a QC of getting a location within 500 m in less than 2 min is a default constraint. However, if the user has access to mobile data (context C5) then a much preciser location can be obtained from the GPS. Under these circumstances, a lock within 500 m may seem like an over-relaxed constraint. For a patient with cardiac arrhythmia (context C10), a more strict QC is needed. Suppose that the system has to ensure that an ambulance reaches

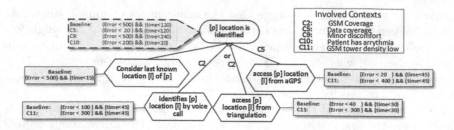

Fig. 2. G_{loc} graphical representation as a pragmatic goal

the patient's home within 5 min. Possibly, in this case, a faster but less precise location would be better suited. The requirements for a minor discomfort (context C9) are also more flexible than those for an arrhythmia (C10). In the three specific contexts, the interpretation must differ from the original baseline.

Figure 2 sums up G_{loc}'s interpretation criteria and presents it as a box connected to the goal itself. However, not only the goal's interpretation may vary according to the context but also the tasks' delivered quality of service (QoS) levels. A task may provide different QoS levels when executed in different contexts. This is also represented in Fig. 2 by the boxes linked to the tasks.

The set of QCs that represent the pragmatic aspects of a goal and the variable delivered QoS levels extend the traditional CGM into a Pragmatic CGM and allow for the reasoning over the achievability of a goal under a certain context.

Achievability of Pragmatic Goals. Pragmatic goals can only be achieved if their provided QoS levels comply with the QCs specified for them, both of which are context-dependent. This means that we extended the basic effect of context on a CGM to cover success and achievement criteria. Such expressiveness enables further analysis for a key adaptation decision: how to reach our goals while respecting the QCs under the current context where the goals' interpretation, the space of applicable alternatives to reach them and the QoS levels provided by the tasks are all context-dependent?

This part also considers situations where it may not be possible to meet the goal's interpretation QoS standards through any of the applicable sub goals, tasks and/or delegations as they deliver not a static but a context-dependent QoS level. In such cases, we classify the goal as unachievable and the reasoning part can explain the reason.

In the example of Fig. 1, if we consider the goal G_{loc} and the contexts impacting on its interpretation, the conclusion is that under a certain context the system may not be able to determine the patient's location with the required precision. This, in practice, does not mean doing nothing. The motivation to do this analysis is because having such knowledge beforehand would allow consideration of other strategies, like adding more alternatives to the same goal to cover a larger range of contexts. At runtime, this conclusion would lead to search for a better variant at a higher-level goal by choosing another branch of an OR-decomposition, which is able to deliver the required quality standard. Therefore, our analysis is both meant for design-time - reasoning to evaluate and validate the comprehensiveness of the solution - and for runtime - searching for the right alternative to reach goals in a specific context.

4   Pragmatic Goal Model

In this section, we concretize our extension to the CGM and elaborate on the new constructs we add as well as their semantics. We mainly enhance the CGM with context-dependent goal interpretations and the expected delivered QoS, which are also context-dependent, for tasks in order to reason about the achievability of the goals for which these tasks are executed.

Figure 3 presents a conceptual model of our extension to the CGM. For the focus of this paper, the CGM could be seen as an aggregation of `Requirements`. A `Requirement` may be specialized into several types: `Tasks`, `Delegations` and `Goals`. A `Delegation` represents when the `Goal` or `Task` (*dependum*) is pursued not by the current but by an external actor (*delegatee*). `Tasks` are performed by the actor in order to achieve a goal. `Tasks` may report the expected delivered quality for each metric through the `providedQuality` method. Goals have a refinements set which define the `Requirements` (subgoals, tasks and/or delegations) that can be used for achieving it as well as a method to distinguish AND- from OR-compositions.

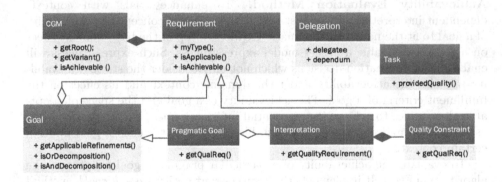

Fig. 3. Conceptual metamodel for our extension to CGMs

`Pragmatic Goals` extend the `Goal` concept with a set of `Interpretations`. A goal `Interpretation` is an abstract concept that has the function of cross referencing a context and the appropriate `Quality Requirement` for that given context. The pragmatic goal is said to be achieved if, and only if, such requirements are met. Otherwise, the goal's delivered QoS is considered inappropriate and the goal is not achieved regardless of achieving one or all of its `Requirements`.

The `Quality Constraints` are expressed in terms of the `applicable Context` in which it holds, the metric that should be considered, the threshold which is a numerical value that represents the scalar value for such metric and the comparison which defines whether the threshold described is the maximum allowed value or the minimum. For instance, to state a quality requirement of at most 250 ms for the execution time when context $C1$ holds, the metric would be "ms", `threshold` would be 250, condition would be "Less or Equal" and applicableContext would be $C1$.

Every `Requirement` inherits the `isAchievable` method. This method can be used either by the final users or by the higher level goals to define whether a particular goal can be achieved for a given quality requirement under the current context. Intermediate goals also have their own predefined interpretation. While this is obviously necessary for the root goal, as the ultimate objective, we also allow certain subgoals to be defined as pragmatic. In principle, actors should be able to impose further constraints on the criteria for achieving any goal within their boundary. The importance of the subgoals quality requirement becomes obvious when dealing with delegation of goals where the external actor may have itself a different, more relaxed or more strict, quality constraint, not necessarily compatible with what the delegator intends.

In this model, both the expectation of delivered quality by the tasks and the quality constraints for the goals, subgoals or delegations are added to the traditional CGM. This is meant to be done by the requirements expert or the domain experts due to the need for specialized knowledge to define such metrics.

Achievability Evaluation Method. To enhance goals with context-dependent interpretation, we must revisit the classical concept of achievability of a goal to fit the nature of Pragmatic CGMs. On top of the basic context effect on a CGM, we enable a higher model expressiveness. Such expressiveness will enable richer adaptation decisions which not only consider the static achievability but also the achievability under the dynamic context and its effect on the fulfillment criteria of a goal. The achievability of a goal and the space of adoptable alternatives to achieve it are essential information to plan adaptation, seen as a selection and enactment of a suitable alternative to reach a goal under a certain context.

To evaluate the achievability of a particular pragmatic goal we present the algorithm in Fig. 1. It implements the `Requirement` entity's *isAchievable* method (Fig. 3) and correlates three context-dependent aspects from the model: (1) the applicable requirements; (2) the goals' interpretations and; (3) the delivered QoS level provided by the tasks.

The algorithm decides whether the root goal is achievable and, if so, lays out an execution plan, i.e., the set of all tasks to be executed, likely to achieve the desired QCs. The algorithm is recursive with a proven linear complexity with respect to the number of refinements in the CGM [9], building on the fact that the CGM is a tree-structured model without loops and that each refinement may be seen as a tree node.

The algorithm considers the root node of the CGM (line 1) and checks whether the root goal is itself applicable under the current context (line 2),

---

Algorithm 1. isAchievable(CGM cgm, Context current, QualityConstraint qualReq)

Require: CGM, current context and desired QCs
 1: Goal root ← cgm.getRoot()
 2: **if** !root.isApplicable(current) **then**
 3:     **return** NULL
 4: **end if**
 5: **if** (root.myType() == task) **then**
 6:     **if** (root.canFulfill(qualReq)) **then**
 7:         **return** new Plan(root)
 8:     **else**
 9:         **return** NULL
10:     **end if**
11: **end if**
12: consideredQualReq ← stricterQC(root.qualReq, qualReq)
13: Plan complete ← NULL
14: deps ← root.getRefinements(cgm, curContext)
15: **for all** Refinement d **in** deps **do**
16:     Plan p ← d.isAchievable(cgm, context, consideredQualReq)
17:     **if** (p != NULL) **then**
18:         **if** (root.isOrDecomposition()) **then**
19:             **return** p
20:         **end if**
21:         **if** (root.isAndDecomposition()) **then**
22:             complete ← addPlanToPlan(p, complete)
23:         **end if**
24:     **else if** (root.isAndDecomposition()) **then**
25:         **return** NULL
26:     **end if**
27: **end for**
28: **return** complete

---

returning NULL if it is not (line 3). In the particular case when the variant's root node is a task (line 5) it can readily decide on the achievability. This is because the task nodes know the expected QoS it can deliver for each metric under the context considered in the CGM. By comparing the delivered QoS and required QCs (line 6), the node can decide whether it is capable or not of delivering such QCs. If it can, it will return a plan consisting only of this task (line 7), otherwise it will return NULL (line 9) and indicate its inability to fulfill the goal's interpretation.

If the root is not a Task, the algorithm will define its quality requirement as the stricter Quality Constraints between its own and the QCs passed on as parameters (line 12) and begin laying out an execution plan to fulfill such QCs (line 13). For each of the applicable refinements, it will evaluate if it is achievable (line 16). If the refinement is achievable then, for OR-decompositions, the algorithm returns this plan immediately (lines 18, 19) and for AND-decompositions it is added to the complete plan (lines 21, 22). Otherwise, if the refinement is unachievable it will immediately return NULL for AND-decompositions (line 25). Finally, for AND-decompositions, should all refinements are achievable it will return the `complete` plan (line 28).

As an outcome, an execution plan is returned for achievable goals. For unachievable goals the NULL value is returned to indicate the inability of fulfilling the required constraints, allowing for alternate means of achieving higher level goals to be explored.

5   Pragmatic Model and Achievability Algorithm Evaluation

To evaluate the need for an algorithmic approach to handle Pragmatic Goals we have conducted a preliminary experiment with volunteers. They were provided with the CGM from Fig. 1 and were asked to check, for a given context set, whether it was achievable or not. The results showed that the volunteers' response had only 26.81 % reliability. For the sake of space, we do not report the results of that experiment in this paper, but they can be accessed on [9]. In this Section, we focused on the evaluation of the proposed model's capability to scale over the Pragmatic CGM size with regard to the amount of goals and contexts and its feasibility as a tool to identify unachievable scenarios.

To do so we used the Goal-Question-Metric (GQM) evaluation methodology [4]. GQM is a goal-oriented approach used throughout software engineering to evaluate products and software processes. It assumes that any data gathering must be based on an explicitly documented logical foundation which may be either a goal or an objective.

GQM's first step is to define high-level evaluation goals. For each goal, a plan consisting of a series of quantifiable questions is devised to specify the necessary measures for duly assessing the evaluation [4]. These questions identify the necessary information to achieve the goals while the metrics define the operational data to be collected to answer each question.

In such a methodology, the main goals of our evaluation are to evaluate: (I) the capability of using our approach for adaptive systems at runtime and; (II) whether it may be used to pinpoint scenarios which render the Pragmatic CGM's root goal unachievable by construction. From these goals, a GQM plan was defined and is presented in Table 1.

Experiment Setup. The experiment setup consisted in evaluating the Pragmatic model and the algorithm's capability to support runtime usage and pinpointing unachievable scenarios. These parts and their evaluations were engineered to provide the metrics demanded by the GQM plan (Table 1).

For the first goal, we evaluated the time to produce an answer, its reliability and a scalability analysis on randomly generated CGMs of different sizes, in terms of goals and contexts amounts. For these time measurements, we have used Java's `nanoTime()` feature, which has the greatest precision, and considered the average of 100 execution repetitions for each model and context set.

Table 1. GQM devised plan

Goal 1: Algorithm's runtime usage capability	
Question	Metric
1.1 How long would it take to come up with an answer?	Execution time
1.2 How reliable are the answers provided?	% of correct answers
1.3 How does it scale over the amount of goals in the model in average?	Execution time
1.4 How does it scale over the amount of contexts in the model in average?	Execution time
1.5 How does the algorithm scale over the amount of goals in the model in the worst case scenario?	Execution time
1.6 How does the algorithm scale over the amount of contexts in the model in the worst case scenario?	Execution time
Goal 2: Pinpoint unachievable context sets	
Question	Metric
2.1 Can it cover all context sets for models increasingly large models?	Context sets coverage

As for the second goal, due to the state explosion problem of possible context sets, we have limited the time amount for each measurement to ten seconds. This way the experiment was faster, and the total amount of time can be easily calculated.

The algorithm from Fig. 1 was implemented[1] using Java OpenJDK 1.7.0_65 and all evaluation tests were performed on a Dell Inspiron 15r SE notebook equipped with a Intel Core i7 processor, 8 GB RAM running Ubuntu 14.10, 64 bits and kernel 3.16.0-29-generic. We also used the EclEmma Eclipse's plugin for ensuring the tests' code coverage.

All experiments to evaluate the correctness and performance of the algorithm were implemented as automated tests under Java's JUnit framework. This guarantees that the evaluation is both effortless and repeatable.

Goal 1: Algorithm's Runtime Usage Capability.

Question 1.1: How long would the algorithm take to come up with an answer? To evaluate the time for the algorithm execution on the CGM of Fig. 1, we executed 1000 iterations of the algorithm for each context set. The results showed that the algorithm took, in average, less than 1 ms to be executed in each of the 4 scenarios.

Question 1.2: How reliable would an answer provided by the algorithm be? To validate its correctness, we implemented tests for all possible context set for Fig. 1.

---

[1] Source code, evaluation mechanisms, and complete result sets are available at https://github.com/felps/PragmaticGoals. Accessed on 2015/03/24.

In each context set we have identified all of the inapplicable tasks - both due to context or quality constraints - and asserted that the outputted execution plan did not contain any of these.

Also, we have implemented over 70 tests for the implementation itself. In total, as reported by the EclEmma plugin[2], this amounted to 95.2 % overall code coverage since EclEmma is unable to cover some code lines due to its implementation. Special consideration was given to the Goal, Pragmatic and Refinement classes as well as to the `isAchievable` method which were extensively tested until achieving 100 % code coverage (Fig. 4). All the tests succeeded thus providing some evidence of the algorithm correctness.

Element	Missed Instructions	Cov.	Missed Branches	Cov.
⊖ Refinement		100%		100%
⊕ Interpretation		100%		100%
⊖ Goal		100%		100%
⊖ Plan		100%		n/a
⊖ Pragmatic		100%		n/a
⊖ CGM		100%		n/a

Fig. 4. Eclipse's EclEmma plugin reporting 100 % code coverage for main classes

One of the purposes of this algorithm is to enable the layout, at runtime, of an execution plan which is able to achieve the CGM's root goal. To do so, the algorithm needs to be able to process models with varying complexities, both in terms of context amount and CGM model size, in a reasonable amount of time so that it will not seriously impact the response time.

Question 1.3 and 1.4: How does the algorithm scale over the amount of goals and contexts in the model in average? To evaluate the algorithm's scalability over the model size in terms of goals and contexts amount, we implemented the following test: for each combination of CGM model size (100–10000 nodes in steps of 100 nodes) and amount of contexts (1–20 contexts), the test would randomly generate 100 CGM models and then the `isAchievable` method was executed 100 times in each model and the average execution time was measured. Finally, it outputs the average execution time for each combination. The resulting average times are presented in Fig. 5a.

Question 1.5 and 1.6: How does the algorithm scale over the amount of goals and contexts in the worst case scenario? Similarly to the previous experiment, to evaluate the algorithm's scalability in the worst case we implemented another test. This time the generated model would be the algorithm's worst case scenario: an achievable Pragmatic CGM composed solely of AND-Decompositions. This forces the algorithm to traverse the whole tree. The test generated random Pragmatic CGMs with sizes varying from 100 to 10000 nodes and, for each model, performed 100 executions. The observed average time per execution is shown in Fig. 5b.

---

[2] http://www.eclemma.org/. Accessed on 2015/03/24.

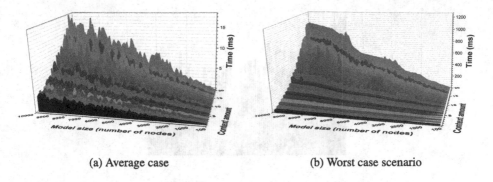

(a) Average case   (b) Worst case scenario

Fig. 5. Algorithm's scalability over the model size, in number of nodes, and context amount

Analysis of the results. With regard to the applicability and reliability, all of our experiments have throughly corroborated the proposed algorithm, with execution times of less than 1 ms and no errors in the 10000 repetitions executed on the CGM represented in Fig. 1. In general, as it can be seen from Fig. 5, the algorithm's execution time grows linearly over the model's amount of goals and contexts, for the average and for the worst-case scenarios. For models consisting of 10000 nodes and 20 contexts the average time to evaluate the random models was about 18 ms and the worst-case scenario took 1081 ms. This was considered a very good result for such large scenario.

Goal 2: Algorithm's Pinpointing Unachievable Context Sets Capability.
To answer question 3.1 (Can it cover all context sets for models increasingly large models?) we implemented one last test which would generate models varying from 100 to 10000 nodes and a fixed set of 15 contexts. For each model size, 10 random models were generated. Finally, on each of these models and for each combination of the 15 contexts, we have executed the algorithm and measured the percentage of possible context sets it was able to sweep, either finding a suitable solution or not, within the first ten seconds. The results can be seen in Fig. 6.

Analysis of the results. As shown in Fig. 6, on smaller models - up to 300 goals - the algorithm was able to fully analyze the 32768 context sets within the 10 s deadline. On larger models - up to 5000 goals - the algorithm was able to analyze around 40 % of the combinations. Even at the limit, with models of 10000 nodes, it was able to cover more than 25 % of the possible combinations. This result suggests that even for models with up to 10000 goals and 20 contexts the complete analysis can be performed within a minute.

6 Related Work

Previous work have tackled similar problems but to the best of our knowledge none has dealt with the dynamic context-dependent interpretation of

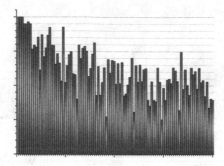

Fig. 6. Percentage of possible context sets analyzed within 10 s vs. model size

requirements and, in particular, of goals. Relevant approaches include the work of: Sebastiani et al., who map the Goal-Model satisfiability problem into a propositional satisfiability (SAT) problem [11]; Souza and Mylopoulos on Awareness Requirement Goals, that define quality objectives for other goals [12,13]; Baresi and Pasquale on Live Goals: goals whose individual behavior change in order to pursue some qualitative objective and bring the system back to a consistent state [2,3]; Dalpiaz et al. on declarative goals: separate goals whose achievement depends on the effects of its refinements on the environment [7]; and the RELAX framework which provides a more rigorous treatment of requirements explicitly related to self-adaptivity [14].

In essence, related work focus on goal-driven adaptation as a system- or model-wise problem. We argue that the notion of pragmatic goals, introduced in our approach, could enrich the rationale of adaptation proposed in those approaches by treating goal-driven adaptation in a case-to-case context-dependent situation.

The Pragmatic Goals' concept differ from the presented work and from traditional softgoals - which do not have clearcut satisfaction criteria [12] - because, unlike [7,12,13], we consider the pragmatic aspect, *i.e.*, the quality objective as an inseparable part from the goal itself: the mere completion of one or all refinements is not enough to achieve a goal, there may be clients' expectations/demands which must be met and which, differently from [14], is itself context-dependent rather than static. We also deal with the identification and reasoning *a priori* and over the CGM as a whole in an effort to keep the system in a consistent state instead of identifying and correcting inconsistent states like [2,3]. Regarding the algorithm itself, our approach enables the algorithm's recursion and achieve linear complexity, by the means of a simplifying assumption: that there are no contributions or denials between different goals. This enabled us to consider the CGM as a tree rather then a generic graph.

Thus, the novelty of our work in comparison to other approaches in requirements-driven adaptation is twofold: (1) The definition of pragmatic goals which means that the satisfaction criteria for goals is context-dependent. (2) The development and implementation of an automated reasoning that can deterministically answer whether the goal is pragmatically achievable and, if it is, point out an execution plan that is likely to achieve it under the current context.

7   Conclusions and Future Work

In this paper we proposed the utilization of a Pragmatic CGM in which the goals' context-dependent interpretation is an integral part of the model. We have also shown why hard goals and softgoals are not enough to grasp some of the real-world peculiarities and context-dependent goal interpretations.

We defined the pragmatic goals' achievability property. A goal's achievability states whether there is any possible execution plan that fulfills the goal's interpretation under a given context. We also proposed, implemented and evaluated an algorithm to decide on the achievability of a goal and lays out an execution plan. We compared the performance of our algorithm to that of a layman's analysis and effectively shown that an algorithmic approach to support the pragmatic goals is needed, considering that human judgment will probably not be fast nor reliable enough. Finally, we performed a scalability analysis on it and shown that it scales linearly over the amount of goals and context amount. We have also shown that, for models up to 10000 nodes and 20 contexts, our algorithm is able to lay out an execution plan in about a second.

For future work, we plan to: (1) integrate this algorithm into a CGM modelling tool; (2) study the possibility of the algorithm to return all task sets instead of a single one and (3) how to enhance the model to integrate task dependencies so that it may represent a context-dependent runtime GM with QoS constraints.

Acknowledgements. This work has been partially funded by CNPq, the EU FP7 Marie Curie Programme through the SOCIAD project and also by Finatec/UnB under call 04/2015, CEPE resolution number 171/2006 and CAPES/PROCAD-grant number 183794.

References

1. Ali, R., Dalpiaz, F., Giorgini, P.: A goal-based framework for contextual requirements modeling and analysis. Requirements Eng. **15**(4), 439–458 (2010)
2. Baresi, L., Pasquale, L.: Adaptive goals for self-adaptive service compositions. In: 2010 IEEE International Conference on Web Services, pp. 353–360, July 2010
3. Baresi, L., Pasquale, L.: Live goals for adaptive service compositions. In: Proceedings of the 2010 ICSE Workshop on Software Engineering for Adaptive and Self-Managing Systems. SEAMS 2010, pp. 114–123 (2010)
4. Basili, V.R., Caldiera, G., Rombach, H.D.: The goal question metric approach. In: Encyclopedia of Software Engineering. Wiley (1994)
5. Bresciani, P., Perini, A., Giorgini, P., Giunchiglia, F., Mylopoulos, J.: Tropos: an agent-oriented software development methodology. Auton. Agents Multi-Agent Syst. **8**(3), 203–236 (2004)

6. Castro, J., Kolp, M., Mylopoulos, J.: Towards requirements-driven information systems engineering: the Tropos project. Inf. Syst. **27**(6), 365–389 (2002)

7. Dalpiaz, F., Giorgini, P., Mylopoulos, J.: Adaptive socio-technical systems: a requirements-based approach. Requirements Eng. **18**(1), 1–24 (2011)

8. Finkelstein, A., Savigni, A.: A framework for requirements engineering for context-aware services. In: STRAW 2001 (2001)

9. Guimarães, F.P., Rodrigues, G.N., Ali, R., Batista, D.M.: Pragmatic Requirements for Adaptive Systems: a Goal-Driven Modelling and Analysis Approach (2015). ArXiv e-prints, http://arxiv.org/abs/1503.07132

10. Mendonça, D.F., Ali, R., Rodrigues, G.N.: Modelling and analysing contextual failures for dependability requirements. In: Proceedings of the 9th SEAMS. pp. 55–64. ACM, New York (2014)

11. Sebastiani, R., Giorgini, P., Mylopoulos, J.: Simple and minimum-cost satisfiability for goal models. In: Persson, A., Stirna, J. (eds.) CAiSE 2004. LNCS, vol. 3084, pp. 20–35. Springer, Heidelberg (2004)

12. Silva Souza, V.E., Lapouchnian, A., Robinson, W.N., Mylopoulos, J.: Awareness requirements for adaptive systems. In: Proceeding of the 6th SEAMS, p. 60. ACM Press, New York, May 2011

13. Souza, V.E.S., Mylopoulos, J.: From awareness requirements to adaptive systems: a control-theoretic approach. In: 2011 2nd International Workshop on Requirements@Run.Time, pp. 9–15, August 2011

14. Whittle, J., Sawyer, P., Bencomo, N., Cheng, B.H., Bruel, J.M.: Relax: Incorporating uncertainty into the specification of self-adaptive systems. In: 17th IEEE RE, pp. 79–88. IEEE (2009)

15. Yu, E.: Modelling strategic relationships for process reengineering. Social Modeling for Requirements Engineering 11 (2011)

16. Yu, E., Mylopoulos, J.: Why goal-oriented requirements engineering. In: Proceedings of the 4th International Workshop on Requirements Engineering: Foundations of Software Quality, pp. 15–22 (1998)

Stress Testing Strategic Goals
with SWOT Analysis

Alejandro Maté[1]([✉]), Juan Trujillo[1], and John Mylopoulos[2]

[1] Lucentia Research Group, Department of Software and Computing Systems,
University of Alicante, Alicante, Spain
{amate,jtrujillo}@dlsi.ua.es
[2] Department of Computer Science, University of Trento, Trento, Italy
jm@cs.toronto.edu

Abstract. Business strategies are intended to guide a company across
the mine fields of competitive markets through the fulfilment of strategic
objectives. The design of a business strategy generally considers a SWOT
operating context consisting of inherent Strengths (S) and Weaknesses
(W) of a company, as well as external Opportunities (O) and potential
Threats (T) that the company may be facing. Given an ever-changing
and uncertain environment, it is important to continuously maintain an
updated view of the operating context, in order to determine whether the
current strategy is adequate. However, traditional SWOT analysis only
provides support for the initial design of business strategy, as opposed to
on-going analysis as new, unexpected factors appear and disappear. This
paper proposes a systematic analysis for business strategy founded on
models of strategic goals and stress test scenarios. Our proposal allows
us to improve decision making by (i) supporting continuous scenario
analysis based on current and future context and, (ii) identifying and
comparing strategic alternatives and courses of action that would lead
to better results.

Keywords: Business intelligence · Strategic goals · SWOT · KPIs ·
Analysis

1 Introduction

A business strategy is intended to fulfill strategic business goals through a specific
course of actions. The achievement of strategic goals is often contingent on internal and external factors that can benefit or harm, make or break business activity.
Due to the importance of such factors, there are multiple conceptualizations for
modelling and analyzing them [5,12], SWOT (Strengths, Weaknesses, Opportunities, and Threats) being one of the most commonly used. SWOT analysis is
applied during business strategy elaboration, both in isolation [3,5,14] as well as
in combination with other techniques such as AHP (Analytic Hierarchy Process)
or ANP (Analytic Network Process) [9,16]. However, during strategic planning,
the impact, or even the presence of many of these factors is uncertain. Indeed,

© Springer International Publishing Switzerland 2015
P. Johannesson et al. (Eds.): ER 2015, LNCS 9381, pp. 65–78, 2015.
DOI: 10.1007/978-3-319-25264-3_5

by taking a quick look at our current context we might identify a number of uncertain factors: Is the price of oil going to keep dropping or will it rise again? Will the euro drop below parity with USD? How long will the sanctions on Russia last? As context evolves, there is increasing uncertainty regarding a business strategy that was initially built for a different context. Unfortunately, current goal modeling and reasoning techniques [4,6,15] are not designed for analyzing and maintaining updated models of a business operating context, nor conducting in-depth stress testing to ensure the viability of currently adopted business strategy.

In this paper, we propose to tackle precisely this problem. Our proposal extends current strategic goal model analysis techniques to support SWOT-based stress testing, consisting of evaluating a business strategy with respect to a series of scenarios. Our proposed analysis takes into account new SWOT factors, in addition to ones considered when a business strategy was designed. We achieve these objectives by means of the following contributions: (i) an extended modelling framework enabling stress testing by using SWOT and scenarios over business strategy models, (ii) a set of reasoning rules using indicators and qualitative values in stress testing, and (iii) a reasoning algorithm that guides the analysis process and returns relevant findings to the analyst. Thanks to these contributions, our approach presents a number of advantages compared to traditional SWOT analysis: (i) it is more flexible and interactive, enabling a user to consider multiple potential futures during stress testing, (ii) it helps us to identify what scenarios lead to success or failure, and estimate the expected probability of occurrence given the current context, and (iii) allows us to comprehensively analyze the impact of external factors on the business strategy, suggesting to decision makers where to focus attention in revising business strategy.

The rest of the paper is structured as follows. Section 2 describes related work, while Sect. 3 presents a running example that we will use throughout the paper. Section 4 discusses modelling extensions required to support stress testing. Section 5 describes the process of stress testing, formalized into an algorithm. Section 6 shows the results of the application of our approach to our running example. Finally, Sect. 7 presents conclusions and sketches directions for future work.

2   Related Work

SWOT analysis and stress testing were introduced to address the need of organizations to anticipate the future and identify external factors that may lead their business strategies to failure. SWOT analysis has been widely applied in different domains [3,5,9,14] as an input for designing business strategies, from universities [3] to forest certification cases [9]. The successful identification of SWOT factors is not a trivial task, and it is usually guided by scenario development [2,3,13,14]. In these scenarios, SWOT factors are key to describe the operating context and help in managing risks. There are multiple ways to develop scenarios [2,13], including future state backcasting, trend analysis, free exploration, etc.

Since using scenarios in an isolated way usually fails to link the results into the strategic planning [13], SWOT factors provide an ideal bridge between scenarios and their stress effects on business strategy.

However, given the uncertainty that surrounds SWOT factors, it can be difficult to prioritize them. A systematic way to prioritize SWOT factors is to pairwise compare them by using the AHP [9]. First, SWOT factors are identified. Then, they are compared, first pairwise within each group (Strengths, Weaknesses, Opportunities, and Threats), and then across groups, in order to determine their relative importance. This way, the company can build a business strategy focusing on the factors that are considered most relevant. However, AHP assumes statistical independence between the factors. Since this is not always the case, ANP is proposed as an alternative [16], allowing SWOT factors to be statistically dependent on each other.

Despite these advances, it may be that (i) not all relevant factors have been identified, and (ii) the impact of each factor may not have been accurately estimated. For example, in the case of Volkswagen during the early 1970s [3], relevant SWOT factors were identified but their impact was underestimated. Generally, much work in the literature makes use of SWOT analysis as input for building the business strategy, but no techniques are provided for carrying forward the analysis into the execution of the business strategy. As a result, it is difficult to understand anomalies [17], and even more challenging to cope with unexpected factors.

While these problems could be alleviated by maintaining an updated model of business goals and external factors, none of the currently available goal modeling languages provides support for stress testing. Several goal modeling languages have been proposed in the literature that allow us to model business goals [1,6,11,15]. Among these models, only the Business Motivation Model (BMM) [11], and the Business Intelligence Model (BIM) [6], are designed specifically for business modeling including external factors, and can be considered as state-of-the-art for business modeling. BMM is oriented towards the definition of textual business strategies, whereas BIM is focused on goal conceptual goal models and reasoning. Thus, we adopted BIM as research baseline for this work.

BIM is founded on the concepts of goal, situation and indicator. Strategic business objectives are modelled as goals, while SWOT factors are modelled as situations.

3   Running Example

The example we present is based on the model presented in [10], where the KPIs of a car manufacturer are modeled on top of the Steel Wheels database. Slight modifications to the BIM-based model have been introduced in order to make the case study more interesting for stress testing. The description has as follows. Steel Wheels is a (fictitious) car manufacturer exploring the impact that future situations may have on its business strategy, shown in Fig. 1. The company is considering two courses of action in order to increase its revenue (main business goal, g1). On one hand, it can focus on creating low-cost designs (g2) by reducing the manufacturing (g3) and distribution costs (g6), thus making its cars affordable for a wider customer base. On the other hand, the company can focus on its high-quality branch and create fancy designs (g9) that attract high-end customers.

Initially, the company considers the low-cost approach as more feasible. However, certain emerging situations, such as an economic downturn (S1) that may reduce the purchasing power of customers, have reinforced the need for a careful analysis of the operating context. During the exploration, experts of the company have identified two scenarios: a future shaped by the effects of economic threats and an alternative one favored by the opening of new markets. The main situations identified using scenario analysis techniques are as follows.

In the economic threats scenario, a continuing economic downturn (S1, threat) is expected to have effects on public policies, leading to a promotion of public transport (S2, threat) in order to alleviate continuing pressure on families. Although the national vehicle fleet is old, and this would favor a renovation plan (S6, opportunity), a promotion of public transport would decrease the likelihood of the government elaborating a renovation plan. Finally, it is also probable that the economic pressure leads to an increase in the cost of raw materials (S3, threat), due to the fluctuating currency values, thus making it more difficult to reduce costs.

On the other hand, with the new opportunities scenario, the negotiation of international trade agreement opens new markets for companies (S4, opportunity). Additionally, the new markets could provide cheaper materials (S5, opportunity), which would reduce manufacturing costs. The combined effect of these opportunities would allow the company to heavily invest in new markets, further expanding its operations, increasing the revenue of Steel Wheels and making it more resilient to the effects of the economic downturn.

These situations have very different impact on the two strategies under study. The low-cost approach is more heavily affected by external factors, whereas the fancy designs strategy is more stable but also introduces a weakness for the company, since expensive cars (S7) are out of scope for public car replacement programs. Nevertheless, this situation also implies that a higher share of the customer base buys expensive cars and thus the company is likely to be less affected by any promotion plan for public transport (S2). Finally, while both approaches are not completely exclusive, adopting one of them hinders the other, as shown by the influence relationships across goals in the diagram.

The results obtained by each approach are quantified by KPIs in the model, that measure the performance of the business with respect to each goal. A green light in Fig. 1 denotes that the target value has been achieved, whereas a yellow light denotes that the performance is only above the threshold. Finally, a red light represents a performance under the specified threshold.

Given this description, the initial question for Steel Wheels is: What would be the likelihood of succeeding if Steel Wheels focused on one strategy or the other? To answer it, we propose to conduct a stress testing analysis.

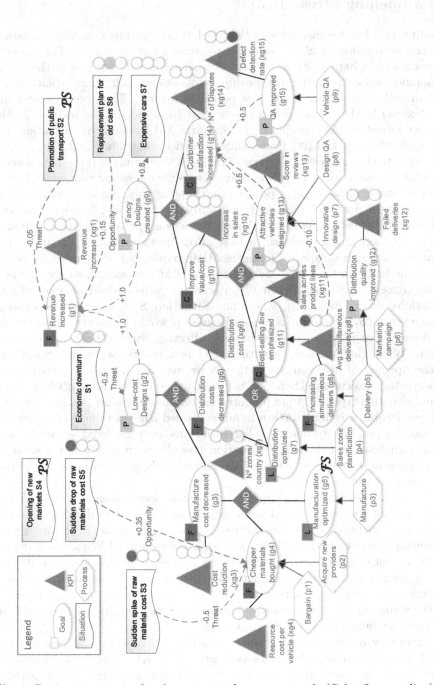

Fig. 1. Business strategies for the car manufacturer example (Color figure online)

4   Modelling Stress Tests

The first problem we encounter in the Steel Wheels strategic model is a lack of constructs to represent all the concepts required for stress testing. For example, we cannot tell what situations are part of each scenario by looking at the model, we cannot model scenario relationships, nor can we tell what situations are certain and what situations are just potential futures. Therefore, the first step is to extend BIM with the constructs needed to support stress testing. To this aim, we introduce the concepts of: (i) Scenario, (ii) Probabilistic Situation, (iii) Material Influence, and (iv) Probability Influence, shown in Fig. 2.

Stress testing is based on scenario analysis. A **Scenario** represents a set of external situations that, together, define a specific operating context and affect other strategic elements in a positive or negative way. Since a scenario is an exploration of potential futures, it will only be active if all the factors that compose it are active. While BIM includes a concept to represent individual situations there are no constructs that allow us to aggregate situations into scenarios, nor can we specify influences that require all of them to be active. Thus, the first construct that we add to the BIM metamodel, shown in Fig. 2, is the concept of Scenario. Formally, a scenario Sc is defined as a set of BIM Situations $Sc = (S_1, ..., S_n)$, where any Situation $S_j \in Sc$ must be external ($Opportunity(S_j)$ or $Threat(S_j)$) as otherwise it would be context-independent.

As we have noted, not all situations affecting Steel Wheels are currently active. While the economic downturn (S1) is a reality, other situations, such as the opening of new markets (S4) may or may not materialize during business strategy execution. Therefore, we need to be able to differentiate between actual situations (reality) from potential situations. Therefore, we introduce the concept **ProbabilisticSituation** in our extended BIM metamodel, and include an isActive property to manage what situations are considered active during a certain analysis. A ProbabilisticSituation has all the properties of a situation in addition to a certain probability (likelihood) of being active during the time span of the business strategy.

In order to estimate the impact that a situation has on the Steel Wheels business strategy, we also need to specify its strength. Indeed, while a sudden drop in raw materials price does not need to become a crash (S5) in order to be beneficial, the extent to which the situation benefits Steel Wheels will depend on the extent of the drop. Related work on business modeling does not define a clear criteria to measure the strength of a situation. In [18] indicators are used to determine only if a situation is active (meets its target) or not, whereas in [6] the whole range [-1,1] of possible values is propagated. In order to avoid this problem, we define the concept of Intensity.

Intensity represents the extent to which a Situation is taking effect, measured according to the value of its associated KPI. KPIs take values within the range [-1,1], are normalized according to their current value (cv) compared to a target value (tv), threshold value (th), and worst value (wv). Negative values for indicators [-1,0) denote that the object measured (a situation in this case) is performing below expectations (th). Intensity takes real values in the range

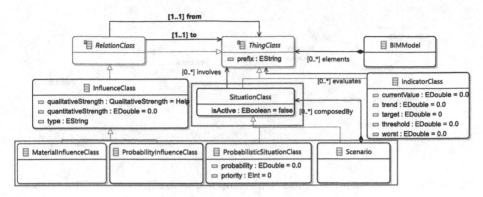

Fig. 2. Extended BIM metamodel

[0,1], from no effect (0) up to full effect (1). According to this description, the Intensity of a Situation is calculated as:

$$Int(S_i) = \begin{cases} (\frac{cv_i - th_i}{tv_i - th_i}) & if\, cv_i \geq th_i \\ 0 & otherwise \end{cases} \quad (1)$$

For example, an economic downturn measured as the GDP drop that is below the threshold value would have an intensity of 0 in terms of how it affects the business strategy or other situations. Once the economic downturn surpasses the established threshold, it would have an increasing effect as the GDP drop increases. For situations that lack an indicator, a correspondence can be established between qualitative and quantitative values, as shown in [6].

Furthermore, the same scenario or situation can have different effects depending on the context [11]. For example, a much stronger currency (not considered on the current analysis) can be a threat (external, negative) to Steel Wheels since it is an international company, and a stronger currency makes its products more expensive in other countries. However, for a domestic competitor the same situation can be an opportunity (external, positive). Situations and scenarios do not only affect the company, but also hinder or magnify other situations and scenarios [9]. For example, by producing expensive cars (S7), Steel Wheels is unable to take advantage of car renovation plans (S6) but also mitigates the effect of public transport promotion (S2). While BIM provides an extension of the Influence class to capture these relationships (Logical influence [6]), this link assumes that both the situation modeled as well as its dual have a symmetric relationship with respect to the business strategy. For example, an economic downturn with a negative indicator would be interpreted as a positive influence for business goals (economic boom). Nevertheless, the assumption of symmetry does not always hold and can lead to undesired results [4]. For example, a successful marketing campaign attracting public attention can have a positive impact on purchases. However, a failed marketing campaign that does not attract any attention simply does not affect purchases.

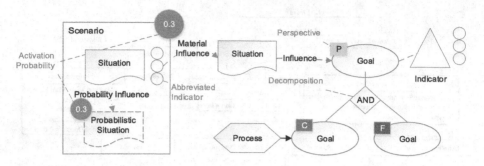

Fig. 3. Notation for new constructs (left) compared to existing ones in BIM (right)

In order to address this lack of expressivity, we introduce a new relationship, Material Influence, which extends the Influence metaclass. A **Material Influence** from a situation S_1 towards a goal or situation S_2 means that the situation will only have an effect on S_2 if S_1 (i) is active and (ii) its intensity is bigger than 0.

In addition to magnifying or mitigating the effects of other situations, situations can also favor or hinder the appearance of other situations. For example, an economic downturn (S1, threat) increases the likelihood of the government promoting public transport (S2, threat) in order to alleviate economic pressures on the population. In turn, adopting such policy diminishes the chances of the government elaborating a renovation plan (S6, opportunity). This is represented as a **Probability Influence** relationship. While BIM provides a similar relationship (Probabilistic Influence), its semantics are slightly different than the ones in our example. Compared to the previous example, a probabilistic link denotes that an economic downturn may have an effect on the strength (performance) of the public transport policies with a certain likelihood, but it does not alter the likelihood of the later happening. Therefore, the properties of the Probability Influence are the same as those included in the InfluenceClass. However, compared to the Probabilistic Influence, it affects the probability of the target situation rather than its performance. As a result, these relationships do not have any effects on normal situations that do not have a likelihood of activating. In order to include this new relationship, we extend the InfluenceClass and add a new type of relationship, the ProbabilityInfluenceClass.

As BIM includes a visual notation, all new concepts have been assigned a compatible a visual representation. The representation of new concepts described in this section is shown on the left hand of Fig. 3, together with the existing elements in BIM shown on the right hand. Moreover, to avoid increasing the complexity of business strategy diagrams, we define a new viewpoint for BIM, the Stress Test viewpoint. This new viewpoint, shown in Fig. 4, includes only situations and the relationships between them, thereby separating stress testing aspect from business strategy concerns.

We can now proceed to analyze the performance of the Steel Wheels strategy for each alternative.

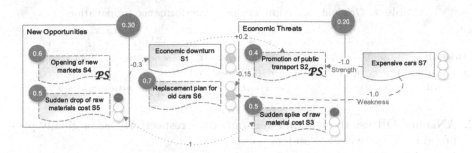

Fig. 4. Stress test viewpoint showing the scenarios in our example

5 Reasoning with a Stress Test

In order to determine how the Steel Wheels strategy will perform depending on the alternative chosen, we need to analyze how each alternative impacts the performance of the overall business strategy, and how the different situations identified affect this performance. For this task, we need to reason and propagate the expected performance values through the BIM model.

There are several reasoning techniques supported by BIM [6], although they are not designed to be used for stress testing. Each reasoning technique differs depending on the amount of information available. Those specifically tailored to work with indicators are based on business formulae and well known relationships or conversion factors across business elements. When formulae and conversion factors between indicators are unknown, BIM falls back to existing goal reasoning techniques [4], treating indicators as quantitative goals.

Goal reasoning techniques are based on label propagation. Label propagation can either be forward, in order to identify the status of a high level goal, or backwards [7], in order to identify potential alternatives that lead to achieving the desired value. While label propagation can also be quantified and extended to indicators by using satisfaction (per^+) and denial (per^-) values, it does not provide adequate support for stress testing. On one hand, stress testing makes use of pre-existing indicator values, which are then stressed to evaluate their performance in different contexts. However, pre-existing values in goal reasoning are considered definitive and will not be altered. On the other hand, existing goal reasoning generates indeterministic values (conflicts [7]) for indicators, where an indicator is both satisfied and denied at the same time, even when all the existing relationships and values in the model have been specified.

In order to address these pitfalls, we adapt the proposal in [6]. The adapted rules are shown in Table 1, where the first two rotes denote the original expressions for satisfaction and denial, while the third row represents the new single value (Intensity) calculated. The most significant modifications are summarized as follows:

1. The value propagated by Material Influences is a single value, Intensity, instead of per^+ and per^- values.

Table 1. Original vs. adapted rules for performance propagation

	$i_2, i_3 \overset{and}{\longmapsto} i_1$	$i_2, i_3 \overset{or}{\longmapsto} i_1$	$i_2 \overset{w+}{\longmapsto} i_1$	$i_2 \overset{w-}{\longmapsto} i_1$
Original(per^+)	$per^+(i_2) \times per^+(i_3)$	$per^+(i_2) + per^+(i_3)$	$w \times per^+(i_2)$	N/A
Original(per^-)	$per^-(i_2) + per^-(i_3)$	$per^-(i_2) \times per^-(i_3)$	N/A	$w \times per^-(i_2)$
Adapted	$min(i_2, i_3)$	$max(i_2, i_3)$	$i_1 + w \times Int(i_2)$	$i_1 - w \times Int(i_2)$

2. AND and OR use the min and max operators respectively, as originally proposed in [4] instead of $\times$ and $+$.
3. Logical operators take precedence against influences.
4. All influences are calculated at the same time, as the sum of all incoming influences on each element. The result is added to the existing value of the target indicator.

It is worth noting that, in general, richer mathematical functions, such as sigmoid, quadratic, etc. can be used during influence propagation if the specific influence relationship formula is known. For simplicity, and assuming a lack of detailed knowledge, we use $F(Int(i_j)) = Int(i_j)$.

We finally need define an efficient way to apply the reasoning rules and gather the results.

Stress testing all potential combinations of active/inactive probabilistic situations individually can be time consuming. Indeed, a naive algorithm requires 2^n passes (n being the number of probabilistic situations) over the goal model in order to test every possible combination of situations. Instead, we propose an algorithm that simplifies the problem and reduces the number of passes required. The algorithm, shown in Algorithm 1, works as follows.

First, we start by assigning values to some of the goal indicators and situations in the model, representing expected values of indicators at the point in time that we are stress testing. These values will be the baseline, and used as input for the algorithm. Values can be quantitative or qualitative, as those used in goal reasoning algorithms. Qualitative values allow us to assign values to goals and situations that do not have a known indicator. Full satisfaction, partial satisfaction, partial denial and full denial are mapped into [1,0.5,-0.5,-1] values respectively [6]. In Figs. 1 and 5 we can see the values for the baseline in our example, composed by the set of goals $(G_4, G_5, G_7, G_8, G_{11}, G_{12}, G_{13}, G_{15})$, and the set of situations $(S_1, ..., S_6)$. Once we have a baseline, the next step is to partition the set of potential situations, obtaining P (line 2).

A partition is a set of probabilistic situations that could be active simultaneously. Partitions are calculated by first removing cycles from the situation graph, obtaining multiple directed acyclic graphs. Then, each resulting set of situations is analyzed as an influence diagram [8] and simplified. The simplification process is performed by assuming that the situation that was previously on a cycle is active, and then discarding those situations that do not influence neither other situations nor goals. In our case, according to Fig. 4, we first obtain two graphs and assume that S3 is active in the first one, while S5 is active in the second one. As S3 and S5 are mutually exclusive, they are removed from the graph where

Data: GoalModel: G, InitialValues: V
Result: R
1 $R = \emptyset$;
2 $P = PartitionSituations(G, V)$;
3 **foreach** $p \in P$ **do**
4      $v = PropagateValues(G, V, p)$;
5      $prob = CalculateProbabilities(G, P)$;
6      $r_v = Normalize(v); r_p = prob$;
7   append(r,R);
8 **end**
9 **return** R;

Algorithm 1. Stress testing algorithm in pseudocode

they are not assumed to be active. Then, as S4 only influences other situations or goals if S5 is active, it is removed from the graph where S3 is active. As a result of partitioning we have two partitions in P: one formed by (S_2, S_3, S_6) and another one formed by (S_2, S_4, S_5, S_6). This allows us to propagate in parallel the values for all the possible combinations of situations within the same partition.

Once we have obtained the partitions, we start the propagation of values (lines 3 to 7). First, we propagate the values through the situations in the partition, assuming all possible combinations of active situations. Then, we start the propagation through the strategic goals as follows. First, we mark each node in the baseline as visited and the rest as unvisited. For each visited node, if a probabilistic situation is influencing it, we propagate its set of values using the rules in Table 1, and add them to the set of values in the node. Then, for each unvisited node, first we calculate its value according to their children. Afterwards, we calculate the combined effect of all influences affecting the node and remove it from the unvisited list. The process is repeated until there are no more nodes to be visited.

Finally, the likelihood of each performance is calculated by using the probabilities of the situations that were considered active for that value (line 5).

After the propagation process is finished, the value of the target nodes is normalized and classified into success, struggling or failing according to their associated indicators, and the resulting model is stored (lines 6,7). The process is repeated for the next partition until no partitions are left.

By following this algorithm, we propagate in parallel all the possible combinations of situations and reduce the number of iterations to $|P|$. In practice, $|P|$ will usually correspond with the number of scenarios explored, as they tend to be exclusive with each other, thus leading to separate partitions.

6 Evaluation

Our algorithm was applied to the extended strategic model, obtained by substituting the situations depicted in Fig. 1 with those from Fig. 4 and updating the relationships between situations and goals. The algorithm was carried out

Low Cost	Initial Performance Values
G4	0,8
G5	1
G7	0,6
G8	1
G12	0,3
G11	0,6
G13	0,3
G15	-0,2

Fancy Designs	Initial Performance Values
G4	-0,1
G5	0,5
G7	0,3
G8	0,2
G12	0,85
G11	0,8
G13	0,7
G15	0,8

Situations	Likelihood	Intensity
S1	100,00%	0,4
S2	40,00%	0,5
S3	50,00%	1
S4	60,00%	0,5
S5	50,00%	1
S6	70,00%	0,7
S7	100,00%	Derived

Fig. 5. Forecasted performance values by focusing on the low-cost (left) or the fancy car design (right) strategies

Table 2. Stressed performance values of the highest level goal

Context/Strategy	Low cost	Fancy designs
Economic threats	Max: 0.75 Avg: 0.44 Min: 0.13	Max: 0.72 Avg: 0.71 Min: 0.7
New opportunities	Max: 1.04 Avg: 0.82 Min: 0.62	Max: 0.77 Avg: 0.73 Min: 0.7

manually using with the aid of an excel spreadsheet. All normalized forecasted KPI values for the different leaf goals and situations are shown in Fig. 5, together with the likelihood of each situation becoming active (100 % denotes active situations). Using these values as an input, the algorithm was able to extract a number of relevant insights for Steel Wheels.

Given space limits, we focus on those related to the highest level goal in the Steel Wheels strategy, shown in Table 2. As we initially expected, the low cost strategy is affected most by the situations, leading to more extreme values. However, it is also the case that it is the only alternative that can achieve complete success.

These results constitute an answer the initial question, but we can provide more insights to make an even better selection for a strategy. For example, what future situations can lead Steel Wheels to success or failure? In order to answer this question we extract the situations that are present only above or below a certain threshold. According to the results obtained, we analyze those situations that result in desired target performance (1) or almost failing (0.25) The analysis shows that the combined appearance of S2, S4, and S6, make the low-cost strategy succeed, whereas the appearance of S3 leads to almost certain failure. However, the likelihood of complete success is only 17,30 % compared to a 50 % chance of sub-par performance. Therefore, this alternative may not be worth the risk and the company would be better focusing on creating fancy designs. Finally, we consider what are the goals most affected by situations? As expected from the strategic model, results point to G1 and G4 as the ones suffering the highest variation depending on the context. With these insights, Steel Wheels can now not only select an adequate strategy, but also prepare actions to mitigate the impact that SWOT factors have on specific goals, for example by looking for alternative providers or materials in order to mitigate S3.

Finally, if one of these -or any other unexpected factor- becomes a reality, Steel Wheels can easily update the stress testing model and re-run the analysis to obtain up-to-date information and make a more informed decision on how to proceed with its strategy and what external factors and goals should be prioritized.

7   Conclusions and Future Work

We have presented a stress testing form of analysis that allows decision makers to conceptualize different scenarios with respect to which they can evaluate their business strategies. This allows decision makers to identify what scenarios potentially lead to success or failure and what strategic alternatives are more tolerant to external factors. We have presented the theoretical foundations of the framework and extended an existing metamodel for business modeling in order to provide a complete framework for stress testing strategic goals. Moreover, we have defined a set of rules for propagating the effects of situations on goals throughout the strategy, and have elaborated an algorithm to apply these rules automatically over a given business strategy. Finally, we have presented the application of our framework by means of an illustrative example.

To the best of our knowledge, this is the first proposal that supports stress testing of strategic goals, and allows decision makers to update the model as context changes, obtaining an updated vision of the future as unexpected factors appear and envisioned situations become a reality or fail to materialize.

Finally, our future work is twofold. On the one hand we intend to better integrate time within the analysis, in order to cope with situations shorter than the business plan time span. On the other hand, we plan to extend current tool support for BIM to also support stress testing of strategic goals using SWOT factors.

Acknowledgments. This work has been partially supported by the GEODAS-BI (TIN2012-37493-C03-03), and by the European Research Council (ERC) through advanced grant 267856, titled "Lucretius: Foundations for Software Evolution" (04/2011/2016) http://www.lucretius.eu. Alejandro Maté is funded by the Generalitat Valenciana (APOSTD/2014/064). Special thanks to Elda Paja for insightful discussions during the development of this work.

References

1. Amyot, D.: Introduction to the user requirements notation: learning by example. Comput. Netw. **42**(3), 285–301 (2003)
2. Duinker, P.N., Greig, L.A.: Scenario analysis in environmental impact assessment: improving explorations of the future. Environ. Impact Assess. Rev. **27**(3), 206–219 (2007)
3. Dyson, R.G.: Strategic development and swot analysis at the university of warwick. Eur. J. Oper. Res. **152**(3), 631–640 (2004)
4. Giorgini, P., Mylopoulos, J., Nicchiarelli, E., Sebastiani, R.: Formal reasoning techniques for goal models. In: Spaccapietra, S., March, S., Aberer, K. (eds.) Journal on Data Semantics I. LNCS, vol. 2800, pp. 1–20. Springer, Heidelberg (2003)

5. Hill, T., Westbrook, R.: Swot analysis: it's time for a product recall. Long Range Plan. **30**(1), 46–52 (1997)
6. Horkoff, J., Barone, D., Jiang, L., Yu, E., Amyot, D., Borgida, A., Mylopoulos, J.: Strategic business modeling: representation and·reasoning. Softw. Syst. Model. **13**(3), 1015–1041 (2014)
7. Horkoff, J., Yu, E.: Finding solutions in goal models: an interactive backward reasoning approach. In: Parsons, J., Saeki, M., Shoval, P., Woo, C., Wand, Y. (eds.) ER 2010. LNCS, vol. 6412, pp. 59–75. Springer, Heidelberg (2010)
8. Howard, R.A., Matheson, J.E.: Influence diagrams. Decis. Anal. **2**(3), 127–143 (2005)
9. Kurttila, M., Pesonen, M., Kangas, J., Kajanus, M.: Utilizing the analytic hierarchy process (AHP) in SWOT analysisa hybrid method and its application to a forest-certification case. Forest Policy Econ. **1**(1), 41–52 (2000)
10. Maté, A., Trujillo, J., Mylopoulos, J.: Conceptualizing and specifying key performance indicators in business strategy models. In: Atzeni, P., Cheung, D., Ram, S. (eds.) ER 2012 Main Conference 2012. LNCS, vol. 7532, pp. 282–291. Springer, Heidelberg (2012)
11. Object Management Group: Business Motivation Model (BMM) 1.2 (2014). http://www.omg.org/spec/BMM/1.2
12. Porter, M.E.: The five competitive forces that shape strategy. Harvard Bus. Rev. **86**(1), 25–40 (2008)
13. Postma, T.J., Liebl, F.: How to improve scenario analysis as a strategic management tool? Technol. Forecast. Soc. Change **72**(2), 161–173 (2005)
14. Schoemaker, P.J., van der Heijden, C.A.: Integrating scenarios into strategic planning at royal dutch/shell. Plann. Rev. **20**(3), 41–46 (1992)
15. Yu, E., Giorgini, P., Maiden, N., Mylopoulos, J. (eds.): Social Modeling for Requirements Engineering. Mit Press, Cambridge (2011)
16. Yüksel, İ., Dagdeviren, M.: Using the analytic network process (ANP) in a SWOT analysis-a case study for a textile firm. Inf. Sci. **177**(16), 3364–3382 (2007)
17. Zoumpatianos, K., Palpanas, T., Mylopoulos, J.: Strategic management for real-time business intelligence. In: Castellanos, M., Dayal, U., Rundensteiner, E.A. (eds.) BIRTE 2012. LNBIP, vol. 154, pp. 118–128. Springer, Heidelberg (2013)
18. Zoumpatianos, K., Palpanas, T., Mylopoulos, J., Maté, A., Trujillo, J.: Monitoring and diagnosing indicators for business analytics. In: CASCON 2013, pp. 177–191. IBM Corporation (2013)

A Method to Align Goals and Business Processes

Renata Guizzardi$^{(\boxtimes)}$ and Ariane Nunes Reis

Ontology and Conceptual Modeling Research Group,
Federal University of Espírito Santo, Vitória, Brazil
rguizzardi@inf.ufes.br, arianenreis@gmail.com

Abstract. Business Process Modeling (BPM) has been for a number of years in the spotlight of research and practice, aiming at providing organizations with conceptual modeling-based representations of the flow of activities that generate its main products and services. It is essential that such flow of activities is engineered in a way to satisfy the organization's goals. However, the work on BPM still makes shy use of goal modeling and the relation between goals and processes is often neglected. In this paper, we propose a method that supports the analyst in identifying which activities in a business process satisfy the organization's goals. Moreover, our method allows reasoning regarding the impact of each of these activities in the satisfaction of the strategic (i.e. top) goals of the organization. The results of this analysis may lead to reengineering, and grant the analyst with the means to design higher quality BPMs. Besides describing the method, this paper presents a preliminary evaluation of the method by the means of an empirical study made in a controlled environment.

1 Introduction

Competitive businesses and an ever-changing market have demanded that current organizations constantly evolve. To achieve that, it becomes necessary to develop a deep understanding of the organizational processes and systems. This motivates an increasing interest in Business Processes Modeling (BPM), the discipline concerned with explicitly capturing, by applying conceptual modeling languages, the flow of activities that generate the main products and services offered by the organization [1].

However, modeling the flow of activities may not be enough to provide the organization with competitive advantage. It is also important to grasp if the current activities and business processes are in line with the goals of the organization. This idea is supported by Rosemann and vom Brocke [2], who claim that strategic alignment is one of the six core elements of BP Management. Although goal modeling is often supported by BPM platforms, it has been regarded by BP practitioners as of secondary importance, and little is explored regarding the relation among processes and goals.

Understanding if and how processes achieve the operational and strategic goals of the organization may guide the decision regarding which activities and processes should become priorities. This may be realized by trying to align (map) activities and business processes to the organization's goals. Moreover, aligning processes and goals may help the analyst to understand if there is any goal being neglected. If a goal is not

© Springer International Publishing Switzerland 2015
P. Johannesson et al. (Eds.): ER 2015, LNCS 9381, pp. 79–93, 2015.
DOI: 10.1007/978-3-319-25264-3_6

aligned to any activity or business process, then one of the following may be happening: (i) the business process is poorly modeled and an activity should be added to make explicit something that is already done in practice; (ii) the business process needs to be reengineered to accommodate one or more new activities, so that the goal is accomplished; or (iii) the goal is no longer important to the organization and should be removed from the goal model.

Many organizations today develop information systems based on their modeled business processes [3]. Thus, aligning goals and business processes may also assist in the development of systems that are more in line with the organization's goals.

Although, recently, some works have focused on the alignment of goals and processes [4–7], a systematic method to support goal and process alignment is still missing. This paper addresses this gap, by describing the results of an ongoing effort towards the alignment of goals and business processes. In a previous work, we have theoretically addressed this topic [8], while this paper focuses on a step-by-step method to support goals and business processes alignment. This method aims at modeling which particular activities or processes achieve the goals of the organization, also providing a reasoning mechanism to verify the impact of the execution (or non-execution) of these activities and processes have on goal satisfaction. Moreover, the method allows one to find inconsistencies in the BP model, supporting the analyst in building models of higher quality and in designing business models more in line with organizational strategies. In this context, goals play the role of drivers of business process improvement.

The remaining of this paper is organized as follows: Sect. 2 presents the applied goal modeling and BPM languages; Sect. 3 describes the proposed method, illustrating its steps with the use of a running example; Sect. 4 discusses a preliminary evaluation conducted by the means of an empirical study; Sect. 5 compares the method with some related works; and finally, Sect. 6 concludes this paper.

2 The Adopted Modeling Languages

2.1 Goal Modeling Language and Reasoning Mechanism

Languages and solutions supporting BPM still make shy use of goal and strategic modeling. This is the case, for example, of the ARIS platform, one of the most used BPM solution in industry. While supporting goal modeling, ARIS proposes a modeling language with very low expressivity, which basically allows relating macro-processes to the leaves of a goal tree, without however distinguishing different kinds of relations between goals and processes and with no reasoning support. Works developed in the context of GORE (Goal-Oriented Requirements Engineering), on the other hand, provide much more powerful modeling languages such as the ones adopted by i* [9] and KAOS [10], for example. Aiming at profiting from this earlier work, we here adopt the Tropos language [11], an i* dialect.

In Tropos, goals are modeled in the perspective of an actor, and often lead to treestructures, where top goals are refined into lower-level ones, mainly by the means of AND and OR decompositions. As one can grasp by intuition, when a top goal is decomposed in two sub-goals by the means of an AND-decomposition, only the satisfaction of the two sub-goals lead to the satisfaction of the top one. If however, a top goal is OR-decomposed into two sub-goals, it is sufficient that only one of them is satisfied in order to satisfy the top goal. Figures 2, 4 and 5 are examples of Tropos models.

Giorgini et al. [12] propose a Tropos-based quantitative approach for verifying goal satisfaction, based on the propagation of values in the goal tree. This approach allows two kinds of propagation, forward and backward propagation. In the forward propagation, initial satisfaction values are attributed to the leaf goals and are propagated to the root goals, thus determining how much evidence there is to the satisfaction of the root goal, given the probability of satisfaction of the leaf goals; while in backward propagation, a desired final satisfaction value is given to the root goal, and then propagated from the root to the leaf goals, so as to determine how much evidence regarding the leaf goals satisfaction is need in order to obtain the desired evidence of satisfaction for the root goal.

To propagate the satisfaction values, it is necessary to take into consideration the labels of the relationships between the goals, as follows:

- AND-decomposition: the decomposed goal is satisfied only with the complete satisfaction of the subgoals. However, even if the evidence of all the subgoals' satisfaction is partial, this evidence propagates to the decomposed goal.
- OR-decomposition: the decomposed goal is satisfied if one or more subgoals are completely satisfied. However, even if the evidence of one of the subgoals satisfaction is partial, this evidence propagates to the decomposed goal.
- Contribution: the contributing goal's satisfaction propagates to the contributed goal, based on a particular contribution (impact) weight. The reasoning approach supports positive and negative contribution weights, both for goal satisfaction and denial. For reasons of space limitation, in this paper we only consider positive contributions, for the case of goal satisfaction.

2.2 Business Process Modeling Notation

For modeling BPs, we adopt BPMN [13], which has become a well-accepted standard, being implemented in many BPM tools and largely adopted, both by researchers and practitioners. BPMN allows the flow of activities to be modeled, by depicting the order in which activities occur. The BPMN elements used in this paper are:

- *pool:* represents a participant (actor) in the process, delimiting a space where the activities performed by that participant are modeled;
- *task (rectangle):* represents a process activity;
- *start event (circle):* triggers the beginning of the process;
- *end event (circle):* determines the process is ended;
- *intermediate event (thick-boarded circle):* models an event that occurs in the middle of the process, usually leading to some decision and altering the flow of activities.

- *exclusive gateway (diamond):* placed in the beginning of a fork of activities, determines that only one of the forked paths is followed.

Figures 1 and 3 present examples of BPMN models.

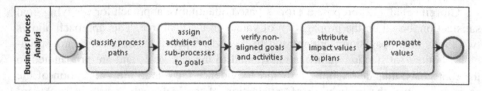

Fig. 1. The process underlying the proposed method

3   A Systematic Method to Align Business Processes and Goals

The proposed method assumes that the organization's goals and processes have been previously modeled, taking both models as entries. While valuing approaches that start with goal modeling and then model processes based on these goals, e.g. [5, 6], we also recognize that, in practice, many organizations have undertaken these modeling tasks separately and today have several non-aligned goal models and BPMs. In this context, it is possible that goal and process model have different granularity level. If that is the case, it may be necessary to refine the models, as our method assumes these models have the same granularity. The general process underlying the proposed method is composed of the five steps depicted in Fig. 1. Aiming at illustrating each of these steps, Subsect. 3.1 describes a running example, further referenced in the remaining sub-sections to demonstrate how each step of the alignment method is carried out.

3.1   Running Example

For the running example, we choose a scenario of conference paper reviewing, for being a well-known and simple enough example to enable the method illustration. Figures 2 and 3 present this scenario's goal and process model respectively.

From now on, we highlight the names of goals and activities from these models, using for them a special font, to facilitate reading.

3.2   Classify Process Paths

In the first step, the analyst classifies the process paths as one of the following three types:

- *Main path:* expected path followed by the process;
- *Secondary path:* forking path whose end leads back to the main path;
- *Alternative path:* forking path that does not return to the main path, hence providing an alternative ending to the process.

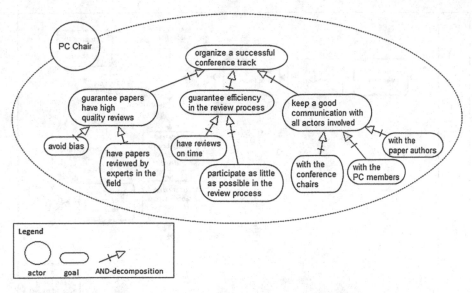

Fig. 2. Conference paper reviewing goal model

This classification is important because, for a given process, some goals may be only satisfied by activities in the main path that, in a certain process execution (i.e. instance), are never actually reached. The opposite situation is also problematic, when goals are only satisfied by activities belonging to secondary or alternative paths.

In the conference paper review process of Fig. 3, the path containing the activity named notify track cancelation to conference chairs is an alternative path, and the path containing the review papers with missing reviews activity is a secondary path. All activities not belonging to these two paths compose the main process path.

3.3   Assign Activities and Sub-processes to Goals

In this step, the analyst takes each leaf goal from the given goal model and assign it to an activity or a sub-process that leads to this goal accomplishment. We call this goal and activity alignment. If the goal is aligned to an activity, it is classified as an *activity goal*; if on the other hand, it is aligned to the process as a whole or to one of its sub-processes, it is classified as a *process goal*. This classification is done according to the taxonomy provided in [8] and it provides for the analyst the information if such a goal is achieved after the execution of solely one activity or if several activities should be taken into account to determine if a goal is successfully accomplished.

Given the aforementioned alignment, the analyst is able to add plans in the goal model, each one representing an activity of the existing process. Plans are linked to goals via the Tropos positive contribution relation, stating that a given plan influences positively the satisfaction of the goal it is linked to. Figure 4 presents the resulting goal

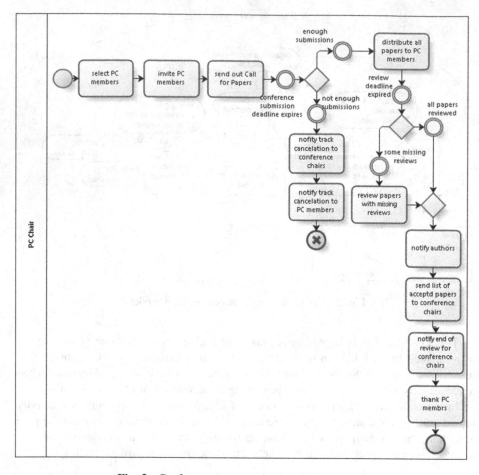

Fig. 3. Conference paper reviewing BPMN model

model, after the alignment and classification. Process goals are depicted in light grey while activity goals are depicted in white.

As can be noted by the alignment of the distribute all papers to PC members activity, it is possible for an activity to contribute to the satisfaction of more than one goal. Many alignments are straightforward, in a way that each activity coming from the BPMN model is designed as a Tropos plan with the same name, in the Tropos model. Some alignments, however, may require some design adaptation, as in the case two or more activities from the process compose a sub-processes aligned to a goal, which normally leads to the creation of a super-plan representing the sub-process. This is the case of the have papers reviewed by experts in the field goal, for which we created the involve PC members plan, which is then AND-decomposed into select PC Members and invite PC Members (the two activities coming from the process). This may also indicate to the analyst that the process model should be changed, to substitute given

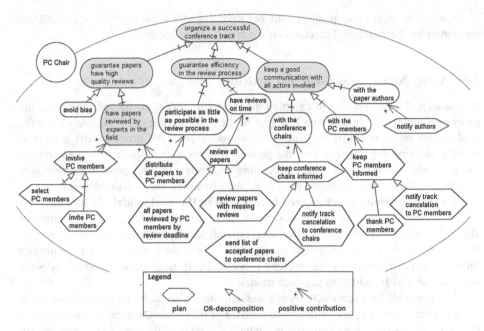

Fig. 4. Goal model after activity alignment

activities by a sub-process[1], so as to create a simpler and easier-to-read process. Another case for design adaptation is related to the path classification discussed in Sect. 3.2. When a goal is accomplished both in the main path and in an alternative or secondary path, the plans representing the activities that achieve such goal should be related via OR-decomposition. This makes sense because only one of the activities will be executed at each process instance, depending if the secondary or alternative path was followed, or if the process executed its main path. This is the case of the have reviews on time goal. This goal is accomplished both if all papers are reviewed by PC Members (the top path in the identified secondary path) or if the PC Chair reviews the missing papers himself (the bottom path in this same path). To model that, an auxiliary plan named review all papers (in time) was created and then decomposed into all papers are reviewed by PC Members and review missing papers, only this last one coming straight from the process model.

In Tropos, there are both positive and negative contributions, respectively meaning positive and negative influences on goal satisfaction. Since our method concerns process and goal alignment, for the sake of exemplifying it, we assume that all activities in the process are executed in order to accomplish the organization's goals, thus we only consider positive contributions. Nevertheless, we are aware that an organization may have conflicting goals. In that case, a process activity may have a positive impact on the satisfaction of one goal, while undermining the accomplishment

---

[1] BPM languages (as BPMN) and frameworks usually offer the option of including as activities in the process model, subprocesses that are also modeled separately, as a modularity strategy.

of another. In such case, positive and negative contributions must be used and are supported by the reasoning mechanism we apply in our method.

3.4   Verify Non-aligned Goals and Activities

In this step, the analyst should examine the model to check: (i) if each activity in the process is aligned to at least one leaf goal in the goal model; (ii) if all goals from the goal model are satisfied by activities or sub-processes in the process; (iii) if there is some goal being neglected in case the process follows a given path. When examining the goal and process model for our running example, we find problems when performing all these three verifications. Let us then examine each of these cases.

A process activity is not aligned to any goal in the goal model. By looking back to the process model of the running example, we realize that there are two activities not aligned to any goal in the goal model: send out call for papers and notify end of review to conference chairs. At this point, we should verify if these activities are simply not relevant and exceeding in the present process, or if new goals satisfied by such activities should be added to the goal model.

The send call for papers activity is essential, otherwise, nobody will find out about the track and, thus, no paper will be submitted. For this reason, we create a new goal in the goal tree, named attract submissions, which can be directly linked to a send call for papers plan, representing this activity.

When performing the step of aligning goals and activities (refer to Sect. 3.3), we realized that the notify end of review to conference chairs activity is actually redundant, because, when we submit the list of accepted papers to the chairs, we are already communicating about the end of the review process. We conclude that such activity is in fact exceeding and exclude it from the process.

A goal is not satisfied by any activity or sub-process. Figure 4 shows that the avoid bias goal has no aligned activity. This means that such goal is never achieved during any execution of the given process. This may be because the goal is not actually important and should, thus, be excluded from the goal model. On the other hand, it may be a clue for the analyst that the process should be reengineered so as to include activities to accomplish such goal. In this case, we believe the latter, so the ask PC Members to perform bidding activity is added in the process, right before the distribute all papers to PC Members activity, thus allowing PC Members to declare conflicts and express their preference regarding papers to review.

Whenever there are missing activities, this may be the result of poor modeling, thus adding activities contributes to enhancing the quality of the process model. In the worst case, missing activities mean that the process must actually be reengineered. In both cases, our method shows that goals may be the drivers to account for such missing activities, supporting the work of the BP analyst.

A goal is neglected in a particular process path. By analyzing the model after path classification, we realize that the keep good communication with paper authors goal is neglected when the process takes the alternative path. In other words, in the first forking, if there are not enough submissions, the conference chairs and the PC members are notified, however, the authors of the submitted papers are never informed regarding the track cancelation. In this case, again, we must add an activity in the process, right after the notify track cancelations to conference chairs activity, named notify track cancelation to authors.

It is important to realize that in the alternative path, many of the goals that are not pertinent to this path are actually neglected. For example, the papers are never going to be reviewed so the have papers reviewed by experts in the field goal and the avoid bias goal will never be achieved. However, this results from an abnormal termination of the process and should not lead to any process change.

Another case in which goals may be correctly achieved only in a secondary path regards the case in which there are two alternative sub-goals (i.e. sub-goals related via OR-decomposition). Suppose, for instance, that the conference track may have either invited submissions or papers submitted in response to a call. In that case, besides the attract submissions goal, there would be an alternative goal named invite submissions. In the process, the send out Call for Papers activity would be forked with a send invitation email to known authors activity. Each of these secondary paths would accomplish only one of the previously mentioned goals, and this would be perfectly natural.

3.5 Attribute Impact Values to Plans

In this step, the analyst (in collaboration with the stakeholder) should attribute weights (i.e. impact values) for the contribution links in the goal model. Given that a plan P contributes to a goal G, a contribution weight means how much impact P has on G's satisfaction. Following the probabilistic model proposed in [12], consider a goal G, which has two contributing plans P_1 and P_2, whose contributions are attributed values of +0,4 and +0,7, respectively. This means that if plan P_1 is completely executed, there is a 40 % chance that goal G is achieved, while if plan P_2 is completely executed, this

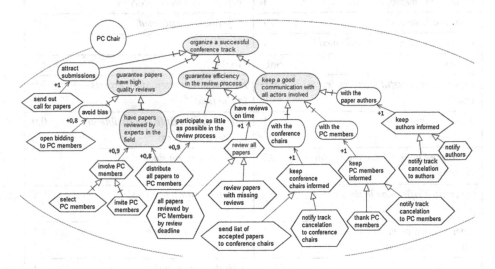

Fig. 5. The goal model after impact values are attributed to goals and plans

represents a greater impact on goal G (70 % chance of achievement). Figure 5 shows the goal model of the running example after the contribution weights attribution.

3.6 Values Propagation

In this step, the analyst should attribute different satisfaction values to the leaf plans of the model and simulate how these values propagate in the goal tree, impacting the middle and root goals. Here, we adapt the proposed approach [12], by considering that for a plan, the satisfaction value consists in the probability of plan execution. For reasons of lack of space, we exemplify the reasoning method only for contribution value propagation in case of satisfaction. Thus, in case the plan is not executed, nothing can be said about the denial of the goal to which the plan contributes. The attributed values should be propagated from the leaves to the top, following the propagation rules proposed in [12] and shown in Table 1. Consider $Sat(G_1)$ as the satisfaction value of G_1 (i.e. the probability that G_1 is satisfied) and w, the contribution weight, explained in Sect. 3.5.

Table 1. Propagation rules

Contribution	$G_2 \overset{w+S}{\rightarrow} G_1$	$Sat(G_1) = Sat(G_2).w$
AND-decomposition	$(G_2, G_3) \overset{and}{\rightarrow} G_1$	$Sat(G_1) = Sat(G_2).Sat(G_3)$
OR-decomposition	$(G_2, G_3) \overset{or}{\rightarrow} G_1$	$Sat(G_1) = Sat(G_2) + Sat(G_3) - Sat(G_2).Sat(G_3)$

Table 2. Propagated values to the plans and goals in the goal model

Plans	Sat	Plans	Sat
involve PC members	1	*keep PC members informed*	1
review all papers	1	*keep authors informed*	1
keep conference chairs informed	1		
Goals	**Sat**	**Goals**	**Sat**
organize successful conference track	0,6	*have papers reviewed by experts in the field*	0,9
attract submissions	1	*participate as little as possible in the review process*	0,9
guarantee papers have high quality reviews	0,7	*have reviews on time*	1
guarantee efficiency in the review process	0,9	*with the conference chairs*	1
Keep a good communication with all actors involved	1	*with the PC members*	1
avoid bias	0,8	*with the paper authors*	1

Suppose that in our example, all plans referring to activities in the main path of the process have been fully executed (thus, having satisfaction value = 1), except the all papers reviewed by PC members by review deadline plan, whose satisfaction value is 0,9. This leads the process to its secondary path, and we will suppose that the review papers with missing review plan is also fully executed (thus, having satisfaction value = 1). Based on these initial values and using the rules of Tables 1, 2 presents the propagated values to the remaining plans and goals of the model.

To exemplify how the values are propagated, let us analyze the evidence of satisfaction of the guarantee efficiency in the review process goal. For that, we start from the leaf plans that are indirectly related to this goal. The all papers reviewed by PC members by review deadline plan (satisfaction value = 0,9) and the review papers with missing review plan (satisfaction value = 1) are related via OR-decomposition to the review all papers plan (satisfaction value = $0,9 + 1 - 0,9 * 1 = 1$). The review all papers plan contributes with a weight of +1 to the have reviews on time goal (satisfaction value = $1 * 1 = 1$). By analogous calculations, we arrive at the value of 0,9 to the satisfaction value of the participate as little as possible in the review process goal. And finally, the participate as little as possible in the review process goal and the have reviews on time goal are related via AND-decomposition to the guarantee efficiency in the review process goal (satisfaction value = $0,9 * 1 = 0,9$). In case there are two plans contributing to a goal (e.g. in the case of the have papers reviewed by experts in the field goal), only the maximum satisfaction value is propagated.

By testing different values assigned to the leaf plans, the process manager is able to reason about the impact of activity execution in the organization's goals. This may help him set priorities and allocate resources to the different process activities, based on their impact on goal achievement. In case the process is automated via a workflow system, after the execution of each activity, the system can automatically update a goal model like the one we are using, providing it with new values of goal satisfaction. This will give the process manager the ability to monitor, in real time, how the middle and the top goal are being achieved.

4 Preliminary Evaluation of the Method

With the aim of evaluating the proposed method, an empirical study was conducted, with the voluntary participation of 14 students from a master course in Computer Science, at the Federal University of Espirito Santo (UFES), in Brazil. For this study, we used real process models from a public regulatory agency. We obtained the real alignment from the employees of such agency and considered it for comparison with the results performed by the empirical study participants. This allowed us to count how many correct or incorrect alignments there were for each participant, as well as the absence of correct alignments.

When quantitatively analyzing the results, we considered the following metrics:

- *Efficiency metrics:* (a) the rate between the number of correct alignments and the time spent performing alignments; (b) the rate between the number of incorrect

alignments and the time spent performing alignments; (c) the rate between the number of absent correct alignments and the time spent performing alignments;
- *Effectiveness metrics:* (d) the rate of correct alignment and the total number of alignments; (e) the rate of incorrect alignment and the total number of alignments; (f) the rate of absent correct alignment and the total number of alignments.

The study was conducted in two rounds and the participants were divided in two groups (A and B), considering their levels of expertise regarding goal and process modeling. Group A aggregated the more experienced participants in both areas, while group B aggregated those less experienced. Information regarding participant's expertise was gathered by the analysis of a profiling form filled by the participants. In the first round, both groups performed the ad-hoc alignment, i.e. each participant had to individually align activities and goals without the use of any particular method. In the second round, the participants of both groups received instructions regarding the proposed method, each one performing a new alignment based on this method. Two cases of similar complexity were chosen for the empirical study. In the first round, group A had case 1, while group B had case 2 (in the second round, this was inverted). Moreover, to enable a qualitative analysis, we designed a form, in which the participants could share their impressions and explicitly give suggestions for the improvement of the method.

For the quantitative analysis, we used boxplot, which allows the visualization of the concentration of the collected data, excluding outliers and supporting the results' comparison. In terms of efficiency, the result showed that the only metric in which the ad hoc alignment performed better is the rate of the number of correct alignment and the time (metric a). This is justifiable, given that the proposed method requires several steps while in the ad hoc method, the alignment is directly made by intuition. Moreover, the fact that all participants were novices in the use of the proposed method may also have influenced the longer time spent in performing alignments. However, in terms of avoiding mistakes (metrics b and c), the proposed method performed better. In terms of effectiveness, the proposed method led to a higher number of correct alignments (metric d) and a lower number of incorrect alignments (metric e). Nevertheless, the ad hoc method performed better in avoiding absence of correct alignments (metric f).

Concerning the qualitative analysis, the participants practically unanimously evaluated the proposed method positively. The biggest contributions of the method, according to the participants is the existence of objective steps that allow them to: (i) systematically identify the goal and activity alignments. With respect to this point, the participants emphasized the usefulness of the process path classification (see Sect. 3.2) and goal classification (see Sect. 3.3) to facilitate the alignment; and (ii) detect discrepancies between the goal and process models (see Sect. 3.4).

Although the results were positive, the participants also highlighted some important limitations: (1) they found the method very complex, being composed of too many steps and sub-steps; and (2) they also claimed that the time dedicated for training was not enough. This possibly explains the bad quantitative results with respect to method efficiency. Moreover it indicates that, when applied in practice, the method requires heavy training and its success relies on the analyst's level of experience.

5 Related Works

The most related research to our own is the GPI method [4]. GPI combines *i** and BPMN, aiming at profiting from each of these languages' dispositions in describing the organization's strategic and operational dimensions, helping to maintain the traceability between goals and processes. Like in our work, the integration between these two languages is obtained through the *i** task element (plan in Tropos). However, differently than in our misalignment detection strategy, goal satisfaction relies on monitoring KPIs related to the resources produced by the BPs that are aligned to the goal. If the given process does not produce an expected KPI, then the goal is not satisfied and there is a misalignment in the organization.

Nagel et al. [5] report on an approach to ensure consistency among goals and BPs, based on the use of KAOS and applying model checking. In their work, the focus is on determining the order in which activities should occur, given by logical and temporal dependencies between goals.

Jander et al. propose Goal-oriented Process Modeling Notation (GPMN) [6], a specific language to model goal-oriented BPs. As in our approach, these authors recognize the importance of explicitly modeling the motivations behind the execution of the processes. However, they propose a different notation instead of using existing goal-modeling languages. We prefer to reuse existing works on goal modeling, so as to build over a big body of work previously done in this research field.

Soffer and Rolland [7] propose a method to combine intention-oriented (i.e. goal modeling) and state-based process modeling. Regarding the modeling languages, instead of Tropos and BPMN, this work adopts two state-based modeling languages. In general, it consists in a more formal work when compared to ours. However, we do share some objectives, such as detecting model incompleteness.

Differently than us, both Nagel et al. [5] and Jander et al. [6] consider that the BP model starts with goal-modeling and systematically proceeds to activities' modeling. Thus they do not pay too much attention to the alignment inconsistencies as described in Sect. 3.4. Moreover, a limitation of the works presented in [5–7] in comparison to ours is not providing any reasoning mechanism regarding goal satisfaction.

6 Conclusion

This paper presents a method for aligning goals and process activities, based on the application of state of the art techniques in goal modeling and BPM, namely Tropos and BPMN.

Goal and process alignment may help the analyst to detect inconsistencies in the goal and process models, thus leading him to enhance the quality of the models and, ultimately, to reengineer. The method also helps the project manager to estimate the impact of the execution of each process activity in goal satisfaction, thus assisting him in setting priorities and making decisions regarding the time, effort and resources spent with each activity.

We are aware that it may be hard to identify the real values of impact each activity has on goal achievement, since these values are subjective and hard to determine.

However, we believe the stakeholders will have at least an idea if the impact is low, medium or high, allowing him to set different values and test them. For the future, we will explore qualitative reasoning mechanisms to evaluate if they are more suitable and practical.

Besides describing the method, the paper presents a preliminary evaluation made with an empirical study. We acknowledge the limitations of this evaluation, especially regarding the low number of subjects. However, we believe that it provides a good indication and a basis for a full experimental evaluation, aimed for the future. Moreover, we aim at performing real case studies, expecting that results from real environments will help us understand the viability of the method, also helping us improve it.

Acknowledgement. This work is partially supported by CAPES/CNPq (grant number 402991/2012-5) and CNPq (grant numbers 461777/2014-2 and 485368/2013-7). The authors are also grateful to Giancarlo Guizzardi, João Paulo Andrade Almeida and Mateus C. Barcellos Costa for the fruitful discussion regarding the proposed method.

References

1. Davies, I., Green, P., Rosemann, M., Indulska, M., Gallo, S.: How do practitioners use conceptual modeling in practice? Data Knowl. Eng. **58**, 358–380 (2006)
2. Rosemann, M., vom Brocke, J.: The six core elements of business process management. In: Rosemann, M., vom Brocke, J. (eds.) Handbook on Business Process Management, 1, pp. 105–122. Springer, Berlin (2015)
3. Dumas, M., van der Aalst, W.M.P., ter Hofstede, A.H.M. (eds.): Process Aware Information Systems: Bridging People and Software Through Process Technology. Wiley, Hoboken (2005)
4. Sousa, H.P., do Prado Leite, J.C.S.: Modeling organizational alignment. In: Yu, E., Dobbie, G., Jarke, M., Purao, S. (eds.) ER 2014. LNCS, vol. 8824, pp. 407–414. Springer, Heidelberg (2014)
5. Nagel, B., Gerth, C., Engels, G., Post, J.: Ensuring consistency among business goals and business process models. In: 17th IEEE International Enterprise Distributed Object Computing Conference. pp. 17–26 (2013)
6. Jander, K., Braubach, L., Pokahr, A., Lamersdorf, W., Wack, K.: Goal-oriented processes with GPMN. Int. J. Artif. Intell. Tools **20**(6), 1021–1041 (2011)
7. Soffer, P., Rolland, C.: Combining intention-oriented and state-based process modeling. In: 24th International Conference on Conceptual Modeling, pp. 47–62 (2005)
8. Cardoso, E., Guizzardi, R., Almeida, J.P.: Aligning objectives and business process models: a case study in the health care industry. Int. J. Bus. Process Integr. Manage. **5**(2), 44–158 (2011)
9. Yu, E., Giorgini, P., Maiden, N., Mylopoulos, J. (eds.): Social Modeling for Requirements Engineering. MIT Press, Cambridge (2011)
10. Dardenne, A., van Lamsweerde, A., Fickas, S.: Goal directed requirements acquisition. In: Selected Papers of the Sixth International Workshop on Software Specification and Design, pp. 3–50. Elsevier Science Publishers (1993)

11. Bresciani, P., Giorgini, P., Giunchiglia, F., Mylopoulos, J., Perini, A.: TROPOS: an agent-oriented software development methodology. Int. J. Auton. Agent. Multi-Agent Syst. **8**(3), 203–236 (2004)
12. Giorgini, P., Mylopoulos, J. Nicchiarelli, E., Sebastiani, R.: Formal Reasoning Techniques for Goal Models. In: Spaccapietra, S. (ed.) Journal on Data Semantics I, pp. 1–21. Springer (2004)
13. OMG: Business process model and notation. http://www.omg.org/spec/BPMN/2.0

Detecting the Effects of Changes on the Compliance of Cross-Organizational Business Processes

David Knuplesch[1]([⊠]), Walid Fdhila[2],
Manfred Reichert[1], and Stefanie Rinderle-Ma[2]

[1] Institute of Databases and Information Systems, Ulm University, Ulm, Germany
{david.knuplesch,manfred.reichert}@uni-ulm.de
[2] Faculty of Computer Science, University of Vienna, Vienna, Austria
{walid.fdhila,stefanie.rinderle-ma}@univie.ac.at

Abstract. An emerging challenge for collaborating business partners is to properly define and evolve their cross-organizational processes with respect to imposed global compliance rules. Since compliance verification is known to be very costly, reducing the number of compliance rules to be rechecked in the context of process changes will be crucial. Opposed to intra-organizational processes, however, change effects cannot be easily assessed in such distributed scenarios, where partners only provide restricted public views and assertions on their private processes. Even if local process changes are invisible to partners, they might affect the compliance of the cross-organizational process with the mentioned rules. This paper provides an approach for ensuring compliance when evolving a cross-organizational process. For this purpose, we construct qualified dependency graphs expressing relationships between process activities, process assertions, and compliance rules. Based on such graphs, we are able to determine the subset of compliance rules that might be affected by a particular change. Altogether, our approach increases the efficiency of compliance checking in cross-organizational settings.

Keywords: Process · Compliance · Process · Change · Cross-organizational · Process

1 Introduction

Ensuring the compliance of their business processes is crucial for enterprises [1]. To cope with this challenge, a variety of approaches are proposed that allow verifying the compliance of business processes with semantic compliance rules (e.g., domain-specific standards, guidelines, and regulations) [2–5]. In this context, only few approaches consider *cross-organizational business processes* (CBP); i.e.,

This work was done within the research project C³Pro funded by the German Research Foundation (DFG), under project number RE 1402/2-1, and the Austrian Science Fund (FWF), under project number I743.

© Springer International Publishing Switzerland 2015
P. Johannesson et al. (Eds.): ER 2015, LNCS 9381, pp. 94–107, 2015.
DOI: 10.1007/978-3-319-25264-3_7

processes involving multiple partners. Ensuring the compliance of a CBP with *global compliance rules* (GCR), however, raises additional challenges. In particular, compliance checking must cope with the fact that the partners do not know all parts of the CBP relevant for a GCR, e.g., due to privacy reasons [6,7]. In this context, we developed techniques to *a priori* ensure compliance of a CBP with global rules [8]. For this purpose, we utilize the *public views* on the processes of the involved partners (i.e. *public process models*) as well as declarative *assertion rules* (AR), provided by each partner on the behavior of its private process. Note that respective views and assertions allow us to approximate the behavior of the partner processes and, hence, the CBP, while satisfying privacy issues. As further shown in [9], CBPs may be subject to change (e.g., a partner may want to change his process or partner interactions shall be changed). Existing work has already addressed issues related to the behavioral correctness (i.e., soundness) of CBPs in the context of such changes [9–12]. In turn, only few approaches address the issue whether or not a changed CBP still complies with a given set of compliance rules [6,7,13]. Unfortunately, these approaches do not provide a solution that considers privacy issues of the involved partners.

An obvious approach to ensure compliance of a changed CBP would be to recheck the former with all imposed compliance rules, e.g., utilizing the approach presented in [8]. However, compliance checking is known to be time-consuming and costly [2,14]. In particular, this applies to *privacy-aware compliance checking* of a CBP, which not only needs to explore the state space of the specified models, but additionally must estimate the effects of activities not visible to all partners due to privacy constraints [8]. Consequently, being able to detect the possible effects, a CBP change may have on the compliance of the CBP with defined rules, would contribute to limit the number of compliance checks to be repeated after having changed the CBP. More precisely, only those compliance rules, which may be affected by the change, should be rechecked (cf. Fig. 1).

Fig. 1. Ensuring compliance of cross-organizational processes after a change

Another *naive approach* would be to solely recheck compliance of those rules referring to activities that are directly affected by the change. However, this approach is not sufficient. On one hand, as known from intra-organizational scenarios [3], it would also recheck compliance rules that become overfulfilled due to a change (*false positive*); e.g., adding a second safety check, although the required one has already been ensured. On the other hand, as opposed to intra-organizational scenarios, the naive approach is unable to identify all compliance rules to be rechecked when the CBP is being changed (*false negative*). This effect occurs when changes not only affect public, but also private elements of partner processes. Although the latter kind of changes are not visible, they could partially be assessed based on direct as well as transitive correlations between changes and assertions. Note that *false positive* cases result in superfluous checks, whereas *false negative* ones give a false sense of security, since they might prevent the detection of compliance violations.

This paper provides a sophisticated approach that enables us to detect and qualify all possible effects any CBP change may have on CBP compliance. To deal with both *false positive* and *false negative* cases, first of all, the dependencies between activities, assertions and compliance rules are analyzed before representing them as *qualified dependency graphs*. Based on the latter, two algorithms are introduced that allow assessing the possible effects a CBP change has on the compliance of the CBP with imposed rules. These algorithms not only enable us to detect possible new compliance violations introduced by the CBP change, but additionally allow determining whether a change has the potential to heal already present violations of a particular rule. The approach is illustrated along a realistic use case and evaluated by a proof-of-concept implementation that was applied to different scenarios.

The remainder of this paper is structured as follows: Sect. 2 provides a running example. Fundamentals are introduced in Sect. 3, whereas Sect. 4 presents the approach. In detail, Sect. 4.1 investigates the dependencies between activities, compliance rules, and assertions. It further introduces the qualified dependency graphs. Section 4.2 then presents algorithms that analyze the dependency graph to detect the effects of CBP changes. A proof-of-concept is provided in Sect. 5. Section 6 discusses related work and Sect. 7 summarizes the paper.

2   Running Example

This section introduces a cross-organizational supply chain scenario as well as a related change (cf. Fig. 2). The scenario highlights the effects of changes on the compliance of a CBP [7]. A CBP involving 6 partners is introduced. It describes a supply chain from the *bulk buyer*'s order of a product batch via the order and provisions of two *intermediate products* a and b by two *suppliers A* and *B* to production and, finally, delivery by the *manufacturer*. Furthermore, the order and delivery of *intermediate product* a not only involve *supplier A* and *manufacturer*, but also a *middleman* and a *special carrier*.

The depicted supply chain process complies with the following 5 global compliance rules (GCR), which reflect regulations and standards to be obeyed:

C1: After production the final test must be performed.
C2: A full test of intermediates is required before starting the production.
C3: Each transport of the intermediate product a requires permission of authority. Furthermore, the special transporter must pass a safety check before.
C4: After a quick test, the parameters of the tests must be compared to ensure validity.
C5: If an intermediate is transported after its full test, a quick test is required after arrival and before production.

The partners only share public views on their processes in order to ensure privacy; e.g., *special carrier* abstracts from activity *safety check* by hiding the latter, whereas *middleman* hides activity *get permission from authority* (cf. Fig. 2). To enable the verification of the GCRs C1-C5, the partners provide the following assertion rules (AR) on the hidden behavior of their private processes:

A1: *Manufacturer* assures that a quick test is performed after the arrival of an intermediate and before processing it, if the manufacturer does not perform a full quality test in this period (A1.1). In turn, if a full quality test is performed after arrival and before production, the manufacturer does not perform a quick test (A1.2).

A2: *Middleman* assures that it gets permission from authority for the special transport before ordering the latter.

A3: *Special carrier* assures to perform a safety check before starting the transport of intermediate a.

A4: *Supplier B* assures that a full quality test of intermediate b is performed before the latter arrives at the *manufacturer* side.

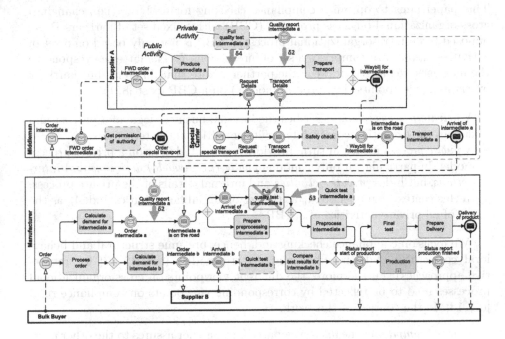

Fig. 2. Example: cross-organizational supply chain process

When utilizing assertions A1-A4, one can successfully verify compliance of the CBP with C1-C5 based on the public process views (cf. [8]).

Change scenario. To decrease costs as well as to optimize the processing of intermediate *a*, *manufacturer* skips the full quality test of intermediate *a*. Instead, the full test shall now be performed by *supplier A* and a quality report be sent to *manufacturer*. In particular the following changes occur (cf. Fig. 2):

δ1: *Manufacturer* skips the *full quality test for intermediate a*.
δ2: Message *quality report for intermediate a* from *supplier A* to *manufacturer* is added.
(δ3: *Manufacturer* adds private activity *quick test for intermediate a*.)
(δ4: *Supplier A* adds private activity *full quality test for intermediate a*.)
δ5: *Supplier A* publishes new *assertion A5*. The latter shall guarantee that supplier A performs a *full quality test for intermediate a* before sending the corresponding *quality report*.

Only δ1, δ2 and δ5 are visible, whereas δ3 and δ4 as well as their effects (i.e., the insertion of private activities) remain hidden from the partners.

Note that it is evident that C2 should be rechecked when considering the public changes, because activity *full quality test for intermediate a* is affected by public change $\delta 1$. By contrast, the public changes do not directly imply the need to recheck C4. However, C4 becomes violated, since a *quick test for intermediate a* occurs, but the parameters of the tests are not compared. Hence, C4 constitutes a *false negative* case as described in the context of the *naive approach*.

3   Fundamentals

This paper aims to optimize compliance checking for evolving (i.e., changing) cross-organizational business processes (CBP) with a fixed set of partners $\mathcal{P}$. As opposed to an intra-organizational process, a CBP is not only based on a set of activities $\mathcal{A}$, but also comprises a set of interactions $\mathcal{I}$. The latter correspond to the messages exchanged between the partners. Note that different, but partially overlapping viewpoints (i.e., process models) on a CBP exist [8,9]:

- A *private process model* describes the internal business logic of a partner and defines the execution constraints for its activities and interactions.
- A *public process model*, in turn, provides a restricted view on a private process model. In particular, it only contains *public (i.e., visible) activities* and *interactions*, but hides *private activities* and internal details of the private process. In this context, we refer to $\mathcal{A}_v$ as the set of all public activities and $\mathcal{A}_h$ as the set of all private activities of a CBP.[1]

We focus on compliance checking and, hence, presume structural and behavioral correctness of both public and private models (see [15–17] for respective approaches). Furthermore, the different viewpoints on cross-organizational processes need to be reflected by corresponding viewpoints on compliance rules [6,8,18]. In the context of this work,

- *asserted compliance* means that a particular partner assures to the other partners that all traces producible on its private process comply with its *assertion rules* (AR), and
- *global compliance* means that all traces virtually producible on the cross-organizational business process (i.e., the concurrent execution of all private processes) comply with all *global compliance rules* (GCR).

We developed the *extended Compliance Rule Graph* (eCRG) language to specify ARs and GCRs [19,20]. The eCRG language is a visual language for modeling compliance rules. It not only focuses on the control flow perspective, but enables integrated support for interactions with business partners as well.[2] The elements of an eCRG may be partitioned into an *antecedence pattern* and a related *consequence pattern*. Both patterns are modeled using *occurrence* and *absence nodes*, which either express the occurrence or absence of certain events related to the execution of a particular activity or the exchange of a particular

---

[1] We assume $\mathcal{A} = \mathcal{A}_v \cup \mathcal{A}_h$, but do not require $\mathcal{A}_v \cap \mathcal{A}_h = \emptyset$.

[2] Note that the eCRG language also addresses the resources, data and time perspectives, but these are not relevant in the context of this paper.

message (i.e., a particular interaction). In turn, eCRG edges are used to specify control flow dependencies (cf. Definition 1). An eCRG considers a process trace as *compliant*, if for each match of the *antecedence pattern* (i.e., *activation* of the eCRG), there exists at least one corresponding match of the *consequence pattern*. *Trivial compliance* refers to the absence of any activation. Figure 3 shows the eCRGs that refer to the ARs and GCRs of the running example.

Note that our approach not depends on eCRG, but can be easily applied to other compliance rule languages as well (e.g. FCL [21]).

Definition 1 (Assertions and Global Compliance Rules).
Let $\mathcal{A}$ be the set of activities and let $\mathcal{I}$ be the set of interactions. Then: a global compliance rule or an assertion rule r is a tuple $r = (A^+, A^-, C^+, C^-, type, \mu^{A^+}, \mu^{A^-}, \mu^{C^+}, \mu^{C^-})$ with

- A^+ (A^-) *being the set of* **antecedence occurrence (absence) nodes,**
- C^+ (C^-) *being the set of* **consequence occurrence (absence) nodes,**
- $type: A^+ \cup A^- \cup C^+ \cup C^- \longrightarrow \mathcal{A} \cup \mathcal{I}$ *mapping each node to its* **activity or message (type),**
- μ^{A^+} (μ^{A^-}) *being the* **antecedence occurrence (absence) sequence flow condition,** *and*
- μ^{C^+} (μ^{C^-}) *being the* **consequence occurrence (absence) sequence flow condition.**

Further, we define $\mathcal{R}_c$ as the set of **global compliance rules** *and $\mathcal{R}_a$ as the set of* **assertions.**

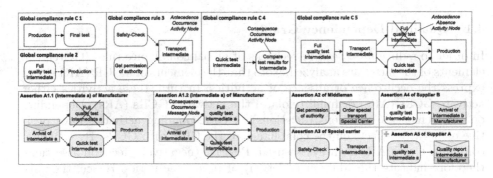

Fig. 3. Global compliance rules C1-C5 and assertions A1-A5 of running example

To verify compliance of an intra-organizational process, model checking can be applied, since the complete state space of a process can be determined. However, in a CBP the partners usually do not publish their *private models*. Hence, the state spaces cannot be determined and global compliance cannot be directly verified. In turn, in [8] we showed that the state space and compliance can be (over-)approximated based on the available information (i.e., the public models, the activities, and the ARs). Figure 4 sketches this approximation using the set of (virtual) traces producible on the CBP in order to characterize the process state space: First, the set of visible traces (i.e. state space) are determined based on the available public models (a). Second, these traces are enriched by including private activities in order to (over-)approximate the behavior of the private processes (b). Third, AR violating traces are filtered out (c). Finally, global compliance is (over-)approximated based on the remaining traces (d).

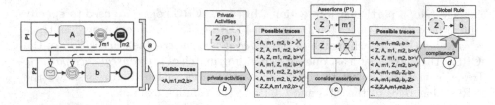

Fig. 4. Approximated global compliance of cross-organizational processes

4 Detecting Effects of Changes on Compliance

This section presents our approach for detecting those compliance rules that need to be rechecked after a change. In order to reduce *false positives*, first of all, different kinds of dependencies are analyzed between activities on one hand and compliance rules and assertions on the other. In particular, these dependencies are represented as qualified dependency graph (cf. Sect. 4.1). Finally, we ensure that there are no *false negative* results by calculating the possible transitive effects of a particular change (cf. Sect. 4.2).

4.1 Qualified Dependency Graph

In order to reduce *false positive* rechecks of global compliance rules (GCR), the elements of the latter are analyzed. Note that the elements of a GCR (AR) either express the occurrence or absence of activities. In turn, these activities either activate GCRs (ARs) or fulfill (violate) the activated GCRs (ARs). Depending on this semantics, additions (deletions) of corresponding activities have positive or negative effects on the compliance of the CBP with GCRs. By contrast, compliance with ARs is always ensured. Positive (negative) effects on assertions therefore indicate the addition (deletion) of private activities. Hence, they are relevant as well. Note that these effects can be both positive and negative (cf. GCR C5 and activity *full quality test intermediate a (b)*). Definition 2 formally introduces different kinds of dependencies, which are then used to express the dependencies between activities on one hand and ARs and GCRs on the other.

Definition 2 (Dependency Qualifications).
Let $\mathcal{Q} := \{\emptyset, +, -, \pm\}$ *denote the* **set of dependency qualifications.** *Thereby, a dependency can be either* $\emptyset$ **independent** *(i.e., there is no dependency),* + **positive**, − **negative**, *or* ± **positive&negative.** *Together with the below operations* **addition** (+) *and* **multiplication** (·), $\mathcal{Q}$ *constructs an idempotent semiring or dioid* $(\mathcal{Q}, (+), \emptyset, (\cdot), +)$, *whereas operations* (+) *and* (·) *are defined on* $\mathcal{Q}$ *as follows:*

(+)	∅	−	+	±		(·)	∅	−	+	±
∅	∅	−	+	±		∅	∅	∅	∅	∅
−	−	−	±	±		−	∅	+	−	±
+	+	±	+	±		+	∅	−	+	±
±	±	±	±	±		±	∅	±	±	±

Based on Definition 2, we can interpret and express the dependencies between activities and rules (i.e., ARs or GCRs) as *qualified dependency graph* (QDG).

The latter constitutes a colored and bipartite graph, whose nodes correspond to activities and rules (i.e., GCRs or ARs). Thereby, positive (solid) and negative (dashed) edges express the dependencies between activities and rules. To construct the QDG, Definition 3 utilizes the partitioning of eCRGs into occurrence and absence nodes of the antecedence and consequence pattern respectively.

Definition 3 (Qualified Dependency Graph).

A qualified dependency graph Φ is a tuple $\Phi = (\mathcal{A}, \mathcal{R}, \vec{d})$, with
- $\mathcal{A}$ being a set of activities and $\mathcal{R}$ being a set of rules, and
- $\vec{d} \subseteq (\mathcal{A} \times \mathcal{R}) \times \{-, +\}$ being the qualified dependency relation, which is defined as follows:
$$\vec{d} := \{(a,r,-)|\exists r \in \mathcal{R}, n \in A_r^+ : type(n) = a\} \cup \{(a,r,+)|\exists r \in \mathcal{R}, n \in A_r^- : type(n) = a\}$$
$$\cup \{(a,r,+)|\exists r \in \mathcal{R}, n \in C_r^+ : type(n) = a\} \cup \{(a,r,-)|\exists r \in \mathcal{R}, n \in C_r^- : type(n) = a\}$$

Further,
- $\Phi_a := (\mathcal{A}, \mathcal{R}_a, \vec{d})$ is the qualified dependency graph between activities and ARs and
- $\Phi_c := (\mathcal{A}, \mathcal{R}_c, \vec{d})$ is the qualified dependency graph between activities and GCRs.

Activities of the antecedence occurrence (A^+) pattern have negative effects $(-)$ on the compliance of the respective rule, since additional activities of the antecedence occurrence (A^+) pattern might trigger additional activations that might be violated. In turn, additional activities of the antecedence absence (A^-) pattern might deactivate existing rule activations and, therefore, increase compliance $(+)$. Further, additional activities of the consequence occurrence (C^+) pattern might fulfill additional activations, and, hence, increase compliance with the respective rule $(+)$. In turn, additional activities of the consequence absence (C^-) pattern might violate of present activations $(-)$. Figure 5 combines the QDGs related to the running example. For the sake of readability, we omitted rules and activities not relevant in the given change scenario.

Note that the partitioning of the eCRG, which is utilized in Definition 3, is not a unique characteristic of the eCRG, but is enabled by other compliance rule languages in a similar way; e.g., FCL [21] distinguishes between *premises* ($\rightarrow$antecedence) and *conclusions* ($\rightarrow$consequence), which both can be *negated* ($\rightarrow$absence).

4.2   Algorithms

Based on the QDGs, we introduce algorithms to determine the direct and transitive effects of CBP changes on private activities (Algorithm 1) as well as their possible influence on the compliance of the CBP with GCRs (Algorithm 2). These algorithms utilize the *public properties of changes*; i.e., the additions and deletions of public activities and assertions (cf. Definition 4). Figure 6 shows the public properties of the changes applied in the context of the running example.

Definition 4 (Public Properties of Changes).
The **public properties of a CBP change** *correspond to a function* $chg : \mathcal{A}_v \cup \mathcal{R}_a \rightarrow \{\emptyset, -, +, \pm\}$ *that states whether activities of a particular type are not affected ($\emptyset$), removed ($-$), added ($+$), or both (i.e. moved ($\pm$)).*

We distinguish between two kinds of effects, a CBP change may have on the compliance of the CBP with a particular GCR. On one hand, a GCR may directly refer to activities affected by the change (i.e., added, deleted or moved).

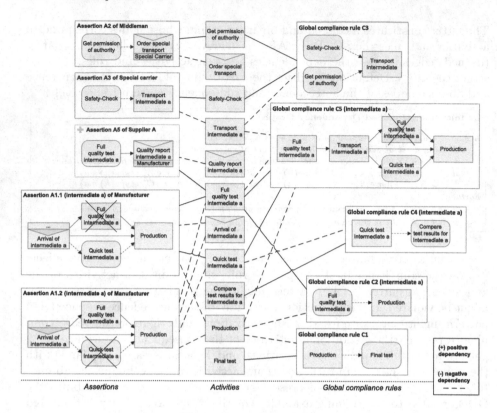

Fig. 5. Dependency graph

$$chg(x) := \begin{cases} -, & \text{if } x \text{ is activity } \textit{Full quality test of intermediate a} & (\delta 1) \\ +, & \text{if } x \text{ is activity } \textit{Quality report intermediate a} & (\delta 2) \\ +, & \text{if } x \text{ is assertion rule A5} & (\delta 5) \\ \emptyset, & \textit{else} \end{cases}$$

Fig. 6. Public properties of the change in the running example

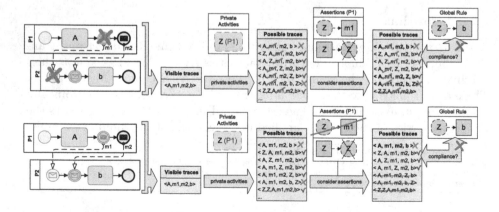

Fig. 7. Effects of local changes on the assertion-based filtering

---

Algorithm 1: Transitive effects of CBP changes

Input:
- Function $chg()$: $\mathcal{A}_v \cup \mathcal{R}_a \rightarrow \{\varnothing, -, +, \pm\}$ specifying the initial CBP change,
- Qualified dependency graph $\Phi_a := (\mathcal{A}, \mathcal{R}_a, \overline{d})$ between activities and ARs.

```
1  begin
2    Q := ∅; //A queue for unhandled, but affected elements.
3    //Initialize the change effects on hidden activities with ∅:
4    foreach a ∈ Aₕ do effects(a) := ∅;
5    //Initialize the change effects on assertions depending on chg
6    //Append them to Q to handle their transitive effects:
7    foreach r ∈ Rₐ do
8      │  effects(r) := chg(r);
9      │  if chg(r) ≠ ∅ and r ∉ Q then Q := Q + r;

10   //Initialize the change effects on visible activities depending on chg:
11   foreach a ∈ Aᵥ do
12     │  if chg(a) ≠ ∅ then
13     │    │  //Calculate the transitive change effects on assertions that depend on activity 'a'
14     │    │  //Append them to Q to handle them:
15     │    │  foreach (a, r, σ) ∈ d̄ do
16     │    │    │  effects(r) := effects(r) − σ · chg(a);
17     │    │    │  if r ∉ Q then Q := Q + r;

18   //As long as there exist unhandled elements in Q, calculate their effects:
19   while Q ≠ ∅ do
20     │  n :← Q; //Remove and store the head of queue Q in variable 'n'.
21     │  //Recalculate the effects on assertions or hidden activities depending on 'n':
22     │  foreach (n, m, σ) or (m, n, σ) ∈ d̄ do
23     │    │  if m ∈ Aₕ ∪ Rₐ then
24     │    │    │  oldEffects := effects(m);
25     │    │    │  effects(m) := effects(m) + effects(n);
26     │    │    │  //Append changed elements to Q to handle them:
27     │    │    │  if oldEffects ≠ effects(m) and m ∉ Q then Q := Q + m;

28   end
```

Output: Function $effects()$ specifying the transitive effects of $chg()$.

Algorithm 2: Effects of CBP changes on global compliance

Input:
- Function $effects()$: $\mathcal{A}_h$ $(\cup \mathcal{R}_a) \rightarrow \{\varnothing, -, +, \pm\}$ specifying the transitive effects of change $chg()$,
- Function $chg()$: $\mathcal{A}_v \cup \mathcal{R}_a \rightarrow \{\varnothing, -, +, \pm\}$ specifying the initial changes,
- Qualified dependency graph $\Phi_c := (\mathcal{A}, \mathcal{R}_c, \overline{d})$ between Activities and GCRs.

```
1  begin
2    //Initialize the change effects on GCRs with ∅:
3    foreach r ∈ Rᴄ do effectsC(r) := ∅;

4    foreach (a, r, σ) ∈ d̄ do
5      │  //Recalculate the direct effects on GCRs based on visual activities:
6      │  if a ∈ Aᵥ then effectsC(r) = effectsC(r) + σ · chg(a);
7      │  //Recalculate the transitive effects on GCRs based on hidden activities:
8      │  if a ∈ Aₕ then effectsC(r) = effectsC(r) + effects(a);

9  end
```

Output: Function $effectsC()$ specifying the effects of $chg()$ on compliance.

On the other, a CBP change can increase (decrease) the activations of assertions or even add (remove) assertions. In turn, the latter might then (no longer) filter out traces that violate the compliance of the CBP with GCRs. Figure 7 illustrates

how the deletion of an activity or an assertion weakens the assertion-based filtering, so that traces can pass and violate compliance. Note that decreasing
activations of assertions weakens the related filtering and, hence, always tends
to decrease compliance of the CBP. In turn, increasing activations of assertions
strengthens the filtering and, hence, always tends to increase compliance of the
CBP. According to this, Algorithm 1 utilizes the given change and the qualified dependencies of the QDG to calculate the increasing (decreasing) effects of
assertions on compliance (cf. Lines 7--9 and Lines 11--17). Note that Lines
8 and 16 change the semantics of qualifications (e.g., + or ±). The original qualifications expressed the additions or deletions of activities and ARs. Afterwards,
in turn, the qualifications express whether ARs and activities have positive or
negative effects on the global compliance of the CBP with GCRs. For public
activities, however, the original semantics is preserved. Since these effects may
transitively spread through private activities and ARs, Lines 19--27 propagate
them based on the QDG. As a result, Algorithm 1 enriches the given change with
its transitive effects on private activities (and ARs respectively).

In turn, Algorithm 2 aggregates the direct and transitive effects of CBP
changes on GCRs based on the dependency relations of the QDG. However, the
qualifications on dependency relations are only relevant in the context of direct
effects; i.e., changed public activities (cf. Line 6). In turn, transitive effects are
directly aggregated (cf. Line 8). Finally, Algorithm 2 returns all possible effects
of a given CBP change on compliance of the CBP with GCRs. Assuming the
latter was ensured before the change, compliance of the CBP with the GCRs,
which are annotated with − or ±, needs to be rechecked, e.g. by applying the
approach presented in [8]. Note that our approach is not limited to this use case.
Additionally, it can be applied to the opposite case; i.e., determining whether a
change has the potential to heal current violations of a particular GCR.

5   Evaluation

In order to evaluate and demonstrate the technical feasibility of the approach,
we implemented a proof-of-concept prototype. The latter is not only able to
construct the QDG and to calculate the results of the presented algorithms,
but also allows visualizing the QDG as well as listing intermediate results of
the algorithms (cf. Fig. 8). Based on these intermediate outputs, we were able
to enhance and optimize the algorithms. In Fig. 8, the prototype applies the
approach to the presented example. In particular, it recommends rechecking 3
GCRs (i.e., C2A, C4A and C5A). Note that the *obvious approach* would recheck
all 8 GCRs, whereas the *naive* one (i.e., to recheck only directly affected GCR)
fails. It rechecks C2A and C5A, but not C4A that is violated.

6   Related Work

In many domains, business processes are subject to laws, regulations, and guidelines [1]. Approaches, methods and techniques ensuring the compliance of a

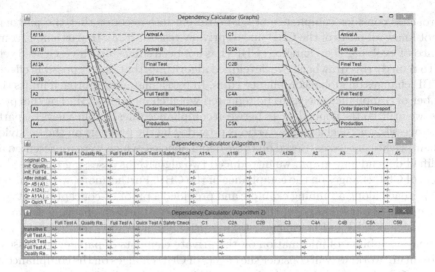

Fig. 8. Proof-of-concept prototype (Screenshots)

process with respective rules and constraints are covered under the term *business process compliance* [22]. In particular, the specification of compliance rules has been addressed by several approaches; e.g., [4,23] provide sets of compliance patterns, whereas [2,19,24] introduce visual notations. Besides the formal specification of compliance rules, their integration along the process lifecycle has been discussed [3,22,25]. Different techniques are applied to *a priori* check the compliance of process models at design time; e.g., [2] applies *model checking*, whereas [26] relies on Mixed-Integer Programming. In turn, [5,27–29] address *compliance monitoring* and *continuous auditing* [30] at runtime. Finally, [4] discusses *a posteriori* compliance checking based on process event logs. However, so far, only little work exists addressing compliance issues in the context of cross-organizational processes (e.g., [27,31]). Furthermore, [13] discusses compliance after changing the set of partners in such settings. However, these approaches have not taken privacy constraints into account yet [6,7].

To remedy this drawback, we investigated *a priori* compliance checking of cross-organizational processes with respect to privacy constraints [8,18]. This paper supplements our previous work by explicitly addressing global compliance in the context of CBP changes.

Note that few approaches deal with structural and behavioral effects of CBP changes and take privacy issues into account [9–12]. However, these approaches do not consider the effects of CBP changes on the compliance of the CBP with imposed global compliance rules.

7   Summary

Ensuring compliance with guidelines, standards and laws is crucial for both intra- and cross-organizational business processes (CBP). However, only few approaches consider compliance of CBPs taking into account that the partners do not know all parts of the CBP due to privacy reasons [6,8]. In particular, compliance of evolving (i.e. changing) CBPs has not been sufficiently investigated yet.

To remedy this drawback, we developed algorithms that detect possible effects of CBP changes on global compliance rules. In particular, our approach limits the number of compliance checks to be repeated when changing a CBP. For this purpose, we utilized the dependencies between compliance rules, public views on partner processes, and declarative assertions provided by the partners on the behavior of their private processes. Based on these dependencies, which were represented as qualified dependency graph, we introduced two algorithms assessing the possible effects of CBP changes on the compliance of the CBP. Furthermore, we illustrated the approach along a running example and provided a proof-of-concept prototype. To the best of our knowledge, there exists no other approach ensuring semantic compliance of CBPs after changes, taking privacy constraints into account (i.e. the non-availability of information on the private elements of partner processes).

In future work, we will consider the effects of CBP changes on the compliance of running CBP instances. In particular, we will investigate, whether these instances can be migrated to new versions of the CBP, without violating compliance. Further, we will improve the approach by taking further information into account (e.g., positions of changed activities, control flow dependencies within compliance rules).

References

1. Governatori, G., Sadiq, S.: The journey to business process compliance. In: Cardoso, J., van der Aalst, W. (eds.) Handbook of Research on BPM, pp. 426–454. IGI Global, Hershey (2009)
2. Awad, A., Decker, G., Weske, M.: Efficient compliance checking using BPMN-Q and temporal logic. In: Dumas, M., Reichert, M., Shan, M.-C. (eds.) BPM 2008. LNCS, vol. 5240, pp. 326–341. Springer, Heidelberg (2008)
3. Ly, L.T., et al.: Integration and verification of semantic constraints in adaptive process management systems. Data Knowl. Eng. **64**(1), 3–23 (2008)
4. Ramezani Taghiabadi, E., Fahland, D., van Dongen, B.F., van der Aalst, W.M.P.: Diagnostic Information for compliance checking of temporal compliance requirements. In: Salinesi, C., Norrie, M.C., Pastor, Ó. (eds.) CAiSE 2013. LNCS, vol. 7908, pp. 304–320. Springer, Heidelberg (2013)
5. Maggi, F.M., Di Francescomarino, C., Dumas, M., Ghidini, C.: Predictive monitoring of business processes. In: Jarke, M., Mylopoulos, J., Quix, C., Rolland, C., Manolopoulos, Y., Mouratidis, H., Horkoff, J. (eds.) CAiSE 2014. LNCS, vol. 8484, pp. 457–472. Springer, Heidelberg (2014)
6. Knuplesch, D., et al.: Towards compliance of cross-organizational processes and their changes. In: BPM 2012 Workshops, pp. 649–661 (2013)
7. Fdhila, W., Knuplesch, D., Rinderle-Ma, S., Reichert, M.: Change and compliance in collaborative processes. In: SCC 2015 (2015)
8. Knuplesch, D., et al.: Ensuring compliance of distributed and collaborative workflows. In: CollaborateCom 2013, pp. 133–142. IEEE (2013)
9. Fdhila, W., et al.: Dealing with change in process choreographies: design and implementation of propagation algorithms. Inf. Syst. **49**, 1–24 (2015)
10. van der Aalst, W.M.P.: Inheritance of interorganizational workflows to enable Business-to-Business E-Commerce. Electron. Commer. Res. **2**(3), 195–231 (2002)

11. Rinderle, S., Wombacher, A., Reichert, M.: Evolution of process choreographies in DYCHOR. In: CoopIS 2006, pp. 273–290 (2006)
12. Mafazi, S., Grossmann, G., Mayer, W., Stumptner, M.: On-the-fly change propagation for the co-evolution of business processes. In: OTM 2013, pp. 75–93 (2013)
13. Comuzzi, M.: Aligning monitoring and compliance requirements in evolving business networks. In: Meersman, R., Panetto, H., Dillon, T., Missikoff, M., Liu, L., Pastor, O., Cuzzocrea, A., Sellis, T. (eds.) OTM 2014. LNCS, vol. 8841, pp. 166–183. Springer, Heidelberg (2014)
14. Knuplesch, D., et al.: On enabling data-aware compliance checking of business process models. In: ER 2010, pp. 332–346 (2010)
15. van der Aalst, W.M.P., et al.: Multiparty contracts: agreeing and implementing interorganizational processes. Comp J. **53**(1), 90–106 (2010)
16. Decker, G., Weske, M.: Behavioral consistency for B2B process integration. In: Krogstie, J., Opdahl, A.L., Sindre, G. (eds.) CAiSE 2007 and WES 2007. LNCS, vol. 4495, pp. 81–95. Springer, Heidelberg (2007)
17. Rouached, M., et al.: Web services compositions modelling and choreographies analysis. Int. J. Web Service Res. **7**(2), 87–110 (2010)
18. Knuplesch, D., Reichert, M., Fdhila, W., Rinderle-Ma, S.: On enabling compliance of cross-organizational business processes. In: Daniel, F., Wang, J., Weber, B. (eds.) BPM 2013. LNCS, vol. 8094, pp. 146–154. Springer, Heidelberg (2013)
19. Knuplesch, D., et al.: Visual modeling of business process compliance rules with the support of multiple perspectives. In: ER 2013, pp. 106–120 (2013)
20. Semmelrodt, F., Knuplesch, D., Reichert, M.: Modeling the Resource perspective of business process compliance rules with the extended compliance rule graph. In: Bider, I., Gaaloul, K., Krogstie, J., Nurcan, S., Proper, H.A., Schmidt, R., Soffer, P. (eds.) BPMDS 2014 and EMMSAD 2014. LNBIP, vol. 175, pp. 48–63. Springer, Heidelberg (2014)
21. Governatori, G., et al.: Detecting regulatory compliance for business process models through semantic annotations. In: BPM 2008 Workshops, pp. 5–17. Springer (2009)
22. Knuplesch, D., Reichert, M.: Ensuring business process compliance along the process life cycle. Technical report 2011–06, Ulm University (2011)
23. Turetken, O., et al.: Capturing compliance requirements: a pattern-based approach. IEEE Softw. **29**, 29–36 (2012)
24. Ly, L.T., Rinderle-Ma, S., Dadam, P.: Design and verification of instantiable compliance rule graphs in process-aware information systems. In: Pernici, B. (ed.) CAiSE 2010. LNCS, vol. 6051, pp. 9–23. Springer, Heidelberg (2010)
25. Koetter, F., et al.: Integrating compliance requirements across business and it. In: EDOC 2014 (2014)
26. Kumar, A., et al.: Flexible process compliance with semantic constraints using mixed-integer programming. INFORMS J. Comput. **25**(3), 543–559 (2013)
27. Berry, A., Milosevic, Z.: Extending choreography with business contract constraints. Coop. Inf. Syst. **14**(2–3), 131–179 (2005)
28. Ly, L.T., et al.: A framework for the systematic comparison and evaluation of compliance monitoring approaches. In: EDOC 2013, pp. 7–16. IEEE (2013)
29. Knuplesch, D., Reichert, M., Kumar, A.: Visually monitoring multiple perspectives of business process compliance. In: BPM 2015 (2015)
30. Alles, M., Kogan, A., Vasarhelyi, M.: Putting continuous auditing theory into practice: lessons from two pilot implementations. Inf. Syst. **22**(2), 195–214 (2008)
31. Governatori, G., et al.: Compliance checking between business processes and business contracts. In: EDOC 2006, pp. 221–232 (2006)

Enhancing Aspect-Oriented Business Process Modeling with Declarative Rules

Amin Jalali[1]([✉]), Fabrizio Maria Maggi[2], and Hajo A. Reijers[3,4]

[1] Stockholm University, Stockholm, Sweden
aj@dsv.su.se
[2] University of Tartu, Tartu, Estonia
f.m.maggi@ut.ee
[3] VU University Amsterdam, Amsterdam, The Netherlands
h.a.reijers@vu.nl
[4] Eindhoven University of Technology, Eindhoven, The Netherlands

Abstract. When managing a set of inter-related business processes, typically a number of concerns can be distinguished that are applicable to more than one single process, such as security and traceability. The proper enforcement of these *cross-cutting* concerns may require a specific configuration effort for each of the business processes involved. Aspect-Oriented Business Process Modelling is an approach that aims at encapsulating these concerns in a model-oriented way. However, state-of-the-art techniques lack efficient mechanisms that allow for the specification of concerns in such a way that they can be executed in *parallel* to other parts of the process. Moreover, existing techniques exclusively focus on the formulation of *mandatory* concerns. To address these limitations, this paper proposes a new approach to encapsulate both optional and mandatory concerns, which can be executed concurrently with other process functionalities. One core element of the new approach is that it extends current Aspect-Oriented Business Process Modelling approaches with declarative rules. Thus, this *hybrid* approach allows for a sophisticated management of cross-cutting concerns.

Keywords: Business process modelling · Aspect orientation · Cross-cutting concerns · Declarative rules

1 Introduction

Separation of concerns is an important strategy for people to deal with the intricate nature of real-life phenomena, in particular, in the development of systems that need to cope with complexity. For this purpose, designers can divide the specifications of the system into smaller individual modules, which can be managed, understood, and changed separately but still integrated into a comprehensive, overall specification.

P. Johannesson et al. (Eds.): ER 2015, LNCS 9381, pp. 108–115, 2015.
DOI: 10.1007/978-3-319-25264-3_8

Business processes, being complex artefacts themselves, usually incorporate many concerns as well. Some of these, e.g., considering security or privacy issues, are not limited to one particular business process, but cross over and are applicable to many business processes. These are known as *cross-cutting concerns*. For example, each time a financial transaction is carried out as a part of a particular business process, certain security protocols are followed.

In modelling business processes, two types of modularization techniques can be distinguished that are regularly employed to encapsulate cross-cutting concerns. Both of these techniques aim at preventing the redundant implementation of a concern, which reduces future maintenance efforts.

With *vertical* modularization, the process logic that implements a concern is "embedded" in the main process model, i.e., it is being implemented as a sub-process. Therefore, the concern has multiple implementations in different process models and, as a result, is scattered over different process models. In this way, if the rules that govern the applicability of a concern change, this will affect all the parts of the process models in which the relevant sub-process appears. In addition, if the implementation of the concern itself changes, again, all the corresponding sub-processes need to be changed. This is called the *scattering problem*.

On the other hand, with *horizontal* modularization, the module that encapsulates the concern is implemented as a separate functionality in a separate process model. Thus, if the concern is changed, only this process model should be changed. However, if *conditions* under which the concern should be addressed are changed, all processes should still be reconsidered to establish whether the module is invoked appropriately, i.e., in accordance with the updated rules. So, even in the case of horizontal modularization, the invocation of the modules that take care of a concern are too closely tangled with the policy module. This problem is referred to as the *tangling problem*.

Aspect-Oriented Business Process Modelling (AO-BPM)[1] aims at separating cross-cutting concerns from process models in a way to solve the scattering and tangling problems [2,3,6]. To solve the scattering problem, each concern is encapsulated in a module called an *advice*. In this way, a change in a concern should only affect a single module. To solve the tangling problem, the relations between the *core* functionality of a business process and the cross-cutting concerns that interact with it are specified separately through a set of *pointcuts*, i.e., positions in the process in which the advice should be applied. In this way, a change in the *relation* between the core of a business process and a concern can be managed without altering the specification of these models. The groups of advices with relevant pointcuts that define how they should relate to the core process model are called *aspects*.

While AO-BPM addresses to a large extent how cross-cutting concerns can be managed, it has a severe limitation that stems from its adoption of concepts and semantics from the field of programming. Specifically, it only allows for

---

[1] Since AOBPM is sometimes used as an acronym for Aspect-Oriented Business Process *Management*, we opt for the hyphenated acronym to minimize confusion.

the definition of cross-cutting concerns that can be executed precisely before, after, or 'around' any specific point in a process. This is similar to a fixed point in a single-threaded piece of executable program code. As a result, AO-BPM approaches are not able to deal with cross-cutting concerns such that these can be fulfilled *in parallel* to some other parts of the business process. For example, as a security concern for the entry process into the United States, the arriving travellers should submit their passport information and customs declaration form prior to Customs and Border Protection (CBP) inspection.[2] A traveller can submit the documents at any time after arrival yet before CBP inspection. This type of concerns cannot be separated through existing AO-BPM approaches since they are not bounded to a single activity in a process model. Indeed, a traveller can do the submission at any time in parallel to the remaining part of the process. In addition, existing AO-BPM approaches cannot specify cross-cutting concerns that are *optional* to follow: they consider each concern as a *mandatory* process part.

To address these limitations, this paper proposes a new approach that extends existing AO-BPM approaches. The new approach supports the definition of both optional and mandatory cross-cutting concerns, which can be defined to be fulfilled in parallel to any other part of the main business process. The major, new element that allows for this is the use of declarative rules that enable a flexible yet precise connection between core process and concerns [10].

The remainder of this paper is structured as follows. Section 2 explains the proposed approach, which is the core idea of this paper. Section 3 discusses related works. Section 4 concludes the paper and explains future directions for research.

2   Approach

This section introduces our proposed approach using a case from the educational domain. Figure 1 depicts four processes, i.e., *Thesis Supervision, Thesis Examination, Course Supervision,* and *Course Examination*. The left-hand side of the figure shows the relation between these business processes and a number of cross-cutting concerns. It is visually expressed that a business process must address a concern if the concern crosses over it.

Each concern is modeled with one or more advices that define its implementation. In Fig. 1, main process, advices and pointcuts are represented by pentagons (annotated by M as an acronym for "Model"). There are four coloured, cascaded pentagons for advices representing the models for the four concerns. For instance, the *Grade Inform* advice can be seen to be taking care of the "Privacy Concern" (see Fig. 2). In this way, a business process model only describes the core functionality of the process, which is the *core concern* or the *main process*. In our setting, the *Course Examination* process is one of the main processes, which is modelled without explicitly addressing any cross-cutting concern (see Fig. 3).

---

[2] http://www.cbp.gov/newsroom/national-media-release/2014-08-11-000000/
new-mobile-passport-control-app-available.

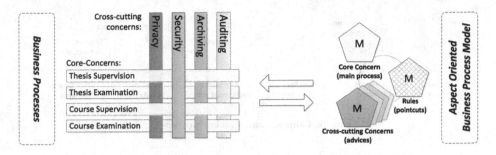

Fig. 1. Aspect-Oriented business process modelling

In this section, we elaborate on how our approach supports a separation of cross-cutting concerns for the *Course Examination* process. This process should comply with three cross-cutting concerns, i.e., privacy, security and archiving in accordance with the concerns that cross over this process in Fig. 1.

2.1 Cross-Cutting Concerns

Figure 2 shows cross-cutting concerns that are applicable for the *Course Examination* process. These concerns are encapsulated as advices and are modelled using Petri nets [1].

As a privacy concern, the *Grade Inform* advice specifies that students should be informed about grades through emails. As a security concern, the *Grade Registration* advice specifies that the administrative staff should provide a grading registration form, which should be signed by the examiner. As an archiving concern, the *Archive Examination* advice specifies that the examined sheets should be scanned by the administrative staff if they are not in digital format. Subsequently, they should be uploaded to the system. Moreover, the *Archive Exam Materials* advice specifies that the examiner can *optionally* archive exam material like questions and correct answers.

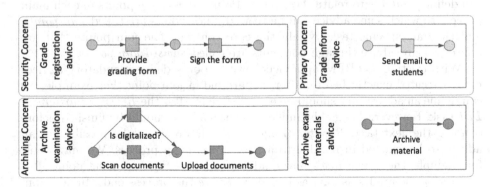

Fig. 2. Cross-cutting concerns for thecourse examination process

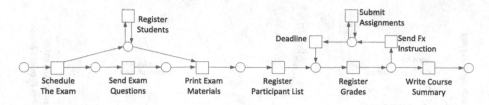

Fig. 3. Course examination process

2.2   Core Concern

The *Course Examination* starts with the *Schedule the Exam* task performed by the education secretary (see Fig. 3). Then, students can register for the exam through the *Register Students* task. In parallel, the *Send Exam Question* task enables the examiner to send the exam questions (at least one week before the exam) to the secretary. After the registration deadline, the secretary prepares the exams through the *Print Exam Materials* task.

After the exam, the administrative personnel registers students who participated in the exam through the *Register Participant List* task. The examiner grades the exam and reports the result through the *Register Grades* task. The grades are A, B, C, D, E, Fx and F. Grades Fx and F are failing grades; however, students who receive an Fx have another chance to improve their grade to E. The examiner should provide instructions (as an assignment) for those who received Fx through the *Send Fx instructions* task. Students can *Submit Assignments* before the specified deadline. Thereafter, the examiner grades the submissions and reports the result through the *Register Grades* task. Finally, the Course Leader (can be the same person as the examiner) writes the course summary.

2.3   Pointcuts

Having both a core concern (the main process) and cross-cutting concerns, we can define *pointcuts* to relate these parts. Pointcuts identify points in each main process to which some advice can be related. Such points are called *join points*, and they are specific tasks within the core concern. The join points that are related to any advice through pointcuts are called *advised join points*.

With existing AO-BPM approaches, we can only define mandatory advices to be executed exactly before, after, or 'around' a join point. For example, we can define an advice that should be executed right after the *Register Participant List* task. However, doing this implies that the advice should be finished before starting the next task, *Register Grades*. Thus, it would not be possible for an advice to be enacted in parallel to a part of the main process. We may want, for example, the *Grade Registration* advice to be executed at *some* point after registering the grades as long as it is done *before* the process ends. In addition, in standard AO-BPM approaches, we cannot define optional advices, like the *Archive Exam Materials* advice in our example. To deal with these issues, we need to extend the definition of pointcuts to:

- support the definition of an *initiator* and a *terminator*. An initiator / terminator pair allows us to define the interval within which an advice can or must be executed. The initiator can refer to the start of a process instance or to an instance of any task in the main process. The terminator can refer to the end of a process instance or to an instance of any task in the main process;
- support the definition of both optional and mandatory advices.

An instance of the initiator can be followed by a terminator with or without other occurrences of the initiator in between. Thus, we can define an advice to be executed: (i) for every occurrence of the initiator until the occurrence of the terminator, or (ii) for the first occurrence of the initiator until the occurrence of the terminator. The first option is called an *alternate response* scenario and the second one a *response* scenario.

Considering that advices can be defined as optional or mandatory, pointcuts can be defined through four declarative rules, called (i) the mandatory conditional response (mcr); (ii) the mandatory conditional alternate response (mcar); (iii) the optional conditional response (ocr); and (iv) the optional conditional alternate response (ocar). The graphical representations of these rules are illustrated in Fig. 4. Note that I, A, and T represent the initiator, the advice name, and the terminator respectively.

Figure 5 shows the application of these rules to our running case. Four declarative rules are defined to relate advices to the main process, which are annotated by numbers in the figure. The first rule indicates that the *Grade Registration* advice should be executed every time the grades are registered, and it should be completed before the process is ended. The second rule indicates that the *Grade Inform* advice should be executed every time the grades are registered, and it should be completed before the process is ended. The third rule indicates that the *Archive Examination* advice should be executed after registering the grades (once for all), and it should be completed before the process is ended. The fourth rule indicates that the *Archive Exam Materials* advice may be executed after registering the participant list, but – if so – it should be finished before the process is ended. The first three rules are mandatory rules, and the last rule is optional. Moreover, the first two rules are of type mcar, because both the security and the privacy concerns should be executed for every change in the grades. The third rule is of type mcr; the fourth rule is an ocr.

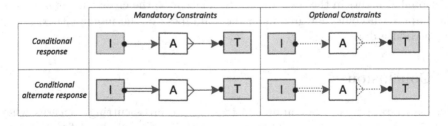

Fig. 4. Proposed approach for definition of pointcuts

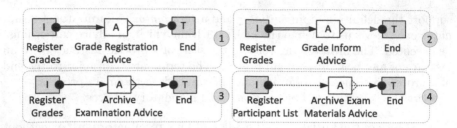

Fig. 5. Definition of pointcuts for the running case

3   Related Work

The separation of cross-cutting concerns has been investigated in business process modelling and enactment by different researchers. *To support modeling*, many works have been proposed in AO-BPM that allow for the definition of a concern for a specific point in a process model [2,3,6,9]. Cappelli et al. [2] introduce an extension for the Business Process Model and Notation (BPMN) [8]. This extension enables the definition of advices "before" or "after" a point in a process model. They also extend the Oryx editor to support this definition. The editor is called CrossOryx. Charfi et al. [3] also introduce an extension for BPMN, called AO4BPMN. This extension supports the definition of advices "before", "after" or "around" a point in a process model. Jalali et al. formalize the AO4BPMN notation for the first time [6]. Patiniotakis et al. [9] introduce bypass and replace elements that enable replacing an advised join point with another activity or skipping it. *To support enactment*, two approaches have been followed by researchers, called static and dynamic weaving. Static weaving defines how aspect oriented business process models should be merged into traditional models at design time, so that they can be executed later [5]. Dynamic weaving defines the operational semantics for executing aspect oriented business process models at runtime [7].

All these contributions are limited to the definition of concerns related to a specific point in a process model. In this paper, we try to support the definition of (possibly optional) concerns that can be executed in parallel to the main process. This problem has been investigated by Wang et al. in [11]. However, their solution does not solve the tangling problem [4], because the authors introduce additional elements in the core concern to relate it to the cross-cutting concerns. Their approach could not guarantee the soundness of the model either, and it can introduce deadlocks.

4   Conclusion

This paper proposes a new approach to separate cross-cutting concerns when modelling business processes. This approach facilitates this by defining declarative rules to relate process models with cross-cutting concerns. In particular, it extends and improves existing AO-BPM approaches by supporting the definition of concerns that can be executed in parallel to the main process. The new approach also enables the definition of optional cross-cutting concerns.

For future work, we aim at defining the formal definition and semantics of this approach. In addition, we plan to support the design and enactment of the proposed approach by implementing artefacts in a Business Process Management System. Finally, we are planning to study the perceived ease of use and perceived usefulness of the approach when modelling a business process. In addition, it would be interesting to investigate how hybrid models can expand the support for flexibility in Business Process Management.

References

1. van der Aalst, W.M.P.: Verification of workflow nets. In: Azma, P., Balbo, G. (eds.) ICATPN 1997. LNCS, vol. 1248, pp. 407–426. Springer, Heidelberg (1997)
2. Cappelli, C., Santoro, F.M., Leite, J.C.S.P., Batista, T., Medeiros, A.L., Romeiro, C.S.C.: Reflections on the modularity of business process models. Bus. Process Manage. J. **16**(4), 662–687 (2010)
3. Charfi, A., Müller, H., Mezini, M.: Aspect-oriented business process modeling with AO4BPMN. In: Khne, T., Selic, B., Gervais, M.-P., Terrier, F. (eds.) ECMFA 2010. LNCS, vol. 6138, pp. 48–61. Springer, Heidelberg (2010)
4. Jalali, A.: Assessing aspect oriented approaches in business process management. In: Johansson, B., Andersson, B., Holmberg, N. (eds.) BIR 2014. LNBIP, vol. 194, pp. 231–245. Springer, Heidelberg (2014)
5. Jalali, A.: Static weaving in aspect oriented business process management. In: Proceedings 34th International Conference on Conceptual Modeling (ER) (2015, to apper)
6. Jalali, A., Wohed, P., Ouyang, C.: Aspect oriented business process modelling with precedence. In: Mendling, J., Weidlich, M. (eds.) BPMN 2012. LNBIP, vol. 125, pp. 23–37. Springer, Heidelberg (2012)
7. Jalali, A., Wohed, P., Ouyang, C., Johannesson, P.: Dynamic weaving in aspect oriented business process management. In: Meersman, R., Panetto, H., Dillon, T., Eder, J., Bellahsene, Z., Ritter, N., De Leenheer, P., Dou, D. (eds.) OTM 2013. LNCS, vol. 8185, pp. 2–20. Springer, Heidelberg (2013)
8. Inc. (OMG) Object Management Group. Business Process Model and Notation (BPMN). Technical report, Object Management Group, Inc. (OMG) (2013)
9. Patiniotakis, I., Papageorgiou, N., Verginadis, Y., Apostolou, D., Mentzas, G.: An aspect oriented approach for implementing situational driven adaptation of BPMN2.0 workflows. In: La Rosa, M., Soffer, P. (eds.) BPM 2012. LNBIP, vol. 132, pp. 414–425. Springer, Heidelberg (2013)
10. Pesic, M.: Constraint-based workflow management systems: shifting control to users. Ph.D. thesis, Eindhoven University of Technology (2008)
11. Wang, J., Zhu, J., Liang, H., Xu, K.: Concern oriented business process modeling. In: IEEE International Conference on e-Business Engineering, ICEBE 2007, pp. 355–358, October 2007

Ontology-Based Modeling

Extending the Foundations of Ontology-Based Conceptual Modeling with a Multi-level Theory

Victorio A. Carvalho[1,2(✉)], João Paulo A. Almeida[1], Claudenir M. Fonseca[1], and Giancarlo Guizzardi[1]

[1] Ontology and Conceptual Modeling Research Group (NEMO),
Federal University of Espírito Santo (UFES), Vitória, ES, Brazil
victorio@ifes.edu.br, jpalmeida@ieee.org,
claudenirmf@gmail.com, gguizzardi@inf.ufes.br
[2] Research Group in Applied Informatics, Informatics Department,
Federal Institute of Espírito Santo (IFES), Colatina, ES, Brazil

Abstract. Since the late 1980s, there has been a growing interest in the use of foundational ontologies to provide a sound theoretical basis for the discipline of conceptual modeling. This has led to the development of ontology-based conceptual modeling techniques whose modeling primitives reflect the conceptual categories defined in a foundational ontology. The ontology-based conceptual modeling language OntoUML, for example, incorporates the distinctions underlying the taxonomy of types in the Unified Foundational Ontology (UFO) (e.g., kinds, phases, roles, mixins etc.). This approach has focused so far on the support to types whose *instances are individuals* in the subject domain, with no provision for *types of types* (or categories of categories). In this paper we address this limitation by extending the Unified Foundational Ontology with the MLT multi-level theory. The UFO-MLT combination serves as a foundation for conceptual models that can benefit from the ontological distinctions of UFO as well as MLT's basic concepts and patterns for multi-level modeling. We discuss the impact of the extended foundation to multi-level conceptual modeling.

Keywords: Ontology · Conceptual modeling · Multi-level modeling

1 Introduction

Conceptual modeling is the activity of formally describing some aspects of the physical and social world around us for the purposes of understanding and communication [1]. It is generally considered a fundamental activity in information systems engineering [2], in which a given subject domain is described independently of specific implementation choices [3]. The main artefact of this activity is a conceptual model, i.e., a specification aiming at representing a conceptualization of the subject domain of interest.

Since the late 1980s, there has been a growing interest in the use of foundational ontologies to provide a sound theoretical basis for the discipline of conceptual modeling [4–6]. The initial hypothesis which was later confirmed by different empirical evidence can be explained by the following arguments: (i) conceptual models are artifacts

© Springer International Publishing Switzerland 2015
P. Johannesson et al. (Eds.): ER 2015, LNCS 9381, pp. 119–133, 2015.
DOI: 10.1007/978-3-319-25264-3_9

produced with the aim of representing a certain portion of reality according to a specific conceptualization; (ii) foundational ontologies describe the categories that are used for the development of these conceptualizations. Therefore, an appropriate conceptual modeling language should provide modeling primitives that reflect the conceptual categories defined in a foundational ontology. This observation has led to the development of approaches for conceptual modeling based on foundational ontologies. An example of such an approach is OntoUML, which is based on the Unified Foundational Ontology (UFO) [3]. In OntoUML, the taxonomy of types of the Unified Foundational Ontology (UFO) has been reflected in the language such that the distinctions of the foundational ontology can be used to provide useful constraints and modeling guidelines, ultimately leading to ontologically well-founded conceptual models. The resulting conceptual models consist of a collection of *types (classes) of individuals in the subject domain* (e.g., the "Person" kind, the "Child" phase, the "Student" role). Each of these domain types instantiate types in the foundational ontology (e.g., kind, subkind, role, phase, etc.).

The approach is so far unable to describe subject domains in which the categorization scheme itself is part of the subject matter. In these subject domains, experts make use of categories of categories in their accounts. For instance, considering the domain of human resource management, organizations are often staffed according to *employee types* (e.g. "Engineer", "Pilot", "Secretary"). They may need to classify those *employee types* giving rise to *types of employee types*. In this case, "Engineer" and "Pilot" could be considered as examples of "Technical Employee Type", as opposed to "Secretary" which is an example of "Administrative Employee Type". Finally, they need to track the allocation of personnel to specific departments (e.g. John is an engineer in the Maintenance Department). Thus, to describe the conceptualization underlying this domain, one needs to represent entities of different (but nonetheless related) classification levels, such as *individual persons* ("John"), *employee types* ("Engineer", "Pilot", "Secretary"), and *types of employee types* ("Technical Employee Type", "Administrative Employee Type").

The need to support the representation of subject domains that deal with multiple classification levels has given rise to what has been referred to as multi-level modeling [7, 8]. In order to address the challenge of multi-level modeling, we have proposed in [9] a theory called MLT. MLT formally characterizes the nature of classification levels, and precisely defines the relations that may occur between elements of different classification levels, encompassing different notions of power type [10, 11]. In this paper, we apply MLT to UFO, in order to extend its applicability to domains that require multiple levels of classification. Conceptual models built with the UFO-MLT combination benefit from the ontological distinctions of UFO as well as the basic concepts and patterns for multi-level modeling of MLT.

This paper is further structured as follows: Sect. 2 presents a fragment of UFO and OntoUML; Sect. 3 presents MLT; Sect. 4 discusses the application of MLT to UFO identifying guidelines for multi-level modeling that arise from the foundational ontology; Sect. 5 discusses the implications of the combined foundations to the practice of multi-level conceptual modeling and finally Sect. 6 presents concluding remarks and topics for further investigation.

2 Ontological Foundations for Conceptual Models

The Unified Foundational Ontology (UFO) is a domain independent system of categories aggregating results from disciplines such as Analytical Philosophy, Cognitive Science, Philosophical Logics and Linguistics. Over the years, UFO has been successfully employed to analyze all the classical conceptual modeling constructs including Object Types and Taxonomic Structures, Part-Whole Relations, Intrinsic and Relational Properties, Weak Entities, Attributes and Datatypes, etc. [3]. Here we present a fragment of UFO that is relevant for this article. An in-depth discussion, formal characterization and discussion regarding empirical support for UFO's categories see [3].

UFO begins with a distinction between *universals* and *individuals*. Universals are patterns of features that can be realized in a number of individuals. For example, John and Mary are individuals that instantiate the universals "Man" and "Woman" respectively. UFO includes a taxonomy of individuals and a taxonomy of universals.

The topmost distinction in the taxonomy of individuals is that between endurants and events. Endurants (as opposed to events) are the individuals said to be wholly present whenever they are present, i.e., they can endure in time, suffering a number of qualitatively changes while maintaining their identity (e.g., a house, a person). Since in this paper we are especially interested in a portion of UFO that accounts for structural (as opposed to dynamic) aspects of conceptual modeling, we focus solely on endurants. Endurants are further classified into *Substantials* and *Moments*. Substantials are existentially-independent endurants (e.g. a person, a forest). A moment, in contrast, is an endurant that *inheres in*, and, therefore, is existentially dependent of, another endurant(s). Moments that are dependent of one single individual are *Intrinsic Moments* (e.g. a person's age) whereas moments that depend on a plurality of individuals are instances of *Relator* (e.g. a marriage, an employment, an enrollment).

These distinctions among individuals are reflected in the taxonomy of universals. Instances of *Intrinsic Moment Universal* apply to intrinsic moments (e.g. "Age"), instances of *Relator Universal* have relators as instances (e.g. "Marriage") and instances of "Substantial Universal" have substantials as instances (e.g. "Person"). The ontological category of Substantial Universal is further specialized according to the ontological notions of identity and rigidity. Substantial universals that carry a uniform principle of identity for their individuals are instances of *Sortal Universal* (e.g., "Person", "Car", "Organization"). In contrast, instances of *Non Sortal Universal* represent an abstraction of properties that are common to instances of various sortals (e.g., the non-sortal "Insurable Item" describes properties that are common to entities of different sortals such as "House", "Car", "Work of Art"). Moreover, a universal is said to be *rigid* if it classifies its instances necessarily (in the modal sense). In other words, if a universal T is rigid, then an instance x of T cannot cease to be an instance of T without ceasing to exist (e.g., "Person", "Organization"). In contrast, a universal is anti-rigid if its instances can move in and out of the extension of that universal without ceasing to exist (e.g., "Student", "Insured Organization"). Rigid sortals that provide a uniform principle of identity to their instances are termed a *Kind* (e.g "Person"). Instances of Kind may be specialized in other rigid sortals, which are instances of a *Subkind* (e.g. "Man"). Anti-rigid sortals are further classified into the categories *Role* or *Phase*. Instances of Role classify substantials through the relational properties they bear in the scope of a relational context (e.g. "Employee", "Husband", "Student") whereas instances of Phase define partitions of a Kind depending on one or more of its intrinsic properties (e.g "Child", "Living Person"). Rigid non-sortals that represent abstractions of properties that apply to instances of different kinds are called *Category* universals (e.g., "Legal Entity" abstracting properties of persons and organizations).

Figure 1 summarizes the discussion so far by depicting a fragment of UFO's taxonomy of universals in the left-hand side ("Endurant Universal" and its specializations) and the taxonomy of individuals in the right-hand side ("Endurant" and its specializations). This fragment of the UFO ontology is presented here as a UML class diagram for presentation-purposes only. The actual representation of the ontology is captured in [3] in a particular type of Intensional Modal Logics with Sortal Quantification.

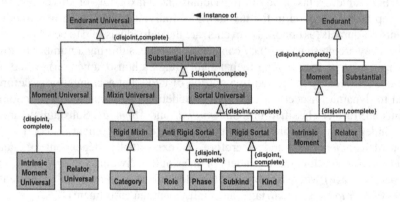

Fig. 1. UFO *endurant individuals* and *universals* taxonomies

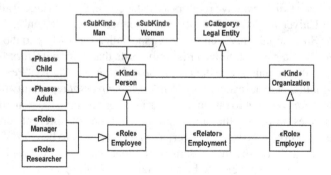

Fig. 2. An OntoUML diagram

In order to support the construction of ontology-driven conceptual models, a UML profile (dubbed OntoUML) was proposed in [3]. OntoUML includes: (i) modeling primitives that reflect ontological distinctions put forth by this ontology (these are represented as stereotypes for each of the leaf ontological categories of the UFO taxonomy of universals, see Fig. 2 for an example); (ii) formal constraints that govern how these constructs can be combined, which are derived from the axiomatization of the ontology. An example of constraint is that since instances of "Subkind", "Role" and "Phase" carry the identity criteria provided by instances of "Kind", OntoUML classes with a «subkind», «role» or «phase» stereotype must specialize (directly or indirectly) exactly one class stereotyped as «kind».

3   MLT: A Theory for Multi-level Modeling

Conceptual domain models constructed in OntoUML are able to express ontological properties of the types that apply to individuals in the subject domain. However, currently, no support is provided to represent domain-specific types of types, since the second-order types of OntoUML are predefined in the profile (as stereotypes). This motivates our investigation into the combination of UFO with a multi-level theory.

We employ a theory for conceptual multi-level called MLT which we introduced in [9]. Similar to UFO, MLT distinguishes between types (universals) and individuals. However, differently from UFO, MLT also considers types that have other types as instances. In order to accommodate these varieties of types, the notion of *type order* is used in MLT. Types having individuals as instances are called *first-order types,* types whose instances are first-order types are called *second-order types* and so on.

In order to link types to the entities that fall under such types, MLT defines a primitive *instance of* relation. This relation is represented by a ternary predicate *iof(e,t,w)* that holds if an entity *e* is instance of an entity *t* (denoting a type) in a world *w*. Indexing the instantiation relation to possible worlds allows MLT to support *dynamic classification,* admitting thus types that apply contingently to their instances (e.g., *John* is an instance of *student* in *w* but not in *w'*, when he has graduated.)

We build up the axiomatic theory defining the conditions for entities to be considered individuals, using the logic constant "Individual". Thus, an entity is an instance of "Individual" iff it does not have any possible instance. The constant "First-Order Type" (or shortly "1stOT") characterizes the type that applies to all entities whose instances are instances of "Individual". Analogously, each entity whose possible extension contains exclusively instances of "1stOT" is an instance of "Second-Order Type" (or shortly "2ndOT"). It follows from this definition that "Individual" is instance of "1stOT" which, in turn, is instance of "2ndOT". We call "Individual", "1stOT" and "2ndOT" the basic types of MLT. According to MLT, every possible entity must be instance of exactly one of its *basic types* (except the topmost type). For our purposes in this paper *first-* and *second-order types* are enough. However, this scheme can be extended to consider as many orders as necessary.

Figure 3 illustrates the elements that form the basis for our multi-level modeling theory, using a notation that is largely inspired in UML. We use the UML class notation to represent the *basic types of* MLT. We use associations as usual to represent relations between instances of the related types (predicates that may be applied to instances of the related types). Since UML does not allow for the representation of links between classes, we use dashed arrows to represent relations that hold between the types, with labels to denote the names of the predicates that apply. This notation is used in all further diagrams in this paper. It is important to highlight here that our focus is not on the syntax of a multi-level modeling language and we use these diagrams to illustrate the concepts intuitively. A complete formalization of MLT can be found in [9].

Fig. 3. Basic foundations of MLT: *basic types* and *instance of* relation.

Some structural relations to support conceptual modeling are defined in MLT, starting with the ordinary specialization between types. A type *t specializes* another type *t'* iff in all possible worlds all instances of *t* are also instances of *t'*. According to this definition every type *specializes* itself. Since this may be undesired in some contexts, we define the *proper specialization* relation as follows: *t proper specializes t'* iff *t specializes t'* and *t* is different from *t'*. Note that the definitions presented thus far guarantee that both *specializations* and *proper specializations* may only hold between types of the same order (these relations are depicted in the upper part of Fig. 4).

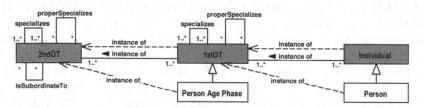

Fig. 4. Illustrating an important basic pattern of MLT and its intra-level structural relations.

Every type that is not a basic type (e.g., a domain type) is an instance of one of the basic higher-order types (e.g., "1stOT", "2ndOT"), and, at the same time proper specializes the basic type at the immediately lower level (respectively, "Individual" and "1stOT"). Figure 4 illustrates this pattern. Since "Person" applies to individuals, it is instance of "1stOT" and proper specializes "Individual". The instances of "Person Age Phase" are specializations of "Person" (e.g. "Child" and "Adult"). Thus, "Person Age Phase" is instance of "2ndOT" and proper specializes "1stOT". In Sect. 4, this pattern will be applied to UFO concepts and will drive patterns for domain models.

In addition to the instantiation and specialization relations, MLT also defines a *subordination* relation. *Subordination* between two higher-order types implies *specializations* between their instances i.e., *t is subordinate to t'* iff every *instance of t proper specializes* an *instance of t'*. Since *subordination* implies *proper specializations* between the instances of the involved types at one order lower, subordination can only hold between higher-order types of equal order. We will use subordination to explain the relation between universals in UFO's taxonomy of universals (e.g., since every "Subkind" must specialize a "Kind", "Subkind" *is subordinate to* "Kind".)

So far, we have only considered intra-level relations, i.e., those that occur between entities of the same order. In addition to that, MLT defines *cross-level structural relations* between types of adjacent orders. These relations support an analysis of the notions of power type in the literature, leading to their incorporation in the theory.

Based on the notion of *power type* proposed by Cardelli [10] (which is founded on the notion of powerset), MLT defines a *power type* relation between a higher-order type and a base type at an order lower: a type *t is power type of* a base type *t'* iff all instances of *t specialize t'* and all possible *specializations of t' are instances of t*. For example, consider a type called "Person Powertype" such that all possible specializations of "Person" are instances of it and, conversely, all its instances specialize "Person". In this case, "Person Powertype" is the *power type* of "Person". Since "Person" is instance of "1stOT", "Person Powertype" is instance of "2ndOT" and specializes "1stOT" (see Fig. 5). (It follows from the definition of *power type* that "1stOT" is *power type of* "Individual". Analogously, "2ndOT" is *power type of* "1stOT".).

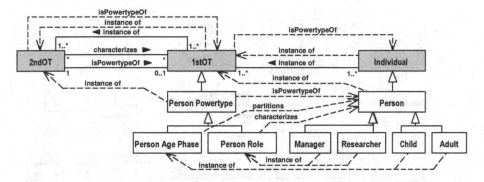

Fig. 5. Illustrating the use of MLT for multi-level conceptual modeling.

Another definition of *power type* that has had great influence in software engineering was proposed by Odell [11]. Inspired on Odell's definition [11], MLT defines the *characterization* relation between types of adjacent levels: a type *t characterizes* a type *t'* iff all *instances of t* are *proper specializations* of *t'*. Note that there may be specializations of the base type *t'* that are not *instances of t*. For instance in Fig. 5, "Person Role" (with instances "Manager" and "Researcher") *characterizes* "Person", but is not a *power type of* "Person", since there are specializations of "Person" that are not instances of "Person Role" ("Child" and "Adult" in the example).

We also define some variations of *characterization*, which are useful to capture further constraints in a multi-level model. We consider that a type *t completelyCharacterizes t'* iff *t characterizes t'* and every instance of *t'* is *instance of, at least*, an *instance of t*. Moreover, iff *t characterizes t'* and every instance of *t'* is *instance of, at most*, one *instance of t* it is said that *t disjointly Characterizes t'*. Finally, a common use for the notion of *power type* in literature considers a *second-order type* that, simultaneously, *completely* and *disjointly characterizes* a *first-order type*. To capture this notion we defined the *partitions* relation. Thus, *t partitions t'* iff each *instance of t* is *instance of exactly one instance of* the base type *t'*. For example of the partitioning relation, consider the second-order type called "Person Age Phase" with instances "Child" and "Adult" (Fig. 5). (This kind of constraint is usually represented in UML through a generalization set, see [9] for a detailed comparison).

4   Combining MLT and UFO

We represent the MLT concepts in the topmost layer of a hierarchy of conceptual models. The basic pattern of MLT is applied in this hierarchy to establish the relation between MLT and UFO, and later to establish the relation between a conceptual domain model and UFO-MLT. More specifically, the concepts of UFO *instantiate* and *specialize* elements of MLT, thereby respecting MLT's axioms and leveraging the use of structural relations and patterns of MLT in UFO. In turn, the concepts of the conceptual domain models *instantiate* and *specialize* concepts of MLT and UFO, respecting all rules and patterns of both MLT and UFO.

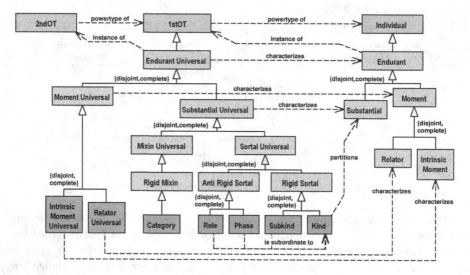

Fig. 6. Applying MLT to UFO taxonomies of endurants

The concepts in UFO's taxonomy of individuals are instances of "1stOT" specializing "Individual". The concepts in the taxonomy of universals are instances of "2ndOT" specializing "1stOT". For each entity in the taxonomy of individuals (e.g., "Endurant", "Substantial", "Moment"), there is a corresponding entity in the taxonomy of universals ("EndurantUniversal", "SubstantialUniversal", "MomentUniversal"). Instances of the entity in the taxonomy of universals specialize the corresponding entity in the taxonomy of individuals. Thus, "EndurantUniversal" *characterizes* "Endurant", "SubstantialUniversal" *characterizes* "Substantial", and so on.

In addition to these general characterization relations, we can also use more specific MLT relations to further explain how the two taxonomies relate according to ground notions in UFO (such as identity). For example, since each instance of "Substantial" is an instance of exactly one instance of "Kind" (the kind that supplies the principle of identity), following MLT, "Kind" *partitions* "Substantial". In addition, since they carry (but do not supply) a principle of identity, instances of "Subkind", "Phase" and "Role" must specialize an instance of "Kind" that supplies such principle. Thus, in MLT terms, "Subkind", "Phase" and "Role" are *subordinate to* "Kind". Figure 6 illustrates the resulting two-layer hierarchy revealing these relations.

In order to benefit from the ontological distinctions of UFO as well as the basic concepts and patterns for multi-level modeling of MLT, conceptual models built with the UFO-MLT combination must adhere to the rules of both theories. Thus, every domain *first-order* type must: (i) instantiate one of the leaf ontological categories of UFO's taxonomy of universals (and, consequently, instantiate MLT's "1stOT"); and (ii) simultaneously, specialize one of the leaf ontological categories of UFO's taxonomy of individuals (and thus, specialize "Individual"). For example, according to the conceptual domain model depicted in Fig. 2, "Legal Entity" is instance of "Category". Since "Category" specializes "1stOT" we conclude that "Legal Entity" is also a first-order type. Considering that "Category" *characterizes* "Substantial", it follows that "Legal Entity" specializes "Substantial" (and, indirectly, specializes "Individual"). Analogously, "Person" and "Organization" are instances of "Kind" and specialize "Legal Entity" (indirectly specializing "Substantial" and "Individual") (Fig. 7).

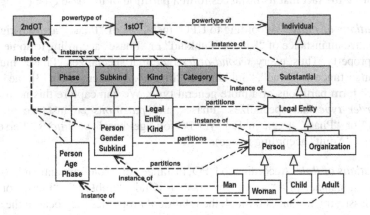

Fig. 7. Using the combination UFO-MLT to create a multi-level conceptual model

The UFO-MLT combination allows us to leverage the UFO taxonomy of universals to provide rules and patterns for *second-order types* in domain models. There are two basic rules for second-order types. First, since every domain first-order type admissible by UFO must be an instance of one of the leaf categories of the taxonomy of universals, *every domain second-order type must specialize one of these categories*. Second, to clarify which first-order type is ultimately instantiated by instances of a second-order type, *every domain second-order type must have an MLT cross-level relation with a first-order type (i.e., characterizes, disjointly characterizes, completely characterizes or partitions)*. These rules are applied below for second-order domain types specializing different leaf categories of UFO's taxonomy of universals.

Specializations of Kind. Considering that instances of "Category" are generalizations of instances of "Kind", a second-order type that specialize "Kind" must *partition* an instance of "Category". For example, considering "Legal Entity" as a "Category" that generalizes properties of different kinds of legal entities, we may define a *second-order type* "Legal Entity Kind" that *specializes* "Kind" and *partitions* "Legal Entity" having as instances "Person" and "Organization". Figure 7 illustrates this scenario.

Specializations of Subkind. Every (instance of) "Subkind" is a rigid universal that specializes an instance of "Kind" according to some intrinsic properties exemplified by their instances. Thus, every *second-order type* that specializes "Subkind" must *characterize* an instance of "Kind". Subkinds are common in taxonomies in which the more specific types form a partition of a more general type distinguishing instances according to immutable intrinsic properties (e.g., "Person" specialized into "Man" and "Woman" according to gender). In this case, a *second-order type* that specializes "Subkind" *partitions* an instance of "Kind" according to immutable intrinsic properties exemplified by their instances (see "Person Gender Subkind" with instances "Man" and "Woman" in Fig. 7). Another example of *second-order type* that can be represented as a specialization of "Subkind" is "Dog Breed", *partitioning* "Dog". Note that MLT does not force the modeler to enumerate the instances of second-order types (such as "Dog Breed"), while still capturing the fact that its instances form a partition of the base type ("Dog").

Specializations of Phase. According to UFO, instances of "Phase" are anti-rigid types that specialize an instance of "Kind", "Subkind" or "Phase" according to some mutable intrinsic property. Thus, every *second-order type* that specializes "Phase" must *characterize* an instance of "Kind", "Subkind" or "Phase". As discussed in [3], the instances of "Phase" form partitions of a more general type. We can capture this notion with a *second-order type* that specializes "Phase" and *partitions* an instance of "Kind", "Subkind" or "Phase" (e.g., in Fig. 7, "PersonAgePhase" *partitions* "Person" into "Child" and "Adult" according to age).

Specializations of Role. According to UFO, instances of "Role" are anti-rigid types that specialize an instance of a "Sortal Universal" ("Kind", "Subkind", "Phase" or another "Role") classifying instances through the relational properties they bear in the scope of a relational context. Thus, every *second-order type* that specializes "Role" must *characterize* an instance of "Sortal Universal". For example, we can define a *second-order type* "Person Role" that *characterizes* "Person" according to roles that persons may play during their lives. Types such as "Employee", "Driver" and "Wife" would be examples of instances of "Person Role". More specific specializations of "Person Role" include: (i) "Woman Role" whose instances specialize "Woman" (an instance of "Subkind") and include those roles that are played exclusively by women, such as "Wife"; (ii) "Adult Role" whose instances specialize "Adult" (an instance of "Phase") and include those roles that are played exclusively by adults, such as "Driver"; and, (iii) "Employee Role" whose instances specialize "Employee" (an instance of "Role") and include those roles that are played exclusively by employees such as "Manager" and "Researcher". These examples of second-order types specializing "Role" are illustrated in Fig. 8.

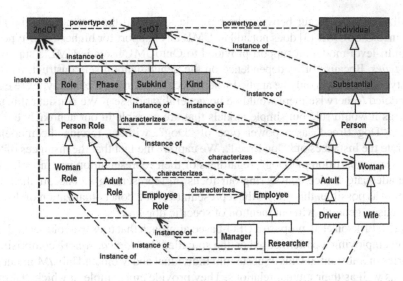

Fig. 8. Some patterns to create second-order types as specializations of "Role"

Note that the strategy that was used previously in OntoUML [3] was one in which the types represented in conceptual models could only instantiate the universals in UFO's taxonomy of universals. These were represented by a fixed set of UML stereotypes, and thus a conceptual model could only have first-order types. In that approach, the axioms of the foundational ontology had to be incorporated into the syntax and semantics of the language profile (e.g., translated into corresponding syntactic rules as shown in [3], or incorporated in a transformation of OntoUML into a logical formalism). This additional step is not necessary here as the structural relations and axioms of UFO-MLT are directly incorporated in the domain ontology. For example, concerning the combinations of the specialization patterns for second-order types discussed in this section, it is inadmissible for a domain second-order type that specializes "Kind" and "Subkind" to be subordinate to a domain second-order type which specializes "Role" or "Phase". This is a consequence of the constraint in UFO that rules out the specialization of an anti-rigid universal by a rigid universal, together with the definition of subordination in MLT.

5   Related Work

Two early attempts to address multi-level modeling, namely *power types* [10, 11] and *materialization* [12], raised from the identification of patterns to represent the relationship between a class of categories and a class of more concrete entities. Despite their different origins, they are based on similar conceptualizations [13] addressing similar concerns. Both approaches establish a relationship between two types such that the instances of one are specializations of another. The power type approach was incorporated in the UML [14], and the language currently includes a *power type association* that relates a classifier (power type) to a generalization set composed by the generalizations that occur between the base classifier and the instances of the power type. Since OntoUML [3] does not add to UML any support for higher-order types, the only multi-level modeling support provided to OntoUML users is UML's support for *power types*. Because of its dependence on the generalization set construct, the UML power type pattern can only be applied when specializations of the base type are explicitly modeled (otherwise there would be no generalization set). We consider this undesirable, as it would rule out simple models that are possible in our approach, e.g., one defining "Dog Breed" as a power type of "Dog", without forcing the modeler to enumerate the instances of "Dog Breed". We capture the fact that the instances of "Dog Breed" form a partition of "Dog", regardless of their representation in a model. This is a more adequate choice considering our focus on representing a conceptualization as accurately as possible; this allows the representation of a conceptualization of dogs and dog breeds in general, without mention of specific dog breeds.

In [15], Erikson et al. propose a UFO-based approach that tries to avoid second-order types by employing a pattern based on the so-called *ontological square* comprising the categories of Substantial Univ./Substantial Individual and Moment Univ./Moment Individual, as well as their mutual relations. They provide an example in which "Horse" is considered a *substantial type,* a horse named "Prancer" is a *substantial object* (instance of "Horse"), "Breed" is a *moment type* and "Shetland Pony" is a *moment object* (instance of "Breed"). Since both *Prancer* and *Shetland Pony* are objects, there is no instance of relation between them. According to the authors, each instance of *Horse* is related to one instance of *Breed* and one instance of *Breed* is related to many instances of *Horse.* Their assumption that the same *moment object* can be related to various *substantial objects* is a misinterpretation of a basic rule of the foundational ontology. In UFO, the relation between *moment objects (individuals)* and *substantial objects (individuals)* is that of *inherence.* In the ontology literature, in general, and in UFO, in particular, it is not possible for a moment to inhere in two different individuals. What the authors seem to intend to represent is actually the relation between a *property* (in the ontological sense) "Shetland Pony" and a number of individuals in which this property is *exemplified* (also in the ontological sense [3]). However, under this interpretation, "Shetland Pony" becomes a universal and "Breed" a second-order universal, defeating what they were trying to accomplish with their approach. Besides this ontological problem, the authors ignore the intuitive mechanisms of defining subtypes of a type according to properties of their instances and the benefits of such mechanisms. For example, using such approach, there is no support to represent properties that inheres only in instances of "Shetland Pony".

Atkinson and Kühne have proposed a deep instantiation based approach [8, 13] as a means to provide for multiple levels of classification whereby an element at some level can describe features of elements at each level beneath that level. The "potency" of an element defines how deep the instantiation chain produced by it may become. A "Mobile Phone Model" class is provided by the authors as an example of a class of potency of 2. The class is given an attribute "IMEI" also with potency of 2 meaning that instances of instances of "Mobile Phone Model" are assigned a value to the "IMEI" attribute. The authors consider that the main benefit of deep instantiation based approach is to support multi-level modeling without the need of introducing what they consider superfluous types (the required base type in the power type pattern) [13]. In the aforementioned example, the concept of "Mobile Phone" would be omitted from the domain model. While the deep instantiation approach can reduce the number of entities represented in a model, this strategy should be used with parsimony. This is because classes that instantiate higher-order classes "inherit" their properties with potency higher than one. In this case, the instantiation relation is overloaded with an implicit specialization relation, and semantic clarity is traded for reduction of model size. Further, by omitting a base type we become unable to express whether the instances of a higher-order type are disjoint types (i.e., we are unable to distinguish which form of *characterization* would apply). We are also prevented from determining metaproperties of the base type (such as e.g., rigidity). Note that the patterns that we have defined in Sect. 4 address these issues specifically. For example, one could define "Mobile Phone Model" as a second-order type that specializes "Subkind" and partitions the first-order type "Mobile Phone" mean that mobile phone cannot instantiate two models and is an instance of the every same model throughout its existence.

The *multi-level objects* (or *m-objects*) [7] is another multi-level modeling approach that applies the notion of *deep instantiation*. This approach is based on the notion of objects that "encapsulate different levels of abstractions that relate to a single domain concept", the so-called *m-objects*. Considering the *mobile phone* example, the modeler could define an m-object named "Mobile Phone Model" with two levels of abstraction, namely *type* and *physical entity* levels. Properties that are characteristics of a mobile phone model, such as *screen size*, should be defined at the *type level* while the ones exemplified by mobile phones, such as *IMEI*, should be defined at the *physical entity level*. The approach defines a *concretize* relationship to associate different m-objects. For example, applying the *concretize relation* to "Mobile Phone Model" we could create an m-object named "IPhone6" attributing values to the properties defined at the type level of "Mobile Phone Model" and at the same time "inheriting" the properties defined at the physical entity level of "Mobile Phone Model" such as "IMEI". Thus, the *concretize* relationship semantically overloads instantiation and specialization. Given that these are relations of different ontological nature, we believe this could affect the understand-ability and usability of the approach. Similarly to Atkinson and Kühne's proposal [8, 13] the approach leads to a model with fewer elements, but prevents us from expressing important aspects of the first- and second-order types.

6   Conclusions and Future Work

In this paper we have extended the Unified Foundational Ontology with the MLT multi-level theory in order to provide foundations for ontology-based multi-level conceptual modeling. MLT is founded on the notion of (ontological) instantiation, which is applied regularly across levels ("orders"). An important basic pattern of the theory has influenced significantly our approach: types *instantiate* a type at an immediately-higher order and *specialize* the basic type of the order to which they belong.

We have shown how the elements of MLT can be used to serve as the topmost layer of a hierarchy of conceptual models, from a foundational ontology to conceptual domain models. The concepts of the foundational ontology *instantiate* and *specialize* elements of MLT, respecting its axioms and using structural relations and patterns of MLT. In turn, the concepts of the conceptual domain model *instantiate* and *specialize* UFO-MLT, respecting MLT and UFO axioms and patterns. The result is an approach to define conceptual domain models that can represent types as well as types of types while adhering to the rules of a foundational ontology.

UFO's original taxonomy of (first-order) universals is leveraged in order to provide patterns for types of types in the domain model. These patterns guide the modeler in the definition of higher-order types and their relations allowing the modeler to express modal properties of instances of higher-order types. To the best of our knowledge, this is the first work to identify patterns and constraints for higher-order types based on a foundational ontology.

Another consequence of employing MLT concerns the engineering of UFO itself. UFO's taxonomies can now be explained in terms of instantiation of higher-order types. Further, as shown in Sect. 4, the relations of MLT (such as characterization) can be used to explain how elements in the taxonomy of universals relate to elements in the taxonomy of individuals.

While we have focused in the definition of domain models, the approach discussed here forms the basis for further extension of UFO itself, as well as to include core ontologies in the hierarchy of models between the foundational ontology and domain models. We will apply this approach to improve the formalization of UFO-based ontologies whose conceptualizations span multiple levels of classifications (e.g., the UFO-S core ontology for services [16] and the O3 organizational ontology [17]).

Finally, we should stress that it is not our intention in this paper to propose a multi-level language, and that our use of a notation inspired in UML has been solely illustrative. As discussed in [3], a reference ontology can be used to inform the revision and redesign of a modeling language, not only through the identification of semantic overload, construct deficit, construct excess and construct redundancy, but also through the definition of modeling patterns and semantically-motivated syntactic constraints. Thus, a natural application for UFO-MLT is to inform the design of a well-founded multi-level conceptual modeling language or to promote the redesign of a language such as UML into a multi-level modeling language.

Acknowledgements. This research is funded by the Brazilian Research Funding Agencies CNPq (grants number 311313/2014-0, 485368/2013-7 and 461777/2014-2) and CAPES/CNPq (402991/2012-5). Victorio A. Carvalho is funded by CAPES.

References

1. Mylopoulos, J.: Conceptual modeling and telos. In: Loucopoulos, P., Zicari, R. (eds.) Conceptual Modeling, Databases and CASE, pp. 49–68. Wiley, New York (1992)
2. Olivé, A.: Conceptual Modeling of Information Systems. Springer, Berlin (2007)
3. Guizzardi, G.: Ontological Foundations for Structural Conceptual Models. University of Twente, Enschede (2005)

4. Wand, Y., Weber, R.: An ontological evaluation of systems analysis and design methods. In: Falkenberg, E., Lingreen, P. (eds.) Information System Concepts: An In-Depth Analysis. Elsevier Science Publishers B.V, Amsterdam (1989)
5. Wand, Y., Weber, R.: On the ontological expressiveness of information systems analysis and design grammars. J. Inf. Syst. **3**, 217–237 (1993)
6. Guarino, N.: Formal ontology and information systems. In Guarino, N. (ed.), Formal Ontology in Information Systems, Proceedings FOIS 1998, pp. 3–15. IOS Press, Amsterdam (1998)
7. Neumayr, B., Grün, K., Schrefl, M.: Multi-level domain modeling with m-objects and m-relationships. In: Proceedings of 6th Asia-Pacific Conference on Conceptual Modeling, New Zealand (2009)
8. Atkinson, C., Kühne, T.: The essence of multilevel modeling. In: Proceedings of the 4th International Conference on the Unified Modeling Language, Toronto, Canada (2001)
9. Carvalho, V.A., Almeida, J.P.A.: Towards a well-founded theory for multi-level conceptual modeling (2015). http://nemo.inf.ufes.br/mlt
10. Cardelli, L.: Structural subtyping and the notion of power type. In Proceedings of the 15th ACM Symposium of Principles of Programming Languages, pp. 70–79 (1988)
11. Odell, J.: Power types. J. Object-Oriented Program. **7**(2), 8–12 (1994)
12. Pirotte, A., Zimanyi, E., Massart, D., Yakusheva, T.: Materialization: a powerful and ubiquitous abstraction pattern. In: Proceedings of the 20th International Conference on Very Large Data Bases, pp. 630–641 (1994)
13. Atkinson, C., Kühne, T.: Reducing accidental complexity in domain models. Softw. Syst. Model. **7**(3), 345–359 (2008). Springer-Verlag
14. OMG : UML superstructure specification – version 2.4.1 (2011)
15. Eriksson, O., Henderson-Sellers, B., Ågerfalk, P.J.: Ontological and linguistic metamodeling revisited: a language use approach. Inf. Softw. Technol. **55**(12), 2099–2124 (2013). Elsevier
16. Nardi, J.C., Falbo, R., Almeida, J.P.A., Guizzardi, G., et al.: A commitment-based reference ontology for services. Inf. Syst. **51**, 1 (2015)
17. Pereira, D., Almeida, J.P.A.: Representing organizational structures in an enterprise architecture language. In: 6th Workshop on Formal Ontologies meet Industry (2014)

Logical Design Patterns for Information System Development Problems

Wolfgang Maaß[1(✉)] and Veda C. Storey[2]

[1] Saarland University, 66123 Saarbrücken, Germany
wolfgang.maass@iss.uni-saarland.de
[2] University Plaza, Georgia State University, Atlanta, 30302, USA
vstorey@gsu.edu

Abstract. Design theories investigate prescriptive and descriptive elements of the activity of design. Central to the descriptive realm are abstract design rules and design goals that are part of the governance of design. On the prescriptive side, models are used on various levels of abstraction for representing different kinds of knowledge for systems engineering. Models for three layers of abstraction are proposed: a business layer, a logical layer, and an implementation layer. At the logical layer, the concept of a logical design pattern is introduced as a natural means for linking business models and technical models as well as design theories and information systems engineering. Ten logical design patterns, extracted from a series of information system development projects, are presented and applied in an example.

Keywords: Design theory · Design rules · Design science · Logical design patterns · Prescriptive and descriptive design knowledge · Design theory · Abstractions · Software engineering

1 Introduction

Designing information systems is traditionally governed by software engineering methods that work directly on the design object using methods from requirements engineering, software development, testing, and organizational change theories [1, 2]. Pure "from-scratch" design approaches have been enhanced by pattern-based [3], domain model approaches [4], frameworks, and reference models. An integrated design theory for prescriptive design knowledge is still needed, however, several concepts have been proposed as key elements, such as design rules, organizational patterns [5], design patterns, and technical design patterns (e.g., [3]). The lack of a prescriptive design theory is only superseded by even more uncertainty of a descriptive design theory [6].

This bottom-up design process is contrasted by top-down approaches based on design theories as proposed in design science research [7–9]. Design theories in information systems focus on products of design and design processes [7]. They investigate sets of goal-oriented, generalized requirements as part of design knowledge that links a design decision with a desired outcome for a given class of design problems [7, 8]. Design principles and design rules, as essential elements of design theories, are either

© Springer International Publishing Switzerland 2015
P. Johannesson et al. (Eds.): ER 2015, LNCS 9381, pp. 134–147, 2015.
DOI: 10.1007/978-3-319-25264-3_10

extracted from existing information systems ex-post [10] or derived from kernel theories, natural or social sciences [7]. In both cases, design theories are top-down approaches for information system design.

This separation between top-down and bottom-up approaches has not always been prevalent. Prior research has focused on the need to create things that serve human purposes such as constructs, models, methods, and implementations. Researchers in software engineering and information systems engineering widely neglect the need for design theories, whereas design theorists separate their results from actual design [7, 11].

Design science investigates knowledge that is used for designing information systems but not for engineering activities as such [8]. Thus, there is a lack of consensus on how design theories and engineering of information systems interact to create effective designs. The objective of this research, therefore, is to propose design rules as higher-level abstractions *of* design (descriptive knowledge in the sense of design science); and to align them with logical design patterns as abstractions *for* design [12] (p. 111) (prescriptive knowledge in the sense of systems engineering). The contributions of doing so are: to aid designers in creating better designs and to connect design theories and engineering practices, in particular on conceptual modeling. To carry out the research, design patterns are proposed as mediators between software engineering and design theories.

2 Related Research

Relationships between design theories and system modeling as explicit descriptions of designs for information systems are sparse [13]. Carlson et al. discuss an initial attempt for bridging design theory and software engineering by describing the "actual form" of a design proposition as "an algorithm, a drawing or picture, a report or a whole book. […] An algorithmic design proposition can in principle guarantee a good (best) outcome […]" [14].

2.1 Design Theories

Research on design theories and design research emphasize how organizations can be improved by enhanced information technologies (e.g., [7, 15]). This is embedded into a more general discussion on the relationships between descriptive and explanatory research, versus prescriptive and exploratory research (e.g., [8, 16]). Van Aken identifies professional designers as being occupied with the creation of three knowledge types [8]: (1) object design, (2) realization-design, and (3) process-design. Object design is the design of the artifact or intervention itself; realization-design is a plan for the implementation; and process-design is a plan for the design process applied in general [8]. Van Aken [8] further emphasizes that design science develops scientific knowledge to support the design of interventions or artifacts but "is not concerned with action itself, but with knowledge to be used in designing solutions, to be followed by design-based action."

Walls et al. [7] employ Nagel's [17] components of a theory. Nagel identified the basic notion of a system and sets of rules that relate empirical contents with abstract calculus. Gregor and Jones [18] emphasize both artifact design and design process. The design process view has often been dominant. Walls et al. [7] focus on the artifact design and production and define a design theory as an explanation of properties an artifact should have or "how an artifact should be constructed." Markus et al. [15] argue that design theory should either target generalized product architectures or generalized methods [18].

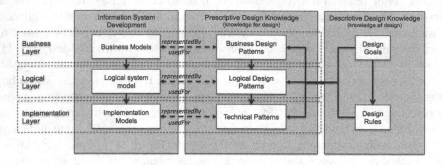

Fig. 1. Design framework

With this separation of "guiding design" and "doing design," design theory artificially distinguishes itself from designing artifacts and interventions. This means, for example, that Bob Kahn's and Vinton Cerf's design principles for the Internet, including minimalism, autonomy, best effort, stateless routers, and decentralized control [19], are beyond the realm of any design theory, but lie within the domain of engineering.

2.2 Requirements Engineering and Modeling

The actual design of information systems depends on prescriptive design knowledge captured by models and patterns (cf. Fig. 1). Models used in the business layer (cf. Fig. 1) shall inform stakeholders about characteristic attributes of an existing or targeted business venture as the system under study (SUS). Business models are used in rapidly changing, complex, and uncertain environments [20] and abstract from implementation issues. Relevant attributes are product and service offerings, information flows, potential benefits, and sources of revenue [21]. Business models focus on value propositions, value exchanges between actors and cost structures, but abstract from technical details (e.g., [20]). Because information systems are mainly developed for domains with virtual goods and services, such as financial industries, e-commerce, and e-government, business models generally abstract from physical real-world entities.

In contrast, technical models in software engineering focus on relevant entities, such as classes, methods, and actors when looking at object-oriented systems as SUS, but abstract from business details and organizational issues (e.g., [22]). Recently, service-oriented systems development use technical models that cut across various parts of business models and technical models with a focus on technical services but abstraction from business values [23]. Service-oriented models consist of a set of models and are, thus, similar to reference models (e.g., [24]).

Information system design distinguishes layers related to business designs from layers related to technical system designs (e.g. [24, 25]). On the business level, conceptual business models consist of: (1) an architecture for product, service and information flows, including a description of business roles; (2) a description of potential benefits; and (3) a description of sources of revenues [21].

On the technical level, software engineering, in particular, model driven engineering, provides explicit knowledge structures in the form of specifications based on different types of modeling languages [26]. Analysis models are abstract specifications of requirements described by (modeling) languages of the domain without direct consideration of technical issues [26]. Analysis models are transformed into technical models, i.e. platform-independent and platform-specific models [26].

In-between, socio-technical models interrelate business models and technical system models. These are called *logical models* because they interrelate abstractions of business models with abstractions of technical models. The modeling language i* [27] and its extension Tropos [28] support the representation of actors, goals, believes, plans, and dependencies. Extensions of i* integrate two additional model classes, i.e. tasks and information resources [29].

This distinction of three levels resembles OMG's definition of model driven architectures (MDA): (1) business layer, (2) logical layer, and (3) implementation layer [30]. Associated with these layers are business and domain models, logical system models, and implementation models. While business and domain models are independent of technical platforms and implementation models focuses on a technical viewpoint, logical system models describe interactions between business entities and/or technical components (cf. Fig. 1).

2.3 Modeling with Patterns

Pattern approaches support efficient reuse of prescriptive design knowledge for gaining models of better quality [4]. Business model pattern have been proposed recently [31] while technical patterns, such as analysis patterns [32] and technical design patterns [3], are well-known and often used. Analysis patterns consist of domain-neutral groups of related, generic objects with attributes, behaviors, and expected interactions. They resemble Alexander's idea of design patterns used in urban planning [33], mixed with an object-oriented view of software development with data structures and methods [34]. Alexander perceives design patterns as practical solutions to a design problem for a domain, such as urban planning. Besides of its use as problem-solving, design patterns also provide an abstracted language that can be shared among designers.

In contrast, logical design patterns have been rarely investigated. Proposals are still driven by technical viewpoints and can rarely be used by domain experts [35]. In adaptation of the definition of technical design patterns [3], a logical design pattern names, abstracts, and identifies the key aspects of a socio-technical design structure that make it useful for creating a reusable design on the logical layer (cf. Fig. 1). In contrast to business models and technical models, logical design patterns cut across business and technical domains without simply combining corresponding models. Logical design patterns abstract from business models and technical models and focus on representing interacting social and technical conceptual entities.

2.4 Design Rules

Design rules can be seen as theory-grounded guidelines for the design process. Hanseth and Lyytinen [10] suggest that a design theory has three elements: (1) a set of generalized design goals; (2) a set of system features that fit these goals; and (3) a set of design principles and rules. Design principles describe broad guidelines and are detailed by design rules that *"formulate in concrete terms how to generate and select desired system features"* (ibid.). Design rules pragmatically constrain design search spaces and, thus, are an essential element of design theories (e.g., [8, 16, 36]). Design rules are generalized guidelines for a class of problems, which can be applied to designing actual information systems. They are part of a governance structure of design processes, particularly, for specification models, such as domain models, business models, and platform models.

The causal character of design rules binds design goals with design activities [7–9]. As part of design theories, design rules are explicitly defined as plans for the task of designing, but not with the activity of designing a solution itself [8]. Most research on design rules that emphasize aspects of design organizations and design processes [8, 15] focus on descriptive knowledge for design teams. Markus et al. [15], for example, propose design rules, such as *"designing for customer engagement by seeking out naïve users"* or van Aken's [8] example *"If you want to realize a large-scale, complex strategic change, use a process of logical incrementalism."* Therefore design rules capture descriptive design knowledge investigating knowledge *of* design [12].

Research in organization science conceptualizes design rules as any coherent set of normative propositions, grounded in the state-of-the-art [37] that are general descriptive for a class of problems [7, 8]. Design rules are perceived as boundary objects that connect practitioners and academics [38].

Some researchers start the discussion of design rules from Bunge's concept of a *technological rule* that is "an instruction to perform a finite number of acts in a given order and with a given aim" [39]. In management theories, a technological rule is defined as *"a chunk of general knowledge, linking an intervention or artifact with a desired outcome or performance in a certain field of application"* [8]. This very practical and prescriptive notion might easily lead to recipes without proper foundations. Therefore, van Aken argues for testing technological rules in various domains and different designer groups and justifying technological rules, either by base theories or by accumulated evidence [8]. Following van Aken's approach, design rules link dependent variables with independent variables. Independent variables *"describe something that can be changed or implemented by the designer,"* whereas dependent variables in design rules represent *"something of value to the organization."* [40].

Walls et al. perceive a set of rules as a means for connecting empirical content with theories [7] and describe design rules as follows: *"If you want to achieve goal X, then make Y happen"* or inversely defined by Goldkuhl: *"Perform act A in order to obtain goal G"* [9]. Van Aken extends this design rule pattern by adding contexts: *"If you want to achieve Y in situation Z, then perform action X"* [8]. Markus et al. argue that, in domains where key knowledge is implicit, if-then rules, cases, or texts can help to make implicit knowledge explicit [15]. Table 1 summarizes definitions of design rules.

Table 1. Definitions of design rules

IF clause (intention)	Then clause	Reference
Achieve goal X	Make Y happen	Walls et al. [7]
Obtain goal G	Perform act A	Goldkuhl [9]
Achieve Y in situation Z	Perform action X	Van Aken [8]

3   Logical Design Patterns as Building Blocks for Logical Models

3.1   Logical Models

According to Adaptive Structuration Theory, social structures serve as templates for planning and accomplishing tasks [41]. Social structures are two-folded. They exist within information technologies and constrain interactions in social situations and are embedded within social actions independent of information technologies (ibid). As a result, social structures are building blocks of socio-technical systems. Logical models focus on socio-technical situations of information systems and abstract from value propositions, technical details, and implementation issues. Therefore, logical models support representation of surface structures of information system designs [42]. They connect information systems theories from social sciences with conceptual modeling that relate models for particular usage situations with implicit or explicit goals [27].

Logical models are consistent with the approach of reusing conceptual models, in particular, analysis patterns [32]. They abstract from business models and technical models and focus on the interaction of entities of both layers. Logical models are condensed representations of a set of designed situations in which social actors, technical services, and physical entities purposefully interact with one another [43]. A basic design goal is that situations ought to be as well-defined as possible. With the application logic on mobile devices, technological objects increasingly obtain service characteristics. Users make deliberate choices on the situated usage of technical objects and co-create solutions for these technical objects by providing context and personal information for a situation. Due to the tendency that real-world environments are increasingly enriched with information technologies, logical models need to be extended with representations of physical aspects with their intrinsic features as well. A user will use a service only if it provides a utility. Utilities are either derived from: (1) direct interaction between a single user and a service; or (2) interaction between users mediated by services. Thus, logical models are differentiated from business models because they abstract from business aspects, such as strategic, value propositional, or financial aspects [20].

Following the theory of social interactionism [44], logical models conceive virtual or real-world contexts with their social structures in which role-taking actors interact with one another by using services provided by technical objects and physical entities for goal-directed behavior. This view on logical models is based on at least four conceptual spaces [43, 45]:

1. *Social system*: (1) set of social roles available with a set of attributes, such as rights, obligations, and prohibitions, interactions that support some tasks, and goals; (2) set of social interactions
2. *Service system*: services are abstractions of functional affordances and support activities for planning and accomplishing tasks. Services are either interface services for direct support of role-taking actors or internal services.
3. *Information sphere*: non-materialized information entities
4. *Physical infrastructure*: information objects carry meaning, i.e. conceive information objects that can be intentionally used in interactions. They are any materialization, including hardware, software, real-world entities, and human beings but also books as materialization of information objects.

In the social system, interactions between social actors are defined. Services and information objects support these descriptions (for details cf. [43]). Whereas entities from social systems, information spheres, and service systems are conceptual by definition, they are materialized by entities in the physical infrastructure. This is accomplished by binding non-material entities to material entities. For instance, social roles are bound to human individuals; poems might be bound to paper.

3.2 Logical Design Patterns

Logical design patterns provide a governance structure for logical models. They connect, on one hand, boundary objects of business models with boundary objects of technical models by obeying guidelines given by design rules. On the other hand, logical design patterns are means for connecting design rules, i.e., descriptive design knowledge, with logical models, prescriptive design knowledge (cf. Figure 1). Thus, logical design patterns capture characteristic, re-occurring structural relationships between entities of prescriptive and descriptive design knowledge.

We define a **logical design pattern** as causal relationships between goals given by a social situation, social actors, services, technical objects, and design rules: *If you want to achieve goal set G_S in a social situation S supported by technical designs T, then derive model M by instantiation of logical design pattern LDP with boundary objects B_S from S and B_T from T governed by design rule D.*

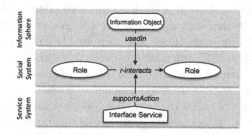

Fig. 2. Example: logical design pattern *role-interaction*

This characterization of logical design patterns replicates the structure of design rules within the context of prescriptive design knowledge. It describes how-to achieve situation-specific goals (G_S) by information system artifacts. Design rule D obtains an analogue structure for descriptive, design theoretic knowledge. Nonetheless design rules require incremental refinement until goals G_D of a design rule D can be directly used as organizational governance structure for a logical design pattern *LDP*.

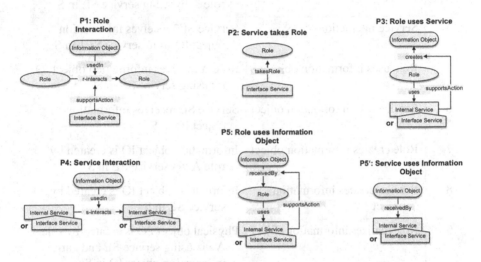

Fig. 3. Logical design pattern (excerpt – for details refer to [45])

An example is the *role-interaction* logical design pattern [45] that links a communication goal G_S with a situation S in which two role-taking actors (elements of B_S) interact with one another by exchanging an information-object (element of B_S) via a technical design T given by a service (element of B_T). In this case, it is an interface service that can be directly used by role-taking actors. How design rules directly impose governmental control on logical design patterns is still under investigation. To appreciate this relationship, one can consider the simple design rule "Runs all the Tests" [46]. The *role-interaction* LDP obeys this design rule by clear causal relationships between goals, situations, boundary objects, and activities. A graphical representation of the model represented by the *role interaction* design pattern is shown in Fig. 2. On the technical level, logical design patterns are formalized by description logic [47] resulting in n-ary relationships transformed into binary relationships [48].

By mining various information system design projects based upon the logical layer domain ontology of one of the authors, 10 logical design patterns were extracted (cf. Fig. 4 and Table 2).

Table 2. Logical design patterns

No.	Model set M of situation S	Goal set G of situation S
1	Role interaction	Role B receives information object IO from role A by service SE in S
2	Service takes role	Service SE is linked with role A in S
3	Role uses service	Information object IO is created by role A by using service SE in S
4	Service interaction	Service SE2 receives information object IO from service SE1 in S
5	Role uses information object	Role A receives information object IO by using service SE in S
6	Service uses information object	Service SE receives information object IO in S
7	Role creates information object	Information object IO is created by role A via service SE in S
8	Service creates information object	Information object IO is created by service SE in S
9	Role realizes information object	Physical object PO is created by role A via using service SE and carries information object IO in S
10	Service creates information object	Physical object PO is created by service SE and carries information object IO in S

3.3   Anchoring Logical Design Patterns in Domain Ontologies

Logical design patterns are used for describing logical models of information systems as a means for translating analysis models into technical implementations. They are also used as a means for building and processing shared understandings between domain experts, technology experts, and other stakeholders. Both purposes are supported by semantically enriching concepts and relationships based on formal ontologies [49]. Embedding logical design patterns into the conceptual framework of formal ontologies supports, for instance, semantic search for existing logical models and logical consistency checks for new logical models.

All logical design patterns have been embedded into DOLCE [50] and OntoUML [51]. In the following, embedding into OntoUML is discussed for the *role-interaction* logical design pattern. Communicative interaction between a sender role and a receiver role is a relational moment with an information object as a kind of social object [52]. Because social relations in the logical layer domain ontology are 4-ary relations, it is decomposed into four binary relations plus introduction of a role *r-relation* [48] (Fig. 5). For this example, roles are instantiated by human beings and relations by speech acts or any other kind of interactions that are compatible with role-taking entities. Any kind of interaction might carry an information object instantiation (Fig. 4).

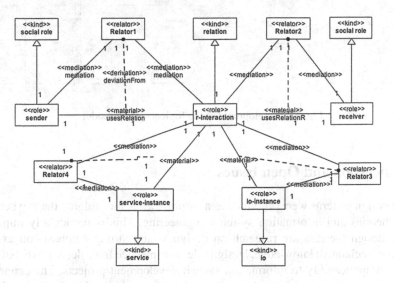

Fig. 4. Representation of the role-interaction pattern in OntoUML

4   Application to Designing Information Systems

In this section, we discuss a small example in more detail. In his seminal book on use cases, Cockburn presents a "get paid for car accident" use case with the claimant's goal of getting paid for a car accident [53]. The scenario is structured as follows:

1. Claimant submits claim with substantiating data
2. Insurance company verifies claimant owns a valid policy
3. Insurance company assigns agent to examine case
4. Agent verifies all details are within policy guidelines
5. Insurance company pays claimant

Logical design patterns were applied in six steps (cf. Fig. 3): (1) *Role creates IO*, (2) *Role interaction*, (3) *Role uses IO*, (4) *Role interaction*, (5) *Roles uses IO*, and (6) *Role interaction*. The formal specification was successfully tested for internal consistency and validity. Details on the method for building logical models with logical design patterns are found in [43, 45].

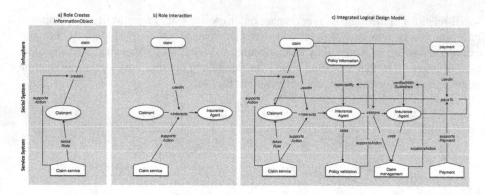

Fig. 5. Development of the logical system model

5   Conclusion and Open Issues

Logical design patterns were presented as a natural means for bridging the gap between design theories and information systems engineering. This is particularly important because design theories are research on design with a strong emphasis on ex-post, normative, declarative knowledge. Design rules are intended to be descriptive, but often lack direct applicability to information system development projects. The connection between design rules and logical design models are considered a means for linking research on design and research of design [12]. Further research will investigate reasoning on formal representations of goals based on work proposed by [54, 55].

In several projects, our proposed approach has proven helpful for guiding the design process as well as building conceptual models. It was particularly helpful for discussing models and goals in heterogeneous development teams, resulting in shared understanding that was better supported than with traditional methods, such as UML [56].

This paper is an initial effort for research on the relationship between design theories and software engineering with design rules, logical models, and logical design patterns as key ingredients for a better understanding. The intent is that design theories might eventually be developed into a knowledge base that is instrumental for high-quality information systems engineering.

References

1. Highsmith, J., Cockburn, A.: Agile software development: the business of innovation. Computer **34**, 120–127 (2001)
2. Jacobson, I., Booch, G., Rumbaugh, J.: The Unified Software Development Process. Addison Wesley Longman, Boston (1998)
3. Gamma, E., Helm, R., Johnson, R., Vlissides, J.: Design Patterns. Elements of Reusable Object-Oriented Software. Addison-Wesley, Boston (1994)
4. Purao, S., Storey, V.C., Han, T.D.: Improving analysis pattern reuse in conceptual design: augmenting automated processes with supervised learning. Inf. Syst. Res. **14**, 269–290 (2003)

5. Kolp, M., Giorgini, P., Mylopoulos, J.: Organizational patterns for early requirements analysis. In: Eder, J., Missikoff, M. (eds.) CAiSE 2003. LNCS, vol. 2681, pp. 617–632. Springer, Heidelberg (2003)
6. Gregor, S., Hevner, A.R.: Positioning and presenting design science research for maximum impact. MIS Q. **37**, 337–356 (2013)
7. Walls, J., Widmeyer, G., El Sawy, O.: Building an information system design theory for viligant EIS. Inf. Syst. Res. **3**, 36–59 (1992)
8. van Aken, J.E.: Management research based on the paradigm of the design sciences: the quest for field-tested and grounded technological rules. J. Manage. Stud. **41**, 219–246 (2004)
9. Goldkuhl, G.: Design theories in information system - a need for multi-grounding. J. Inf. Technol. Theory Appl. (JITTA) **6**, 59–72 (2004)
10. Hanseth, O., Lyytinen, K.: Design theory for dynamic complexity in information infrastructures: the case of building internet. J. Inf. Technol. **25**, 1–19 (2010)
11. Gregor, S.: The nature of theory in information systems. Mis Q. 611–642 (2006)
12. Protzen, J.-P., Harris, D.: The Universe of Design: Horst Rittel's Theories of Design and Planning. Routledge, London (2010)
13. Dey, D., Storey, V.C., Barron, T.M.: Improving database design through the analysis of relationships. ACM Trans. Database Syst. (TODS) **24**, 453–486 (1999)
14. Carlsson, S.A.: Developing knowledge through IS design science research. Scand. J. Inf. Syst. **19**, 75–86 (2007)
15. Markus, M.L., Majchrzak, A., Gasser, L.: A design theory for systems that support emergent knowledge processes. MIS Q. **26**, 179–203 (2002)
16. Romme, A.: Making a difference: organization as design. Decis. Support Syst. **14**, 558–573 (2003)
17. Nagel, E.: The structure of science. Am. J. Phys. **29**, 716 (1961)
18. Gregor, S., Jones, D.: The anatomy of a design theory. J. Assoc. Inf. Syst. **8**, 312–335 (2007)
19. Leiner, B.M., Cerf, V.G., Clark, D.D., Kahn, R.E., Kleinrock, L., Lynch, D.C., Postel, J., Roberts, L.G., Wolff, S.: A brief history of the internet. ACM SIGCOMM Comput. Commun. Rev. **39**, 22–31 (2009)
20. Osterwalder, A.: The business model ontology: a proposition in a design science approach. Faculty of Business and Economics, Doctor. University of Lausanne (2004)
21. Timmers, P.: Business models for electronic markets. Electron. Markets **8**, 3–8 (1998)
22. Seidwitz, E.: What models mean. IEEE Softw. **20**, 26–32 (2003)
23. Papazoglou, M.P.: Service-oriented computing: concepts, characteristics and directions. In: Web Information Systems Engineering, WISE 2003, Proceedings of the Fourth International Conference on, pp. 3–12. IEEE (2003)
24. Scheer, A.W.: ARIS Business Process Modeling. Springer, Berlin (2000)
25. Zachman, J.A.: The Zachman framework: a primer for enterprise engineering and manufacturing (electronic book) (2003)
26. Aßmann, U., Zschaler, S., Wagner, G.: Ontologies, meta-models, and the model-driven paradigm. In: Calero, C., Ruiz, F., Piattini, M. (eds.) Ontologies for Software Engineering and Software Technology, pp. 249–273. Springer, Berlin (2006)
27. Yu, E.: Modelling strategic relationships for process reengineering. Soc. Model. Requirements Eng. **11**, 2011 (2011)
28. Bresciani, P., Perini, A., Giorgini, P., Giunchiglia, F., Mylopoulos, J.: Tropos: an agent-oriented software development methodology. Auton. Agents Multi-Agent Syst. **8**, 203–236 (2004)

29. Gregoriades, A., Shih, J.-E., Sutcliffe, A.: Human-centred requirements engineering. In: Requirements Engineering Conference, 2004, Proceedings, 12th IEEE International, pp. 154–163. IEEE (2004)
30. Model Driven Architecture (MDA), MDA Guide rev. 2.0. Object Management Group (2014)
31. Telang, P.R., Singh, M.P.: Specifying and verifying cross-organizational business models: an agent-oriented approach. IEEE Trans. Serv. Comput. **5**, 305–318 (2012)
32. Fowler, M.: Analysis Patterns, Reusable Object Models. Addison-Wesley, Longman, Boston (1997)
33. Alexander, C.: A Pattern Language: Towns, Buildings, Construction (1978)
34. Meyer, B.: Object-Oriented Software Construction. Prentice-Hall, New York, London (1988)
35. Fowler, M.: Patterns of Enterprise Application Architecture. Addison-Wesley Longman Publishing Co., Inc., Boston (2002)
36. Baskerville, R., Pries-Heje, J.: Explanatory design theory. Bus. Inf. Syst. Eng. **5**, 271–282 (2010)
37. Romme, A.G.L., Endenburg, G.: Construction principles and design rules in the case of circular design. Organ. Sci. **17**, 287–297 (2006)
38. Pascal, A., Thomas, C., Romme, G.A.: An integrative design methodology to support an inter-organizational knowledge management solution. In: Proceedings of the ICIS Conference (2009)
39. Bunge, M.: Scientific Research II: the Search for Truth. Springer, Berlin (1967)
40. van Aken, J.E.: Management research as a design science: articulating the research products of mode 2 knowledge production in management. Br. J. Manage. **16**, 19–36 (2005)
41. DeSanctis, G., Poole, M.: Capturing the complexity in advanced technology use: adaptive structuration theory. Organ. Sci. **5**, 121–147 (1994)
42. Wand, Y., Weber, R.: On the deep structure of information systems. Inf. Syst. J. **5**, 203–223 (1995)
43. Maaß, W., Varshney, U.: Design and evaluation of ubiquitous information systems and use in healthcare. Decis. Support Syst. **54**, 597–609 (2012)
44. Mead, G.H.: Mind, Self and Society from the Standpoint of A Social Behaviorist. [Edited and with an Introduction by Charles W. Morris]. University of Chicago Press, Chicago, London, 18, 1972 (1934)
45. Maass, W., Janzen, S.: Pattern-based approach for designing with diagrammatic and propositional conceptual models. In: Jain, H., Sinha, A.P., Vitharana, P. (eds.) DESRIST 2011. LNCS, vol. 6629, pp. 192–206. Springer, Heidelberg (2011)
46. Martin, R.C.: Clean Code: a Handbook of Agile Software Craftsmanship. Pearson Education, New York (2009)
47. Baader, F., Calvanese, D., McGuinness, D., Nardi, D., Patel-Schneider, P. (eds.): The Description Logic Handbook: Theory, Implementation and Applications. Cambridge University Press, Cambridge (2003)
48. Noy, N., Rector, A., Hayes, P., Welty, C.: Defining n-ary relations on the semantic web. W3C Working Group Note 12 April 2006
49. De Nicola, A., Missikoff, M., Navigli, R.: A software engineering approach to ontology building. Inf. Syst. **34**, 258–275 (2009)
50. Masolo, C., Borgo, S., Guarino, N., Oltramari, A.: The WonderWeb Library of Foundational Ontologies. WonderWeb (2003)
51. Guizzardi, G.: Ontological Foundations for Structural Conceptual Models. CTIT, Centre for Telematics and Information Technology (2005)
52. Wagner, G.: The agent–object-relationship metamodel: towards a unified view of state and behavior. Inf. Syst. **28**, 475–504 (2003)

53. Cockburn, A.: Structuring Use Cases with Goals1 (1997)
54. Singh, S.N., Woo, C.: Investigating business-IT alignment through multi-disciplinary goal concepts. Requirements Eng. **14**, 177–207 (2009)
55. Yu, E.: Modelling Strategic Relationships for Process Reengineering. Department of Computer Science, Ph.D. University of Toronto (1995)
56. Maass, W., Storey, V.C.: Recall of concepts and relationships learned by conceptual models: the impact of narratives, general-purpose, and pattern-based conceptual grammars. In: Yu, E., Dobbie, G., Jarke, M., Purao, S. (eds.) ER 2014. LNCS, vol. 8824, pp. 377–384. Springer, Heidelberg (2014)

A Middle-Level Ontology for Context Modelling

Oscar Cabrera[(✉)], Xavier Franch, and Jordi Marco

Universitat Politècnica de Catalunya – BarcelonaTech, GESSI – UPC,
c/Jordi Girona 1-3, 08034 Barcelona, Spain
{ocabrera,franch}@essi.upc.edu,
jmarco@lsi.upc.edu

Abstract. Context modelling is one of the stages conducted during the context life cycle. It has the aim of giving meaning and structure to the collected context's raw data. Although there are different context models proposed in the literature, we have identified some gaps that are not fully covered, particularly related to the reusability of the models themselves and the lack of consolidated and standardized ontological resources. To tackle this problem, we adopt a three-layered context ontology perspective and we focus on this paper in the middle layer, which is defined following a prescriptive process and structured in a modular way for supporting reuse.

Keywords: Context modelling · Context ontology · Ontology reuse

1 Introduction

Context is a term widely used in many areas of conceptual modelling. Dey defines it as *"any information that can be used to characterize the situation of an entity. An entity is a person, place, or object that is considered relevant to the interaction between a user and an application"* [1]. From this perspective, *context modelling* is the research topic in which the notion of context has been structured in different formalisms to represent context information, which affects positively or negatively an entity. One of its major applications is concentrated on context-aware infrastructures, i.e. areas such as Smart Cities, Pervasive Computing and Internet of Things [2].

In a previous work, we analyzed the state of the art on context modelling [3] identifying certain gaps that have motivated this work. One of the most important issues is the difficulty of reusing the context models proposed in the literature mainly due to the lack of homogeneity among their elements, as well as the shortage of their definitions. This problem calls for efforts to consolidate the context knowledge already available and to specify a clear schema of knowledge reutilization.

To contribute solving these issues, in this paper we adopt a three-level ontology approach for context modelling:

This work is partially supported by the Spanish project TIN2013-44641-P.

© Springer International Publishing Switzerland 2015
P. Johannesson et al. (Eds.): ER 2015, LNCS 9381, pp. 148–156, 2015.
DOI: 10.1007/978-3-319-25264-3_11

- The *upper-level* provides a basic taxonomy of context classes that represent general context concepts. We have presented this upper level in a previous work [3].
- The *middle-level* supports reusing and extending ontological resources of existing context models and other consolidated ontologies from a modular perspective.
- The *lower-level* includes a set of detailed classes highly dependent on the domain.

The focus of this paper will be the middle-level. We propose reusing a set of ontological resources that represent structured modules selected using different strategies.

The rest of the paper is organized as follows. Sections 2 and 3 present the background and antecedents of our work. Section 4 describes in depth the proposed model focused on the middle-level ontology. Finally, Sect. 5 presents the conclusions.

2   Background

2.1   Context Modelling Approaches

Recently, Perera et al. [2] presented a comparison of the six most popular context modelling techniques and concluded that the most appropriate technique to manage context is *ontology-based modelling*. According to Sudhana et al. [4], Noy [5] and Chen et al. [6], ontologies are a key feature in the making of context-aware distributed systems because they support knowledge sharing, reasoning and interoperability. For all these reasons, we are adopting ontology-based modelling in our work.

2.2   Classification of Ontologies

Different designs and structures of ontologies have been proposed so far. Two usual criteria are: generality and expressiveness. Generality has the purpose of specifying general classes towards top levels and more specific classes towards lower levels [7]. This criterion supports the adoption of a layered view of ontologies [8, 9]. Expressiveness indicates the level of detail of an ontology. Usually they are classified into lightweight and heavyweight [10]. In this paper we adopt a 3-level view of abstraction with an expressivity closer to heavyweight ontologies since we want to express axioms and constraints more than only concepts, taxonomies and relationships.

2.3   Reuse in Methodologies for Developing Ontologies

A large number of methodologies have been proposed to conduct the ontology building process (e.g., [11, 12]). Generally, these methodologies specify an activity based on the reuse of existing knowledge. According to Pinto et al. [13] there are two different reuse processes: merge and integration. In a merge process, it is usually difficult to identify regions in the resulting ontology that were taken from the merged ontologies and that were left more or less unchanged. In an integration process source ontologies are aggregated, combined and assembled together, possibly after reused ontologies have suffered some changes, such as extension, specialization or adaptation. We propose integration because we are more interested on unifying modules than complete ontologies. We will apply the integration process defined by Pinto and Martins [14] since they have compiled integration activities from different methodologies.

3   Antecedents

In a previous paper [3] we included a preliminary state of the art in context. This study: (1) compiled different gaps reported by researchers in the context modelling area; (2) identified gaps through the analysis and evaluation conducted in the contributions; and (3) established a basic taxonomy of high level classes intended to serve as basis of the abstract level of a context model consolidating all these proposals. The consolidated resources encompass all the perspectives already provided in context modeling, especially, resources regarding context information vocabulary, properties and terminology definitions.

As next step, we deepened this state of art by conducting a systematic mapping study according to the guidelines proposed by Kitchenham and Charters [15].

The results obtained from the review were used to evolve the taxonomy presented in [3] as depicted in Fig. 1. To populate the schema presented in the figure we considered different aspects from the surveyed models such as the most addressed classes in the surveyed proposals, definitions provided clustering those that represent a generic description of the class, common patterns identified through the proposed schemas, and alignment with foundational ontologies that were partially reused by the contributions. Hence, we adopt this taxonomy as the upper-level ontology that frames the middle-level that we are proposing in this paper.

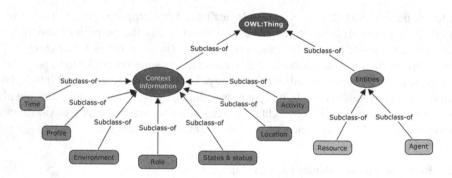

Fig. 1. Upper-level ontology.

4   Middle-Level Ontology for Context Modelling

The main objective of a middle-level ontology is to provide a set of modules easy to reuse and extend. In addition, we aim at building a proposal aligned with existing context ontologies, reusing the ontological resources presented in Fig. 1. To this purpose, we adopt the integration process defined by Pinto and Martins [14] which defines several tasks that are applied below. For the sake of brevity, we will present some of the tasks together.

Identify Integration Possibility. In this first step, Pinto and Martins propose to select the framework being used to build the ontology. Following the criteria provided by Su and Ilebrekke [17], we have selected Protégé as ontology development tool. Particularly, we are interested on importing specific modules from existing ontologies in an easy way providing a clear schema of reutilization, and connection with the upper level classes. Moreover, in cases where the ontologies selected are provided in a different framework, we will translate the selected modules into the semantic of Protégé. The selection of this tool is greatly influencing the proposal given. This is unavoidable because we want an ontology that can be used in an engineering context in order to provide tangible value in the development of contextual software and services.

Identify Modules and Knowledge to be Represented in Each of Them. As starting point, we associate a module to each of the leaves of the class hierarchy established in the upper-level ontology (see Fig. 1). We have consolidated the knowledge coming from the selected ontologies and as a result, these classes are defined as follows:

- Time. Temporal concepts and properties common to any formalization of time [6].
- Profile. Biographical sketch [18].
- States and Status. A state at a particular time (e.g., a condition or state of disrepair, the current status of the arms negotiations) [18].
- Environment. Environment in which the user interacts [19].
- Role. Role of an agent can be used to characterize the intention of the agent [20].
- Location. By location context, we mean a collection of dynamic knowledge that describes the location of an agent [20].
- Activity. Represents a set of actions [6].
- Resource. Resources describe anything used to perform the activity [21].
- Agent. Both computational entities and human users can be modeled as agents [6].

Identify and Get Candidate Ontologies. According to [14], this task first identifies candidate ontologies that could be used as modules of our middle-level ontology. We did this through the mapping study reported in Sect. 3. Next, we selected 64 of the context models as possible candidates to be integrated in the modules of the middle-level ontology[1].

To obtain the candidate ontologies in an adequate form, we analyzed their knowledge and implementation levels as well as the documentation available. We realized serious problems in the detailed expression of the knowledge and the coverage of the implementation. Still, for each model we aimed at identifying and retrieving the ontological resources that we considered relevant to create or complement modules

---

[1] For the sake of brevity we do not specify the 64 references of the ontologies selected and other resources of the mapping study that were used in this work; you may find them at [16].

of the middle-level. It is worth to mention that we considered not only the 64 context ontologies but also other 12 ontologies that they reused. We decided to establish a common base of candidate ontologies acting as reference point to structure the modules required. To carry out this task, from these ontologies, we selected those ones more referenced by existing proposals of context modelling: CONON [9], SOUPA [6], SUMO [8], OpenCyc [22], FOAF (xmlns.com/foaf/spec), CCPP (www.w3.org/TR/CCPP-struct-vocab2/), OWL-Time (www.w3.org/TR/owl-time) and OWL-S (www.w3.org/Submission/OWL-S).

Study and Analysis of Candidate Ontologies. In this task the candidate ontologies are analyzed to identify possible problems in the integration process. We applied the SEQUAL evaluation framework formulated by Hella and Krogstie [23]. There are 7 quality categories used to evaluate the reusability of the ontologies. For each category, they propose some values that we have mostly kept:

- Physical (Phy). The ontology should be physically available and it should be possible to make changes to it. Available ($\checkmark$); available, presenting some problems to open in Protégé ($\checkmark$-); available, but too big to open ($\checkmark$–); not available (X).
- Empirical (Emp). If a visual representation of the ontology is provided it should be intuitively and easy to understand. Satisfactory ($\checkmark$); less satisfactory ($\checkmark$-).
- Syntactic (Syn). The ontology should be represented according to the syntax of a preferred machine readable language. OWL full ($\checkmark$); partial OWL ($\checkmark$-); RDF ($\checkmark$–).
- Semantic (Sem). The ontology should cover the area of interest. Overlap, satisfactory validity ($\checkmark\checkmark$); partial overlap but not complete, satisfactory validity (-$\checkmark$); partial overlap but not complete, poor validity (–); not overlapping (X). Since this category is too coarse to be applied globally, Table 2 shows its evaluation for the modules identified in the middle-level ontology.
- Pragmatic (Prag). It should be possible to understand what the ontology contains, and being able to use it for our purpose. Satisfactory ($\checkmark$); not satisfactory (X).
- Social (Soc). The ontology should have a relatively large group of users. Mature and widely used ($\checkmark\checkmark$); assumed mature, not specified how much it is used (–); not mature, but referenced (-$\checkmark$).
- Organizational (Org). The ontology should be freely available, accessible, maintained and supported. Free, accessible, and stable ($\checkmark\checkmark\checkmark$); free, accessible, and probably stable ($\checkmark\checkmark$-); free, not accessible, and probably stable ($\checkmark$x-).

Hella and Krogstie already provided in [23] the evaluation of some of the candidate ontologies selected in the previous task, concretely FOAF, OpenCyc and SUMO. We reused this evaluation reviewing the current status of the ontologies in order to check that the results obtained remain consistent; the only change has already been reported above (currently SUMO can be opened in Protégé). The rest of ontologies were evaluated from scratch. The results obtained are depicted in Table 1.

Table 1. Evaluation of candidate ontologies

Ontologies	Phy	Emp	Syn	Prag	Soc	Org
CONON	X	✓	✓-	✓	-✓	✓x-
SOUPA	✓	✓-	✓-	✓	-✓	✓✓-
SUMO	✓-	✓	✓	X	–	✓✓-
OpenCyc	✓–	✓-	✓	X	–	✓✓-
FOAF	✓	✓	✓	✓	✓✓	✓✓✓
OWL-Time	✓	✓-	✓	✓	–	✓✓-
CCPP	✓	✓	✓–	✓	–	✓✓-
OWL-S	✓	✓-	✓	✓	–	✓✓-

Table 2. Semantic evaluation of candidate ontologies organized by modules

Ontologies	Agent	Resource	Activity	Time	Environment	Location	Profile	Role	Status
CONON	-✓	-✓	-✓	X	X	-✓	–	X	X
SOUPA	-✓	–	-✓	✓✓	X	-✓	-✓	–	–
SUMO	-✓	-✓	-✓	✓✓	–	-✓	–	–	–
OpenCyc	–	–	–	–	–	–	–	–	–
FOAF	X	X	X	X	X	X	-✓	X	X
OWL-Time	X	X	X	✓✓	X	X	X	X	X
CCPP	X	X	X	X	X	X	-✓	X	X
OWL-S	X	X	X	X	X	X	-✓	X	X

Choosing Source Ontologies. Given the study and analysis of candidate ontologies the final choices must be made in this task. Pinto and Martins propose two stages. In a first stage, a critical look to the characteristics analyzed in the previous task is made[2].

Although the schema presented by SOUPA and CONON for modelling context is widely referenced in the academic research, they present major drawbacks on the availability, completeness and maintenance of the resources provided. Still, the design of the model presented here is partially inspired by the modular schema of SOUPA and the intuitive visual representation of CONON.

For the rest of the ontologies, all of them are physically available. However, SUMO and OpenCyc are big ontologies difficult to import into ontology editors. SUMO, FOAF and CCPP provide a visual representation of their schema that is easy to understand. In the semantic and pragmatic qualities, SUMO and OpenCyc are upper ontologies providing an extensive vocabulary; although it can be used for purposes of context modelling, a large set of this vocabulary is irrelevant for this purpose. The rest of the

---

[2] Unlike Pinto and Martins, we base the selection on the analysis previously made.

foundational ontologies are more concrete and provide a smaller set of vocabulary partially covering the context of an entity.

On the basis of this assessment, the final decision is made in a second stage. Our aim is to select the parts of each candidate ontology that cover satisfactorily a module identified from the upper-level ontology; also, we consider the overall ontology evaluation to decide whether to include it in the result or not. Table 2 provides details on the analysis that support our choice of middle-level ontology, presented in Fig. 2. As it can be seen, in the middle-level of the model we propose different modules associated to the corresponding high level classes of the upper-level ontology. These modules are selected from the candidate ontologies by means of the following considerations: (1) integrate modules fulfilling the conceptualization of a given entity or context information; (2) otherwise, a new module combining ontological resources from different sources is proposed.

Several situations have been found when selecting. For instance, the *Object* module is selected from SUMO since it provides the overlap required to conceptualize this module. However, this ontology does not provide at all the required resources to conceptualize a computational entity, so we complement it with resources from CONON.

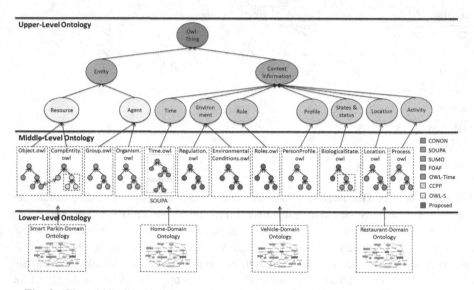

Fig. 2. The middle-level ontology and its relationships with the upper and lower levels.

Apply Integration Operations. Once the candidate ontologies have been filtered, the final task is to perform the integration. For the sake of space we cannot provide the full process. We just illustrate it one example. The upper-level class *Time* can be structured by using the semantics of SUMO, SOUPA or OWL-Time as it is depicted in the semantic evaluation of Table 2. However, according to the evaluation of Table 1 it is difficult to identify certain resources from SUMO. Then, we take as a basis the semantics of time given by OWL-Time because it is particularly focused on modelling time and for other features also evaluated in Table 1. Then, we adopted the following integration operations: (1) we integrated in a module the OWL-Time as it is and then we make some modifications in the structure and vocabulary taking into account the next operation; (2) we identified the equivalent resources among the vocabulary and patterns presented in ontologies assessed where time was also modelled in order to consolidate, standardize and minimize semantic inconsistencies.

5 Conclusions

In this paper we presented a middle-level ontology with the purpose of consolidating the context knowledge already available from a modular perspective yielding a clear schema of knowledge reutilization. The main contribution has been the effort of analyzing, selecting and combining many useful vocabularies from different existing proposals. To do so, we gathered parts of different ontologies to be integrated into modules. From this perspective, we face the gaps of a generic context model allowing the instantiation of existing context knowledge in a unique and simple model easy to be extensible in the establishment of new knowledge. The implemented resources are available at: https://github.com/ocabgit/Three-LevelContextOntology.git.

References

1. Dey, A.: Understanding and using Context. Pers. Ubiquit. Comput. **5**(1), 4–7 (2001)
2. Perera, C., Zaslavsky, A., Christen, P., Georgakopoulos, D.: Context aware computing for the internet of things: a survey. IEEE Commun. Surv. Tutorials **16**(1), 414–454 (2014)
3. Cabrera, O., Franch, X., Marco, J.: A context ontology for service provisioning and consumption. In: IEEE RCIS (2014)
4. Sudhana, K.M., Raj, V.C., Suresh, R.M.: An ontology-based framework for context-aware adaptive e-learning system. In: IEEE ICCCI (2013)
5. Noy, N.F.: Semantic integration: a survey of ontology-based approaches. ACM SIGMOD Rec. **33**(4), 65–70 (2004)
6. Chen, H., Perich, F., Finin, T., Joshi, A.: SOUPA: standard ontology for ubiquitous and pervasive applications. In: IEEE MOBIQUITOUS (2004)
7. Guarino, N.: Formal ontology and information systems. In: FOIS (1998)
8. Niles, I., Pease, A.: Towards a Standard Upper Ontology. In: FOIS (2001)
9. Wang, X.H., Zhang, D.Q., Gu, T., Pung, H.K.: Ontology based context modeling and reasoning using OWL. In: IEEE PERCOMW (2004)
10. Corcho, O., Fernández-López, M., Gómez-Pérez, A.: Methodologies, tools and languages for building ontologies. Where is their meeting point? Data Knowl. Eng. **46**(1), 41–64 (2003)
11. Fernández-López, M., Gómez-Pérez, A., Juristo, N.: METHONTOLOGY: from ontological art towards ontological engineering. In: AAAI 1997 Spring Symposium Series (1997)
12. Uschold, M., King, M.: Towards a methodology for building ontologies. In: IJCAI 1995 Workshop on Basic Ontological Issues in Knowledge Sharing (1995)
13. Pinto, H.S., Gómez-Pérez, A., Martins, J.P.: Some issues on ontology integration. In: IJCAI 1999 Workshop on Ontologies and Problem Solving Methods (1999)
14. Pinto, H.S., Martins, J.P.: A Methodology for Ontology Integration. In: K-CAP (2001)
15. Kitchenham, B., Charters, S.: Guidelines for Performing Systematic Literature Reviews in Software Engineering, version 2.3. EBSE Technical Report. EBSE-2007-01 (2007)

16. Cabrera, O., Franch, X., Marco, J.: Appendix of: a Middle-Level Ontology for Context Modelling. http://gessi.lsi.upc.edu/threelevelcontextmodelling/ (2015)
17. Su, Xiaomeng, Ilebrekke, Lars: A comparative study of ontology languages and tools. In: Pidduck, A.B., Mylopoulos, J., Woo, C.C., Ozsu, M.T. (eds.) CAiSE 2002. LNCS, vol. 2348, p. 761. Springer, Heidelberg (2002)
18. Miller, G.A.: WordNet: a lexical database for English. Commun. the ACM **38**(11), 39–41 (1995)
19. Preuveneers, D., Van den Bergh, J., Wagelaar, D., Georges, A., Rigole, P., Clerckx, T., Berbers, Y., Coninx, K., Jonckers, V., De Bosschere, K.: Towards an extensible context ontology for ambient intelligence. In: Markopoulos, P., Eggen, B., Aarts, E., Crowley, J.L. (eds.) EUSAI 2004. LNCS, vol. 3295, pp. 148–159. Springer, Heidelberg (2004)
20. Chen, H., Finin, T., Joshi, A.: An ontology for context-aware pervasive computing environments. Knowl. Eng. Rev. **18**(3), 197–207 (2003)
21. Prekop, P., Burnett, M.: Activities, context and ubiquitous computing. Comput. Commun. **26**(11), 1168–1176 (2003)
22. Curtis, J., Baxter, D., Cabral, J.: On the application of the Cyc ontology to word sense disambiguation. In: FLAIRS (2006)
23. Hella, L., Krogstie, J.: A structured evaluation to assess the reusability of models of user profiles. In: Bider, I., Halpin, T., Krogstie, J., Nurcan, S., Proper, E., Schmidt, R., Ukor, R. (eds.) BPMDS 2010 and EMMSAD 2010. LNBIP, vol. 50, pp. 220–233. Springer, Heidelberg (2010)

Ontology Patterns

An Ontology Design Pattern to Represent False-Results

Fabrício Henrique Rodrigues[✉], José Antônio Tesser Poloni, Cecília Dias Flores,
and Liane Nanci Rotta

Universidade Federal de Ciências da Saúde de Porto Alegre, Porto Alegre, Brazil
titofhr@gmail.com, jatpoloni@yahoo.com.br,
{dflores,lnrotta}@ufcspa.edu.br

Abstract. Observations are an important aspect of our society. Arguably, great part of them is captured by means of sensors. Despite the importance of the matter, the ontology of observations and sensors is not well developed, with few efforts dealing with the fundamental questions about their nature. As a result, an important aspect of sensors is overlooked: sensors may fail, producing false-results (i.e. false-positives and false-negatives). The lack of a proper representation of this aspect prevents us from communicating and reasoning about sensor failures, making it harder to assess the correctness of observations and to treat possible errors. In view of this problem, we propose an ontology design pattern (ODP) to represent false-results of sensors. It covers a special case of sensor that exclusively produces positive or negative results regarding the presence of the type of entity the sensor is designed to perceive. The paper introduces the ODP structure as well as its ontological commitments, bringing an example from the biomedical field. Discussion and further research opportunities of research are posed at the end of the paper.

Keywords: Ontology · Ontology design patterns · Sensors · False-positive · Knowledge representation

1 Introduction

Observations constitute the first step in collecting data, allowing us to acquire the knowledge about our surroundings both to learn about our condition on this world as well as to test our hypotheses about nature. Moreover, some would even say that they are our only connection with reality. In any case, our practical life is significantly based on them, both in personal and societal sphere (e.g. from the temperature of our rooms to the levels of rivers and oceans).

Arguably, great part of our observations is carried out by means of sensors – or even the totality of them, depending on the adopted meaning for 'sensor'. Indeed, there are a variety of definitions for the term: it is seen as specifically as a device that responds to a physical stimulus and transmits a resulting impulse [1] and as a physical object that performs observations (i.e. detects stimuli related to properties of a feature of interest and transforms them into another representation) [2]; or as generalized as a source producing a value within a value space representing a phenomenon in a given domain of discourse [3] and as the role of an observer [4].

© Springer International Publishing Switzerland 2015
P. Johannesson et al. (Eds.): ER 2015, LNCS 9381, pp. 159–172, 2015.
DOI: 10.1007/978-3-319-25264-3_12

Despite the importance of the matter, the ontology of observation is somewhat unexplored, focusing in particular applications rather than fundamental questions about its nature [4]. The same seems to occur with the ontology of sensors, with most of the efforts concerned with technical issues (ranging from classification of sensor types [5–7] and their components and structure [3, 8] to measurements [7, 9–12] and organization of sensor data [7, 9–11], managing sensors to enable the Semantic Sensor Web [13]) and just a few research works covering fundamental aspects of sensing process (e.g. [2, 4]).

With no doubt, all these efforts have introduced valuable representations. Even so, all of them assume sensors as foolproof tools, in the sense that sensor results are understood as always related to the real value of what are being observed. Some models (e.g. [2, 3]) include representation of conditions in which the sensors are expected to work properly, but this approach just narrows the impact of the issue rather than addressing it – i.e. placing sensor function in a safe area, but still not dealing with problems in the results when they arise. However, as real-world artifacts, it is reasonable to expect that sensors may have flaws. Indeed, their observations are said to be liable to effects of the sensor itself as well as those related to other parameter of the sensing process [14]. Such condition may originate false-results – here understood as the union of false-positive (i.e. claim something as positive when that is not the case) and false-negative (i.e. claim something as negative when it is, in fact, positive) results [15]. Nevertheless, even though this is a real issue, no known effort cover this aspect.

Considering that, this work presents an ontology design pattern (ODP) [16] to represent sensors in the perspective of occurrence of false-results. The proposed pattern covers a special case of sensors that sense the environment and gives a Boolean result indicating the presence or absence of a given type of entity. Such sort of sensor will be hereafter referred to as detectors.

The ODP was built employing the Unified Foundational Ontology (UFO) [17], which provides a set of meta-types described in terms of philosophically well-defined meta-properties. Then, the universals (i.e. a notion equivalent to that of concept or class) of a domain can be classified into those meta-types, what favors their analysis according to the referred meta-properties. Among such meta-properties we can highlight rigidity, identity criterion and relational dependence.

A concept is rigid if all its possible instances must instantiate such concept in all possible worlds (i.e. an instance of a rigid concept cannot stop being an instance of this concept without ceasing to exist). On the other hand, a concept is anti-rigid if entities can freely instantiate it and stop being instances of such concept. Identity criteria determine which properties must be observed to assess if two entities (or an entity in different times) are the same or not (e.g. what makes it possible to say that Paul McCartney now and Paul McCartney from 30 years ago are the same person). We say that a concept carry an identity criteria if it makes all its instances be differentiable by the same identity criteria. Finally, a concept A is relationally dependent on a concept B if an entity can only instantiate A while establishing certain relationship with an instance of B (e.g. Student is a concept relationally dependent on the concept of Educational Institution, since an individual is instance of Student only while it is enrolled in an instance of Educational Institution) [18].

Lastly, UFO divides concepts in two broad categories: those comprising endurants (i.e. individuals that are wholly present whenever they are present) and those comprising perdurants (i.e. individuals that extend in time accumulating temporal parts).

The remainder of this paper is structured as follows: Sect. 2 presents the proposed ODP, Sect. 3 shows an application of such pattern in urinalysis domain, Sect. 4 briefly discuss some aspects of the work and Sect. 5 brings final remarks.

2 False-Results ODP

This work proposes an ODP to cover false-results for detectors, with the purpose of answering to some ontological questions such as:

- "What is a false-result? What does it mean to have a false-result?"
- "How a false-result happens?"
- "What does participate in a false-result?"

It builds upon ideas from [2] and [4] and uses similar structure and ontological commitments for the core concepts related to sensors and observations, which make up the basic structure for the ODP (Fig. 1). The definitions of such concepts are presented below.

Observable Entity. An entity that can be perceived by a sensor through a detectable event (i.e. a stimulus) that acts as proxy for such entity. Some observable entities can be perceived by a variety of stimulus (e.g. smoke is perceived due to the optical interference it causes in photoelectric detectors as well as due to the electrical interference it causes in ionization detectors). Being an observable entity is a role played when the entity is liable to be perceived by some sensor. As entities with diverse identity criteria can play such role, this concept can be classified as a UFO's Role Mixin universal (i.e. a dispersive, anti-rigid, relationally dependent universal that does not carry any identity principle and aggregates properties common to different roles [17]).

Stimulus. An event that represents a change in the environment, that is causally related to the presence of an observable entity (or several ones), and that can be perceived by some sensor. The *proxyFor* relationship captures the idea that the observable entity contributes to the stimulus occurrence in such degree that the stimulus can be regarded as a sign of the presence of the observable entity. Considering that, *Stimulus* is classified as an UFO's Event universal, which comprises perdurants that are regarded as transformations in the state of affairs from one situation to another [19].

Sensor. An entity able to detect a particular class of stimuli and to produce a result from it. This ability is presented by *detects* – a formal relationship relying on some detectable property of the stimulus and the disposition of the sensor in sensing such property. As in [4], we understand that being a sensor is analogous to the role of an observer, being played by an entity when it is observing the environment in search for the type of observable entity it is intended to identify. As entities with different identity criteria can act as sensors (e.g. an electronic device that senses light and reports its intensity, a scout who observes footprints and reports which type of animal is present in that region), this concept is also an UFO's Role Mixin universal.

Detector. A sensor subtype that exclusively produces positive or negative results concerning the presence of a particular type of observable entity. Such concept is also classified as an UFO Role Mixin universal.

Result. Entity that acts as a symbol representing a mapping between given class of stimulus (or the absence of it) and some information about the corresponding observable entity. It may stand for any type of value representation (e.g. continuous numbers, discrete symbols, categories). As any type of object can play such a role, depending on the nature of the sensor producing the result (e.g. a symbol depicted on a screen, an emitted particle, a note written by an human observer), this concept is also an UFO Role Mixin universal.

Positive Result. A result that indicates the detection of some stimulus and that is regarded as a sign of the presence of the type of observable entity the detector is designed to perceive. As a type of result, it is also an UFO Role Mixin universal.

Negative Result. A result that indicates absence of the stimulus and that is regarded as a sign that the intended observable entity is also absent. It is also an UFO Role Mixin universal.

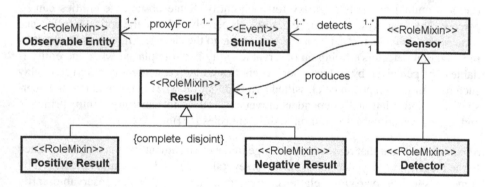

Fig. 1. Basic structure of the ODP.

These concepts cover the main aspects of the sensing process that were object of related work and that are particularly necessary to explain the idea of this article. Next sections deepen the discussion seeking to address the false-result problem.

2.1   False-Result Problem

Sometimes detectors fail [14]. Detector failures consist of production of false (wrong) results: either giving a positive result when the observable entity that the detector is designed to detect is absent (i.e. false-positive) or giving a negative result when the observable entity is indeed present (i.e. false-negative). In spite of that, current models of detectors (and sensors in general) do not provide means to represent the situation as so. Thereby, either the false-results are represented and stored as if they were correct results (e.g. as if the intended observable entity was actually detected when in fact it was a false-positive), or they are deleted and possibly replaced by the correct results, if available.

Not being able to correctly represent false-results prevents us from appropriately communicate or reason about them, what hampers the process of assessing the correctness of observations and results that detectors produce. For instance, if some detection process keeps failing due to the presence of some entity x, there is no way to formal and explicitly represent such failure history – and so, it is not possible to investigate what may be wrong (i.e. what may be this entity x affecting the process). For these reasons, next session proposes an ODP to solve this problem.

2.2 Proposed Solution

Even though detectors do not directly perceive their intended observable entities – relying on proxy stimuli – it is clear that their goal is identifying the observable entities themselves rather than their related stimuli. This notion is reinforced by [2] and [4], which poses the stimulus solely as a mean to information about what is observed reach the sensor (or detector, in our particular case). All in all, if stimulus was the objective of a detector, why should one bother in representing an observable entity?

Behind this conclusion lies an important idea: *intention is part of the nature of detectors*. Here, we are not referring to intention as some purpose the sensor would have as an agentive entity. Instead, intention is understood as the deliberated assertion (made by who has designed and/or deployed the detector) that a positive result means that the detector has identified its intended observable entity – and a negative result means such observable entity was not perceived. In other words, each detector is designed to identify a specific type of observable entity and its result is expected to be related to the presence of such type of entity. It allows detectors to provide meaningful results (i.e. identifying the occurrence of one possibility and ruling out all others). Having no intention, a positive result would give no more information than the self-evident fact that the detector has produced a positive result. Moreover, without an intention, it is impossible to have a false-result, since there is no way of a result be wrong if it does not make a clear assertion to be checked against an underlying reality. Evidence that supports this idea is present in [4], where it is suggested that choosing a type of observable entity (i.e. determining the type of entity a sensor will observe) is part of the observation process. Thus, if choosing is part of observation, the intention that supports such choice must be part of the nature of the detector.

Besides the notion that a detector is defined by its intention, another fact is important: *a detector relies on a proxy stimulus to identify its intended observable entity*. This way, its observations are greatly affected by matters related to such stimulus. For example, as a detector get to know about its intended observable entity by perceiving a particular type of stimulus, anything that produces an equivalent type of stimulus would also be observable by such detector. Then, given the intention of the detector, it is possible to divide all the things it can observe in two major categories: the intended observable entities (hereafter referred to as *analytes*) and those not intended to be observed (hereafter referred to as *confounders*). Another implication is that, as the only way a detector can identify its analyte is through a proxy stimulus, anything able to suppress its effects – either by preventing the analyte from causing the stimulus or by preventing the detector from perceiving it – would make the detector blind for such observable entity. Things with such capability will be referred to as *inhibitors*.

Considering the exposed so far, we can define the nature of false-results. A result is defined in relation to its detector intention, so that a positive result represents the presence of what is intended to be observed (i.e. analyte) and a negative result represents its absence. Thus, false-results are situations in which a detector produces a result that, owing to failures in observation process, does not depict the correct state regarding analyte presence. Following that, false-positives occur when a detector correctly detects the stimulus related to its analyte, but the analyte is not present – which means that the stimulus came from a confounder. Likewise, false-negatives take place when the analyte is present, but the detector is not able to detect its related stimulus due to an inhibitor (which prevents stimulus from happening, changes its effects, or reduces detector sensibility to it). Finally, true-positives and true-negatives are situations in which the result produced by the detector agrees with the underlying reality with respect to the presence of the analyte: the former being a positive result when analyte is indeed present; the latter, a negative result when analyte is absent.

Given such nature of false-result phenomenon, we devised an ODP to account for this problem (Fig. 2), expanding the structure from Fig. 1 with the following concepts:

Analyte. The type of observable entity that the detector is designed to identify. Coupled with the *hasAnalyte* formal relationship, it reveals the intention of such detector. Analyte is a role played in relation to a detector by entities with different identity criteria, being classified as an UFO's Role Mixin universal.

Confounder. Anything whose related stimulus makes a detector produce a positive result without being its analyte. This notion is captured by the *hasConfounder* formal relationship, so that each confounder must be related to (i.e. have the disposition of activating) at least one detector, whereas detectors do not necessarily have confounders. As well as analyte, this concept is also an UFO's Role Mixin universal.

Inhibitor. Anything able to prevent an observable entity from causing a stimulus or to prevent the detector from detecting it. This notion is represented by the *hasInhibitor* formal relationship, so that each inhibitor must be related to (i.e. have the disposition of inhibiting the activation of) at least one detector, whereas detectors do not necessarily have inhibitors. This concept is also an UFO's Role Mixin universal.

False-Positive. A whole formed by a positive result referring to a single sample (i.e. the portion of the environment that the detector observes in search for its analyte) and the absence of the analyte in that particular sample, at a given time. In other words, it is the situation in which a single positive result is present, but the corresponding analyte is not. Each new positive result produced in absence of the corresponding analyte originates a new false-positive situation. Considering the described nature of false-results, some confounder must also be present. Being a whole with its own meaning, this concept is classified as an UFO's Situation universal (i.e. a state of affairs, a type of configuration of a part of reality that can be understood as a whole [19]).

False-Negative. Situation in which exactly one negative result is present, as well as at least one analyte (i.e. a negative result is obtained for a sample that has an analyte for the corresponding detector). Each new negative result produced in presence of analyte originates a new false-negative situation. In such situations, some inhibitor must also be present. Like false-positive, this concept is also an UFO's Situation universal.

True-Positive. Situation in which exactly one positive-result is present as well as at least one analyte for the detector that produced the result. Each new positive result obtained in presence of an analyte brings about a new true-positive. Since results are correct solely against analyte presence, it does not matter whether some confounder or inhibitor is present or not. This concept is also an UFO's Situation universal.

True-Negative. Situation in which a single negative-result is present and the corresponding analyte is not. Each new negative result produced in absence of analyte makes a new true-negative situation. Similarly to true-positive, it does not matter whether some confounder or inhibitor is present or not. It is also an UFO's Situation universal.

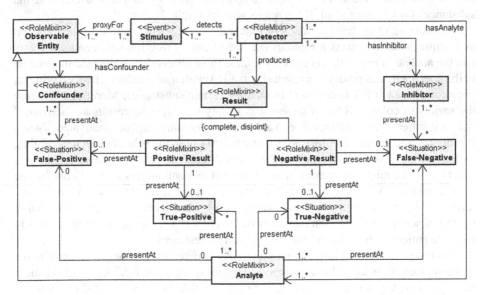

Fig. 2. ODP structure.

As an additional constraint, even though a confounder must be present at false-positive situation, the conditions to recognize such situation are just the presence of a positive result and the absence of analytes. So, the solely presence of a positive result brings into existence either a true-positive or false-positive situation, only depending on the presence of the analyte. Given that, a positive result will always be present at one of those situations and never at both of them simultaneously. Similar reasoning can be applied to negative results and their corresponding situations.

Applying this pattern is a matter of arranging the concepts of the desired domain following the scheme depicted in Fig. 2. Alternatively, this model can be used as a top ontology, providing detector-related semantics for domain-specific concepts. Next section presents an example of application.

3 Case Study: Physicochemical Analysis of Urine

Urinalysis is the urine test, which provides valuable information about patient's urinary and renal conditions, as well as about major metabolic functions [20]. It begins with a physicochemical analysis of a urine specimen (looking for substances present in it and measuring its pH and specific gravity), followed by a microscopic analysis (searching for formed elements – e.g. cells, crystals, casts).

Usually, physicochemical analysis is carried out using urinary dipsticks. A urinary dipstick is a plastic strip with reactive areas (RAs), each of them giving an approximate estimation about the presence of a specific chemical substance (e.g. hemoglobin, glucose), pH or specific gravity of the specimen. When exposed to the specimen, reagents in each RA react with the substance whose presence the RA is intended to measure, changing the RA color according to the intensity of the substance. If the substance is not present at all, RA remains with its original color.

Considering the definition of sensor elaborated in this work, a RA can be regarded as a sensor, since it detects a stimulus (i.e. the chemical reaction between its reagents and the analyte in urine) that is causally related to an observable entity (i.e. the analyte in the specimen) and produces a result (i.e. the RA with a new color), clearly having an intention (i.e. each RA is designed to identify certain substance). Moreover, ignoring the variety of colors a RA can present and simply regarding its results as "activated" (when it suffers a color change due to any analyte amount) and "non-activated" (when it remains with its initial color), RAs can perfectly play the role of detectors. In fact, understanding RAs as detectors is relevant and of great help. There are several known substances that might be present in urine and act as confounders or inhibitors for the most common RAs that compose urinary dipsticks – what leads to usual false-results. Such occurrences must be observed and checked by the biochemist who is performing the test, since such information integrates the test report, which is used by physicians to evaluate patient's clinical condition and prescribe treatments.

In view of illustrating how representing RAs as detectors can help to cope with the problem, we will consider the case of three of the most important RAs in urinary dipstick: hemoglobin RA, glucose RA and leukocyte esterase RA. Initially, Fig. 3 shows glucose RA representation according to the ODP. In the white boxes, we have concepts from urinalysis domain and from ODP structure (here employed as a top ontology). Concepts in gray boxes are those created to suit glucose RA representation to the ODP – and that, if taken alone, reveal a structure analogous to that of Fig. 2.

Glucose RA plays the role of glucose detector while exposed to urine (when it is able to get in touch with the substances in the specimen). Glucose plays the role of analyte for glucose RA when in urine (where it is liable to be perceived by a glucose

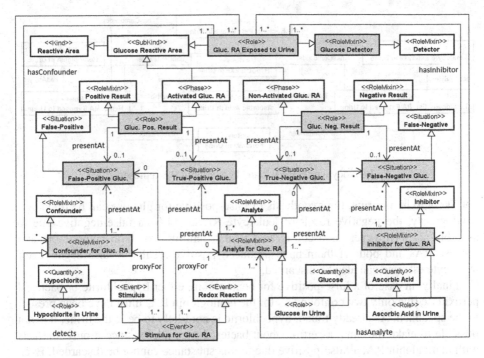

Fig. 3. Glucose reaction area representation.

RA). Likewise, hypochlorite, when in urine, is a confounder for glucose RAs, since it causes the same type of chemical reaction (i.e. redox reaction) that activates such RAs (i.e. causes the same stimulus). Finally, ascorbic acid acts as inhibitor for this type of glucose detector since it reduces glucose RA sensibility to the referred chemical reaction (i.e. prevents detector from detecting the stimulus).

Hemoglobin and leukocyte esterase RAs are also liable to the effects of confounders and inhibitors (illustrated in a simpler way in Fig. 3, due to space matters): ascorbic acid plays the role of inhibitor and hypochlorite acts as confounder for both RAs, bacterial peroxidase (i.e. an enzyme secreted by some types of bacteria) acts as confounder for hemoglobin RA, and formalin is inhibitor for hemoglobin RA as well as confounder for leukocyte esterase RA.

This representation helps in assessing urinalysis correctness, allowing raising hypothesis about false-results and providing means to test them. For example, lets imagine a urine specimen for which glucose and hemoglobin RAs have given positive results while leukocyte esterase RA have produced a negative one. In this case, a false-positive for glucose may be considered. By the definition of false-positive, we can infer that, if such situation is true, some confounder for this RA must be present in the urine. According to Fig. 3, the only known confounder for glucose RA is hypochlorite. In accordance with Figs. 3 and 4, hypochlorite also acts as confounder for the other two RAs. Since leukocyte esterase RA has produced a negative result, hypochlorite is not likely to be present in the specimen (otherwise it would have made this RA produce a positive result). Then, as hypochlorite is the only confounder for glucose RA and it is

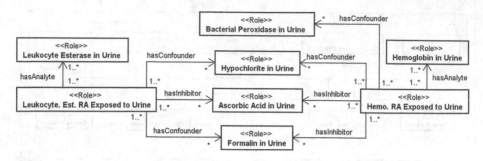

Fig. 4. Leukocyte esterase RA and hemoglobin RA representations.

not likely to be in urine, the hypothesis of false-positive for glucose can be ruled out. Analogously, the negative result of leukocyte RA could be a false-negative due to ascorbic acid. However, as ascorbic acid is also inhibitor for both hemoglobin and glucose RAs and both of them have produced positive results, a false-negative for leukocyte esterase can also be discarded.

Finally, in case of a false-positive for hemoglobin, either hypochlorite or bacterial peroxidase (the only two confounders for the hemoglobin RA) must have been present in the specimen. As already found, hypochlorite is not likely to be in the urine. It is not possible to make the same assertion about bacterial peroxidase: since it only interacts with hemoglobin RA, a false-positive due to this substance cannot be discarded. Even so, just raising such hypothesis is of considerable help, since it would lead the biochemist to search for other signs to confirm or dismiss the positive result for hemoglobin (e.g. finding red blood cells in microscopy, which would justify the presence of some hemo-globin that could have activated the RA).

Emphasizing the generality of the proposed solution, we can briefly introduce exam-ples in other fields. For instance, a human presence detector is intended to detect the presence of a person (analyte), generally relying on infrared emissions from her/his body (stimulus). However, a large dog (also with warm body) or a device emitting infrared rays could also originate similar stimulus (being confounders), causing the detector to produce a positive result in a false-positive situation. Moreover, a person wearing a coverage able to block infrared rays (an inhibitor) would pass by the sensor without activating it – leading to a false-negative situation.

Another example would be an optical smoke detector – with smoke regarded as "the gaseous products of burning materials" (http://www.merriam-webster.com/dictionary/smoke). This type of detector consists of a light source and a photoelectric receiver to sense such light. Smoke is detected due to its optical interference in this light sensing process (stimulus). However, steam or dust can also cause a similar interference that activates the detector (being confounders). Withal, though not a usual occurrence, an external strong light aiming detector's photoelectric receiver could cancel smoke inter-ference (being an inhibitor).

4   Discussion

Sometimes, knowing about false-result occurrences may allow permanently preventing them, choosing a detector that implements a better procedure to carry out the observation, without interferences. In these situations, false-results are just a temporary issue, which may not worth a representation of their own. Nevertheless, there certainly are cases in which such situation is unavoidable (e.g. alternative procedures are not known or their implementation is unfeasible). So, having means to represent such cases would be useful in order to better dealing with them. This work should be considered under this perspective.

Thereto, this work does not deny the approaches adopted in previous works. In fact, it builds upon them, adding an "intention layer" (i.e. what the result of a detector is intended to stand for) in order to allow the representation of an aspect of sensors (though still restricted to the particular case of detectors) not covered by those efforts (i.e. false-results). Covering such new aspect, this work provides interesting capabilities. For example, knowing that a false-result happened and having a complete knowledge about it, our pattern makes it possible to represent such find – indicating which detector has failed, what result it has produced, what such result means (i.e. which type of entity it asserts as present/absent) and what was really causing it (i.e. which type of entity acted as confounder/inhibitor). This would allow storing and communicate such situations, while providing answers to questions such as "what has caused this false-positive?" or "what was expected to be observed in this case?".

Beyond that, representing detectors using the ODP enables some important reasoning tasks. They come in two main forms: given the result of a detector, one can get interfering entities (confounders or inhibitors) that may explain the result; given interfering entities, one can get the detectors that may be affected by them. For example, suspecting about the correctness of a result, it is possible to identify which confounders or inhibitors may be causing it. This allows seeking alternative means to detect such interfering entities and confirm or refute the suspicions. Moreover, knowing about the presence of some type of entity in the environment in which a group of detectors is deployed, it is possible to identify which of them may be suffering interference of such entity – either as confounder or inhibitor (e.g. if detector D has C as confounder, knowing about the presence of C may raise suspicions about the result of D if it is positive). Finally, having results of a set of detectors deployed in the same environment, we can assess the coherence among them by identifying entities that interact with more than one detector and checking if such detectors behave as expected in view of the presence of such types of entity (e.g. consider detectors A and B, the first having no known confounders and whose analyte is confounder for B, which has no known inhibitors; in this scenario, we would have an incoherence if A gives a positive result and B and a negative one).

These reasoning possibilities could help in dealing with the challenge of characterizing and managing the quality of sensor data, as exposed in [21]. Implementing it in a knowledge representation language (e.g. OWL) would aid in inspecting such reasoning procedures. Yet, as currently OntoUML editors have no support to Situation and Event universals [22] (and thus cannot generate logical specifications for our ODP), it would require the development of a mapping strategy to represent the concepts with the primitives of the chosen language, what is out of the scope of this work.

This work has also some noticeable limitations. For example, it is reasonable to argue that there may be some detectors designed to observe endurants – either objects (e.g. metallic objects carried by a person) or tropes (e.g. fever in a patient) – and others to observe perdurants (e.g. the occurrence of a storm in a given region). Still, though the general ODP scheme can be applied to all the cases, the ODP constrains the concept of observable entity to a subtype of endurant (i.e. an UFO's RoleMixin), preventing the representation of tropes and perdurants under this concept. Albeit there is real need to represent such types of observable entities, the description of an observable entity as a role played in relation to a detector is just too suitable to be readily put away – and even has ground in previous works [23]. Thereby, describing the dynamics portrayed by the ODP in a way that generalizes the concept of observable entity to admit tropes and perdurants remains an open question.

Another weakness is that, as implied by [14], detector results (in fact, sensor results in general) are not only liable to interferences of external entities (i.e. confounders and inhibitors), but also to intrinsic aspects of detectors themselves. It means some internal failures may cause detectors to produce positive/negative results, regardless the occurrence or not of some stimulus. Probably, the nature of such failures is more subtle than that of interferences in the stimulus detection (i.e. it does not seem to be a matter of an entity interacting with detectors or their observable entities, but a combination of conditions and/or complex events involving structural components, which may vary among different types of detector) - what will require further research to identify possible invariance that can enable a faithful description of such aspect.

Finally, along with that, another open line of research is transposing the ODP to the case of sensors in general. We believe that good results may come if we regard the action of confounders as that of reinforcing (i.e. increasing the value of) sensor response – and consider the action of inhibitors as that of attenuating such response.

5   Final Remarks

This work presented an ODP to represent detectors and other related concepts involved in false-results occurrence. Despite its current limitations, the pattern allows answering the questions it was designed for (e.g. those posed in Sects. 2 and 4) and includes the necessary concepts for a rich description of the issue. Given the examples introduced in this article, the ODP seems sufficiently general to be applied in a wide range of domains. Also, its grounding in the meta-types of UFO provides good evidence of logical consistency.

The pattern is already in use in the development of a knowledge-based system for urinalysis. Even so, as work in progress, major improvements may be achieved by exploring several lines of research, including: (i) modeling false-results due to internal issues, (ii) representing tropes and perdurants as observable entities and (iii) extend the ODP to the general sensor case.

Acknowledgements.   Thanks to Mara Abel and Joel L. Carbonera for the great help. This work was partially supported by *CAPES* (project *Pró-Ensino na Saúde - n 39*).

References

1. Ceruti, M.G.: Ontology for level-one sensor fusion and knowledge discovery. In: International Knowledge Discovery and Ontology Workshop (KDO), pp. 97–102, Pisa, Italy (2004)
2. Janowicz, K., Compton, M.: The stimulus-sensor-observation ontology design pattern and its integration into the semantic sensor network ontology. In: Taylor, K., Ayyagari, A., De Roure, D. (eds.) The 3rd International workshop on Semantic Sensor Networks 2010 (SSN10) in Conjunction with the 9th International Semantic Web Conference (ISWC 2010), Shanghai, China (2010)
3. Neuhaus, H., Compton, M.: The semantic sensor network ontology: a generic language to describe sensor assets. In AGILE Workshop: Challenges in Geospatial Data Harmonisation (2009)
4. Kuhn, W.: A functional ontology of observation and measurement. In: Janowicz, K., Raubal, M., Levashkin, S. (eds.) GeoS 2009. LNCS, vol. 5892, pp. 26–43. Springer, Heidelberg (2009)
5. Calder, M., Morris, R., Peri, F.: Machine reasoning about anomalous sensor data. In: International Conference on Ecological Informatics (2008)
6. Matheus, C.J., Tribble, D., Kokar, M.M., Ceruti, M.G., McGirr, S.C.: Towards a formal pedigree ontology for level-one sensor fusion. In: 10th International Command and Control Research and Technology Symposium (2005)
7. Russomanno, D., Kothari, C., Thomas, O.: Sensor ontologies: from shallow to deep models. In: 37th Southeastern Symposium on System Theory (2005)
8. Witt, K.J., Stanley, J., Smithbauer, D., Mandl, D., Ly, V., Underbrink, A., Metheny, M.: Enabling sensor webs by utilizing SWAMO for autonomous operations. In: 8th NASA Earth Science Technology Conference (2008)
9. Avancha, S., Patel, C., Joshi, A.: Ontology-driven adaptive sensor networks. In: 1st Annual International Conference on Mobile and Ubiquitous Systems, Networking and Services (2004)
10. Eid, M., Liscano, R., Saddik, A.E.: A universal ontology for sensor networks data. In: IEEE International Conference on Computational Intelligence for Measurement Systems and Applications (2007)
11. Kim, J., Kwon, H., Kim, D., Kwak, H., Lee, S.: Building a service-oriented ontology for wireless sensor networks. In: 7th IEEE/ACIS International Conference on Computer and Information Science (2008)
12. Amato, F., Casola, V., Gaglione, A., Mazzeo, A.: A semantic enriched data model for sensor network interoperability. In: Simulation Modelling Practice and Theory, pp. 1–16. Elsevier, Amsterdam (2010)
13. Sheth, A., Henson, C., Sahoo, S.: Semantic sensor web. In: IEEE Internet Computing, p. 78–83 (2008) [NÃO MAIS USADA – DELETAR !!!]
14. Henry, M.P., Clarke, D.W.: The self-validating sensor: rationale, definitions and examples. Control Eng. Pract. 1(4), 585–610 (1993). Pergamon Press, GB
15. Sharma, D., Yadav, U.B., Sharma, P.: The concept of sensitivity and specificity in relation to two types of errors and its application in medical research. J. Reliab. Stat. Stud. 2(2), 53–58 (2009)
16. Gangemi, A.: Ontology design patterns for semantic web content. In: Gil, Y., Motta, E., Benjamins, V.R., Musen, M.A. (eds.) ISWC 2005. LNCS, vol. 3729, pp. 262–276. Springer, Heidelberg (2005)
17. Guizzardi, G. Ontological Foundations for Structural Conceptual Models. PhD thesis, University of Twente, The Netherlands (2005)

18. Lozano, J., Carbonera, J.L., Abel, M., Pimenta, M.: Ontology view extraction: an approach based on ontological meta-properties. In: Proceedings of ICTAI 2014 (2014)
19. Guizzardi, G., Wagner, G., de Almeida Falbo, R., Guizzardi, R.S.S., Almeida, J.P.A.: Towards ontological foundations for the conceptual modeling of events. In: Ng, W., Storey, V.C., Trujillo, J.C. (eds.) ER 2013. LNCS, vol. 8217, pp. 327–341. Springer, Heidelberg (2013)
20. Strasinger, S.K., Di Lorenzo, M.S.: Urinalysis and Body Fluids, 5th edn. F.A. Davis Company, Philadelphia (2008)
21. Corcho, O., García-Castro, R.: Five challenges for the semantic sensor web. Semantic Web 1, 121–125 (2010). IOS Press
22. Benevides, A.B., Guizzardi, G.: A model-based tool for conceptual modeling and domain ontology engineering in OntoUML. In: Filipe, J., Cordeiro, J. (eds.) Enterprise Information Systems. LNBIP, vol. 24, pp. 528–538. Springer, Heidelberg (2009)
23. Probst, F.: Ontological analysis of observations and measurements. In: Raubal, M., Miller, H.J., Frank, A.U., Goodchild, M.F. (eds.) GIScience 2006. LNCS, vol. 4197, pp. 304–320. Springer, Heidelberg (2006)

Ontology Engineering by Combining Ontology Patterns

Fabiano B. Ruy[1,2](✉), Cássio C. Reginato[1], Victor A. Santos[1],
Ricardo A. Falbo[1], and Giancarlo Guizzardi[1]

[1] Ontology and Conceptual Modeling Research Group (NEMO),
Computer Science Department,
Federal University of Espírito Santo, Vitória, Brazil
{fabianoruy, cassio.reginato, victor.amsantos, falbo,
gguizzardi}@inf.ufes.br
[2] Informatics Department,
Federal Institute of Espírito Santo, Campus Serra, Serra, Brazil

Abstract. Building proper reference ontologies is a hard task. There are a number of methods and tools that traditionally have been used to support this task. These include foundational theories, reuse of domain and core ontologies, development methods, and software tool support. In this context, an approach that has gained increased attention in recent years is the systematic application of ontology patterns. This paper discusses how Foundational and Domain-related Ontology Patterns can be derived, and how they can be applied in combination for building more consistent ontologies in a reuse-centered process.

Keywords: Ontology patterns · Conceptual ontology patterns · Ontology reuse · Ontology engineering

1 Introduction

Although nowadays ontology engineers are supported by a wide range of Ontology Engineering (OE) methods and tools, building proper reference domain ontologies is still a difficult task even for experts [1]. Besides the domain knowledge, the ontology engineer needs to apply ontological foundations in order to develop well-founded ontologies.

According to their generality level, ontologies can be classified into Foundational, Core and Domain ontologies [2]. At the highest level, foundational ontologies span across many fields and model the very basic and general concepts and relations that make up the world [3]. Domain ontologies, in turn, describe the conceptualization related to a specific domain [3]. Core ontologies are located between foundational and domain ontologies, and provide a definition of structural knowledge in a specific field that spans across different application domains in this field [2]. The generality levels are not a discrete classification, but a continuum [4] from foundational ontologies, totally domain-independent, to domain ontologies, for a very particular domain. Core ontologies, though more general than domain ontologies, are domain-dependent.

© Springer International Publishing Switzerland 2015
P. Johannesson et al. (Eds.): ER 2015, LNCS 9381, pp. 173–186, 2015.
DOI: 10.1007/978-3-319-25264-3_13

Reuse is pointed out as a promising approach for OE, since it enables a speeding up of the ontology development process. Higher level ontologies can be used to support the development of lower level ontologies, e.g., foundational ontologies can be used to support the development of core and domain ontologies, and core ontologies can be the basis for developing domain ontologies. However, ontology reuse, in general, is a hard research issue, and one of the most challenging and neglected areas of OE [5]. The problems of selecting the right ontologies to reuse, extending them, and composing several ontology fragments have not been properly addressed yet [6].

Ontology Patterns (OPs) are an emerging approach that favors reuse of encoded experiences and good practices. OPs are modeling solutions to solve recurrent ontology development problems [7]. There are many different types of OPs that can be used in different phases of the OE process. In this paper, we are interested in Conceptual OPs (COPs), since our focus is on developing reference ontologies. A reference ontology is constructed with the goal of making the best possible description of the domain in reality, representing a model of consensus within a community, regardless of its computational properties [8]. In other words, when developing a reference ontology, the focus is on expressivity of the representation and truthfulness to the domain being represented (domain appropriateness), even if computational properties such as tractability and decidability have to be sacrificed. In summary, in the view employed here, an ontology is a particular kind of conceptual model, namely, a reference conceptual model capturing the shared consensus of a given community. As such, although our discussion is somehow focused on domain reference ontologies, the approach advanced here should be beneficial to ontology-driven conceptual modeling in general [9].

COPs are modeling fragments extracted from either foundational ontologies (Foundational OPs - FOPs) or core/domain ontologies (Domain-Related OPs - DROPs). They are to be used during the ontology conceptual modeling phase, and focus only on conceptual aspects, without any concern with the technology or language [10]. An OP extracted from a higher level ontology can be used to support the development of lower level ontologies.

We argue that if FOPs and DROPs are systematically applied in combination, reuse is maximized, making the OE process more productive, and improving the quality of the resulting domain ontologies. In this paper, we discuss how FOPs and DROPs can be derived, and how they can be used to develop core and domain ontologies. The FOPs discussed here are derived from the Unified Foundational Ontology - UFO [8]. The DROPs, in turn, are related to the Enterprise domain [11]. Moreover, we show a tool supporting the application of COPs for developing ontologies. Our main goal is to show how the combined application of COPs could be useful for OE.

This paper is organized as follows. Section 2 briefly presents the Unified Foundational Ontology (UFO), the source for the discussed FOPs. Section 3 discusses how FOPs are derived from UFO, how DROPs are extracted from a core ontology, and how COPs can be used in combination for developing domain ontologies. Section 4 briefly presents a software tool supporting COPs definition and application. Section 5 discusses related works. Finally, Sect. 6 presents our final considerations.

2   The Unified Foundational Ontology - UFO

UFO [8] is a foundational ontology that has been developed based on a number of theories from Formal Ontology, Philosophical Logics, Philosophy of Language, Linguistics and Cognitive Psychology. The aspects we present here cover object-type (endurant) distinctions in UFO, namely, the distinction between sortal types, non-sortal types and relators. For an in depth presentation, formal characterization and discussion about empirical support for UFO, the interested reader should refer to [8].

By referring to a number of formal and ontological meta-properties, UFO proposes a number of distinctions among object types. Within these, sortal types are types that either provide or carry a uniform principle of identity for their instances. Within sortal types, we have the distinction between rigid and anti-rigid sortals. A rigid type is a type that classifies its instances necessarily (in the modal sense), i.e., the instances of that type cannot cease to be an instance of that type without ceasing to exist. Anti-rigidity, in contrast, characterizes a type whose instances can move in and out of the extension of that type without altering their identity. For instance, contrast the rigid type *Person* with the anti-rigid types *Student* or *Husband*. While the same individual *John* never ceases to be instance of *Person*, he can move in and out of the extension of *Student* or *Husband*, once he enrolls in/finishes college or marries/divorces, respectively.

Kinds are sortal rigid types that provide a principle of identity for their instances. Subkinds are rigid types that carry the principle of identity supplied by a Kind. Concerning anti-rigid sortals, we have the distinction between Roles and Phases. Phases are relationally independent types defined as a partition of a sortal. This partition is derived based on an intrinsic property of the type (e.g., *Child* is a phase of *Person* while she has less than 18 years). Roles are relationally dependent types, capturing that entities play roles when related to other entities (e.g., *Student, Husband*). Since the principle of identity is provided by a unique Kind, each sortal hierarchy has a unique Kind in the top [8].

Non-Sortals are categorizations that aggregate properties of distinct Sortals. *Furniture* is an example of Non-Sortal that aggregates properties of *Table, Chair* and so on. Non-Sortals do not provide principle of identity; instead, they just classify things that share common properties but that obey different principles of identity. Rigidity and anti-rigidity also applies to non-sortals. Rigid non-sortals are termed Categories (e.g., *Physical Object*); Anti-rigid sortals are RoleMixins (e.g., *Customer, Crime Weapon*). However, non-sortals can also exhibit a meta-property termed semi-rigidity. An object type is semi-rigid if it is rigid for some of its instances and anti-rigid for others. A semi-rigid non-sortal is termed a *Mixin* (e.g., *Insurable Item*).

For capturing these distinctions between object types put forth by UFO, Guizzardi [8] proposed a UML profile called OntoUML. Over the years, OntoUML has been successfully employed to build conceptual models and domain ontologies in a number of complex domains (e.g., [12, 13]). Table 1 shows the OntoUML stereotypes.

Concerning relationships, UFO's theory of relations makes a fundamental distinction between two main types of relationships, namely: *formal* and *material* relations. Whilst the former holds directly between two entities without any further intervening individual, the latter is induced by the presence of mediating entities called Relators. Relators are individuals with the power of connecting entities. For example,

Table 1. OntoUML object type distinction

Stereotype	Identity	Rigidity	Allowed supertypes
Kind	Provide	Rigid	Category, RoleMixin, Mixin
Subkind	Carry	Rigid	Kind, Subkind, Category, RoleMixin, Mixin
Role	Carry	Anti-Rigid	Kind, Subkind, Role, Phase, Category, RoleMixin, Mixin
Phase	Carry	Anti-Rigid	Kind, Subkind, Role, Phase, Category, RoleMixin, Mixin
Category	–	Rigid	Category
Mixin	–	Semi-Rigid	Mixin, RoleMixin
Role mixin	–	Anti-Rigid	Mixin, RoleMixin

an *enrollment* connects a *student* with an *educational institution*, and an *employment* connects an *employee* with an *employer organization* [8]. OntoUML has a construct for modeling relator universals. Every instance of a relator universal is existentially dependent on at least two distinct entities. The formal relations that take place between a relator universal and the object classes it mediates are termed *mediation* relations (a particular type of existential dependence relation).

3   Conceptual Ontology Patterns

In the Ontology Engineering process, pattern reuse can occur in many possible ways [10]. Our focus is on FOPs and DROPs, how they are extracted, and how they are applied for building core and domain ontologies. Figure 1 shows, on the left side, the ontology generality levels, and on the right, the corresponding types of COPs.

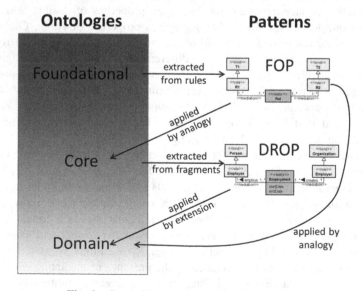

Fig. 1. Generality levels and ontology patterns.

FOPs are extracted from the foundations and rules of a foundational ontology. A FOP, capturing foundational structural content, can be reused **by analogy** [14], reproducing their structure in the ontology being developed, by matching the referenced concepts of the pattern to the corresponding ones in the domain. The result is an ontology fragment with the FOP structure shaping the structures at the level of domain concepts. FOPs can be applied for building both core and domain ontologies.

On the other hand, DROPs capture the core knowledge of the target domain, and can be extracted directly from core/domain ontologies fragments. It is worth to note that when a DROP is extracted from a core/domain ontology modeled already reusing FOPs, we have as result a richer DROP, carrying both structural and domain knowledge, characterizing a chained COP application at the domain level. DROP reuse occurs **by extension** [15], i.e., their concepts and relations become part of the domain ontology and they are typically extended, giving rise to new more specific concepts and relations.

Both DROPs and FOPs can be applied for engineering domain ontologies. When developing domain ontologies, for each aspect to be modeled, the ontology engineer should first try to find an applicable DROP, which captures the core knowledge, and possibly a foundational structure, suitable for the modeling problem at hand. If there is no DROP satisfying the domain requirements, she needs to build a new ontology fragment. In this case, she can apply a FOP and reuse its structure.

Of course, modeling domain ontologies is not limited to the direct application of patterns. Domain ontology fragments created from COPs are interrelated and need to be put together. To do that, the ontology engineer can look for related DROPs, and also use FOPs for combining the structure inherent to the different fragments.

3.1   Deriving and Applying Foundational Ontology Patterns (FOPs)

FOPs pack the structure and rules of some foundational aspects recurrently applied when building core and domain ontologies. A FOP is not a foundational ontology fragment; instead, it is a self-contained set of related foundational rules and constraints that is applied to solve a common modeling problem independently of domain. By self-contained we mean that the pattern satisfies some foundational ontology constraints imposed to the elements involved, not contradicting any foundational ontology rule.

We distinguish between two main types of FOPs, namely Structural FOPs and Derivation FOPs. A Structural FOP captures a template structure that comes from the application of certain foundational ontology aspects that enforce this structure to take place. In order to exemplify, let us consider the following foundational rules from UFO: (r1) A *relator* mediates at least two distinct entities; (r2) A *mediation* is a bidirectional mandatory relation between a *relator* and an *endurant*; (r3) A *role* is an anti-rigid type and must inherit the identity principle of exactly one *kind*; and (r4) A *role* is externally dependent on a *relator*. A combination of (r1) and (r2) gives rise to the Relator FOP, whilst a combination of (r3) and (r4) gives rise to the Role FOP. Since the Relator FOP is generic, allowing mediations with different types of *endurants* (*kinds, roles, phases, mixins*, etc.), it enables some variations combining other ontological rules or even other FOPs. It is the case of the Role-Relator FOP, which combines the Relator and the Role FOPs (or the four foundational rules mentioned above) to represent the case where a *relator* mediates two *roles*. These patterns are shown in Fig. 2, as templates. They can be applied during the development of core or domain ontologies by analogy, i.e., by reproducing their structure in the ontology being developed [10], and completing the structure with the specific knowledge. For instance, to represent the relation of a Person booking a Room, the *relator* Reservation mediates two *roles*: Customer as a *role* of a Person *kind*, and Reserved Room as a *role* of a Room *kind*. The Role-Relator FOP can be applied to many situations relating things, such as in a marriage, in an employment, etc.

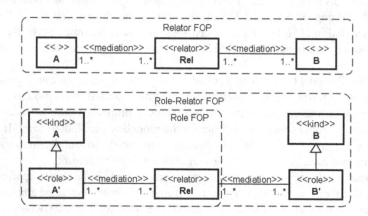

Fig. 2. The relator FOP.

Another recurrent applied variation of the generic Relator FOP is the Kind-Relator FOP. It takes place when the *relator* mediates a *role* and a *rigid sortal* (*kind* or *subkind*). This FOP is useful when the ontology engineer needs to represent a *relator* between two entities but the *role* that could be assumed by the instances of the *rigid type* is not representative in the domain (e.g., a *role* "University with Students", since it is supposed that all University has Students), and can be discarded. Box D in Fig. 6 shows an example of the application of the Kind-Relator FOP.

The Role Mixin FOP [8], which addresses the problem of modeling roles with multiple allowed types is another example of Structural FOP. As Fig. 3 shows, the abstract class C is the *role mixin* that covers different *role* types. Classes A' and B' are the disjoint subclasses of C that can have direct instances, representing the *roles* that carry the principles of identity that govern the individuals that fall in their extension. Classes A and B are the ultimate sortals (*kinds*) that supply the principle of identity carried by A' and B', respectively. The material relation R represents the common specialization condition for A' and B', which is represented in C. Finally, class D represents a *endurant* (such as *kind, role, phase*) that C is *relationally dependent on*.

Another important class of FOPs regards derived concepts [16]. Derivation regards how someone can derive concepts from operations applied to other concepts. For instance, in [17], Guizzardi demonstrates that ontological meta-properties can be derived from derived Object-Types. For instance, any object-type derived by union from two

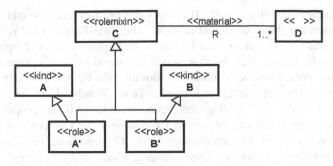

Fig. 3. Role Mixin FOP [8]

Table 2. Derivation FOPs: union and intersection

Type 1	Operation	Set of types	Result
Kind	∪	Kind, ..., Kind	Category
Category	∪	Category, ..., Category	Category
Subkind	∪	Subkind, ..., Subkind	Subkind, Kind
Role	∩	Role, ..., Role	Role

rigid Object-Types is also a rigid type. In this paper, we approach two cases of derivation: derivation by union and derivation by intersection. Table 2 shows four possible derived concepts obtained by union and intersection, and thus four Derivation FOPs.

In order to exemplify the use of FOPs, we choose the Enterprise domain. Figure 4 depicts a portion of an Enterprise Core Ontology (adapted from [11]), answering the following competency questions: (CQ1) Who are the employees of an Organization? (CQ2) What is the time period of an Employment? (CQ3) How is an Organization structured in terms of Organizational Units? (QC4) Which are the Projects of an Organization? (CQ5) Which are the parties involved in a Project?

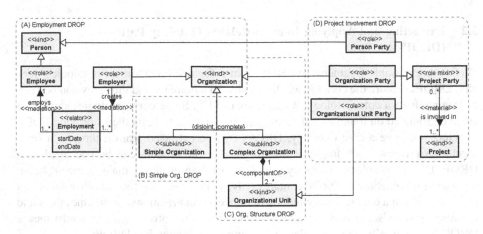

Fig. 4. Enterprise Core Ontology fragmented in DROPs (partial version, adapted from [11]).

The first competence question (CQ1) points the need to relate people to an organization, representing the employment relation. The Role-Relator FOP is suitable for this case. By analogy, we can say that Person (a *kind*) plays the *role* Employee while participating in the Employment *relator*, which is created by an Employeer, a *role* of the Organization *kind*. In the same Enterprise domain, the Relator FOP applies several times, being useful to represent, for instance, Team Membership. The competency question CQ2 is solved by adding the *startDate* and *endDate* properties to the Employment concept. CQ3 can be solved by applying the Composition FOP, considering two subtypes of Organization: Simple Organization, representing those organizations not broken down into Organizational Units; and Complex Organization, representing organizations composed of two or more Organizational Units (satisfying the so-called *weak supplementation* [18] axiom for compositional structures). CQ4 can be answered by the same fragment of CQ5, since the Organization itself is a party involved in its Projects. The parties involved in Projects have diverse nature, such as Organizations, Organizational Units and People. Since each party type assumes a role applying different principles of identity, the Role Mixin FOP applies. Thus, Project Party is a *role mixin* with a material relation with the Project *kind*.

To exemplify the application of a derivation pattern, let us take a fragment of the Enterprise Core Ontology, where Team, Organizational Unit and Organization give rise by union to the concept of Institutional Agent. Since these three concepts are classified as *kind* in the ontology, the resulting derived by union type (Institutional Agent) should be a *category* (see the first line of Table 2), as Fig. 5 shows.

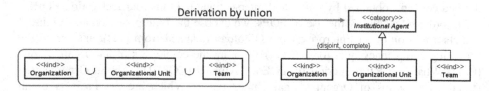

Fig. 5. The kind union FOP.

3.2   Extracting and Applying Domain-Related Ontology Patterns (DROPs)

DROPs are reusable fragments extracted from reference core/domain ontologies. Ideally, DROPs capture the core knowledge related to a domain, and thus they can be seen as fragments of a core ontology of that domain [10]. Since core ontologies should be created grounded on foundational ontologies, it is natural to build them applying FOPs.

Once we have a core ontology, DROPs can be extracted from it through a fragmentation process. Each fragment meaningful for the domain can be packaged as a DROP. DROP complexity can vary greatly depending on the domain fragment being represented. Sometimes a DROP contains only two related concepts; in other situations they can contain a complex combination of concepts and relations. Sometimes the same fragment gives rise to two (or more) variant and alternative patterns; sometimes a DROP is structurally open in order to be completed by another DROP.

Regarding the DROP derivation process, while FOPs are focused on foundational and structural aspects, when deriving a DROP, domain aspects come first. The main rule for a DROP is to represent a recurrent fragment in the field, regardless of its foundational structure. Thus, while FOPs tend to be generally applied, DROPs for a specific field are very interrelated [4]. For this reason, it is usual to apply many DROPs in combination or in a sequence for engineering domain ontologies. Ontology Pattern Languages (OPLs) [4], for instance, are used to organize DROPs in a guided application process.

With a core ontology in hands, fragments can be extracted to form DROPs. In order to illustrate the extraction of DROPs from core ontologies, in Fig. 4, the dotted boxes show four DROPs: (A) Employment, (B) Simple Organization, (C) Organizational Structure, and (D) Project Involvement. Although each DROP represents a distinct aspect of the core domain knowledge, they can be related in different ways. For instance, the DROPs Simple Organization (B) and Organizational Structure (C) are very related from the foundational point of view, since they model two complementary *subkinds* of the same *kind*. However, they can also be applied in isolation, depending on the application domain. Another interesting aspect is that different DROPs partially overlap between them. This is the case of the Organization concept, which is shared by three distinct DROPs. This characteristic shows that DROPs have a more strict relation and are often applied in combination or even in sequence. Additional examples of DROPs of the Enterprise domain and possible combinations between them can be found in [11].

In order to illustrate the combined application of COPs for building a domain ontology, Fig. 6 presents a portion of a domain ontology for Universities. This example explores some different situations of COP reuse at the domain level. Since University is a specific type of Enterprise, the Enterprise DROPs are applicable. Three competency questions were defined for this portion of the domain ontology: (CQ1) Who are the employees of a University? (CQ2) Who are the students of a University? (CQ3) Which are the parties involved in a Research Project? As we can notice, CQ1 and CQ3 are, indeed, specializations of the competency questions of the Enterprise core ontology presented before. Two DROPs of Fig. 4 applies to answer CQ1 (Employment DROP) and CQ3 (Project Involvement DROP).

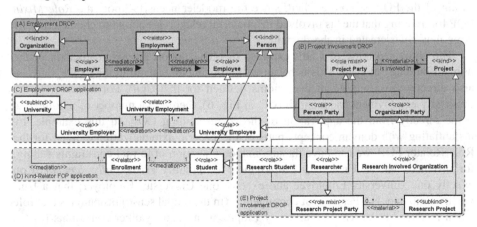

Fig. 6. Building a Domain Ontology with different types of COPs.

CQ1 can be solved directly by applying the Employment DROP. This DROP is added to the domain model (box A in Fig. 6) and its concepts and relations are extended (in box C): University is an Organization, University Employment is an Employment, and the *roles* Employer and Employee are specialized into the specific *roles* University Employer and University Employee, respectively. The relations "creates" and "employs" are also specialized. This reuse by extension mechanism allows some modifications, such as adding new properties to the concepts, restricting cardinalities and adding new axioms. It is worthwhile to point out that the Employment DROP inherits the structure of the Role-Relator FOP. Thus, the university employment fragment being modeled (box C) reuses these two COPs, from different levels.

For CQ2, there is not a DROP available, since Student is a notion of the University domain, and is not modeled for Enterprises in general. However, the question treats a relation between a specific role assumed by people and a University. Thus, the Kind-Relator FOP applies. Reusing this FOP structure, the *role* Student extends Person, and the *relator* Enrollment establishes the participation of Students in a University (box D). It is important to observe that only two new concepts, Student and Enrollment, were added to the model, taking the advantage of reusing the concepts of Person and University already modeled.

The third question, CQ3, addresses the different parties involved in a (Research) Project, and can be solved applying the Project Involvement DROP. As in the case of CQ1, the DROP is added to the model (box B) and its concepts and relations are extended (box E). Research Project is a Project, and Research Project Party is a specialization of the Project Party *role mixin*. The University domain considers three types of involved parties: Researchers, a more specific *role* of University Employee; Research Student, a specialization of the Student *role*; and Research Involved Organization, a *role* for Organizations involved in research projects. The first two types, have their identity principle provided by the Person *kind*, thus extend Person Party. The last one is an Organization, and extends Organization Party. This third fragment is another case of combined reuse of COPs, since the fragment of the Project Involvement DROP reuses the structure of the *Role Mixin* FOP. In this case, the joint application of the FOP is more explicit, since the modeler needs to know the *Role Mixin* FOP for assuring that the "is involved in" material relation is enough to characterize the three new *roles* created in the domain.

The combined reuse of COPs can also directly support the axiomatization of the domain ontology at hand. This is because, by reusing and adapting the COPs, one can also directly reuse and adapt the axioms defined for that COP. For instance, in the university employment fragment (box C in Fig. 6), there are a number of axioms that come directly from the application of the corresponding COP. In this example, by instantiating with domain concepts parts of the generic axiomatization defined for the Role-Relator FOP, we can have the automatic generation of formal constraints stating, for instance, that an University Employment is multiply *existentially dependent* on exactly one University Employee and exactly one University Employer; that a University Employer is a Person that contingently (in the modal sense) instantiates that role when mediated by at least a University Employment, among other constraints [8].

4 Tool Support for Ontology Pattern Application in OLED

For the definition and application of FOPs and DROPs, we extended the OLED tool. OLED[1] is an ontology framework that supports the OntoUML language [8], providing a number of features such as construction, evaluation, simulation and transformations of OntoUML models. In order to provide support for pattern-based engineering of OntoUML models, OLED allows the definition of FOPs and DROPs libraries. These libraries can be created in the tool by defining each pattern as a model. Then, the suitable libraries can be imported and applied to build core and domain ontologies. This feature also turns possible the combination of the ontology patterns aforementioned, since the OLED extension enables to build a core ontology applying FOPs libraries and then saving fragments of this core ontology in a DROP library. These DROPs (and FOPs), in turn, can be reused to build domain ontologies.

Figure 7 illustrates these pattern supporting features for creating a domain ontologies with a DROP library. First, the Employment DROP is selected for reuse in the pallet (left side of Fig. 7a). A new window for COP configuration opens (Fig. 7b), where some adaptations can be made (changing names, picking up from the model, adding new classes, etc.). Next, the DROP classes are included in the current model (highlighted classes), and can be specialized with specific concepts and relations from the application domain.

5 Related Work

There are few works addressing the combined application of OPs. Most of them focuses on reusing patterns of the same generality level, or of the same type.

Uschold and colleagues [19] consider the use of an implementation guide for building domain ontologies. This process has been applied, for instance, in the Gist ontology [20]. Gist has the main purpose of being used as a catalog of generic concepts for the enterprise domain. The Gist's intents are similar to the idea of DROPs, but restricted to the enterprise domain. Gist can be considered an enterprise core ontology that establishes a method to apply its concepts. Hence, it would be plausible to consider the Gist's fragments as DROPs. As Gist, others trustful generic ontologies are capable to be a source of DROPs for specific domains. One of our main concerns here is to show that core ontologies can take advantage of using foundational ontologies and consequently FOPs.

In [5], Gangemi and Presutti present six categories of patterns (namely: structural, content, lexico-syntactic, reasoning, presentation and correspondence) for ontologies at the design and implementation levels. Their Content Patterns are quite similar to our DROPs. However, they do not consider foundational patterns nor discuss how patterns can be combined.

Falbo and colleagues [10] extended Gangemi and Presutti's pattern classification, by considering DROPs and FOPs as COPs. They also briefly show how these types of

---

[1] OLED: http://nemo.inf.ufes.br/projects/oled.

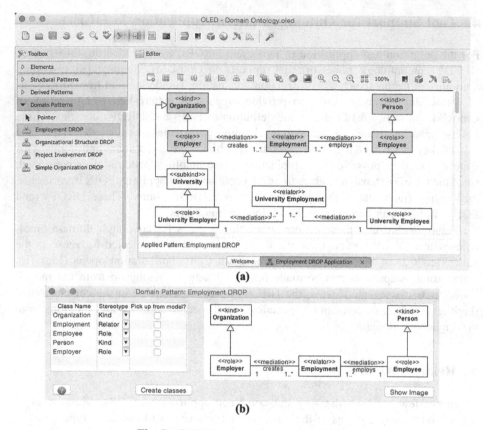

Fig. 7. OLED's support for reusing COPs.

COPs can be used for building ontologies (reuse by analogy and by extension). In fact, we took this work as a baseline. With respect to that work, we define here a new type of Foundational OP, the Derivation FOPs, and discussed how the mechanism for deriving FOPs and DROPs works. We also discuss, in details, how FOPs are applied by analogy and DROPs by extension. Finally, we present a software tool supporting the definition and application of these COPs.

In another work of the same group [4], Falbo and colleagues discussed how core ontologies can be organized as Ontology Pattern Languages (OPLs). An OPL provides guidance for the application of DROPs to build domain ontologies. OPLs are a relevant example of COPs joint application, when DROPs are reused in combination, or even in sequence. However, OPLs explore only DROP application. As contributions, we presented how FOPs, combined with DROPs, are useful for supporting the development of domain ontologies. This combination of different level COPs is useful especially for (i) applying a FOP when there is no DROP suitable for meeting a domain requirement; (ii) applying a DROP enriched by a FOP background (combined application of a sequence of DROPs and FOPs); (iii) combining different DROP applications in structurally valid ontology configurations; and (iv) combining DROPs from different related OPLs. Considering all these situations, we claim that OPLs should also consider including FOPs for building domain ontologies.

6 Final Considerations

Ontology Patterns have been recognized as a beneficial approach for Ontology Engineering [5, 6]. This paper discussed how FOPs and DROPs can be derived, respectively from foundational and core ontologies, and how they can be applied in combination for creating ontology-driven conceptual models, in general, and domain ontologies, in particular. The combined application of these two different levels of COPs enriches the model development process in many ways: (i) FOPs can complement a domain ontology fragment by adding the necessary concepts to make the model consistent; (ii) FOPs can be used when there is no suitable DROP for the problem at hands, and (iii) the combined reuse of COPs can enrich a domain ontology fragment with structural and domain knowledge. Besides the advantages for modeling, COPs can also contribute for reusing competency questions and axioms from foundational and core ontologies. The key point is that when developing domain ontologies, foundational aspects count as much as the domain aspects. Thus, for building well-founded and domain compliant ontologies, it is essential to reuse both aspects. This reuse can be achieved by applying foundational and domain-related pattern in combination.

As future work, we are studying a proper way to include FOPs into OPLs, improving the language guidance for building domain ontologies. Some efforts are also invested on evolving some aspects of the OLED tool, mainly for integrating different patterns libraries and improving the model construction usability. We are also planning an empirical study to demonstrate the advantages of this pattern-based approach in which FOPs and DROPs are combined, both in terms of the productivity gained and in terms of the cognitive tractability of the resulting models.

Acknowledgements. This research is funded by the Brazilian Research Funding Agency CNPq (Processes 485368/2013-7 and 461777/2014-2).

References

1. Noppens, O., Liebig, T.: Ontology patterns and beyond towards a universal pattern language. In: WOP (2009)
2. Scherp, A., Saathoff, C., Franz, T., Staab, S.: Designing core ontologies. Appl. Ontology **6**, 177–221 (2011)
3. Guarino, N.: Formal ontology and information systems. In FOIS 1998, vol. 46 (1998)
4. de Almeida Falbo, R., Barcellos, Monalessa Perini, Nardi, J.C., Guizzardi, Giancarlo: Organizing ontology design patterns as ontology pattern languages. In: Cimiano, P., Corcho, O., Presutti, V., Hollink, L., Rudolph, S. (eds.) ESWC 2013. LNCS, vol. 7882, pp. 61–75. Springer, Heidelberg (2013)
5. Gangemi, A., Presutti, V.: Ontology design patterns. In: Staab, S., Studer, R. (eds.) Handbook on Ontologies, Second, pp. 221–243. Springer, Berlin (2009)

6. Blomqvist, E., Gangemi, A., Presutti, V.: Experiments on pattern-based ontology design. In: Proceedings of 5th International Conference on Knowledge Capture, K-CAP 2009 (2009)
7. Presutti, V., Daga, E., Gangemi, A., Blomqvist, E.: eXtreme design with content ontology design patterns. In: Proceedings of Workshop on Ontology Patterns (WOP 2009) (2009)
8. Guizzardi, G.: Ontological Foundations for Structural Conceptual Models. Telematica Instituut Fundamental Research Series, Enschede (2005)
9. Guizzardi, G.: Ontological patterns, anti-patterns and pattern languages for next-generation conceptual modeling. In: Yu, E., Dobbie, G., Jarke, M., Purao, S. (eds.) ER 2014. LNCS, vol. 8824, pp. 13–27. Springer, Heidelberg (2014)
10. Falbo, R.A., Guizzardi, G., Gangemi, A., Presutti, V.: Ontology patterns: clarifying concepts and terminology. In: Proceedings of the 4th Workshop on Ontology and Semantic Web Patterns, Sidney, Australia (2013)
11. Falbo, R.A., Ruy, F.B., Guizzardi, G., Barcellos, M.P., Almeida, J.P.A.: Towards an enterprise ontology pattern language. In: Proceedings of the 29th Annual ACM Symposium on Applied Computing - SAC 2014, pp. 323–330 (2014)
12. U.S. Department of Defense (DoD), Data Modeling Guide (DMG) For An Enterprise Logical Data Model, V2.3, 15 March 2011
13. Guizzardi, G., Lopes, M., Baião, F., Falbo, R.: On the importance of truly ontological distinctions for ontology representation languages: an industrial case study in the domain of oil and gas. In: Halpin, T., Krogstie, J., Nurcan, S., Proper, E., Schmidt, R., Soffer, P., Ukor, R. (eds.) Enterprise, Business-Process and Information Systems Modeling. LNBIP, vol. 29, pp. 224–236. Springer, Heidelberg (2009)
14. Maiden, N.A., Sutcliffe, A.G.: Exploiting reusable specifications through analogy. Commun. ACM 35(4), 55–64 (1992)
15. Mattson, M., Bosch, J., Fayad, M.: Framework integration problems, causes, solutions. Commun. ACM 42(10), 80–87 (1999)
16. Olivé, A.: Conceptual Modeling of Information Systems. Springer Science & Business Media, Berlin (2007)
17. Guizzardi, G.: Ontological meta-properties of derived object types. In: Ralyté, J., Franch, X., Brinkkemper, S., Wrycza, S. (eds.) CAiSE 2012. LNCS, vol. 7328, pp. 318–333. Springer, Heidelberg (2012)
18. Varzi, A.: Mereology. In: Zalta, E.N. (ed.) The Stanford Encyclopedia of Philosophy, Spring 2015 (2015)
19. Uschold, M., King, M., Moralee, S., Zorgios, Y.: The enterprise ontology. Knowl. Eng. Rev. 13(01), 31–89 (1998)
20. Uschold, M., McComb, D.: Introduction to Gist. IAOA. http://iaoa.org/isc2014/uploads/Whitepaper-Uschold-IntroductionToGist.pdf. Accessed 2013

Towards a Service Ontology Pattern Language

Glaice K. Quirino[1(✉)], Julio C. Nardi[2], Monalessa P. Barcellos[1], Ricardo A. Falbo[1],
Giancarlo Guizzardi[1], Nicola Guarino[3], Mario Bochicchio[4], Antonella Longo[4],
Marco Salvatore Zappatore[4], and Barbara Livieri[4]

[1] Federal University of Espírito Santo, Vitória, Brazil
{gksquirino,monalessa,falbo,gguizzardi}@inf.ufes.br
[2] Federal Institute of Espírito Santo, Campus Colatina, Colatina, ES, Brazil
julionardi@ifes.edu.br
[3] Laboratory for Applied Ontology, Institute of Cognitive Sciences and Technologies,
Trento, Italy
nicola.guarino@cnr.it
[4] University of Salento, Lecce, Italy
{mario.bochicchio,antonella.longo,marcosalvatore.zappatore,
barbara.livieri}@unisalento.it

Abstract. In this paper we partially present an initial version of an Ontology
Pattern Language, called S-OPL, describing the core conceptualization of services as a network of interconnected ontology modeling patterns. S-OPL builds on
a commitment-based core ontology for services (UFO-S) and has been developed
to support the engineering of ontologies involving services in different domains.
S-OPL patterns address problems related to the distinction of general kinds of
customers and providers, service offering, service negotiation and service
delivery. In this paper, we focus on the first two. The use of S-OPL is demonstrated
in a real case study in the domain of Information and Communication Technology
services.

Keywords: Service · Ontology pattern · Ontology pattern language

1 Introduction

Ontologies have been recognized as a useful instrument for reducing conceptual ambiguities and inconsistencies [1–3]. Ontology solutions are gaining importance in all
activities that require a deep understanding of an enterprise and the business sector in
which it operates [4]. Thus, ontologies, as reference models, may be useful to share the
conceptualization inherent to business models within the company (e.g., among divi-
sions and departments, favoring communication between Business/IT), as well as, to
some extent, outside the company, when business models are communicated to third
parties (e.g., in communication to investors and business partners).

Building ontologies, however, has not been considered an easy task. In this context,
reuse is currently recognized as an important practice for Ontology Engineering, and
pattern-oriented approaches are promising for supporting ontology reuse [5].

© Springer International Publishing Switzerland 2015
P. Johannesson et al. (Eds.): ER 2015, LNCS 9381, pp. 187–195, 2015.
DOI: 10.1007/978-3-319-25264-3_14

An ontology pattern describes a particular recurring modeling problem that arises in specific ontology development contexts and presents a well-proven solution [6].

Ontology Pattern Languages (OPLs) are networks of interconnected ontology modeling patterns that provide holistic support for solving ontology development problems in a given field. An OPL is not a mere catalogue of patterns. It offers a set of interrelated patterns, plus a process guiding how to use these patterns, how to combine them in a specific order and suggesting one or more patterns for a given modeling problem. An OPL addresses several types of relationships among patterns, such as: dependence, temporal precedence of application, and mutual exclusion [7]. Thus, an OPL provides explicit guidance on how to reuse and integrate related patterns into a concrete ontology conceptual model. In summary, an OPL gives concrete guidance for developing ontologies, addressing at least the following issues [7]: (i) What are the key problems to solve in the domain of interest? (ii) In what order should these problems be tackled? (iii) What alternatives exist for solving a given problem? (iv) How should the dependencies between problems be handled? (v) How to resolve each individual problem most effectively in the presence of its correlated problems?

In this paper, we present part of the initial version of S-OPL (Service OPL), an OPL that can be used for building service domain ontologies. S-OPL's ontology patterns were extracted from UFO-S [2], a commitment-based core ontology for services. As a *core ontology* [8], UFO-S favors ontology patterns extraction [9], since it presents general concepts that span across several applications domains (e.g., education, transportation, and insurance) and can be reused when ontologies are developed to these domains. UFO-S is grounded on the Unified Foundational Ontology (UFO) [10], and it is represented in OntoUML [10], an UML profile that incorporates the foundational distinctions of UFO. By being based on UFO-S, the patterns of S-OPL are also grounded on such foundational distinctions. Moreover, by using OntoUML, it is possible to count on a well-maintained ontology engineering apparatus that can be applied for building service-related domain ontologies. Such apparatus includes model verification and validation via visual simulation [11], as well as model transformation to languages such as OWL (Web Ontology Language) [12] and support for ontology patterns application [9].

In UFO-S, service relations are characterized by the commitments and claims established between service participants, which drives the actions in service delivery. Following the modular structure of UFO-S, S-OPL comprises patterns covering four main areas: (i) *Service Offering*, which includes patterns to model a service offering to a target community; (ii) *Provider and Target Customer*, which deals with defining types of service providers and target customers; (iii) *Service Negotiation*, which deals with the negotiation between provider and customer in order to get an agreement; and (iv) *Service Delivery*, which models aspects related to the actions performed for fulfilling a service agreement. In this paper, we focus on the first two groups of patterns.

The remainder of this paper is organized as follows. Section 2 partially presents S-OPL. Section 3 demonstrates the use of S-OPL by discussing a real case study in the Information and Communication Technology (ICT) domain. Section 4 presents the final considerations of the paper.

2   S-OPL: A Service Ontology Pattern Language

S-OPL comprises a set of ontology patterns plus a process describing how to combine them in order to build a service domain ontology (an ontology about services in a specific application domain). Such patterns are organized in four groups: *Service Offering, Provider and Target Customer, Service Negotiation and Agreement,* and *Service Delivery.* Due to space limitation, only the patterns of the two first groups are presented.

The *Service Offering* group consists of five patterns, as Fig. 1 shows. The main pattern in this group is *SOffering,* which deals with the problem of modeling a service offering made by a service provider (e.g., a car rental company "XCompany") to a *Target Customer Community* (e.g., community of people and organizations that can rent cars). *Target Customer Community* is the collective of agents that constitute the community to which the service is being offered. *Target Customer,* in turn, is the role played by the agents when they become members of a target customer community. The *TCCMembership* pattern addresses this membership relation between *Target Customer Community* and *Target Customer.* Finally, the *SDescription* pattern allows describing a service offering by means of *Service Offering Descriptions,* such as folders, registration documents in a chamber of commerce, and so on.

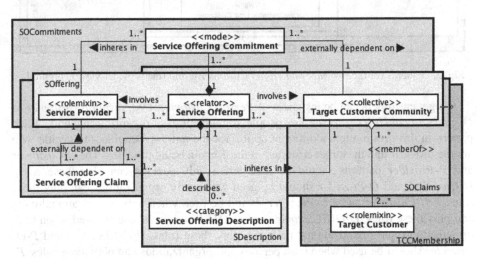

Fig. 1. *Service Offering* group.

A service offering is composed of *Service Offering Commitments* from the service provider towards the target customer community (e.g., "to grant temporary use of a vehicle to the customer") and *Service Offering Claims* from target customer community towards the service provider (e.g., "having a car available with a tank full of fuel"). *SOCommitments* and *SOClaims* patterns address, respectively, these aspects.

Service Provider and *Target Customer* are roles (in fact, technically, they are *role mixins,* i.e., roles that are played by entities of different *kinds* [10]). They can be played by a *Person,* an *Organization,* or an *Organizational Unit.* These different types of providers and target customers are addressed by the patterns of the *Provider and Target*

Customer group. Figure 2 shows the patterns of this group that describe the types of *Target Customer*. Since the patterns addressing types of providers are analogous to the ones addressing target customer types, we omit them here. The prefix of pattern names indicates the types of agents that play the roles of *Provider* or *Target Customer*: *P = Person, O = Organization, OU = Organizational Unit*.

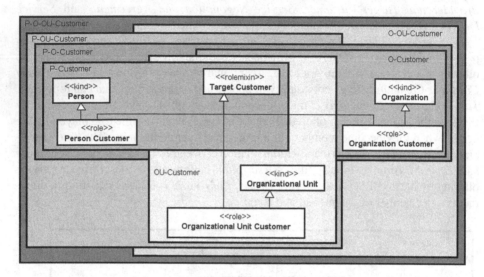

Fig. 2. Patterns for the *Provider and Target Customer* group (the provider view is omitted).

Target customers can be instances of *Person Customer, Organization Customer* or *Organizational Unit Customer*, i.e., people, organizations or organizational units. Each pattern in this group offers a different option for the ontology engineer to decide who are the provider and the target customer in the domain being modeled. The *P-Customer* and *P-Provider* patterns should be used when only persons can play these roles. *O-Customer* and *O-Provider* should be used when only organizations can play these roles. *OU-Customer* and *OU-Provider* should be used when only organizational units can play these roles. *O-OU-Customer* and *O-OU-Provider* should be used when both organizations and organizational units can play these roles. *P-O-Customer* and *P-O-Provider* should be used when both persons and organizations can play these roles. *P-OU-Customer* and *P-OU-Provider* should be used when both persons and organizational units can play these roles. Finally, *P-O-OU-Customer* and *P-O-OU-Provider* should be used when any of these types of entities (persons, organizations and organizational units) can play these roles. The patterns *P-Customer, O-Customer, OU-Customer, O-OU-Customer, P-O-Customer, P-OU-Customer* and *P-O-OU-Customer* are alternatives, i.e., the ontology engineer should select and use only one of them. The same occurs with the corresponding patterns related to *Provider*.

As previously discussed, an OPL includes, besides the patterns, a process suggesting an order in which they can be applied. Figure 3 presents the fragment of the S-OPL process used in this paper, which is represented by means of an extended UML activity diagram as proposed in [7]. In that figure, patterns are represented by action nodes (the labeled rounded rectangles); patterns groups are delimited by blue lines; initial nodes (solid circles) are used to represent entry points in the OPL, i.e., patterns in the language that can be used without using other patterns first. Control flows (arrowed lines) represent the admissible sequences in which patterns can be used. Fork nodes (line segments with multiple output flows) are used to represent independent and possibly parallel paths. Join nodes (line segments with multiple input flows) are used to represent path junctions. Dotted lines delimit variant patterns, i.e., a set of patterns from which it is necessary to select only one to be used. Patterns in grey and thick black lines are the patterns used and paths followed in the example discussed in Sect. 3.

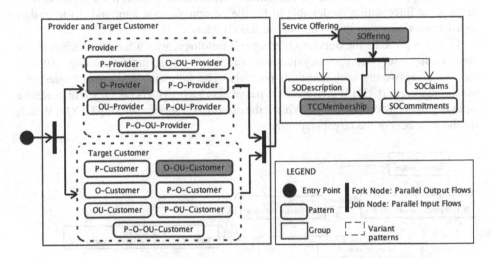

Fig. 3. The fragment of S-OPL process used in this paper (Color figure online).

As Fig. 3 shows, S-OPL has only one entry point. Thus, the ontology engineer must start developing the specific service domain ontology by deciding which types of providers and target customers are involved in the service being modeled. As previously discussed, providers and target customers can be people, organizations or organizational units. Therefore, the ontology engineer must select one of the patterns of the *Provider* sub-group and one of the patterns of the target customer sub-group. Once the types of providers and target customers are modeled, the ontology engineer can focus on the service offering. The next pattern to be used is *SOffering*. Next, the following patterns can be used: *TCCMembership*, for modeling the members of the Target Customer Community; *SOCommitments* and *SOClaims*, if the ontology engineer is interested in modeling service offering commitments and claims, respectively; and *SODescription*, if the ontology engineer is interested in the problem of describing the service offering by means of a service offering description.

3   Applying S-OPL: A Case Study

In this section, we shortly present a case study applying S-OPL to develop a service ontology in a particular application domain: the Email Service Ontology (ESO). The case is concerned with an Email Service internally delivered to a big Italian company with more than 5000 employees spread out into more than 100 offices all over the country. The IT Department of the company is responsible for providing this Email Service. For doing this, the department hires two underpinning ICT (Information and Communication Technology) services provided by two different organizations: the "emailbox service" and the "networking service". ESO was developed for a two-fold reason: (i) to integrate and make consistent the stakeholders' perspectives on the service, defining an information model able to fill the communication gap; and (ii) to design appropriate views over the model for the stakeholders.

Figure 4 presents the Service Offering sub-ontology, which has been achieved by first eliciting a set of competency questions and then by selecting the ontology patterns capable of answering such questions properly. The following competency questions were considered: (CQ1) Which are the types of service providers? (CQ2) Which are the types of target customers? (CQ3) What is the established service offering? (CQ4) Which are the members of the target community?

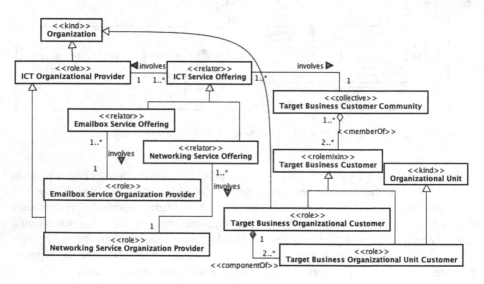

Fig. 4. Service offering sub-ontology.

The first competency question (CQ1) refers to the available types of service providers. Emailbox and networking services are offered only by organizations. Thus, the *O-Provider* pattern was used, originating the following types of service providers: *Emailbox Service Organization Provider* and *Networking Service Organization Provider*, both subtypes of *ICT Organizational Provider*.

Regarding target customers (CQ2), we used the *O-OU-Customer*, since in this specific case, offerings of ICT services are made only to organizations and organizational units (e.g., departments). Thus, *Target Business Organizational Customer* and *Target Business Organizational Unit Customer* are both *Target Business Customers*.

In order to model the service offering established between the service provider and the target customer community (CQ3), the *S-Offering* pattern was applied. Thus, we defined that the *Emailbox Service Organization Provider* is involved in the *Emailbox Service Offering*, whilst the *Networking Service Organization Provider* is involved in the *Networking Service Offering*. *Emailbox Service Offering* and *Networking Service Offering* are types of *ICT Service Offerings*. An *ICT Services Offering* is a relator involving an *ICT Organizational Provider* and a *Target Business Customer Community*.

Finally, for answering the fourth competency question (CQ4), we used the *TCCMembership pattern*. By applying this pattern, we captured that *Target Business Customers* are members of a *Target Business Customer Community*.

4   Final Considerations

The main contribution of this paper is to present part of the initial version of Service Ontology Pattern Language (S-OPL). More than an additional concrete experience with the approach proposed in [7], S-OPL is a significant contribution in itself, given the importance of the service field.

To the best of our knowledge, S-OPL is the first proposal for an Ontology Pattern Language (OPL) for service modeling. Moderately related to our approach, however, we can discuss works concerning service modeling. Service Ontology of Oberle et al. [13] proposes a number of core modules that can be extended by new ontology models addressing different applications domains (e.g., healthcare and automotive). Differently from S-OPL, this approach does not propose a service modeling pattern language, with a set of high-granularity primitives (patterns) and a modeling process. There are also in the literature proposals for ontology-based modeling of services. These include the OBELIX Service Ontology and its related tools for graphical modeling of services and for knowledge-based configuration of service bundles [14]. OBELIX takes the *service as a whole* as its reuse unit, i.e., as building blocks that can be combined to form service bundles. Our proposal, in contrast, uses self-contained building blocks (organized as fine-grained patterns and guided by the S-OPL process) that address the representation of different aspects of service relations (e.g., service offering, and service agreement). These patterns are to be used in tandem for building a service ontology in a specific application domain.

From the case study, we noticed that the use of S-OPL, whose patterns were extracted from a well-founded core ontology (UFO-S), tends to bring the following benefits: (i) the resulting ontology tends to contain fewer inconsistency problems given that many of the potentially recurring source of inconsistencies tend to be solved by the basic patterns of the core ontology; (ii) the development process of the derived domain-specific service ontologies tends to be accelerated by the massive reuse of modeling fragments; and (iii) S-OPL guides pattern selection, also facilitating their combination.

Despite that, some limitations were identified, such as: (a) the presence of only one entry point for the OPL and (b) the insufficient account of resources, which constitute a strategic aspect for some services (e.g., in cloud services dynamic resource allocation is a key aspect) and (c) the lack of distinction between contractual relationship and factual relationship, as a way to consider the relationships both at the contractual layer and at the operative layer. Moreover, (d) there is the need to extend the service lifecycle in order to account for service discovery and service dismission. Finally, (e) the concept of core action, as defined in [15], is missing.

The latter point gains importance in the case of instrumental services. For instrumental services, the offer of the provider does not consist in an action (e.g., cutting your hair), but rather in allowing the possibility for the user to perform a given action, which constitute the *core action* of the service. For instance, in our case study, the providers offer the Internet connection and the mailbox application, with whom they guarantee to the users that they can send/receive or manage emails. In this frame, the provider performs *supporting actions* apt at enabling the core service consumption [16]. In other words, although the actions are guaranteed by the provider, they are executed by the user.

As ongoing future works, we intend to enlarge S-OPL by adding new patterns (e.g., patterns that deal with relationships between services and business goals) and to define new entry points in S-OPL process for making it more flexible. Moreover, new applications and evaluations of S-OPL are intended to be conducted to provide relevant feedback for further improvements.

Acknowledgments. This research is funded by the Brazilian Research Funding Agency CNPq (Processes 485368/2013-7 and 461777/2014-2).

References

1. Mora, M., Raisinghani, M., Gelman, O., Sicilia, M.A.: Onto-servsys: a service system ontology. In: Demirkan, H., Spohrer, J.C., Krishna, V. (eds.) The Science of Service Systems, pp. 151–173. Springer, Berlin (2011)
2. Nardi, J.C., de Almeida Falbo, R., Almeida, J.P.A., Guizzardi, G., Pires, L.F., Van Sinderen, M., Guarino, N.: Towards a commitment-based reference ontology for services. In: 17th IEEE International Enterprise Distributed Object Computing Conference (EDOC), pp. 175–184 (2013)
3. Milosevic, Z., Bojicic, S., Rossman, M., Walker, M.: Healthcare SOA ontology (release 1) (2013)
4. Case, G.: ITIL v3. Continual service improvement (2007)
5. Buschmann, F., Henney, K., Schimdt, D.: Pattern-Oriented Software Architecture: On Patterns and Pattern Language. Wiley, New York (2007)
6. de Almeida Falbo, R., Guizzardi, G., Gangemi, A., Presutti, V.: Ontology patterns: clarifying concepts and terminology. In: Proceedings of the 4th Workshop on Ontology and Semantic Web Patterns, Sydney, Australia (2013)
7. de Almeida Falbo, R., Barcellos, M.P., Nardi, J.C., Guizzardi, G.: Organizing ontology design patterns as ontology pattern languages. In: Cimiano, P., Corcho, O., Presutti, V., Hollink, L., Rudolph, S. (eds.) ESWC 2013. LNCS, vol. 7882, pp. 61–75. Springer, Heidelberg (2013)

8. Scherp, A., Saathoff, C., Franz, T., Staab, S.: Designing core ontologies. Appl. Ontol. **6**, 177–221 (2011)
9. Ruy, F.B., Reginato, C.C., Santos, V.A., de Almeida Falbo, R., Guizzardi, G.: Ontology engineering by combining ontology patterns. In: 34th International Conference on Conceptual Modeling (ER 2015), Stockholm, Sweden (2015)
10. Guizzardi, G.: Ontological Foundations for Structural Conceptual Models. Universal Press, The Netherlands (2005)
11. Guizzardi, G.: Ontological patterns, anti-patterns and pattern languages for next-generation conceptual modeling. In: Yu, E., Dobbie, G., Jarke, M., Purao, S. (eds.) ER 2014. LNCS, vol. 8824, pp. 13–27. Springer, Heidelberg (2014)
12. Guizzardi, G., Zamborlini, V.: Using a trope-based foundational ontology for bridging different areas of concern in ontology-driven conceptual modeling. Sci. Comput. Program. **96**, 417–443 (2014)
13. Oberle, D., Bhatti, N., Brockmans, S., Janiesch, C.: Countering service information challenges in the internet of services. J. Bus. Inf. Syst. Eng. **1**(5), 370–390 (2009)
14. Akkermans, H., Baida, Z., Gordijn, J., Pena, N., Altuna, A., Laresgoiti, I.: Value webs: using ontologies to bundle real-world services. IEEE Comput. Soc. **19**, 57–66 (2004)
15. Ferrario, R., Guarino, N.: Commitment-based modeling of service systems. In: Snene, M. (ed.) IESS 2012. LNBIP, vol. 103, pp. 170–185. Springer, Heidelberg (2012)
16. Ferrario, R., Guarino, N., Fernández-Barrera, M.: Towards an ontological foundation for services science: The legal perspective. In: Sartor, G., Casanovas, P., Biasiotti, M., Fernandez-Barrera, M. (eds.) Approaches to Legal Ontologies, pp. 235–258. Springer, Netherlands (2011)

Constraints

Incremental Checking of OCL Constraints with Aggregates Through SQL

Xavier Oriol[(✉)] and Ernest Teniente

Department of Service and Information System Engineering,
Universitat Politcnica de Catalunya – BarcelonaTech, Barcelona, Spain
{xoriol,teniente}@essi.upc.edu

Abstract. Valid states of data are those satisfying a set of constraints. Therefore, efficiently checking whether some constraint has been violated after a data update is an important problem in data management. We tackle this problem by incrementally checking OCL constraint violations by means of SQL queries. Given an OCL constraint, we obtain a set of SQL queries that returns the data that violates the constraint. In this way, we can check the validity of the data by checking the emptiness of these queries. The queries that we obtain are incremental since they are only executed when some relevant data update may violate the constraint, and they only examine the data related to the update.

Keywords: Constraints Checking · SQL · OCL · Aggregates

1 Introduction

A conceptual schema is the formal specification of an information system in terms of the structure of the data to be stored together with the operations applicable to modify such data. The specification of the data structure includes several integrity constraints, i.e., some conditions that any state of the data should satisfy in order to be valid.

One of the most used languages for specifying conceptual schemas is UML, a standard language maintained by the OMG [1]. In UML, the structure of the data is mainly defined by a taxonomy of classes/associations (i.e., a class diagram), complemented with several textual integrity constraints written in OCL [2]. This leads to the problem of efficiently checking whether the current data of a running information system truly satisfies the OCL constraints defined in its UML schema. This is an important problem in data management since any violation of an integrity constraint would indicate an invalid state of the data.

Consider, for instance, the UML schema in Fig. 1. This schema specifies an information system storing data about *movies* and the *people* participating in them as *cast members*, where each cast member exercises a *role* (e.g. director, actor, actress, etc.).

We are clearly interested to ensure that data over such schema satisfies a set of integrity constraints. For instance, there should not be any cast member

© Springer International Publishing Switzerland 2015
P. Johannesson et al. (Eds.): ER 2015, LNCS 9381, pp. 199–213, 2015.
DOI: 10.1007/978-3-319-25264-3_15

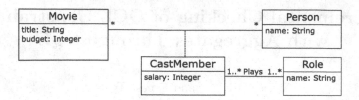

Fig. 1. Simplified UML class diagram for the IMDb Information System

playing the role of *actor* and *actress* at the same time; the sum of any movie budget should be higher than the salaries of its cast members; and there should be at least one cast member playing the role of *director* for each movie. These constraints are defined in OCL in Fig. 2. Note that aggregate operations such as sum or count are used in the constraint definitions.

```
context CastMember inv NotActorAndActress:
self.role.name->excludes('actor') or self.role.name->excludes('actress')
context Movie inv BudgetIsHigher:
self.budget >= self.castMember.salary->sum()
context Movie inv HasSomeDirector:
self.castMember.role.name->count('director') > 0
```

Fig. 2. OCL Constraints for the simplified UML class diagram of IMDb

Several techniques have been proposed so far to efficiently checking the violation of OCL constraints. Some of them handle the problem by translating the constraints into SQL [3,4]. For example, the work in [3] builds some SQL views that return the data violating the constraints. Thus, a constraint is violated if its corresponding view is not empty. Other techniques follow a different approach aimed at incrementally checking the OCL constraints [5–7]. That is, assuming that no OCL constraint is violated for the current data, they determine which OCL constraints should be checked, and for which data, after some update is applied to the information system.

The method we present in this paper follows a combination of these approaches and so we take advantage of both of them. On the one hand, the idea is to translate OCL constraints into SQL queries, thus, benefiting from DBMS optimizations such as query planners, different join algorithms, indexes, caches, etc. but stating the query in terms of current data and data being inserted/deleted. In this way, the query we define is only executed when a data update matches it, hence, when the update potentially violates its corresponding constraint. On the other hand, the query only searches for violations in the current data that joins with the update. In this way, for example, we do not need to recheck all the constraints for all the movies when some update is applied in only a few of them.

As far as we know, there is only another proposal following a similar app-roach to ours [8]. However, the core of this work is based on the RETE algorithm, which encompasses materializing any relational algebra operation performed by the queries. In this sense, the RETE algorithm has been criticized because of its combinatorially explosive materialized data growth and the execution time it might take to maintain such materialization [9]. On the contrary, our pro-posal only materializes the aggregated values accessed by the constraints (e.g. sum, size), avoiding in this way the inefficiencies caused by such intermediate materializations.

This paper extends our previous work in [10], where we outlined how to perform incremental checking of OCL constraints for a specific fragment of OCL, in the following directions:

- *Dealing with Aggregation Operations.* We extend our approach to be able to deal with distributive aggregation functions of OCL (i.e., sum, count, size).
- *Expressiveness Analysis.* We analyze the expressiveness of the OCL our method can deal with and we show that, because of dealing with aggrega-tion, we can handle constraints beyond the OCL$_{FO}$ fragment of OCL [11].
- *Experimentation.* We have made several experiments using real data extracted from the public interface of IMDb[1] information system. With these experi-ments we show the scalability of our approach in realistic conditions.

2 Basic Concepts and Notation

Terms, Atoms and Literals. A *term* t is either a variable or a constant. An *atom* is formed by a n-ary *predicate* p together with n terms, i.e., $p(t_1, ..., t_n)$. We may write $p(\bar{t})$ for short. If all the terms $\bar{t}$ of an atom are constants, we call the atom to be *ground*. A literal l is either an atom $p(\bar{t})$, a negated atom $\neg p(\bar{t})$, or a built-in literal $t_i \; \omega \; t_j$, where ω is an arithmetic comparison (i.e., $<,\leq,=,\neq$).

Derived/Base/Aggregate Predicates. A predicate p is said to be *derived* if the boolean evaluation of an atom $p(\bar{t})$ depends on some derivation rules, otherwise, it is said to be *base*. A *derivation rule* has the form: $\forall \bar{t}. \; p(\overline{t_p}) \leftarrow \phi(\bar{t})$ where $\overline{t_p} \subseteq \bar{t}$. In the formula, $p(\overline{t_p})$ is an atom called the *head* of the rule and $\phi(\bar{t})$ is a conjunction of literals called the *body*. We restrict all derivation rules to be safe (i.e., any variable appearing in the head or in a negated or built-in literal of the body also appears in a positive literal of the body) and non-recursive. Given several derivation rules with predicate p in its head, $p(\bar{t})$ is evaluated to true if and only if one of the bodies of such derivation rules is evaluated to true.

An aggregate predicate p_a (aka aggregate query/rule [12–14]) is a predi-cate defined over some predicate p that aggregates one of the terms of p with some aggregation function f. An aggregate predicate is defined by means of a rule: $\forall \bar{t}. \; p_a(\overline{t_p}, f(x)) \leftarrow p(\bar{t})$ where $\overline{t_p} \subseteq \bar{t}$ and $x \in \bar{t}$. An atom $p_a(\overline{t_p}, x_f)$ eval-uates to true if and only if x_f equals to aggregating all values x in $p(\bar{t})$ by

---

[1] www.imdb.com.

means of f. E.g. given the aggregate predicate $sumSalaries(e, x)$ defined by $sumSalaries(e, sum(s)) \leftarrow salary(e, s)$, $sumSalaries(e, x)$ evaluates to true if and only if x is equal to the sum of all salaries s such that $salary(e, s)$.

We also extend the notion of base/derived/aggregate predicate to atoms and literals. I.e., when the predicate of some atom/literal is base/derived/aggregate, we say that such atom/literal is base/derived/aggregate respectively.

Logic Formalization of the UML Schema. As proposed in [15] we formalize each class C in the class diagram with attributes $\{A_1, \ldots, A_n\}$ by means of a base atom $c(Oid)$ together with n atoms of the form $cA_i(Oid, A_i)$, each association R between classes $\{C_1, \ldots, C_k\}$ by means of a base atom $r(C_1, \ldots, C_k)$, and, similarly, each association class R between classes $\{C_1, \ldots, C_k\}$ with attributes $\{A_1, \ldots, A_n\}$ by means of a base atom $r(Oid, C_1, \ldots, C_k)$ together with n atoms $rA_i(Oid, A_i)$.

Structural Events. A *structural event* is an elementary change in the population of a class or association of the schema [16]. That is, a change in the contents of the data. We consider six kinds of structural events: class instance insertion/deletion, association instance insertion/deletion and attribute instance insertion/deletion. Attribute updates are simulated by means of a simultaneous deletion and insertion of the old and new value respectively.

We denote insertions by ι and deletions by δ. Given a base atom $p(\overline{x})$, insertion structural events are formally defined by $\forall \overline{x}. \ \iota p(\overline{x}) \leftrightarrow p^n(\overline{x}) \wedge \neg p(\overline{x})$, while deletion structural events by $\forall \overline{x}. \ \delta p(\overline{x}) \leftrightarrow p(\overline{x}) \wedge \neg p^n(\overline{x})$, where p^n stands for predicate p evaluated in the new data state, i.e., the one obtained after applying the change.

Aggregate Events. An *aggregate event* is a change in the value of some aggregate predicate. Aggregate events are determined by the structural events applied in a data state. Similarly as before, we denote increases of aggregate events by means of ι and decreases by means of δ.

Dependencies. A *tuple-generating dependency* (TGD) is a formula of the form $\forall \overline{x}, \overline{z}. \ \varphi(\overline{x}, \overline{z}) \rightarrow \exists \overline{y}. \ \psi(\overline{x}, \overline{y})$. A *denial constraint* is a special type of TGD of the form $\forall \overline{x}. \ \varphi(\overline{x}) \rightarrow \bot$, in which the conclusion only contains the $\bot$ atom, which cannot be made true, and the premise may contain positive, negated and built-in literals. An *event dependency constraint* is a special type of denial constraint containing at least one positive event atom.

3   Our Approach

We follow a two-steps approach to translate, in compilation time, a set of OCL constraints to SQL queries. In the first step, each OCL constraint is translated into a set of event dependency constraints (EDCs). An EDC is a logic rule identifying a particular situation where some structural events applied to a certain state of the data will cause the violation of the OCL constraint.

In the second step, each EDC is translated into a different SQL query. We assume a mapping from the EDC predicates to SQL tables for this purpose.

Roughly speaking, base predicates representing UML classes/associations are mapped to SQL tables containing their instances, structural event predicates are mapped to auxiliary SQL tables containing the structural events being applied, and aggregate predicates are mapped to auxiliary SQL tables containing a materialization of the relevant aggregated values. Given this mapping, the translation from an EDC to SQL results into a query joining the three: current data, current aggregated values, and structural events.

At runtime, our method works as follows: (1) an actor places the structural events he/she wants to apply in the auxiliary structural event tables; (2) our method executes the queries for checking the violation of OCL constraints; (3) if the queries return the empty set, there is no violation and the method commits the structural events in the tables representing the data, and it incrementally updates the materialized aggregated values. On the contrary, if the query returns some data, there is some constraint violation and thus, the update is rejected.

The key for incrementality is the join in the SQL queries between structural events, current data and current aggregated values. First of all, any SQL query joining a structural event which is not applied (i.e., whose SQL table is empty) is immediately discarded. Therefore, we only check those constraints that can be violated according to the ongoing structural events. Second, the data considered by an SQL query during its execution is necessarily the data joining the structural events applied, thus, avoiding to look through all the database. Third, the materialized aggregated values avoid the need to recompute from scratch the necessary aggregations required to check a constraint, which might be expensive. Finally, our method guarantees that the materialized aggregated values can be updated incrementally, thus, with negligible time penalty.

In our previous work [10] we tackled efficient integrity checking for OCL_{UNIV} constraints[2], a specific fragment of OCL. We extend here our approach to handle OCL distributive aggregates, which extends the expressiveness of the OCL we can deal with beyond OCL_{FO} [11]. For the sake of self-containment of the paper, we start by reviewing the translation from OCL_{UNIV} constraints to SQL queries. Then, we extend our approach dealing with aggregates and we show how this extension allows us to deal with any OCL_{FO} constraint and beyond.

3.1 OCL_{UNIV} Translation into SQL

First, we encode each OCL_{UNIV} constraint as a logic denial according to [15]. Logic denials are written over the logic formalization of the UML schema which has been defined in Sect. 2. The logic formalization corresponding to our UML schema in Fig. 1 is:

$$movie(m),\ mTitle(m,t),\ mBudget(m,b),\ person(p),\ pName(p,n),$$
$$castMember(c,m,p),\ cmSalary(c,s),\ plays(c,r),\ role(r),\ rName(r,n)$$

---

[2] The name is due to the observation that any OCL expression in OCL_{UNIV} can be written as a logic formula where all variables are universally quantified.

We assume, without loss of generality, that instances of *Role* and *Person* can be identified by its *name*. Thus, we will use the name attribute as their OIDs. In this way, we can omit the *rName* and *pName* predicates.

Given the previous logic formalization, the translation of the OCL$_{\text{UNIV}}$ constraint *NotActorAndActress* is encoded into the following denial constraint:

$$plays(c, Actor) \land plays(c, Actress) \rightarrow \bot$$

In the rest of this section, we explain how to obtain the EDCs from denial constraints and how to transform them into SQL queries.

Obtaining EDCs for OCL$_{\text{UNIV}}$ Constraints. An event dependency constraint (EDC) identifies a particular situation in which the original OCL$_{\text{UNIV}}$ constraint would be violated in a data state D^n resulting from applying some set of structural events to some initial data state D. Therefore, each denial constraint obtained in the previous step will be translated into several EDCs, each one corresponding to a different way in which the constraint may be violated.

The main idea for obtaining the EDCs is to replace each literal in the denial constraint by the expression that evaluates it in the new state D^n. Positive and negative literals in the denial are handled in a different way according to the following formulas:

$$\forall \overline{x}.\ p^n(\overline{x}) \leftrightarrow (\iota p(\overline{x})) \lor (\neg \delta p(\overline{x}) \land p(\overline{x})) \tag{1}$$
$$\forall \overline{x}.\ \neg p^n(\overline{x}) \leftrightarrow (\delta p(\overline{x})) \lor (\neg \iota p(\overline{x}) \land \neg p(\overline{x})) \tag{2}$$

Rule 1 states that an atom $p(\overline{x})$ will be true in the new state D^n if its insertion structural event has been applied or if it was already true in the initial state D and its deletion structural event has not been applied. In an analogous way, rule 2 states that $p(\overline{x})$ will not hold in D^n if it has been deleted or if it was already false and it has not been inserted.

By applying the substitutions above, we get a set of EDCs that states all possible ways to violate a constraint by means of the possible structural events of the schema. From the previous denial constraint we get:

$$plays(c, Actor) \land \neg \delta plays(c, Actor) \land \iota plays(c, Actress) \rightarrow \bot \tag{3}$$
$$\iota plays(c, Actor) \land plays(c, Actress) \land \neg \delta plays(c, Actress) \rightarrow \bot \tag{4}$$
$$\iota plays(c, Actor) \land \iota plays(c, Actress) \rightarrow \bot \tag{5}$$

EDCs are grounded on the idea of *event rules* which were defined in [17] to perform integrity checking in deductive databases. In general, we will get $2^k - 1$ EDCs for each denial constraint dc, where k is the number of literals in dc.

Translating EDCs into SQL. Now, we translate EDCs into SQL. For this purpose, we require a mapping from logic predicates to SQL tables. Each literal from the EDC is translated into a table reference placed in the FROM clause of the query. We define an SQL JOIN for a table reference when the literal is positive and it has some variable in common with another previously translated literal. In contrast, we define an SQL *antijoin* (by means of a LEFT JOIN together a IS NULL condition) for negative literals. Built-in literals and constant bindings are translated in the WHERE clause. Following our previous example, we would translate EDC 3 as:

SELECT P_1.*cast*
FROM *Plays* AS P_1
    LEFT JOIN *del_Plays* AS dP_1 ON (P_1.*cast* = dP_1.*cast* and P_1.*role* = dP_1.*role*)
    JOIN *ins_Plays* AS iP_2 ON (iP_2.*cast* = P_1.*cast*)
WHERE P_1.*role* = '*actor*' and dP_1.*cast* IS NULL and iP_2.*role* = '*actress*'

Intuitively, the previous SQL query looks for those cast members in the table *Plays* such that: (1) their role is *actor*, (2) their role *actor* is not being deleted by the structural events, (3) some structural events are inserting the new role *actress* to them. When executing this query, the query planner specifies its start from the *ins_Plays* table, rather than *Plays*, since the cardinality of *ins_Plays* is expected to be much lower. In this way, the DBMS does not look through all current data (i.e., all data in *Plays*), but only to the data that joins the update. Moreover, if *ins_Plays* has no tuples, and thus the original OCL constraint cannot be violated, the query returns the empty set without accessing data because any join with no tuples trivially returns the empty set. In this way, our queries behave incrementally since they only look to the data related to the update, and only when the update may cause a violation.

3.2 OCL Aggregation Translation into SQL

We extend now the previous approach to deal with aggregates. There exist several kinds of aggregates according to the complexity to incrementally update them when some structural event is applied [18]. In this work, we focus on *distributive* aggregation. Intuitively, distributive aggregates are those that can be updated by taking into account the current aggregated value of the data, the aggregated value of the data inserted, and the aggregated value of the data deleted. The distributive aggregates of OCL are: sum, size and count.

As we did before, we first translate any OCL constraint into a logic denial constraint; then, we translate this denial constraint into several EDCs and, finally, we translate each EDCs into an SQL query.

OCL Aggregation Translation into Logic Denials. Any OCL aggregation expression is defined by means of a *source* (i.e., a navigation) and an *aggregation operation* (e.g. sum), where the resulting aggregate value is normally used in some arithmetic comparison. We translate the *source* of the expression following the same lines as [15], and from there, we use an *aggregate predicate* to aggregate the required value. Once we obtain the aggregated required value, we can define the built-in literal encoding the arithmetic comparison.

For instance, given the *BudgetIsHigher* constraint, the source of the OCL aggregation expression is *self. castMember. salary*. This expression is translated as the following conjunction of literals:

$$castMember(c, m, p) \land cmSalary(c, s)$$

From there, we can aggregate the salaries (i.e., the s term) by means of defining an aggregate predicate:

$$sumSalaries(m, sum(s)) \leftarrow castMember(c, m, p) \land cmSalary(c, s)$$

In this way, the atom *sumSalaries(m, x)*, indicates that the sum of salaries of the movie m is x. Thus, we can use x to check whether the sum of the movie salaries is greater than its budget:

$$mBudget(m, b) \land sumSalaries(m, x) \land b < x \rightarrow \perp$$

Obtaining EDCs for Denial Constraints with Aggregates. The most important issue for obtaining the EDCs in the presence of aggregate predicates relies on how to compute the aggregate value in the new data state D^n. We make use of two aggregate event predicates for this purpose: one for computing the aggregated value x_ι for the data being inserted, and another one for computing the aggregated value x_δ for the data being deleted. Since we focus on OCL distributive aggregation functions, we can guarantee that the aggregated value in the new state D^n equals to the current aggregated value x plus x_ι minus x_δ.

For instance, the previous denial constraint would be translated as:

$$\iota mBudget(m, b) \land sumSal(b, x) \land \iota sumSal(m, x_\iota) \land \delta sumSal(m, x_\delta) \land$$
$$b < x + x_\iota - x_\delta \rightarrow \perp \tag{6}$$
$$mBudget(m, b) \land \neg \delta mBudget(m, b) \land sumSal(m, x) \land \iota sumSal(m, x_\iota) \land$$
$$\delta sumSal(m, x_\delta) \land x_\iota \neq x_\delta \land b < x + x_\iota - x_\delta \rightarrow \perp \tag{7}$$

The first rule captures those violations occurring when newly inserted movie budgets with possibly some cast member salary updates are lower than the aggregated salaries in the new data sate. The second captures those violations that occur when updating the cast member salaries of those movies for which no budget update is applied.

These EDCs result from, first, replacing any base literal with the rules defined in 1 and 2 as before to obtain their evaluation in D^n. Then, we add, for any aggregate literal, two additional aggregate event literals computing the aggregated value x_ι and x_δ for the data being inserted/deleted respectively. Afterwards, we replace any occurrence of the aggregated value x for its value in D^n, that is $x + x_\iota - x_\delta$. Finally, we add a new built-in literal $x_\iota \neq x_\delta$ in the rule with no structural event literals for *mBudget* to ensure that there is, at least, some update in the aggregated value.

Now, we need to define the aggregate event predicates $\iota sumSal$ and $\delta sumSal$. Intuitively, for $\iota sumSal$ we want to sum the new salaries being added to the source $self.\,castMember.\,salary$. Again, we can compute the new instances added to the *source* by replacing their literals according to the formulas 1 and 2:

$$\iota sumSal(m, sum(s)) \leftarrow castMember(c, m, p) \wedge \neg \delta castMember(c, m, p) \wedge \iota cmSalary(c, s)$$
$$\iota sumSal(m, sum(s)) \leftarrow \iota castMember(c, m, p) \wedge cmSalary(c, s) \wedge \neg \delta cmSalary(c, s)$$
$$\iota sumSal(m, sum(s)) \leftarrow \iota castMember(c, m, p) \wedge \iota cmSalary(c, s)$$

Similarly, we can define $\delta sumSal$. In this case, we have to replace insertions by deletions since we are looking for instances which are deleted from the *source*:

$$\delta sumSal(m, sum(s)) \leftarrow castMember(c, m, p) \wedge \neg \delta castMember(c, m, p) \wedge \delta cmSalary(c, s)$$
$$\delta sumSal(m, sum(s)) \leftarrow \delta castMember(c, m, p) \wedge cmSalary(c, s) \wedge \neg \delta cmSalary(c, s)$$
$$\delta sumSal(m, sum(s)) \leftarrow \delta castMember(c, m, p) \wedge \delta cmSalary(c, s)$$

Note that, in both cases, the different rules form a partitioning of the instances being inserted/deleted in the *source* expression. In this way, we can compute the total aggregated value x_ι and x_δ by the sum of the aggregated values obtained from the various rules.

Translating EDCs with Aggregated Events to SQL. Translating EDCs with aggregates into SQL queries follows the same principles as before: each literal is translated as a table reference in the FROM clause possibly with a JOIN condition. For example, the rule EDC 6 is translated as:

SELECT $M.movie\_id, M.budget, sumSalaries.X+ins\_sumSalaries.X-del\_sumSalaries.X$
FROM $ins\_mBudget$ AS M
    LEFT JOIN $sumSalaries$ ON($M.movie\_id$ = $sumSalaries.movie\_id$)
    LEFT JOIN $ins\_sumSalaries$ ON($M.movie\_id$ = $ins\_sumSalaries.movie\_id$)
    LEFT JOIN $del\_sumSalaries$ ON($M.movie\_id$ = $del\_sumSalaries.movie\_id$)
WHERE $M.budget$ <$sumSalaries.X+ins\_sumSalaries.X-del\_sumSalaries.X$

where $sumSalaries$ is a table containing the materialized aggregation of the cast member salaries of the different movies, and $ins\_sumSalaries$ and $del\_sumSalaries$ are two views computing the aggregation of the cast member salaries being inserted and deleted for the different movies. Note that we need to use LEFT JOIN instead of JOIN in order not to lose those *movies* for which we do not have any of these aggregate values.

Now, we need to define the SQL views $ins\_sumSalaries$ and $del\_sumSalaries$. Such views are defined by means of translating into SQL the different definition rules of the predicates $\iota sumSalaries$ and $\delta sumSalaries$ specified in the EDCs. For instance, the first definition rule of $\iota sumSalaries$ would be translated as:

```
CREATE VIEW ins_sumSalaries1 AS
SELECT CM.movie_id, SUM(iCMS.salary) AS X
FROM castMember as CM
    LEFT JOIN del_castMember as dCM ON (CM.id = dCM.id)
    LEFT JOIN ins_cmSalary as iCMS ON (CM.id = iCMS.id)
WHERE dCM.id IS NULL
GROUP BY CM.movie_id
```

These views are obtained by translating the body of the rule into SQL following the same principles as before, then applying a GROUP BY with the attributes corresponding to the terms of the rule's head, and finally aggregating the corresponding term.

At this point, we only need to define a view combining all the different salary sums corresponding to the different definition rules. In the case of $\iota sumSalaries$ it would be:

```
CREATE VIEW ins_sumSalaries AS
SELECT iSS1.movie_id, iSS1.X + iSS2.X + iSS3.X AS X
FROM ins_sumSalaries1 as iSS1
    FULL OUTER JOIN ins_sumSalaries2 as iSS2 ON (iSS1.movie_id = iSS2.movie_id)
    FULL OUTER JOIN ins_sumSalaries3 as iSS3 ON (iSS1.movie_id = iSS3.movie_id)
```

Note the usage of FULL OUTER JOIN in order not to lose any of the aggregated values in the different $ins\_sumSalariesX$ views.

Note also that we use the views computing the aggregated value being inserted x_ι and the aggregated value being deleted x_δ to incrementally update the materialized aggregated value x. In our example, we use the views $ins\_sumSalaries$ and $del\_sumSalaries$ to update the table $sumSalaries$ in case we finally commit the structural events.

3.3 Expressiveness of OCL$_{UNIV}$ with Aggregation

The expressiveness of the OCL$_{UNIV}$ language is already determined by a formal grammar in [19]. Incorporating distributive aggregation in OCL$_{UNIV}$ extends the expressiveness of the language beyond the newly supported aggregate operations sum, size, count since there are many OCL operations beyond OCL$_{UNIV}$ that can be rewritten in terms of aggregation (e.g. notEmpty). To see to what extent we are improving the expressiveness of the OCL fragment, we first discuss the expressiveness in terms of logics, and then, go back to OCL.

From the point of view of logics, dealing with aggregate predicates allows us to handle negative derived literals, that is, denial constraints with the form $\phi(\overline{x}) \wedge \neg d(\overline{x}) \rightarrow \perp$ where d is a derived predicate. Note that negative derived literals can be encoded as an aggregate predicate counting the number of instances satisfying the body of the derived literal, and comparing such number to 0.

Since the work in [15] defines a translation from OCL_{FO} [11] into the language of denial constraints with derived negative literals, our method can deal with any OCL_{FO} constraint. OCL_{FO} is the OCL fragment limited to first order constructs, in other words, it encompasses almost any OCL operation in exception of aggregates (min, max, sum, count, size) and transitive closure (closure).

Nevertheless, with the method proposed here we are also able to deal with some of these aggregates. In particular, we can deal with sum, count, and size because they are distributive. In this manner, the expressiveness of the constraints we deal with is beyond OCL_{FO}.

4 Experiments

We have conducted some experiments to show the scalability of our approach. We have loaded the IMDb public interface available data (about $13*10^4$ movies and $2.5*10^6$ cast members) into an SQL schema corresponding to the UML schema shown in Fig. 1. Then, we have measured the execution time of our method to check the OCL constraints of Fig. 2 in three different scenarios: adding new movies, modifying salaries of cast members, and deleting movie directors. All these experiments have been conducted in MySQL 5.6 running on Windows 8 in an Intel i7-4710HQ up to 3.5 GHz machine with 8 GB of RAM.

For each scenario, we have executed our method several times increasing the number of structural events applied in each case. Note that inserting a movie requires several structural events: inserting the movie, its budget, its cast members, etc.; updating a salary requires two events: deleting the old salary and inserting the new one; and deleting a director from a movie requires three: deleting its cast membership, its role and its salary. In Table 1 we show the execution times in seconds for checking each constraint of the example regarding to the number of structural events applied to each scenario.

From these results we can see that the time to check any constraint increases with the number of movie insertions. This is because all three constraints can be violated in this scenario. Insertions of 1000 movie had better response times than those of inserting 500 due to the cache memories of MySQL. When analyzing salary updates, we see that only the constraint *BudgetIsHigher* gets worse results when increasing the number of events considered, while the other two remain almost constant. This is because it is impossible to violate them when updating salaries. The same phenomena occurs with the constraint *NotActorAndActress* in the third scenario since it is impossible to violate it by deleting cast members.

It is worth noting that most of the experiments took less than one second and that only one of them was over 30 s. Moreover, the cache memories improved the results of the last experiments with the largest number of structural events. In the case with most number of data changes (22,037 structural events), it took 12.37 s to check one constraint in a database with more than 3 million rows.

We also show in Table 2 the execution time in seconds required to update the materialized aggregates for each scenario once the constraint check has been performed. Note that updating the materialized aggregates does not suppose any scalability problem since none of them takes more than 0.5 s.

To show the benefits of our incremental approach, we have measured the execution times of the SQL queries obtained from the OCLDresden tool [3], which translates each constraint into an SQL query, but without following an incremental approach. In this case, the execution time to check *NotActorAndActress* was 21.47 s, while checking *HasSomeDirector* and *BudgetIsHigher* did not finish within two hours. We could improve these last execution times after manually rewriting the automatic translation provided by the tool, but their results were still hight: 238.33 s and 79.44 s. Note that, since this method is not incremental, its execution time is independent of the events applied, thus, it takes these times even when the events applied cannot violate any of the constraints.

Table 1. Time in seconds to check constraints

#Movie Insertions	1	5	10	50	100	500	1000
#Structural Events	28	149	272	1254	2059	10876	22037
NotActorAndActress	0.36	0.75	5.31	3.51	5.54	9.39	9.13
HasSomeDirector	0.28	0.08	0.33	0.41	0.56	1.42	0.38
BudgetIsHigher	0.41	0.90	5.37	5.54	9.82	30.34	12.37
#Salary Updates	**1**	**5**	**10**	**50**	**100**	**500**	**1000**
#Structural Events	2	10	20	100	200	1000	2000
NotActorAndActress	0.14	0.05	0.05	0.03	0.05	0.03	0.06
HasSomeDirector	0.31	0.00	0.00	0.02	0.02	0.00	0.00
BudgetIsHigher	0.17	0.09	0.30	0.69	1.16	1.00	1.44
#Director Deletions	**1**	**5**	**10**	**50**	**100**	**500**	**1000**
#Structural Events	3	15	30	150	300	1500	3000
NotActorAndActress	0.19	0.20	0.17	0.13	0.16	0.70	0.12
HasSomeDirector	0.45	0.53	0.95	1.79	2.07	13.09	3.93
BudgetIsHigher	0.30	0.47	0.37	0.44	2.38	0.60	0.48

Table 2. Time in seconds to update the materialized aggregates

	1	5	10	50	100	500	1000
Movie Insertions	0.05	0.14	0.12	0.15	0.11	0.48	0.35
Salary Updates	0.08	0.08	0.12	0.08	0.06	0.11	0.10
Director Deletions	0.05	0.05	0.08	0.06	0.42	0.06	0.14

5   Related Work

OCL Constraints to SQL. Similarly to our method, the work of [3] is based on translating each OCL constraint into an SQL view that will be empty if and only if the constraint is satisfied. Likewise, the work of [4] defines a translation from OCL expressions into MySQL queries/procedures that return the evaluation of the OCL expression. Both translations are able to deal with aggregates, but they are not incremental since whenever a data update occurs, the overall queries need to be recomputed from scratch. In a different way, [20] offers a translation from OCL to SQL queries that are triggered by those data updates that might violate the constraints. However, in this method queries might still look for violations trough the overall data instead of the limited data related to the update.

Incremental OCL Constraints Checking. The work in [6] is based on, for each OCL constraint, mapping the different context elements for which the constraint must be evaluated to the related data required to perform such evaluation. Thus, when any of these relevant data is modified, the corresponding OCL constraint is reevaluated for such context element. However, this strategy might be prohibitive due to excessive memory usage depending on the kind of constraints involved [21]. Instead of storing all these data, the works of [5,7] are able to compute, given a data update, the constraints that might be violated together the context element for which they should be reevaluated. However, none of these approaches is fully incremental since they only compute the relevant context element for which to evaluate the constraint, but not the relevant data for that context element for which to perform the evaluation.

OCL Translation to Graph Patterns. The work of [8] consists in translating OCL constraints into graph patterns to benefit from graph-pattern incremental queries. Nevertheless, such method uses the RETE algorithm to achieve the incremental behavior, which materializes every relational algebra operation performed by the queries. Such materialization has already been criticized because its huge memory usage and the penalization time required to maintain such materialization. To solve such issues, other algorithms have been suggested such as TREAT [9]. However, TREAT does not handle aggregates.

6 Conclusions

Checking the satisfaction of constraints over some data state is an essential problem in order to ensure data validity. In this paper, we have proposed a method based on SQL to incrementally check the satisfaction of OCL constraints. That is, we build several SQL queries that return the data violating the constraints. Thus, we can check data validity by means of checking SQL query emptiness. Moreover, the queries we propose are defined in a way that perform incrementally. In other words, the SQL queries are only executed when some data update may have violated a constraint, and only examine the data related to the update.

Our method is based on translating each OCL constraint into a set of logic rules that we call event dependency constraints (EDCs). Each EDC is a rule that captures a different combination of events (i.e., data updates) causing the violation of a constraint given the current data state. Each EDCs is then translated into SQL. These SQL queries make use of some auxiliary tables containing the events being applied in the information system, and some other tables containing a materialization of the relevant aggregates required to check the constraint. Thus, the SQL query is a join between the three: events, current data and aggregate values. This join is the key for incrementality since it forces the query to only search for possible violations with the data related to the update.

We have shown that the expressiveness of the OCL constraints we can deal with is beyond OCL_{FO}, the first order fragment of OCL, because of dealing with the distributive OCL aggregation operations (sum, count, size). Furthermore, we have shown the scalability of our method by means of some experiments with real data examining both the time to execute the queries and the time to maintain the materialized aggregated data. As further work, we plan to extend our approach to deal with OCL transitive closure and non-distributive aggregate operations such as min and max.

Acknowledgements. This work has been partially supported by the Ministerio de Economía y Competitividad, under project TIN2014-52938-C2-2-R and by the Secreteria d'Universitats i Recerca de la Generalitat de Catalunya under 2014 SGR 1534 and a FI grant.

References

1. Object Management Group (OMG): Unified Modeling Language (UML) Superstructure Specification, version 2.4.1 (2011). http://www.omg.org/spec/UML/
2. Object Management Group (OMG): Object Constraint Language (UML), version 2.4 (2014). http://www.omg.org/spec/OCL/
3. Heidenreich, F., Wende, C., Demuth, B.: A framework for generating query language code from OCL invariants. ECEASST **9**, 1–10 (2008)
4. Egea, M., Dania, C., Clavel, M.: MySQL4OCL: a stored procedure-based MySQL code generator for OCL. ECEASST **36**, 1–16 (2010)
5. Uhl, A., Goldschmidt, T., Holzleitner, M.: Using an OCL impact analysis algorithm for view-based textual modelling. ECEASST **44**, 1–20 (2011)
6. Groher, I., Reder, A., Egyed, A.: Incremental consistency checking of dynamic constraints. In: Rosenblum, D.S., Taentzer, G. (eds.) FASE 2010. LNCS, vol. 6013, pp. 203–217. Springer, Heidelberg (2010)
7. Cabot, J., Teniente, E.: Incremental integrity checking of UML/OCL conceptual schemas. J. Syst. Softw. **82**(9), 1459–1478 (2009)
8. Bergmann, G.: Translating OCL to graph patterns. In: Dingel, J., Schulte, W., Ramos, I., Abrahão, S., Insfran, E. (eds.) MODELS 2014. LNCS, vol. 8767, pp. 670–686. Springer, Heidelberg (2014)
9. Miranker, D.P.: TREAT: A better match algorithm for AI production systems. In: Proceedings of the 6th National Conference on Artificial Intelligence, vol. 1, AAAI. pp. 42–47. AAAI Press (1987)
10. Oriol, X., Teniente, E.: Incremental checking of OCL constraints through SQL queries. In: Proceedings of the 14th International Workshop on OCL and Textual Modelling, pp. 23–32 (2014)
11. Franconi, E., Mosca, A., Oriol, X., Rull, G., Teniente, E.: Logic foundations of the OCL modelling language. In: Fermé, E., Leite, J. (eds.) JELIA 2014. LNCS, vol. 8761, pp. 657–664. Springer, Heidelberg (2014)

12. Afrati, F.N., Chirkova, R.Y.: Selecting and using views to compute aggregate queries. In: Eiter, T., Libkin, L. (eds.) ICDT 2005. LNCS, vol. 3363, pp. 383–397. Springer, Heidelberg (2005)

13. Cohen, S., Nutt, W., Serebrenik, A.: Rewriting aggregate queries using views. In: Proceedings of the 18th ACM SIGMOD-SIGACT-SIGART Symposium on Principles of Database Systems, PODS 1999, pp. 155–166. ACM, New York (1999)

14. Consens, M.P., Mendelzon, A.O.: Low complexity aggregation in graphlog and datalog. In: Abiteboul, S., Kanellakis, P.C. (eds.) ICDT 1990. LNCS, vol. 470, pp. 379–394. Springer, Heidelberg (1990)

15. Queralt, A., Teniente, E.: Verification and validation of UML conceptual schemas with OCL constraints. ACM TOSEM **21**(2), 13 (2012)

16. Olivé, A.: Conceptual Modeling of Information Systems. Springer, Heidelberg (2007)

17. Olivé, A.: Integrity constraints checking in deductive databases. In: Proceedings of the 17th International Conference on Very Large Data Bases (VLDB), pp. 513–523 (1991)

18. Gray, J., Chaudhuri, S., Bosworth, A., Layman, A., Reichart, D., Venkatrao, M., Pellow, F., Pirahesh, H.: Data cube: a relational aggregation operator generalizing group-by, cross-tab, and sub-totals. Data Min. Knowl. Discov. **1**(1), 29–53 (1997)

19. Oriol, X., Teniente, E., Tort, A.: Fixing up non-executable operations in UML/OCL conceptual schemas. In: Yu, E., Dobbie, G., Jarke, M., Purao, S. (eds.) ER 2014. LNCS, vol. 8824, pp. 232–245. Springer, Heidelberg (2014)

20. Al-Jumaily, H.T., Cuadra, D., Martínez, P.: OCL2Trigger: deriving active mechanisms for relational databases using model-driven architecture. J. Syst. Softw. **81**(12), 2299–2314 (2008)

21. Falleri, J., Blanc, X., Bendraou, R., da Silva, M.A.A., Teyton, C.: Incremental inconsistency detection with low memory overhead. Softw. Pract. Exper. **44**(5), 621–641 (2014)

Probabilistic Cardinality Constraints

Tania Roblot and Sebastian Link[✉]

Department of Computer Science, University of Auckland, Auckland, New Zealand
{tkr,s.link}@auckland.ac.nz

Abstract. Probabilistic databases address well the requirements of an increasing number of modern applications that produce large collections of uncertain data. We propose probabilistic cardinality constraints as a principled tool for controlling the occurrences of data patterns in probabilistic databases. Our constraints help organizations balance their targets for different data quality dimensions, and infer probabilities on the number of query answers. These applications are unlocked by developing algorithms to reason efficiently about probabilistic cardinality constraints, and to help analysts acquire the marginal probability by which cardinality constraints hold in a given application domain. For this purpose, we overcome technical challenges to compute Armstrong PC-sketches as succinct data samples that perfectly visualize any given perceptions about these marginal probabilities.

Keywords: Data and knowledge visualization · Data models · Database semantics · Management of integrity constraints · Requirements engineering

1 Introduction

Background. The notion of cardinality constraints is fundamental for understanding the structure and semantics of data. In traditional conceptual modeling, cardinality constraints were already introduced in Chen's seminal paper [3]. They have attracted significant interest and tool support ever since. Intuitively, a cardinality constraint $card(X) \leq b$ stipulates for an attribute set X and a positive integer b that a relation must not contain more than b different tuples that all have matching values on all the attributes in X. For example, bank customers with no more than 5 withdrawals from their bank account per month may qualify for a special interest rate. Traditionally, cardinality constraints empower applications to control the occurrences of certain data, and have applications in data cleaning, integration, modeling, processing, and retrieval among others.

Example. Relational databases target applications with certain data, such as accounting, inventory and payroll. Modern applications, such as data integration, information extraction, scientific data management, and financial risk assessment produce large amounts uncertain data. For instance, RFID (radio frequency identification) is used to track movements of endangered species of animals, such as wolverines. Here it is sensible to apply probabilistic databases. Table 1 shows a

P. Johannesson et al. (Eds.): ER 2015, LNCS 9381, pp. 214–228, 2015.
DOI: 10.1007/978-3-319-25264-3_16

probabilistic relation (p-relation) over TRACKING = $\{rfid, time, zone\}$, which is a probability distribution over a finite set of possible worlds, each being a relation. Data patterns occur with different frequency in different worlds. That is, different worlds satisfy different cardinality constraints. For example, the cardinality constraint $c_1 = card(time, zone) \leq 1$ holds in the world w_1 and $\{time, zone\}$ is therefore a key in this world, and

Table 1. Probabilistic relation

w_1 ($p_1 = 0.75$)			w_2 ($p_2 = 0.15$)			w_3 ($p_3 = 0.1$)		
rfid	time	zone	rfid	time	zone	rfid	time	zone
w2	06	z1	w1	08	z4	w1	08	z4
w2	07	z1	w1	08	z5	w1	08	z5
w3	15	z7	w1	08	z6	w1	08	z6
w3	16	z8	w2	05	z1	w2	05	z1
w3	17	z9	w2	06	z1	w2	06	z1
w10	10	z11	w2	07	z1	w2	07	z1
w11	10	z12	w4	11	z3	w4	11	z3
w12	10	z13	w5	12	z3	w5	12	z3
w4	11	z3	w6	13	z3	w6	13	z3
w5	12	z3	w7	14	z3	w7	14	z3
w6	13	z3	w8	09	z2	w8	09	z2
w7	14	z3	w9	09	z2	w9	09	z2
						w0	09	z2

$c_2 = card(time, zone) \leq 2$ holds in the world w_1 and w_2. Typically, the likelihood of a cardinality constraint to hold in a given application domain, i.e. the constraint's degree of meaningfulness, should be reflected by its marginal probability. In the example above, c_1 and c_2 have marginal probability 0.75 and 0.9, respectively, and we may write $(card(time, zone) \leq 1, \geq 0.75)$ and $(card(time, zone) \leq 2, \geq 0.9)$ to denote the *probabilistic cardinality constraints* (pCCs) that c_1 holds at least with probability 0.75 and c_2 holds at least with probability 0.9.

Applications. PCCs have important applications. *Data quality:* Foremost, they can express desirable properties of modern application domains that must accommodate uncertain data. This raises the ability of database systems to enforce higher levels of consistency in probabilistic databases, as updates to data are questioned when they result in violations of some pCC. Enforcing hard constraints, holding with probability 1, may remove plausible worlds and lead to an incomplete representation. The marginal probability of cardinality constraints can balance the consistency and completeness targets for the quality of an organization's data. *Query estimation:* PCCs can be used to obtain lower bounds on the probability by which a given maximum number of answers to a given query will be returned, without having to evaluate the query on any portion of the given, potentially big, database. For example, the query

SELECT *rfid* FROM Tracking WHERE *zone*='z2' AND *time*='09'

asks for the rfid of wolverines recorded in zone z2 at 09am. Reasoning about our pCCs tells us that at most 3 answers will be returned with probability 1, at most 2 answers will be returned with minimum probability 0.9, and at most 1 answer will be returned with minimum probability 0.75. A service provider may return these numbers, or approximate costs derived from them, to a customer, who can make a more informed decision whether to pay for the service. The provider, on the other hand, does not need to utilize unpaid resources for querying the potentially big data source to return the feedback.

Contributions. The applications motivate us to stipulate lower bounds on the marginal probability of cardinality constraints. The main inhibitor for the uptake of pCCs is the identification of the right lower bounds on their marginal probabilities. While it is already challenging to identify traditional cardinality constraints which are semantically meaningful in a given application domain, identifying the right probabilities is an even harder problem. Lower bounds appear to be a realistic compromise here. Our contributions can be summarized as follows. *(1) Modeling:* We propose pCCs as a natural class of semantic integrity constraints over uncertain data. Their main target is to help organizations derive more value from data by ensuring higher levels of data quality and assist with data processing. *(2) Reasoning:* We characterize the implication problem of pCCs by a simple finite set of Horn rules, as well as a linear time decision algorithm. This enables organizations to reduce the overhead of data quality management by pCCs to a minimal level necessary. For example, enforcing $(card(rfid) \leq 3, \geq 0.9)$, $(card(zone) \leq 4, \geq 0.9)$ and $(card(rfid, zone) \leq 3, \geq 0.75)$ would be redundant as the enforcement of $(card(rfid, zone) \leq 3, \geq 0.75)$ is already implicitly done by enforcing $(card(rfid) \leq 3, \geq 0.9)$.

(3) Acquisition: For acquiring the right marginal probabilities by which pCCs hold, we show how to visualize concisely any given system of pCCs in the form of an Armstrong PC-sketch. Recall that every p-relation can be represented by some PC-table. Here, we introduce Armstrong PC-sketches as finite semantic representations of some possibly

Table 2. PC-sketch of Table 1

card	rfid	time	zone	ι
3	w1	08	*	2,3
2	w2	*	z1	1,2,3
1	w2	*	z1	2,3
2	*	09	z2	2,3
1	*	09	z2	3
3	w3	*	*	1
3	*	10	*	1
4	*	*	z3	1,2,3

ι	$\Pi(\iota)$
1	.75
2	.15
3	.1

infinite p-relation which satisfies every cardinality constraint with the exact marginal probability by which it is currently perceived to hold. Problems with such perceptions are explicitly pointed out by the PC-sketch. For example, Fig. 2 shows a PC-sketch for the p-relation from Table 1, which is Armstrong for the pCCs satisfied by the p-relation. The sketch shows which patterns of data must occur in how many rows (represented in column *card*) in which possible worlds (represented by the world identifiers in column ι). The symbol * represents some data value that is unique within each world of the p-relations derived from the sketch. Π defines the probability distribution over the resulting possible worlds. Even when they represent finite p-relations, PC-sketches are still more concise since they only show patterns that matter and how often these occur.

Organization. We discuss related work in Sect. 2. PCCs are introduced in Sect. 3, and reasoning tools for them are established in Sect. 4. These form the foundation for computational support to acquire the correct marginal probabilities in Sect. 5. We conclude and outline future work in Sect. 6. Due to lack of space, all proofs have been made available in the technical report [16].

2   Related Work

Cardinality constraints are one of the most influential contributions conceptual modeling has made to the study of database constraints. They were already present in Chen's seminal paper [3]. It is no surprise that today they are part of all major languages for data and knowledge modeling, including UML, EER, ORM, XSD, and OWL. Cardinality constraints have been extensively studied in database design [4–9, 12, 14, 15, 18]. For a recent survey, see [19].

Probabilistic cardinality constraints $card(X) \leq b$, introduced in this paper, subsume the class of probabilistic keys [2] as the special case where $b = 1$.

For possibilistic cardinality constraints [10], tuples are attributed some degree of possibility and cardinality constraints some degree of certainty saying to which tuples they apply. In general, possibility theory is a qualitative approach while probability theory is a quantitative approach to uncertainty. Our research thereby complements the qualitative approach to cardinality constraints in [10] by a quantitative approach.

Our contribution extends results on cardinality constraints from traditional relations, which are covered by our framework as the special case where the p-relation consists of only one possible world [1, 6]. As pCCs form a new class of integrity constraints, their associated implication problem and properties of Armstrong p-relations have not been investigated before.

There is also a large body of work on the discovery of "approximate" business rules, such as keys, functional and inclusion dependencies [13]. Here, approximate means that almost all tuples satisfy the given rule; hence allowing for very few exceptions. Our constraints are not approximate since they are either satisfied or violated by the given p-relation or the PC-sketch that represents it.

3   Cardinality Constraints on Probabilistic Databases

Next we introduce some preliminary concepts from probabilistic databases and the central notion of a probabilistic cardinality constraint. We use the symbol $\mathbb{N}_1^\infty$ to denote the positive integers together with the symbol ∞ for infinity.

A *relation schema* is a finite set R of attributes A. Each attribute A is associated with a domain $dom(A)$ of values. A tuple t over R is a function that assigns to each attribute A of R an element $t(A)$ from the domain $dom(A)$. A *relation* over R is a finite set of tuples over R. Relations over R are also called *possible worlds* of R here. An expression $card(X) \leq b$ with some non-empty subset $X \subseteq R$ and $b \in \mathbb{N}_1^\infty$ is called a *cardinality constraint over R*. In what follows, we will always assume that a subset of R is non-empty without mentioning it explicitly. A cardinality constraint $card(X) \leq b$ over R is said to *hold* in a possible world w of R, denoted by $w \models card(X) \leq b$, if and only if there are not $b + 1$ different tuples $t_1, \cdots, t_{b+1} \in W$ such that for all $1 \leq i < j \leq b+1$, $t_i \neq t_j$ and $t_i(X) = t_j(X)$.

A *probabilistic relation* (p-relation) over R is a pair $r = (W, P)$ of a finite non-empty set W of possible worlds over R and a probability distribution $P :$ $W \to (0, 1]$ such that $\sum_{w \in W} P(w) = 1$ holds.

Table 1 shows a p-relation over relation schema $\text{WOLVERINE} = \{rfid, time,$ $zone\}$. World w_2 satisfies the CCs $card(rfid) \leq 3$, $card(time) \leq 3$, $card(zone) \leq 4$, $card(rfid, time) \leq 3$, $card(rfid, zone) \leq 3$, and $card(time, zone) \leq 2$ but violates the CC $card(time, zone) \leq 1$.

A cardinality constraint $card(X) \leq b$ over R is said to *hold with probability* $p \in [0, 1]$ in the p-relation $r = (W, P)$ if and only if $\sum_{w \in W, w \models card(X) \leq b} P(w) = p$. In other words, the probability of a cardinality constraint in a p-relation is the marginal probability with which it holds in the p-relation. We will now introduce the central notion of a cardinality constraint on probabilistic databases.

Definition 1. *A probabilistic cardinality constraint, or pCC for short, over relation schema R is an expression $(card(X) \leq b, \geq p)$ where $X \subseteq R$, $b \in \mathbb{N}_1^\infty$ and $p \in [0, 1]$. The pCC $(card(X) \leq b, \geq p)$ over R is said to* hold *in the p-relation r over R if and only if the probability with which the cardinality constraint $card(X) \leq b$ holds in r is at least p.*

Example 1. In our running example over relation schema WOLVERINE, the p-relation from Table 1 satisfies the set Σ of the following pCCs $(card(rfid) \leq 3, \geq 1)$, $(card(time) \leq 3, \geq 1)$, $(card(zone) \leq 4, \geq 1)$, $(card(time, zone) \leq 2, \geq 0.9)$, $(card(rfid, time) \leq 1, \geq 0.75)$, $(card(rfid, zone) \leq 2, \geq 0.75)$, as well as $(card(time, zone) \leq 1, \geq 0.75)$. It violates the pCC $(card(rfid, time) \leq 1, \geq 0.9)$.

4   Reasoning Tools

When enforcing sets of pCCs to improve data quality, the overhead they cause must be reduced to a minimal level necessary. In practice, this requires us to reason about pCCs efficiently. We will now establish basic tools for this purpose.

Implication. Let $\Sigma \cup \{\varphi\}$ denote a finite set of constraints over relation schema R, in particular Σ is always finite. We say Σ *(finitely) implies* φ, denoted by $\Sigma \models_{(f)} \varphi$, if every (finite) p-relation r over R that satisfies Σ, also satisfies φ. We use $\Sigma_{(f)}^* = \{\varphi \mid \Sigma \models_{(f)} \varphi\}$ to denote the *(finite) semantic closure* of Σ. For a class $\mathcal{C}$ of constraints, the (finite) $\mathcal{C}$-implication problem is to decide for a given relation schema R and a given set $\Sigma \cup \{\varphi\}$ of constraints in $\mathcal{C}$ over R, whether Σ (finitely) implies φ. Finite implication problem and implication problem coincide for the class of pCCs, and we thus speak of *the* implication problem.

Axioms. We determine the semantic closure by applying *inference rules* of the form $\dfrac{\text{premise}}{\text{conclusion}}$. For a set $\mathfrak{R}$ of inference rules let $\Sigma \vdash_{\mathfrak{R}} \varphi$ denote the *inference* of φ from Σ by $\mathfrak{R}$. That is, there is some sequence $\sigma_1, \ldots, \sigma_n$ such that $\sigma_n = \varphi$ and every σ_i is an element of Σ or is the conclusion that results from an application of an inference rule in $\mathfrak{R}$ to some premises in $\{\sigma_1, \ldots, \sigma_{i-1}\}$. Let $\Sigma_{\mathfrak{R}}^+ = \{\varphi \mid \Sigma \vdash_{\mathfrak{R}} \varphi\}$ be the *syntactic closure* of Σ under inferences by $\mathfrak{R}$. $\mathfrak{R}$ is *sound* (*complete*) if for every set Σ over every $(R, \mathcal{S})$ we have $\Sigma_{\mathfrak{R}}^+ \subseteq \Sigma^*$ ($\Sigma^* \subseteq \Sigma_{\mathfrak{R}}^+$). The (finite) set $\mathfrak{R}$ is a (finite) *axiomatization* if $\mathfrak{R}$ is both sound and complete. In the set $\mathfrak{P}$ of inference rules from Table 3, R denotes the underlying relation schema, X and Y form attribute subsets of R, $b, b' \in \mathbb{N}_1^\infty$, and p, q as well as $p + q$ are probabilities. Due to lack of space we omit the soundness and completeness proof of the following theorem, see [16].

Table 3. Axiomatization $\mathfrak{P} = \{\mathcal{D}, \mathcal{Z}, \mathcal{U}, \mathcal{S}, \mathcal{B}, \mathcal{P}\}$

$(card(R) \leq 1, \geq 1)$ (Duplicate-free, $\mathcal{D}$)	$(card(X) \leq b, \geq 0)$ (Zero, $\mathcal{Z}$)	$(card(X) \leq \infty, \geq 1)$ (Unbounded, $\mathcal{U}$)
$\dfrac{(card(X) \leq b, \geq p)}{(card(XY) \leq b, \geq p)}$ (Superset, $\mathcal{S}$)	$\dfrac{(card(X) \leq b, \geq p)}{(card(X) \leq b + b', \geq p)}$ (Bound, $\mathcal{B}$)	$\dfrac{(card(X) \leq b, \geq p + q)}{(card(X) \leq b, \geq p)}$ (Probability, $\mathcal{P}$)

Theorem 1. $\mathfrak{P}$ *forms a finite axiomatization for the implication of probabilistic cardinality constraints.*

Example 2. The set Σ of pCCs from Example 1 implies $\varphi = (card(rfid, time) \leq 4, \geq 0.8)$, but not $\varphi' = (card(rfid, time) \leq 1, \geq 0.8)$. In fact, φ can be inferred from Σ by applying $\mathcal{S}$ to $(card(rfid) \leq 3, \geq 1)$ to infer $(card(rfid, time) \leq 3, \geq 1)$, applying $\mathcal{B}$ to this pCC to infer $(card(rfid, time) \leq 4, \geq 1)$, and then applying $\mathcal{P}$.

If a data set is validated against a set Σ of pCCs, then the data set does not need to be validated against any pCC φ implied by Σ. The larger the data set, the more time is saved by avoiding redundant validation checks.

Algorithms. In practice it is often unnecessary to determine all implied pCCs. In fact, the implication problem for pCCs has as input $\Sigma \cup \{\varphi\}$ and the question is whether Σ implies φ. Computing Σ^* and checking whether $\varphi \in \Sigma^*$ is hardly efficient. Indeed, we will now establish a linear-time algorithm for computing the maximum probability p, such that $\varphi = (card(X) \leq b, \geq p)$ is implied by Σ. The following theorem provides the foundation for the algorithm [16].

Theorem 2. *Let $\Sigma \cup \{(card(X) \leq b, \geq p)\}$ denote a set of pCCs over relation schema R. Then Σ implies $(card(X) \leq b, \geq p)$ if and only if (i) $X = R$ or (ii) $p = 0$ or (iii) $b = \infty$ or (iv) there is some $(card(Z) \leq b', \geq q) \in \Sigma$ such that $Z \subseteq X$, $b' \leq b$, and $q \geq p$.*

Example 3. Continuing Example 2, we can apply Theorem 2 directly to see that Σ implies $\varphi = (card(rfid, time) \leq 4, \geq 0.8)$. Indeed, the pCC $(card(rfid) \leq 3, \geq 1) \in \Sigma$ satisfies the sufficient conditions of Theorem 2 to imply φ, since $\{rfid\} \subseteq \{rfid, time\}$, $3 \leq 4$, and $1 \geq 0.8$.

Theorem 2 motivates the following algorithm that returns for a given cardinality constraint $card(X) \leq b$ the maximum probability p by which $(card(X) \leq b, \geq p)$ is implied by a given set Σ of pCCs over R: If $X = R$ or $b = \infty$, then we return probability 1; Otherwise, starting with $p = 0$ the algorithm scans all input pCCs $(card(Z) \leq b', \geq q) \in \Sigma$ and sets p to q whenever q is larger than the current p, X contains Z and $b' \leq b$. $\|\Sigma\|$ denotes the total number of attributes together with the logarithm of the integer bounds in Σ. Here, we assume without loss of generality that ∞ does not occur.

Theorem 3. *On input $(R, \Sigma, card(X) \leq b)$ our algorithm returns in $\mathcal{O}(\|\Sigma \cup \{(card(X) \leq b, \geq p)\}\|)$ time the maximum probability p with which $(card(X) \leq b, \geq p)$ is implied by Σ.*

Example 4. Continuing Example 1, we can apply our algorithm to the schema WOLVERINE, pCC set Σ, and the cardinality constraint $card(rfid, time) \leq 4$, which gives us the maximum probability 1 for which it is implied by Σ.

Theorem 3 allows us to decide the associated implication problem efficiently, too. Given $R, \Sigma, (card(X) \leq b, \geq p)$ as an input to the implication problem, we use our algorithm to compute $p' := \max\{q : \Sigma \models card(X) \leq b, \geq q\}$ and return an affirmative answer if and only if $p' \geq p$.

Corollary 1. *The implication problem of probabilistic cardinality constraints can be decided in linear time.*

Example 5. Continuing Example 4 we can see directly that Σ implies the pCC $\varphi = (card(rfid, time) \leq 4, \geq 0.8)$ since our algorithm returned 1 as the maximum probability for which $card(rfid, time) \leq 4$ is implied by Σ. Since the given probability of 0.8 does not exceed $p = 1$, φ is indeed implied.

5   Acquiring Probabilistic Cardinality Constraints

Data quality, and therefore largely the success of data-driven organizations, depend on the ability of analysts to identify the semantic integrity constraints that govern the data. For cardinality constraints $(card(X) \leq b, \geq p)$ the "right" marginal probability p and the "right" upper bound b must be identified for a given set X of attributes. Choosing p too big or b too small prevents the entry of clean data, resulting in a lower level of data completeness. Choosing p too small or b too high can lead to the entry of dirty data, resulting in a lower level of data consistency. Analysts benefit from computational support to improve upon their ad-hoc perceptions on an appropriate probability p and bound b.

Goal. Armstrong relations are a useful tool for consolidating the perception of analysts about the cardinality constraints (CCs) of a given application domain. Starting with a set Σ, the tool creates a small relation r that satisfies Σ and violates all CCs not implied by Σ. This property makes r a perfect sample for Σ: any CC is satisfied by the relation if and only if it is implied by Σ.

Our goal is to develop the tool of Armstrong p-relations for a given set Σ of pCCs: the marginal probability by which a traditional constraint $card(X) \leq b$ holds on the Armstrong p-relation is the maximum probability p by which the pCC $(card(X) \leq b, \geq p)$ is implied by Σ. So, if an analyst wants to check for an arbitrary pCC $(card(X) \leq b, \geq p)$ whether it is implied by Σ, she can compute the marginal probability p' by which the CC $card(X) \leq b$ holds on the Armstrong p-relation and verify that $p \geq p'$. For the remainder of this section, we will review Armstrong relations, add new results, and then devise our construction of Armstrong p-relations and more concise representations thereof.

Armstrong Relations. An Armstrong relation w for a given set Σ of CCs over relation schema R violates all CCs $card(X) \leq b$ over R which are not implied by Σ. However, $\Sigma \models card(X) \leq b$ if and only if $X = R$ or $b = \infty$ or there is some $card(Z) \leq b' \in \Sigma$ where $Z \subseteq X$ and $b' \leq b$. Hence, if $\Sigma \not\models card(X) \leq b$, then $X \neq R$, $b < \infty$ and for all $card(Z) \leq b' \in \Sigma$ where $Z \subseteq X$ we have $b' > b$. Our strategy is therefore to find for all subsets X, the smallest upper bound b_X that applies to the set X. In other words, $b_X = \inf\{b \mid \Sigma \models card(X) \leq b\}$. Moreover, if $b_{XY} = b_X$ for some attribute sets X, Y, then it suffices to violate $card(XY) \leq b_{XY} - 1$. For this reason, the set $dup_\Sigma(R)$ of *duplicate sets* is defined as $dup_\Sigma(R) = \{\emptyset \subset X \subset R \mid b_X > 1 \wedge (\forall A \in R - X(b_{XA} < b_X))\}$. For each duplicate set $X \in dup_\Sigma(R)$, we introduce b_X new tuples $t_1^X, \ldots, t_{b_X}^X$ that all have matching values on all the attributes in X and all have unique values on all the attributes in $R - X$. An Armstrong relation for Σ is obtained by taking the disjoint union of $\{t_1^X, \ldots, t_{b_X}^X\}$ for all duplicate sets X.

Example 6. For a probability p and a given set Σ of pCCs let $\Sigma_p = \{card(X) \leq b \mid \exists p' \in (0, 1](card(X) \leq b, \geq p') \in \Sigma\}$. Continuing Example 1 consider the sets $\Sigma_{0.75}$, $\Sigma_{0.9}$ and Σ_1 of traditional cardinality constraints on WOLVERINE. The attribute subsets which are duplicate with respect to these sets are illustrated in Fig. 1, together with their associated cardinalities. The worlds w_1, w_2 and w_3 in Table 1 are Armstrong relations for $\Sigma_{0.75}$, $\Sigma_{0.9}$ and Σ_1, respectively.

Armstrong Sketches. While this construction works well in theory, a problem occurs with the actual use of these Armstrong relations in practice. In some cases, the Armstrong relation will be infinite and therefore of no use. These cases occur exactly if there is some attribute $A \in R$ for which $b_A = \infty$, in other words, if there is some attribute for which no finite upper bound has been specified. For a practical solution we introduce Armstrong sketches, which are finite representations of possibly infinite Armstrong relations.

Let R_* denote a relation schema resulting from R by extending the domain of each attribute of R by the distinguished symbol $*$. A *sketch* $\varsigma = (card, \omega)$ over R consists of a finite relation $\omega = \{\tau_1, \ldots, \tau_n\}$ over R_*, and a function $card$ that

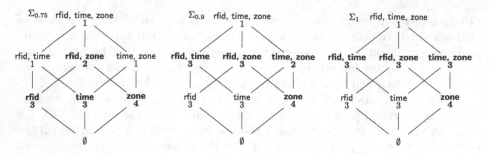

Fig. 1. Duplicate sets X in bold font and their cardinalities b_X for Example 6

maps each tuple $\tau_i \in \omega$ to a value $b_i = card(\tau_i) \in \mathbb{N}_1^\infty$. An *expansion* of ς is a relation w over R such that

- $w = \bigcup_{i=1}^n \{t_i^1, \ldots, t_i^{b_i}\}$,
- (preservation of domain values) for all $i = 1, \ldots, n$, for all $k = 1, \ldots, b_i$, for all $A \in R$, if $\tau_i(A) \neq *$, then $t_i^k(A) = \tau_i(A)$,
- (uniqueness of values substituted for $*$) for all $i = 1, \ldots, n$, for all $A \in R$, if $\tau_i(A) = *$, then for all $k = 1, \ldots, b_i$, for all $j = 1, \ldots, n$, and for all $l = 1, \ldots, b_j$ (where $l \neq k$, if $j = i$), $t_i^k(A) \neq t_j^l(A)$.

We call ς an *Armstrong sketch* for Σ, if every expansion of ς is an Armstrong relation for Σ. The following simple algorithm can be used to construct an Armstrong sketch $\varsigma = (card, \omega)$ for Σ: for each duplicate set $X \in dup_\Sigma(R)$ we introduce a tuple τ_X into ω such that, for all $A \in X$, $\tau_X(A)$ has some unique domain value from $dom(A) - \{*\}$, and for all $A \in R - X$, $\tau_X(A) = *$, and $card(\tau_X) = b_X$. The main advantage of Armstrong sketches over Armstrong relations is their smaller number of tuples. In fact, this number coincides with the number of duplicate sets which is guaranteed to be finite. In contrast, if some $b_X = \infty$, then every Armstrong relation must be infinite.

Example 7. Continuing Example 6 the following tables show Armstrong sketches (A-sketches) for the sets $\Sigma_{0.75}$, $\Sigma_{0.9}$, and Σ_1, which have expansions w_1, w_2, and w_3 as shown in Table 1, respectively.

A-sketch for $\Sigma_{0.75}$				A-sketch for $\Sigma_{0.9}$				A-sketch for Σ_1			
card	*rfid*	*time*	*zone*	*card*	*rfid*	*time*	*zone*	*card*	*rfid*	*time*	*zone*
2	w2	*	z1	3	w1	08	*	3	w1	08	*
3	w3	*	*	3	w2	*	z1	3	w2	*	z1
3	*	10	*	2	*	09	z2	4	*	*	z3
4	*	*	z3	4	*	*	z3	3	*	09	z2

Armstrong p-sketches. An *Armstrong p-relation* for a set Σ of pCCs over R is a p-relation r over R such that for all pCCs φ over R the following holds:

$\Sigma \models \varphi$ if and only if r satisfies φ. As relations are the idealized special case of p-relations in which the relation forms the only possible world of the p-relation, there are sets of pCCs for which no finite Armstrong p-relation exists, i.e., the Armstrong p-relation contains some possible world that is infinite. For this reason we introduce probabilistic sketches and their expansions, as well as Armstrong p-sketches which are guaranteed to be finite p-relations.

A *probabilistic sketch* (p-sketch) over R is a probabilistic relation $s = (\mathcal{W}, \mathcal{P})$ over R_* where the possible worlds in $\mathcal{W}$ are sketches over R. A *probabilistic expansion* (p-expansion) of s is a p-relation $r = (W, P)$ where W contains for every sketch $\varsigma \in \mathcal{W}$ a single expansion w over R of ς, and $P(w) = \mathcal{P}(\varsigma)$.

An *Armstrong p-sketch* for a set Σ of pCCs over R is a p-sketch over R such that each of its p-expansions is an Armstrong p-relation for Σ.

Example 8. Continuing Example 1 the following table shows an Armstrong p-sketch s for the given set Σ of pCCs.

$\varsigma_1(p_1 = 0.75)$				$\varsigma_2(p_2 = 0.15)$				$\varsigma_3(p_3 = 0.1)$			
$card_1$	rfid	time	zone	$card_2$	rfid	time	zone	$card_3$	rfid	time	zone
2	w2	*	z1	3	w1	08	*	3	w1	08	*
3	w3	*	*	3	w2	*	z1	3	w2	*	z1
3	*	10	*	4	*	*	z3	4	*	*	z3
4	*	*	z3	2	*	09	z2	3	*	09	z2

A p-expansion of s is the finite Armstrong p-relation of Table 1.

Naturally the question arises whether Armstrong p-sketches exist for any given set of pCCs over any given relation schema. The next theorem shows that every distribution of probabilities to a finite set of cardinality constraints, that follows the inference rules from Table 3, can be represented by a single p-relation which exhibits this distribution in the form of marginal probabilities [16].

Theorem 4. *Let* $l : 2^R \times \mathbb{N}_1^\infty \to [0,1]$ *be a function such that the image of* l *is a finite subset of [0,1],* $l(R,1) = 1$ *and for all* $X \subseteq R$, $l(X, \infty) = 1$, *and for all* $X, Y \subseteq R$ *and* $b, b' \in \mathbb{N}_1$, $l(X, b) \leq l(XY, b + b')$ *holds. Then there is some p-sketch* s *over* R *such that every p-expansion* r *of* s *satisfies* $(card(X) \leq b, \geq l(X, b))$, *and for all* $X \subseteq R$, $b \in \mathbb{N}_1^\infty$ *and* $p \in [0,1]$ *such that* $p > l(X, b)$, r *violates* $(card(X) \leq b, \geq p)$.

We say that pCCs *enjoy* Armstrong p-sketches, if for every relation schema R and for every finite set Σ of pCCs over R there is some p-sketch over R that is Armstrong for Σ [16].

Theorem 5. *Prob. cardinality constraints enjoy Armstrong p-sketches.*

Armstrong PC-sketches. Probabilistic databases can have huge numbers of possible worlds. It is therefore important to represent and process probabilistic data concisely. Probabilistic conditional databases, or short PC-tables [17] are a popular system that can represent any given probabilistic database. Considering our aim of finding concise data samples of pCCs, we would like to compute Armstrong p-sketches in the form of Armstrong PC-sketches.

For this purpose, we first adapt the standard definition of PC-tables [17] to that of PC-sketches. A *conditional sketch* or *c-sketch*, is a tuple $\Gamma = \langle \varsigma, \iota \rangle$, where $\varsigma = (card, \omega)$ is a sketch (where ω may contain duplicate tuples), and ι assigns to each tuple τ in ω a finite set ι_τ of positive integers. The set of *world identifiers* of Γ is the union of the sets ι_τ for all tuples τ of ω. Given a world identifier i of Γ, the possible world sketch $\varsigma_i = (card_i, \omega_i)$ associated with i is $\omega_i = \{\tau | \tau \in \omega$ and $i \in \iota_\tau\}$ and $card_i$ is the restriction of $card$ to ω_i. The *representation* of a c-sketch $\Gamma = \langle \varsigma, \iota \rangle$ is the set $\mathcal{W}$ of possible world sketches ς_i where i denotes some world identifier of Γ. A *probabilistic conditional sketch* or *PC-sketch*, is a pair $\langle \Gamma, \Pi \rangle$ where Γ is a c-sketch, and Π is a probability distribution over the set of world identifiers of Γ. The *representation* of a PC-sketch $\langle \Gamma, \Pi \rangle$ is the p-sketch $s = (\mathcal{W}, \mathcal{P})$ where $\mathcal{W}$ is the set of possible world sketches associated with Γ and the probability $\mathcal{P}$ of each possible world sketch $\varsigma_i \in \mathcal{W}$ is defined as the probability $\Pi(i)$ of its world identifier i.

It is simple to see that every p-sketch can be represented as a PC-sketch [16].

Theorem 6. *Every p-sketch can be represented as a PC-sketch.*

A PC-sketch is called an *Armstrong PC-sketch* for Σ if and only if its representation is an Armstrong p-sketch for Σ.

Example 9. Table 2 shows a PC-sketch $\langle \Gamma, \Pi \rangle$ that is Armstrong for the set Σ of pCCs from Example 1.

Algorithm 1 contains the pseudo-code and comments how to compute an Armstrong PC-sketch for any given set Σ of pCCs over any given relation schema R. In particular, line (5) uses the definition of the cardinality $b_X^i := \inf\{b \mid card(Y) \leq b \in \Sigma_{p_i} \wedge Y \subseteq X\}$ to compute them.

Theorem 7. *For every set Σ of pCCs over relation schema R, Algorithm 1 computes an Armstrong PC-sketch for Σ.*

Finally, we derive some bounds on the time complexity of finding Armstrong PC-sketches. Since the relational model is subsumed there are cases, where the number of tuples in every Armstrong PC-sketch for Σ over R is exponential in $||\Sigma||$. Such a case is given by $R_n = \{A_1, \ldots, A_{2n}\}$ and $\Sigma_n = \{(card(A_{2i-1}, A_{2i}) \leq 1, \geq 1) \mid i = 1, \ldots, n\}$ with $||\Sigma_n|| = 2 \cdot n$. Indeed, every Armstrong PC-sketch for Σ_n must feature 2^n different tuples to accommodate the 2^n different duplicate sets X with associated cardinality $b_X^1 = \infty$, and there is only one possible world. Algorithm 1 was designed with the goal that the worst-case time bound from the traditional relational case does not deteriorate in our more general setting. This is indeed achieved, as the computationally most demanding part of Algorithm 1

Algorithm 1. Armstrong PC-sketch

Require: R, Σ
Ensure: Armstrong PC-sketch $\langle\langle(card, \omega), \iota\rangle, \Pi\rangle$ for Σ
 1: Let $p_1 < \cdots < p_n$ be the probabilities in Σ;   ▷ If $p_n < 1$, $n \leftarrow n + 1$ and $p_n \leftarrow 1$
 2: $p_0 \leftarrow 0$; $\Pi \leftarrow \emptyset$;
 3: **for** $i = 1, \ldots, n$ **do**   ▷ Process one possible world sketch at a time
 4:     $\Pi \leftarrow \Pi \cup \{(i, p_i - p_{i-1})\}$;   ▷ World i has probability $p_i - p_{i-1}$
 5:     Compute $\{b_X^i \mid X \subseteq R\}$;   ▷ Smallest upper bound for each X in world i
 6:     $dup_i \leftarrow$ Set of duplicate sets for Σ_{p_i};   ▷ Duplicate sets to realize in world i
 7: $\omega \leftarrow \emptyset$; $k \leftarrow 0$;
 8: $dup \leftarrow \{(X, \{i \mid X \in dup_i\}) \mid X \in dup_i$ for some $i\}$;
 9: **for all** $(X, W) \in dup$ **do** ▷ For each X that is a duplicate set in every world in W
10:     $b \leftarrow 0$; $j \leftarrow k + 1$;
11:     **for** $i = 1, \ldots, n$ **do**   ▷ Add some τ_k that realizes X in every world in W
12:         **if** $X \in dup_i$ **and** $b_X^i > b$ **then**   ▷ if there are any remaining cardinalities
13:             $k \leftarrow k + 1$;
14:             **for all** $A \in R$ **do**   ▷ Define τ_k with...
15:                 **if** $A \in X$ **then**
16:                     $\tau_k(A) \leftarrow j$;   ▷ ...fixed values on X
17:                 **else**
18:                     $\tau_k(A) \leftarrow *$;   ▷ ...and unique values outside of X
19:             $\omega \leftarrow \omega \cup \{\tau_k\}$;   ▷ Add new tuple
20:             $card(\tau_k) \leftarrow b_X^i - b$;   ▷ Stipulate remaining cardinality
21:             $\iota(\tau_k) \leftarrow W - \{1, \ldots, i - 1\}$;   ▷ Worlds that require this cardinality
22:             $b \leftarrow b_X^i$;   ▷ Mark cardinalities as already realized
23: **return** $\langle\langle(card, \omega), \iota\rangle, \Pi\rangle$;

---

is the computation of the cardinalities in line (5) which is achieved in time exponential in $max(||\Sigma||, |R|)$, where $|R|$ denotes the number of attributes in R.

Theorem 8. *The time complexity to find an Armstrong PC-sketch for a given set Σ of pCCs over schema R is precisely exponential in $max(||\Sigma||, |R|)$.*

There are also cases where the number of tuples in some Armstrong PC-sketch for Σ over R is logarithmic in $||\Sigma||$. Such a case is given by $R_n = \{A_1, \ldots, A_{2n}\}$ and $\Sigma_n = \{(card(X_1 \cdots X_n) \leq 1, \geq 1) \mid X_i \in \{A_{2i-1}, A_{2i}\}$ for $i = 1, \ldots, n\}$ with $||\Sigma_n|| = n \cdot 2^n$. There is an Armstrong PC-sketch for Σ that contains only one tuple for each of the n duplicate sets $X = R - \{A_{2i-1}, A_{2i}\}$ with associated cardinality $b_X^1 = \infty$.

In practice we recommend to use both representations of business rules: one in the form of the set Σ of pCCs itself and one in the form of an Armstrong PC-sketch for Σ. This is always possible by our results. We think Armstrong PC-sketches help identify bounds b that are too low or probabilities p that are too high, while the set Σ helps identify bounds b that are too high or probabilities p that are too low.

Graphical User Interface. We have implemented Algorithm 1 in the form of a graphical user interface (GUI) called *Fortuna*[1]. A user can enter some attributes and specify probabilistic cardinality constraints using any combination of these. The GUI shows an Armstrong PC-sketch for the specified input, sketches of the possible

PC–Sketch				
card	rfid	time	zone	W
4	*	*	v_zone,1	1, 2, 3
2	v_rfid,2	*	v_zone,2	1, 2, 3
1	v_rfid,2	*	v_zone,2	2, 3
3	v_rfid,3	*	*	1, 2, 3
3	*	v_time,4	*	1, 2, 3
2	*	v_time,5	v_zone,5	2, 3
1	*	v_time,5	v_zone,5	3
3	v_rfid,6	v_time,6	*	2, 3

Probability Distribution over Worlds			Possible World W1			
Index	P		card	rfid	time	zone
1	0.75		4	*	*	v_zone,1
2	0.15		2	v_rfid,2	*	v_zone,2
3	0.1		3	v_rfid,3	*	*
			3	*	v_time,4	*

Fig. 2. Screenshot of the GUI *Fortuna*

worlds can be brought up, and their individual tuples can be expanded at will. Figure 2 shows a partial screenshot of our GUI *Fortuna* with some outputs for our running example.

6 Conclusion and Future Work

Probabilistic cardinality constraints were introduced to stipulate lower bounds on the marginal probability by which a maximum number of the same data pattern can occur in sets of uncertain data. As shown in Fig. 3 the marginal probability can be used to balance the consistency and completeness targets for the quality of data, enabling organizations to derive more value from it.

Axiomatic and algorithmic tools were developed to reason efficiently about probabilistic cardinality constraints. This can help minimize the overhead in using them for data quality purposes or deriving probabilities on the maximum number of query answers without querying any data. These applications are effectively unlocked by developing computational support in the form of probabilistic Armstrong samples for identifying the right marginal probabilities by which cardinality constraints should hold in a given application domain. Analysts and domain experts can jointly inspect Armstrong samples which point out any flaws in the current perception

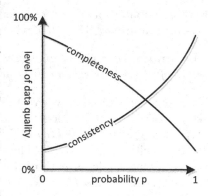

Fig. 3. Control mechanism p

of the marginal probabilities. Our tool *Fortuna* can be used to generate Armstrong samples for any input, and to explore the possible worlds it represents.

Our results constitute the core foundation for probabilistic cardinality constraints, which can be extended into various directions in future work. It will

---

[1] Available for download at https://www.cs.auckland.ac.nz/~tkr/fortuna.html.

be interesting to raise the expressivity of probabilistic cardinality constraints by allowing the stipulation of lower bounds on the number of the same data patterns, and/or upper bounds on the marginal probabilities, for examples. For a given PC-table it would be interesting to develop efficient algorithms that compute the marginal probability by which cardinality constraints hold on the data the table represents. Experiments with our implementation are expected to provide further insight into the average case performance of Algorithm 1 in relationship to the worst- and best-cases discussed. Finally, it would be interesting to conduct an empirical investigation into the usefulness of our framework for acquiring the right marginal probabilities of cardinality constraints in a given application domain. This will also require us to extend empirical measures from certain [11] to probabilistic data sets. Particularly intriguing will be the question which of Armstrong PC-sketches and Armstrong p-sketches are actually more useful. While Armstrong PC-sketches are more concise, they may prove to be too concise to draw the attention of analysts and domain experts to critical constraint violations.

Acknowledgement. This research is supported by the Marsden fund council from Government funding, administered by the Royal Society of New Zealand.

References

1. Beeri, C., Dowd, M., Fagin, R., Statman, R.: On the structure of Armstrong relations for functional dependencies. J. ACM **31**(1), 30–46 (1984)
2. Brown, P., Link, S.: Probabilistic keys for data quality management. In: Zdravkovic, J., Kirikova, M., Johannesson, P. (eds.) CAiSE 2015. LNCS, vol. 9097, pp. 118–132. Springer, Heidelberg (2015)
3. Chen, P.P.: The Entity-Relationship model - toward a unified view of data. ACM Trans. Database Syst. **1**(1), 9–36 (1976)
4. Currim, F., Neidig, N., Kampoowale, A., Mhatre, G.: The CARD system. In: Parsons, J., Saeki, M., Shoval, P., Woo, C., Wand, Y. (eds.) ER 2010. LNCS, vol. 6412, pp. 433–437. Springer, Heidelberg (2010)
5. Ferrarotti, F., Hartmann, S., Link, S.: Efficiency frontiers of XML cardinality constraints. Data Knowl. Eng. **87**, 297–319 (2013)
6. Hartmann, S., Köhler, H., Leck, U., Link, S., Thalheim, B., Wang, J.: Constructing Armstrong tables for general cardinality constraints and not-null constraints. Ann. Math. Artif. Intell. **73**(1–2), 139–165 (2015)
7. Hartmann, S., Link, S.: Efficient reasoning about a robust XML key fragment. ACM Trans. Database Syst. 34(2) (2009)
8. Hartmann, S., Link, S.: Numerical constraints on XML data. Inf. Comput. **208**(5), 521–544 (2010)
9. Jones, T.H., Song, I.Y.: Analysis of binary/ternary cardinality combinations in entity-relationship modeling. Data Knowl. Eng. **19**(1), 39–64 (1996)
10. Koehler, H., Link, S., Prade, H., Zhou, X.: Cardinality constraints for uncertain data. In: Yu, E., Dobbie, G., Jarke, M., Purao, S. (eds.) ER 2014. LNCS, vol. 8824, pp. 108–121. Springer, Heidelberg (2014)

11. Langeveldt, W.D., Link, S.: Empirical evidence for the usefulness of Armstrong relations in the acquisition of meaningful functional dependencies. Inf. Syst. **35**(3), 352–374 (2010)
12. Liddle, S.W., Embley, D.W., Woodfield, S.N.: Cardinality constraints in semantic data models. Data Knowl. Eng. **11**(3), 235–270 (1993)
13. Liu, J., Li, J., Liu, C., Chen, Y.: Discover dependencies from data - a review. IEEE Trans. Knowl. Data Eng. **24**(2), 251–264 (2012)
14. McAllister, A.J.: Complete rules for n-ary relationship cardinality constraints. Data Knowl. Eng. **27**(3), 255–288 (1998)
15. Queralt, A., Artale, A., Calvanese, D., Teniente, E.: OCL-Lite: Finite reasoning on UML/OCL conceptual schemas. Data Knowl. Eng. **73**, 1–22 (2012)
16. Roblot, T., Link, S.: Probabilistic cardinality constraints. Tech. Rep. 481. https://www.cs.auckland.ac.nz/research/groups/CDMTCS/researchreports/ (2015)
17. Suciu, D., Olteanu, D., Ré, C., Koch, C.: Probabilistic Databases, Synthesis Lectures on Data Management. Morgan and Claypool Publishers, San Rafael (2011)
18. Thalheim, B.: Entity-Relationship Modeling. Springer, Heidelberg (2000)
19. Thalheim, B.: Integrity constraints in (conceptual) database models. In: Kaschek, R., Delcambre, L. (eds.) The Evolution of Conceptual Modeling. LNCS, vol. 6520, pp. 42–67. Springer, Heidelberg (2011)

SQL Data Profiling of Foreign Keys

Mozhgan Memari, Sebastian Link[✉], and Gillian Dobbie

Department of Computer Science, University of Auckland,
Auckland, New Zealand
{m.memari,s.link,g.dobbie}@auckland.ac.nz

Abstract. Referential integrity is one of the three inherent integrity rules and can be enforced in databases using foreign keys. However, in many real world applications referential integrity is not enforced since foreign keys remain disabled to ease data acquisition. Important applications such as anomaly detection, data integration, data modeling, indexing, reverse engineering, schema design, and query optimization all benefit from the discovery of foreign keys. Therefore, the profiling of foreign keys from dirty data is an important yet challenging task. We raise the challenge further by diverting from previous research in which null markers have been ignored. We propose algorithms for profiling unary and multi-column foreign keys in the real world, that is, under the different semantics for null markers of the SQL standard. While state of the art algorithms perform well in the absence of null markers, it is shown that they perform poorly in their presence. Extensive experiments demonstrate that our algorithms perform as well in the real world as state of the art algorithms perform in the idealized special case where null markers are ignored.

Keywords: Data profiling · Foreign key · Null · Semantics in data · SQL

1 Introduction

Motivation. Domain, entity, and referential integrity form the three inherent integrity constraints in relational database systems [5]. Entity integrity is enforced by primary keys to ensure that entities of an application domain are uniquely represented within a database [12]. Referential integrity is enforced by foreign keys to ensure that the intrinsic links between entities of an application domain are correctly represented within a database [12]. In the TPC-C database, the ORDER table contains the foreign key o_c_id, o_d_id, o_w_id, which references the primary key c_id, c_d_id, c_w_id of the CUSTOMER table. The primary key uniquely identifies a customer c_id within a district c_d_id of a warehouse c_w_id. The foreign key references the unique customer in a district of a warehouse who placed the order. Primary and foreign keys are fundamental building blocks of database system, and are indispensable for most database applications [5,12].

In practice, designers fail to specify primary and foreign keys for various reasons. For example, they do not comprehend well enough the intrinsic links between data in the application domain, it is infeasible to enforce integrity due

© Springer International Publishing Switzerland 2015
P. Johannesson et al. (Eds.): ER 2015, LNCS 9381, pp. 229–243, 2015.
DOI: 10.1007/978-3-319-25264-3_17

to inconsistencies arising from data integration or database evolution over time, or because integrity enforcement will inhibit data acquisition [17]. Databases at enterprise level frequently contain hundreds of tables and thousands of columns, and are often insufficiently documented. In such situations it difficult to identify foreign keys. As a consequence, the discovery of primary and foreign keys is an important yet challenging core activity of data profiling [13]. This is due to the large number of candidates and the fact that the given data source may not even satisfy the meaningful primary and foreign keys; since these are not always enforced. Previous work on profiling primary and foreign keys has exclusively focused on purely relational databases, in which null markers are treated naively as any other domain value [17]. In the real world, null markers occur frequently and the SQL standard recommends a designated treatment for them.

Goal. This background motivates our objective to establish algorithms that efficiently profile unary and multi-column foreign keys from dirty and incomplete data. Efficiency refers to the following factors: (i) there should be a different algorithm for each of the semantics for foreign keys as proposed by the SQL standard, i.e. simple, full, and partial; (ii) the precision and recall of the algorithms should be similar to those of the state of the art algorithm for the idealized special case in which null markers are treated naively, and (iii) the profiling time of our algorithms should be similar to that of the state of the art algorithm.

Key Idea. Let X denote the sequence of distinct foreign key columns on table R_1 and Y the equal-length sequence of distinct primary key columns on R_2. Then we write $R_1[X] \subseteq R_2[Y]$ to denote the foreign key on R_1. For example, ORDER[o_c_id, o_d_id, o_w_id] $\subseteq$ CUSTOMER[c_id, c_d_id, c_w_id] denotes the foreign key on ORDER from the TPC-C benchmark. For the projection $r[Z]$ of a table instance r over table R onto the columns in sequence Z over R, the *inclusion coefficient* between $r_1[X]$ and $r_2[Y]$ has been defined as $\sigma(r_1[X], r_2[Y]) = |r_1[X] \cap r_2[Y]|/|r_1[X]|$ where $|S|$ denotes the number of elements in S. Intuitively, the inclusion coefficient $\sigma(r_1[X], r_2[Y])$ measures the proportion of values in $r_1[X]$ that are also in $r_2[Y]$. The higher $\sigma(r_1[X], r_2[Y])$ the more likely it is that $R_1[X] \subseteq R_2[Y]$ forms a true foreign key. However, the inclusion coefficient has only been defined for relations in which no null markers occur [14,17]. The SQL standard recommends three different options for the semantics of foreign keys, which can be declared in the MATCH clause as simple, full, and partial [12]. The semantics differ in how they treat tuples with occurrences of NULL in foreign key columns. Simple semantics regards such tuples as non-offensive, while full semantics regards them as offensive. Finally, partial semantics regards tuples $t \in r_1[A_1, \ldots, A_n]$ only as offensive if there is no tuple $t' \in r_2[B_1, \ldots, B_n]$ that *subsumes* t, i.e., where $t[A_i] = t'[B_i]$ on all attributes A_i where $t[A_i] \neq$ NULL. We say that a tuple t is X-total on a sequence X of distinct attributes, if $t[A] \neq$ NULL on all $A \in X$. Furthermore, we call the tuple t over table R total if it is R-total, and a relation r over R total if all tuples in r are total. We thus obtain the notions of simple σ_s, full σ_f, and partial inclusion coefficients σ_p by defining

Table 1. TPC-C foreign key and its different inclusion coefficients

ORDER				CUSTOMER		
... o_c_id	o_d_id	o_w_id ...		... c_id	c_d_id	c_w_id ...
1	2	1		1	2	1
NULL	2	2		1	2	2
3	3	3		3	3	3
NULL	5	3		2	5	1
2	3	2		2	5	2
3	6	3		3	6	3
3	NULL	3		4	9	2
4	2	2		4	2	2
4	NULL	2		7	9	3
2	5	NULL		6	5	1

$$\sigma_f = 4/10 = 0.4$$
$$\sigma_s = 9/10 = 0.9$$
$$\sigma_p = 8/10 = 0.8$$

- $r_1[X] \cap_s r_2[Y] := \{t \in r_1[X] \mid$ if t is X-total, then $t \in r_2[Y]\}$,
- $r_1[X] \cap_f r_2[Y] := \{t \in r_1[X] \mid t$ is X-total and $t \in r_2[Y]\}$, and
- $r_1[X] \cap_p r_2[Y] := \{t \in r_1[X] \mid \exists t' \in r_2[Y]$ such that t is subsumed by $t'\}$.

Indeed, when $r_1[X]$ and $r_2[Y]$ are total relations, simple σ_s, full σ_f, and partial inclusion coefficients σ_p all coincide with the standard inclusion coefficient.

Example 1. For illustration consider the foreign key *FK* ORDER[o_c_id, o_d_id, o_w_id] $\subseteq$ CUSTOMER[c_id, c_d_id, c_w_id] of the TPC-C schema. Table 1 shows sample data that violates the foreign key under all three semantics, since the fifth tuple - which is total - is not present in CUSTOMER. An intriguing question is if *FK* can still be discovered as a good foreign key candidate. For total relations this question has been answered affirmatively [17], but our data features NULL.

Indeed, $\cap_f$ consists of only four tuples since full semantics regards incomplete tuples as offensive, $\cap_s$ consists of nine tuples since simple semantics regards incomplete tuples as non-offensive, and $\cap_p$ consists of eight tuples since the fourth and fifth ORDER tuples are not subsumed by any CUSTOMER tuple.   □

Example 1 illustrates that the three null marker semantics, proposed by the SQL standard, require different techniques to profile foreign keys.

Contributions. Firstly, we demonstrate that current methods for foreign key profiling over complete data are not suitable for incomplete data. These methods already exhibit poor precision and recall when few null markers occur. Secondly, based on the recommended semantics of foreign keys by the SQL standard, we propose three notions of inclusion coefficients. Each notion leads to a different algorithm for profiling SQL foreign keys in dirty and incomplete data. We focus on foreign keys that reference a primary key, the most common case in practice. Thirdly, experiments with benchmark data demonstrate that our algorithms for simple and partial semantics perform as well for dirty and incomplete data as the current state of the art algorithm performs in the idealized special case where incomplete data is absent. Our results for precision and recall are robust under different rates of incompleteness. The penalty for sustaining strong profiling results for dirty and incomplete data is only marginal when comparing our profiling time to that of the state of the art for complete data.

Organization. In Sect. 2 we recall the three semantics the SQL standard proposes for foreign keys. Our profiling algorithms are proposed in Sect. 3. The algorithms are evaluated in Sect. 4. Their combination with schema-driven techniques is evaluated in Sect. 5. Related work is discussed in Sect. 6 before we conclude and comment on future work in Sect. 7.

2   Referential Integrity and the SQL Standard

Inclusion dependencies, and foreign keys in particular, are referential integrity constraints that enforce the intrinsic links between data items of an application domain within a database system [8,11]. An *inclusion dependency* (IND) over a database schema $\mathbf{R}$ is an expression of the form $R_1[X] \subseteq R_2[Y]$, where R_1, R_2 are relation schemata of $\mathbf{R}$, and X and Y are equal-length sequences of distinct columns from R_1 and R_2, respectively. Having $X = f_1, \ldots, f_n$ and $Y = k_1, \ldots, k_n$, the inclusion dependency holds in the given database over $\mathbf{R}$ if for every tuple t in the relation over R_1 there is some tuple t' in the relation over R_2 such that $t[f_i] = t'[k_i]$ for $1 \leq i \leq n$. An inclusion dependency $R_1[X] \subseteq R_2[Y]$ is a *foreign key* if Y forms a key of R_2 [11].

Match Options to Handle Nulls. An exception to the semantics of foreign keys are occurrences of the null marker NULL. If some foreign key column features an occurrence of NULL, then the semantics of a foreign key is defined by the MATCH clause of the SQL standard [12]. Available options are full, simple and partial, each proposing different ways to handle occurrences of NULL.

Under *full* semantics, the foreign key $R_1[f_1, \ldots, f_n] \subseteq R_2[k_1, \ldots, k_n]$ is satisfied if every tuple $t \in r_1$ over R_1 is $\{f_1, \ldots, f_n\}$-total and there is some tuple $t' \in r_2$ over R_2 such that $t[f_i] = t'[k_i]$ for all $i = 1, \ldots, n$. Hence, full semantics does not permit any occurrences of NULL on any of the foreign key columns.

Under *simple* semantics, the foreign key $R_1[f_1, \ldots, f_n] \subseteq R_2[k_1, \ldots, k_n]$ is satisfied if for every tuple $t \in r_1$ over R_1, either $t[f_i] = $ NULL for some $i \in \{1, \ldots, n\}$, or there is some tuple $t' \in r_2$ over R_2 such that $t[f_i] = t'[k_i]$ for all $i = 1, \ldots, n$. Hence, simple referential integrity is never violated by tuples that are partially defined on the foreign key columns.

Under *partial* semantics, the foreign key $R_1[f_1, \ldots, f_n] \subseteq R_2[k_1, \ldots, k_n]$ is satisfied if for every tuple $t \in r_1$ over R_1 there is some tuple $t' \in r_2$ over R_2 such that $t[f_1, \ldots, f_n]$ is subsumed by $t'[k_1, \ldots, k_n]$. Hence, partial referential integrity may also be violated by tuples that are partially defined on the foreign key columns. It can be understood as a compromise that balances the pessimistic full semantics and the optimistic simple semantics.

In Table 1, only tuples 1, 3, 6, and 8 of the ORDER table satisfy full semantics, while all tuples except for tuple 5 satisfy simple semantics, and all tuples except for tuples 4 and 5 satisfy partial semantics.

3   Profiling Foreign Keys from Dirty and Incomplete Data

We propose two-step algorithms for profiling full, simple, and partial foreign keys from dirty and incomplete data. In phase one, we apply full, simple, and partial inclusion coefficients to identify good candidates for foreign keys. In phase two, a randomness test validates whether the distribution of the foreign key values is similar to that of the key values.

Inclusion Coefficients for Incomplete Data. The first challenge is to find reasonable candidates from data that may violate meaningful foreign keys. Similar to [2,6,17], a foreign key is a candidate when its inclusion coefficient meets a user-defined threshold θ. We introduce three different inclusion coefficients, resulting from the SQL standard semantics for null markers.

Example 2. When $\theta = 0.8$, 20 % inconsistencies in the given data are considered acceptable. The simple and partial coefficients of the foreign key *FK* from Example 1 meet θ, but the full inclusion coefficient does not.   □

Exact computations of inclusion coefficients are infeasible for big data sizes, so we approximate them by bottom-k sketches [17]. A bottom-k sketch consists of those values of a given data set that have the k smallest ranks assigned to them by a hash function. For a bottom-k sketch of a multi-column key, the hash function is applied to the concatenation of the values on each column [17].

Algorithm 3.1. DISCOVERY($\mathbf{C}, \mathbf{T}, \mathbf{P_u}, \mathbf{P_m}, \theta, O$)

---

main
    for all $C \in \mathbf{C} : \hat{C} \leftarrow BottomK[C]$
    for all $P = \{C_1, \ldots, C_n\} \in (\mathbf{P_u} \cup \mathbf{P_m})$
      ⎧ **for** $p \leftarrow 1$ **to** n
      ⎪   ⎧ **for all** $\hat{C}_f \in \mathbf{C}$
do ⎨ **do** ⎨   ⎧ $\sigma :=$ ⎰ **output** (INCLUSION-S($\hat{C}_f, \hat{C}_p$)), **if** $O = \mathbf{s}$ **or** $O = \mathbf{p}$
      ⎪   ⎪ **do**⎨   ⎱ **output** (INCLUSION-F($\hat{C}_f, \hat{C}_p$)), **if** $O = \mathbf{f}$
      ⎪   ⎩   ⎨ **if** $\sigma \geq \theta$
      ⎩   ⎩ **then** ⎰ $\mathbf{F_u} \leftarrow (C_p, C_f)$, **if** $n = 1$
                       ⎱ $S[P, C_p] \leftarrow C_f$, **if** $n > 1$
    for all $P = \{C_1, \ldots, C_n\} \in \mathbf{P_m}$
      ⎧ **for all** $\mathbf{T} \in schema$
      ⎪   **do** $F_m \leftarrow (\{\{C'_1, \ldots, C'_n\}|C'_i \in S[P, C_i] \cap T\}, P)$
      ⎪ **for all** $F = (\{C'_1, \ldots, C'_n\}, P) \in F_m$
      ⎪   ⎧ $\hat{F} \leftarrow BottomK[F]$
do ⎨   ⎪ $\hat{P} \leftarrow BottomK[P]$
      ⎪   ⎪   ⎧ **output** (INCLUSIONS($\hat{F}, \hat{P}$)), **if** $O = \mathbf{s}$
      ⎪ **do**⎨ $\sigma :=$ ⎨ **output** (INCLUSIONF($\hat{F}, \hat{P}$)), **if** $O = \mathbf{f}$
      ⎪   ⎪   ⎩ **output** (INCLUSIONP($\hat{F}, \hat{P}$)), **if** $O = \mathbf{p}$
      ⎩   ⎩ **if** $\sigma(\hat{F}, \hat{P}) \geq \theta$
             then $\mathbf{F_m} \leftarrow (F, P)$
    return $(\mathbf{F_u} \cup \mathbf{F_m})$

Algorithm 3.1 shows the pseudo-code of our proposed algorithm. Like in [17] we assume that a set of unary ($\mathbf{P_u}$) and multi-column ($\mathbf{P_m}$) primary keys has been obtained already. $\mathbf{C}$ and $\mathbf{T}$, respectively, refer to the set of columns and tables in the given schema. The option O for the MATCH clause of the SQL standard are denoted by $\mathbf{f}$ for full, $\mathbf{s}$ for simple, and $\mathbf{p}$ for partial. $\hat{C}$ refers to the bottom-k sketch of column C. For $O \in \{\mathbf{f}, \mathbf{s}, \mathbf{p}\}$, INCLUSION-O($\hat{F}, \hat{P}$) computes an approximation of the full, simple, or partial inclusion coefficient by counting the number of foreign key tuples in the current bottom-k sketch $\hat{F}$ that meet the requirements of full, simple, or partial semantics, respectively, and dividing this number by the total number of foreign key tuples in $\hat{F}$. If f_j denotes the jth foreign key tuple in $\hat{F}$, then f_j meets the requirements of (i) full semantics, when $f_j[C'] \neq$ NULL for all columns C' and f_j occurs in $\hat{P}$; (ii) simple semantics, when either $f_j[C'] =$ NULL for some column C' or f_j occurs in $\hat{P}$; and (iii) partial semantics, when there is some $p \in \hat{P}$ that subsumes f_j.

Algorithm 3.1 starts by profiling unary inclusion dependencies as in [6,17]. The novel idea is to profile foreign keys under different SQL semantics. For unary candidates, simple and partial semantics coincide. Thus, INCLUSION-S computes unary inclusion coefficients under simple and partial semantics.

All candidates for unary inclusion dependencies are stored in memory for profiling multi-column foreign keys. If a unary inclusion dependency references some $P \in \mathbf{P}_u$, it is stored as a unary foreign key in $\mathbf{F}_u$. If there is some multi-column primary key, Algorithm 3.1 calculates the inclusion coefficient for a set of unary inclusion dependencies which occur in one table and pair-wisely reference the columns of this primary key. Finally, all pairs (F, P) which pass the inclusion test are stored in $\mathbf{F}_m$. Algorithm 3.1 returns a set $\mathbf{F_u} \cup \mathbf{F_m}$ of unary and multi-column candidate foreign keys.

Example 3. Consider the TPC-C data set from Example 1 and a threshold of $\theta = 0.9$. Algorithm 3.1 calculates inclusion coefficients for unary candidates first:

	$|\text{ORDER}[o\_c\_id, o\_d\_id, o\_w\_id]|$	σ_p	σ_s	σ_f
$O[o\_c\_id] \subseteq C[c\_id]$	5	1	1	0.9
$O[o\_d\_id] \subseteq C[c\_d\_id]$	5	1	1	0.9
$O[o\_w\_id] \subseteq C[c\_w\_id]$	4	1	1	0.75
FK	10	0.8	0.9	

In particular, $\sigma_f(o\_w\_id, c\_w\_id) = 0.75 < \theta$, which means this unary inclusion dependency does not meet the requirements of full semantics. Therefore, our FK ORDER$[o\_c\_id, o\_d\_id, o\_w\_id] \subseteq$ CUSTOMER$[c\_id, c\_d\_id, c\_w\_id]$ is not further considered under full semantics. Final results show that this foreign key is profiled under simple but not under partial semantics, since $\sigma_p = 0.8 < \theta$. □

Testing Randomness on Dirty and Incomplete Data. The randomness test from [14,17] checks if the distinct values of the referencing column set X have the same distribution as the distinct values in the referenced column set Y. This test is good for eliminating false positives [14,17]. For complete data the Earth Mover's Distance (EMD) represents the least amount of work required to move the set of values in the candidate foreign key to the set of values in the referenced primary key. The smaller the EMD the closer the distributions for the candidate foreign and primary key values. We extend the randomness test of [14,17] from complete to incomplete data. Our method measures the likelihood of the candidate pair (F, P) dependent on either full, simple, or partial semantics.

Algorithm 3.2 returns a set of candidates ranked in increasing order of their EMDs. Firstly, Algorithm 3.2 applies the extended randomness test to further prune the candidates found by Algorithm 3.1. As in [17], we apply quantile histograms to approximate EMDs. The histogram is calculated by a user-defined quantile constant for every column of a primary key. For multi-column keys, a quantile grid is constructed from the quantile histogram of each column. The grid for a candidate foreign key is constructed by populating the associated primary key grid with the values in its sketch. The approximate EMD is computed from the distance between the quantile grids of primary and foreign key. We empower this method to deal with real world data by imputing null markers with actual domain values that are consistent with simple and partial semantics, respectively. The procedure SIMPLE replaces the incomplete bottom-k sketch $\hat{F}$ of the given foreign key F by the completed sketch $\hat{F}_{total}$ in which NULL occurrences in $\hat{F}$ have been imputed with randomly chosen domain values from the referenced primary key sketch $\hat{P}$. This is consistent with the simple semantics. The procedure PARTIAL imputes NULL occurrences in *subsumed* foreign key tuples t with randomly chosen domain values from tuples of the primary key sketch that subsume t. This is consistent with the partial semantics. Based on the completed sketches, Algorithm 3.2 proceeds by approximating the EMD of candidate foreign keys based on quantiles of $\hat{F}_{total}$.

Algorithm 3.2. RANDOMNESS TEST($\mathbf{C}, \mathbf{T}, \mathbf{P_u}, \mathbf{P_m}, \theta, O$)

---

procedure PARTIAL($\hat{F}, \hat{P}$)
$F_{total} \leftarrow \emptyset$
for all $j \leftarrow 1$ **to** $|\hat{F}|$
$\quad$**do** $\begin{cases} \textbf{if } \{C'_i \in C' | \hat{f}_j[C'_i] = \text{NULL}\} \neq \emptyset \textbf{ and } \exists \hat{p} \in \hat{P}(\hat{p} \text{ subsumes } \hat{f}_j) \\ \quad \textbf{then } \hat{f}_j[C'] \leftarrow \hat{p}[C] \\ F_{total} \leftarrow F_{total} \cup \{\hat{f}_j[C']\} \end{cases}$
return (F_{total})

procedure SIMPLE($\hat{F}, \hat{P}$)
$F_{total} \leftarrow \emptyset$
for all $j \leftarrow 1$ **to** $|\hat{F}|$
$\quad$**do** $\begin{cases} \textbf{if } \exists C'_i(\hat{f}_j[C'_i] = \text{NULL}) \\ \quad \textbf{then } \hat{f}_j[C'] \leftarrow \hat{p}[C] \\ F_{total} \leftarrow F_{total} \cup \{\hat{f}_j[C']\} \end{cases}$
return (F_{total})

main
$\mathbf{F_u} \cup \mathbf{F_m} \leftarrow$ **output** (DISCOVERY($\mathbf{C}, \mathbf{T}, \mathbf{P_u}, \mathbf{P_m}, \theta, O$))
for all $P = \{C_1, \ldots, C_n\} \in \{F_m, P_m\} \cup \{F_u, P_u\}$
$\quad$**do** $\begin{cases} Q[\hat{P}] \leftarrow Quantile[\hat{P}] \\ \textbf{for all } F \in \{F_m, P_m\} \\ \quad \textbf{do} \begin{cases} \hat{F}_{total} := \begin{cases} \textbf{output } (\text{PARTIAL}(\hat{F}, \hat{P})), \textbf{if } O = \mathbf{p} \\ \textbf{output } (\text{SIMPLE}(\hat{F}, \hat{P})), \textbf{if } O = \mathbf{s} \\ \hat{F}, \textbf{if } O = \mathbf{u} \end{cases} \\ Q[\hat{F}_{total}] \leftarrow Quantile[\hat{F}_{total}] \\ L \leftarrow (F_m, EMD_n(Q[\hat{F}_{total}], Q[\hat{P}])) \end{cases} \\ \textbf{for all } F \in \{F_u, P_u\} \\ \quad \textbf{do} \begin{cases} Q[\hat{F}] \leftarrow Quantile[\hat{F}] \\ L \leftarrow (F_u, EMD_n(Q[\hat{F}], Q[\hat{P}])) \end{cases} \end{cases}$
return (L)

Example 4. Applying PARTIAL from Algorithm 3.2 to Example 1 may yield the following completed sketch: $\{(1,2,1), (1,2,2), (3,3,3), (\text{NULL},5,3), (2,3,2), (3,6,3),$

(4,2,2), (4,9,2), (2,5,2)}. An alternative is {(1,2,1), (4,2,2), (3,3,3), (NULL,5,3), (2,3,2), (3,6,3), (2,5,1)}. In each completion, all partial tuples are replaced by complete tuples that subsume them. Tuple (NULL,5,3) is not subsumed by any primary key tuple and violates partial semantics. Applying SIMPLE produces a completed sketch in which all NULL occurrences of (NULL,2,2), (NULL,5,3), (3,NULL,3), (4,NULL,2), (2,5,NULL) are imputed by random domain values from the referenced primary key, for example by (1,2,2), (3,5,3), (4,5,2), (2,5,3). Even (NULL,5,3) is replaced, in consistency with simple semantics.   □

4   Evaluation of Data-Driven Profiling Techniques

We evaluate the data-driven profiling of foreign keys from real world data sets.

Characteristics of Experiments. Our algorithms are evaluated on two benchmark databases TPC-C and TPC-H[1]. The characteristics of the data sets are summarized as follows:

Data set	#tables	#rows	Data set size	#unary FKs	#multi-column FKs
TPC-C	9	2.41M	0.39G	3	7
TPC-H	9	9.42M	1.43G	7	1

Algorithms were implemented in C++ and run on an Intel Core i5 CPU 3.3 GHz with 8 GB RAM. The operating system was 64-bit Windows 7 Enterprise, Service pack 1. The database management system we used was MySQL version 5.5. Our algorithms are evaluated in terms of accuracy and time with respect to increasing levels of incompleteness in the data, starting with complete data. We use three well-known measures of accuracy: precision, recall and f-measure. For each data set, we consider the constraints which are explicitly declared in the schema as the golden standard for determining the accuracy of our algorithms. This ensures that we can compare our results to those of [17] who followed the same approach. In TPC-C, there are three single-column, three 2-column ($Cu \subseteq Di$, $Hi \subseteq Di$, $OL \subseteq St$) and four 3-column ($Hi \subseteq Cu$, $Or \subseteq Cu$, $OL \subseteq Or$, $NO \subseteq Or$) foreign keys (FKs). In TPC-H, 7 of the 8 foreign keys are single-column and the only composite foreign key has 2-columns. We randomly generated null marker occurrences in the foreign key columns. Starting from the given complete data set, null marker occurrences were randomly generated for foreign key columns, and in different percentages of tuples to pinpoint the impact of different levels of incompleteness on our measures. The names of the resulting data sets originate from the original data set augmented by the percentage of nullified tuples. For example, TPC-C2 results from TPC-C where 2 % of the tuples had null markers in their foreign key columns.

Profiling Accuracy and Time. We applied Algorithm 3.2 to the benchmark data sets with increasing levels of incompleteness under simple, partial and full

---

[1] http://www.tpc.org/tpcc/ and http://www.tpc.org/tpch/.

Table 2. Data profiling of SQL foreign keys from incomplete benchmark variants

	TPC-C									
	Simple				Partial				Full		
	t_p f_p	$T_{3.1}$	$T_{3.2}$	t_p f_p	$T_{3.1}$	$T_{3.2}$	t_p f_p	$T_{3.1}$	$T_{3.2}$
Dataset		hr:mn	hr:mn		hr:mn	hr:mn		hr:mn	hr:mn
TPC-C	9 165	0:47	3:09	9 165	0:47	3:09	9 165	0:34	2:44
TPC-C2	9 166	0:41	3:29	9 162	0:58	3:47	3 24	0:33	1:14
TPC-C5	9 170	0:40	3:32	9 164	0:41	3:31	1 21	0:26	0:56
TPC-C10	8 160	0:45	3:09	8 155	1:02	3:24	1 21	0:29	0:42
TPC-C25	9 149	0:41	2:41	9 145	0:59	3:03	1 20	0:23	0:35
TPC-C50	9 167	0:36	3:22	9 162	0:52	3:38	1 20	0:30	0:57
TPC-C75	9 150	0:48	2:55	7 145	1:05	3:12	1 20	0:27	0:45

	TPC-H									
	Simple				Partial				Full		
	t_p f_p	$T_{3.1}$	$T_{3.2}$	t_p f_p	$T_{3.1}$	$T_{3.2}$	t_p f_p	$T_{3.1}$	$T_{3.2}$
Dataset		hr:mn	hr:mn		hr:mn	hr:mn		hr:mn	hr:mn
TPC-H	7 40	0:50	1:36	7 40	0:50	1:36	7 40	0:49	1:35
TPC-H2	7 52	1:12	2:09	6 46	1:24	2:12	6 46	0:56	1:44
TPC-H5	7 46	1:14	2:04	6 40	1:22	2:07	5 39	0:55	1:40
TPC-H10	7 46	1:17	2:09	6 40	1:25	2:11	5 39	0:56	1:43
TPC-H25	7 46	1:13	2:04	6 40	1:25	2:06	5 39	0:54	1:40
TPC-H50	5 45	0:59	1:46	4 42	1:07	1:49	3 37	0:51	1:30
TPC-H75	5 39	1:02	1:48	4 38	1:06	1:44	3 37	0:51	1:28

semantics, respectively. As in [17] we applied $\theta = 0.9$, bottom-256 sketches, 256-quantiles for unary, and 16-quantiles for multi-column foreign key candidates. The results are summarized in Fig. 1 and Table 2. Here, t_p denotes the number of true positives, f_p the number of false positives, $T_{3.2}$ the overall time of running Algorithm 3.2 in hours and minutes (hr:mn), and $T_{3.1}$ indicates the time spent on Algorithm 3.1 as part of running Algorithm 3.2. *Partial-PK* denotes an optimization technique discussed in Sect. 5.

Our first observation is that on the original data sets TPC-C and TPC-H, respectively, all accuracy measures agree under all 3 different semantics. The sweet spot for balancing precision and recall is given by the top-20 % ranked foreign keys discovered from TPC-C, which yield an f-measure of 0.36. For TPC-H, the f-measure is around 0.4 when the sweet spots of top-55 % or -60 % ranked foreign keys are considered, see Table 3.

Our second observation is that full semantics is inadequate for SQL data profiling of foreign keys. Since this is the semantics applied by [17], their good performance for complete data does not carry over to incomplete data. The poor performance is more pronounced for TPC-C, since there are more multi-column foreign keys. Just having null markers present in 2 % of all tuples, lowers the recall for full semantics from 0.9 to 0.3. Introducing null markers in 5 % of all tuples brings the recall down to 0.1, and this foreign key no longer features amongst the top-25 %. On TPC-H with 7 unary foreign keys and 1 binary foreign key, the

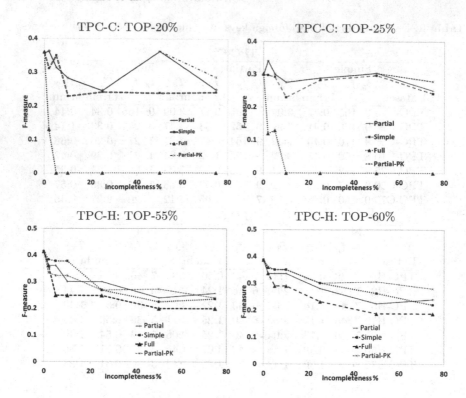

Fig. 1. F-measure: TPC-C and TPC-H

f-measure under full semantics also declines as soon as null markers occur. In particular, the binary foreign key is never profiled under full semantics on any data set with some level of incompleteness.

Our third observation is that both simple and partial semantics perform very well in profiling foreign keys. Their measures of accuracy remain robust under increasing levels of incompleteness and are competitive with the corresponding measures on complete data, as Fig. 1 and Table 2 show.

Our final observation is that the robustness of simple and partial semantics under incomplete data comes at a very reasonable increase in profiling time, when compared to the state of the art algorithm for complete data. It is natural that the profiling time increases with the recall. However, spending 3hr:47mn - which is already the worst case - instead of 2hr:44mn is well worth the increase in accuracy, considering the purpose of data profiling.

Comparing Precision of Simple and Partial Semantics. While simple semantics outperforms partial semantics in recall, the roles change for precision and f-measure. Table 3 shows for each data set and semantics, the lowest percentage that captures all discovered foreign keys from the golden standard. Hence, partial semantics ranks meaningful foreign keys higher.

Table 3. Partial semantics ranks meaningful foreign keys higher

	Partial			Simple			Full		
	t_p	X%	EMD	t_p	X%	EMD	t_p	X%	EMD
TPC-C2	9	25	0.026	9	29	0.028	3	37	0.004
TPC-C5	9	31	0.04	9	36	0.047	1	27	0.033
TPC-C10	8	35	0.05	8	37	0.05	1	32	0.001
TPC-C25	9	33	0.059	9	37	0.07	1	29	0.001
TPC-C50	9	35	0.05	9	36	0.05	1	29	0.001
TPC-C75	7	36	0.054	9	38	0.062	1	38	0.001

Table 4. Semantics with lower EMD for TPC-C multi-column foreign keys

	TPC-C2	TPC-C5	TPC-C10	TPC-C25	TPC-C50	TPC-C75
$Cu \subseteq Di$	S	P	P	S	P	P
$Hi \subseteq Di$	S	S	P	P	S	P
$OL \subseteq St$	S	S	-	S	S	S
$Hi \subseteq Cu$	P	P	P	P	P	P
$Or \subseteq Cu$	P	P	P	P	P	P
$OL \subseteq Or$	P	P	S	P	S	P

Table 5. Profiling benchmarks with different sample sizes under partial semantics

	Sample Size	t_p	f_p	$T_{3.1}$ hr:mn		Sample Size	t_p	f_p	$T_{3.1}$ hr:mn
TPC-C75	128	5	131	0:41	TPC-H75	128	3	36	0:36
TPC-C75	256	7	145	1:05	TPC-H75	256	4	38	1:06
TPC-C75	512	7	152	1:12	TPC-H75	512	4	38	1:53
TPC-C75	1024	7	156	1:20	TPC-H75	1024	4	42	2:33
TPC-C75	10K	8	162	19:16	TPC-H75	10K	5	45	19:27

Table 4 shows for each of the meaningful multi-column foreign keys of TPC-C, which of simple (S) and partial (P) semantics resulted in a lower EMD. In about two thirds of the cases (23 out of 35) this was partial semantics.

Size of Sketches. As in [17] we observe that bottom-256 sketches are a sweet spot for the experiments. Bottom-128 sketches lead to drops in recall, and bottom-k sketches for $k = 512, 1024, 10K$ lead to minor improvements in recall, more false positives, and higher profiling time, see Table 5 for example.

Other Experiments. We have applied our algorithms to the CM data set about employer costs for employee compensation from US government data http://catalog.data.gov/. It consists of 11 tables featuring 11 one-attribute foreign keys and two 3-column foreign keys. On total data, we obtain a recall of 0.85 and

Table 6. TPC-C and TPC-H with partial-PK, 256-bottom sketch

	Partial					Partial-PK			
	t_p	f_p	$T_{3.1}^a$	Recall	F-measure[b]	t_p	f_p	$T_{3.1}$	Recall	F-measure[b]
TPC-C75	7	145	1:05	0.7	0.25	9	152	1:07	0.9	0.28
TPC-H75	4	38	1:06	0.5	0.24	5	41	1:06	0.63	0.28

[a]hr:mn: Running time (hour & minute) of Algorithm 3.1
[b]In top-20 % for TPC-C and top-60 % for TPC-H

precision of 0.3 under all semantics. Adding 2 %, 5 % and 10 % null values, respectively, simple and partial semantics maintain these rates, while full semantics drops to a recall of 0.31 and precision of 0.2.

5   Optimization Strategies

We propose and evaluate two strategies for improving our algorithms.

Increasing Recall. Recall under partial semantics can be improved to the level of recall for simple semantics. Under partial semantics, our algorithm is looking for a matching reference in the bottom-256 sketch of the primary key. If a match does not exist in the sketch, it may still exist in the whole table. Referring to this strategy as *Partial-PK*, Fig. 1 shows its increase in f-measure in comparison to scanning only the bottom-256 sketch of the referenced primary key. Moreover, Table 6 gives an indication of the very minor penalty in profiling time, resulting from the presence of the index for the referenced primary key.

Schema-Driven Strategies. We add two schema-driven techniques leading to significant improvements of our measures, as in [17].

The first technique concerns the golden standard. So far, it comprises only those foreign keys that were explicitly specified in the benchmark documentation. In reality, these foreign keys logically imply other foreign keys [3], which our current algorithms treat (incorrectly) as false positives. On the TPC-C schema, for example, the two foreign keys ORDER[o_c_id, o_d_id, o_w_id] $\subseteq$ CUSTOMER[c_id, c_d_id, c_w_id] and CUSTOMER[c_d_id, c_w_id] $\subseteq$ DISTRICT[d_id,d_w_id] logically imply the foreign key ORDER[o_d_id, o_w_id] $\subseteq$ DISTRICT[d_id,d_w_id]. Adding all implied foreign keys to the golden standard is more realistic and leads to an improvement of our accuracy measures.

Our second technique applies a pruning strategy from [17] regarding the similarity of column names. Finding good measures for string similarity in database schemata is an open problem and can easily lead to false positives [17]. However, the documentation for the benchmark databases is extensive and explains the naming conventions. For TPC-C, for example, we can apply string identity after removing table name prefixes from column names. The resulting names are identical only if the candidate is an implied foreign key.

Adding these techniques to our algorithms results in a perfect precision score of 1, and more than doubles our f-measure in comparison to applying the data-driven algorithms only, see Fig. 2.

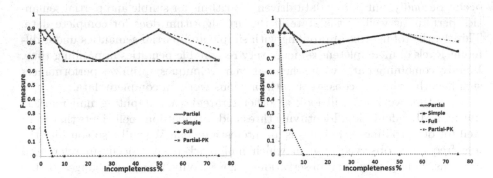

Fig. 2. F-measures after adding schema-driven techniques to TPC-C using the Top-20 % of candidates on the left and the Top-25 % of candidates on the right

6   Related Work

Previous algorithms for profiling referential integrity constraints do not consider incomplete data. The state of the art algorithm that considers dirty data is presented in [17]. The authors established a randomness test that reduces a large number of false positives and can profile foreign keys efficiently from complete data. Our evaluation on the original TPC-C data set, without introducing null markers, provides further evidence for the efficiency of those techniques [17], which did not consider TPC-C despite its high number of multi-column foreign keys compared to TPC-H and TPC-E.

State of the art algorithms for efficiently profiling uniqueness constraints are described in [7,16]. Profiling inclusion dependencies is an effective pre-cursor for profiling foreign keys [14,17]. In fact, the discovery of unary foreign keys from unary inclusion dependencies has been studied by several authors [2,6,8,11]. Data summaries and random sampling, cliques, min-hash sketches and Jaccard coefficient all constitute pruning techniques from previous work [2,4,6,9,17]. Lopes et al. [10] profile foreign keys from given SQL query workloads. Approximate inclusion dependencies constitute a different popular approach to accommodate dirty data within the data profiling framework [2,6,17]. The profiling of data quality semantics is important for big data [15].

7   Conclusion and Future Work

We have proposed the first algorithms for profiling foreign keys from dirty and incomplete data. Our techniques target the three different semantics for null marker occurrences as proposed by the SQL standard. On complete data, all three semantics coincide with the strictly relational semantics. Our detailed evaluation demonstrates several insights. Firstly, full semantics is not suitable for profiling foreign keys from incomplete data, even when only a few null markers occur. Secondly, our purely data-driven algorithms for simple and partial semantics perform as well as the state of the art algorithm does for complete data. Thirdly, the robustness of profiling with simple and partial semantics under different levels of incompleteness incurs very reasonable penalties in profiling time. Finally, combining data- with schema-driven techniques optimizes performances significantly, which is consistent with previous work on complete data.

In future work we will look at other strategies for imputing null markers during EMD calculation. Identifying thresholds of inclusion coefficients that work well for our profiling techniques is an interesting goal. We will also consider the less frequently observed case in which null markers also occur in referenced columns. Our preliminary observations suggest only marginal changes to the overall trend we have described. The extension of our techniques to profiling conditional foreign keys from dirty and incomplete data is an interesting goal for data cleaning. So far, all approaches to this problem do not address dirty data or different semantics for null markers [1].

Acknowledgement. This research is supported by the Marsden fund council from Government funding, administered by the Royal Society of New Zealand.

References

1. Bauckmann, J., Abedjan, Z., Leser, U., Müller, H., Naumann, F.: Discovering conditional inclusion dependencies. In: Chen, X., Lebanon, G., Wang, H., Zaki, M.J. (eds.) 21st ACM International Conference on Information and Knowledge Management, CIKM 2012, Maui, HI, USA, October 29–November 02, 2012, pp. 2094–2098. ACM (2012)
2. Bauckmann, J., Leser, U., Naumann, F., Tietz, V.: Efficiently detecting inclusion dependencies. In: Chirkova, R., Dogac, A., Özsu, M.T., Sellis, T.K. (eds.) Proceedings of the 23rd International Conference on Data Engineering, ICDE 2007, The Marmara Hotel, Istanbul, Turkey, 15–20 April 2007, pp. 1448–1450. IEEE (2007)
3. Casanova, M.A., Fagin, R., Papadimitriou, C.H.: Inclusion dependencies and their interaction with functional dependencies. J. Comput. Syst. Sci. **28**(1), 29–59 (1984)
4. Chen, Z., Narasayya, V.R., Chaudhuri, S.: Fast foreign-key detection in Microsoft SQL server PowerPivot for Excel. Proc. VLDB **7**(13), 1417–1428 (2014)
5. Codd, E.: A relational model of data for large shared data banks. Commun. ACM **13**(6), 377–387 (1970)
6. De Marchi, F., Lopes, S., Petit, J.M.: Unary and n-ary inclusion dependency discovery in relational databases. Int. J. Intell. Syst. **32**(1), 53–73 (2009)
7. Heise, A., Quiané-Ruiz, J., Abedjan, Z., Jentzsch, A., Naumann, F.: Scalable discovery of unique column combinations. Proc. VLDB **7**(4), 301–312 (2013)
8. Kantola, M., Mannila, H., Räihä, K.J., Siirtola, H.: Discovering functional and inclusion dependencies in relational databases. Int. J. Intell. Syst. **7**(7), 591–607 (1992)

9. Koeller, A., Rundensteiner, E.A.: Heuristic strategies for the discovery of inclusion dependencies and other patterns. In: Spaccapietra, S., Atzeni, P., Chu, W.W., Catarci, T., Sycara, K. (eds.) Journal on Data Semantics V. LNCS, vol. 3870, pp. 185–210. Springer, Heidelberg (2006)

10. Lopes, S., Petit, J., Toumani, F.: Discovering interesting inclusion dependencies: application to logical database tuning. Inf. Syst. **27**(1), 1–19 (2002)

11. Mannila, H., Toivonen, H.: Levelwise search and borders of theories in knowledge discovery. Data Min. Knowl. Disc. **1**(3), 241–258 (1997)

12. Melton, J., Simon, A.R.: SQL:1999: Understanding Relational Language Components. Morgan Kaufmann, Boston (2001)

13. Naumann, F.: Data profiling revisited. SIGMOD Rec. **42**(4), 40–49 (2013)

14. Rostin, A., Albrecht, O., Bauckmann, J., Naumann, F., Leser, U.: A machine learning approach to foreign key discovery. In: 12th International Workshop on the Web and Databases, WebDB 2009, Providence, RI, USA, 28 June 2009 (2009)

15. Saha, B., Srivastava, D.: Data quality: the other face of big data. In: Cruz, I.F., Ferrari, E., Tao, Y., Bertino, E., Trajcevski, G. (eds.) 30th International Conference on Data Engineering, ICDE 2014, Chicago, IL, USA, 31 March–4 April 2014, pp. 1294–1297. IEEE (2014)

16. Sismanis, Y., Brown, P., Haas, P.J., Reinwald, B.: GORDIAN: efficient and scalable discovery of composite keys. In: Dayal, U., Whang, K., Lomet, D.B., Alonso, G., Lohman, G.M., Kersten, M.L., Cha, S.K., Kim, Y. (eds.) Proceedings of the 32nd International Conference on Very Large Data Bases, Seoul, Korea, 12–15 September 2006, pp. 691–702. ACM (2006)

17. Zhang, M., Hadjieleftheriou, M., Ooi, B.C., Procopiuc, C.M., Srivastava, D.: On multi-column foreign key discovery. Proc. VLDB **3**(1), 805–814 (2010)

Normalization

From Web Tables to Concepts:
A Semantic Normalization Approach

Katrin Braunschweig[✉], Maik Thiele, and Wolfgang Lehner

Technische Universität Dresden, 01062 Dresden, Germany
{katrin.braunschweig,maik.thiele,wolfgang.lehner}@tu-dresden.de

Abstract. Relational Web tables, embedded in HTML or published on data platforms, have become an important resource for many applications, including question answering or entity augmentation. To utilize the data, we require some understanding of what the tables are about. Previous research on recovering Web table semantics has largely focused on simple tables, which only describe a single semantic concept. However, there is also a significant number of de-normalized multi-concept tables on the Web. Treating these as single-concept tables results in many incorrect relations being extracted. In this paper, we propose a normalization approach to decompose multi-concept tables into smaller single-concept tables. First, we identify columns that represent keys or identifiers of entities. Then, we utilize the table schema as well as intrinsic data correlations to identify concept boundaries and split the tables accordingly. Experimental results on real Web tables show that our approach is feasible and effectively identifies semantic concepts.

Keywords: Web tables · Conceptualization · Normalization · Semantics

1 Motivation

The Web has developed into a comprehensive resource not only for unstructured or semi-structured data, but also for relational data. Millions of relational tables embedded in HTML pages or published in the course of *Open Data/Open Government* initiatives provide extensive information on entities and their relationships from almost every domain. To fully utilize this data in applications such as question answering or entity augmentation, we need to *understand* the content of these tables. However, when the data is published on the Web, there is generally no conceptual model or formal description provided from which we can derive the semantics. Automatically recovering the semantics of Web tables has attracted some attention in the research community, leading to a number of approaches, which map the tables to large ontologies or knowledge bases to identify the most likely concepts described in the data. However, most of these approaches, as well as other applications which make use of Web tables, are based on a *single concept assumption*. Motivated by the fact that many HTML tables on the Web are small, they follow the simplifying assumption that each table covers only a single semantic concept. However, there are also a substantial number

© Springer International Publishing Switzerland 2015
P. Johannesson et al. (Eds.): ER 2015, LNCS 9381, pp. 247–260, 2015.
DOI: 10.1007/978-3-319-25264-3_18

City	Mayor	Elevation(m)	Country	Area (km²)	Population (Mio.)
London	B. Johnson	35	England	130,395	53
Manchester	N. u. Hassan	38	England	130,395	53
Dublin	O. Quinn	-	Ireland	70,273	4.59
Berlin	K. Wowereit	34	Germany	357,021	80.5
Paris	B. Delanoë	35	France	551,695	63.4

City

Name	Mayor	Elevation(m)	Country
London	B. Johnson	35	England
Manchester	N. u. Hassan	38	England
Dublin	O. Quinn	-	Ireland
Berlin	K. Wowereit	34	Germany
Paris	B. Delanoë	35	France

Country

Name	Area (km²)	Population (Mio.)
England	130,395	53
Ireland	70,273	4.59
Germany	357,021	80.5
France	551,695	63.4

Fig. 1. Example of a multi-concept table, with two corresponding single-concept tables.

of much larger and more complex tables on the Web, especially on Open Data platforms. These tables often describe multiple concepts as well as the relationships between them. Figure 1 shows an example of such a multi-concept table. It is easy to see that treating this table as a single-concept table will lead to incorrect information being taken from the data, for instance that the columns *area* and *population* are attributes of *cities* instead of *countries*. Also, matching this de-normalized schema to concepts and attributes in a knowledge base to understand its semantics is much more complex and possibly inconclusive.

In this paper, we propose a *semantic normalization* approach to identify concept boundaries in these wide, multi-concept tables, and split them into multiple single-concept tables, which can then be processed as before. Semantic normalization, which has the objective to identify concepts and relationships in the tables [11], is closely related to traditional normalization techniques in relational databases. However, in contrast, semantic normalization is a heuristic approach to support table understanding and does not necessarily provide the same formal guarantees.

Outline and Contributions. The remaining paper is organized as follows: in Sect. 2, we formally define the problem that is addressed by our approach. In Sect. 3, we analyze the unique challenges that arise when dealing with heterogeneous public datasets from the Web as well as their implications on the semantic normalization task. We then present our approach in detail in Sect. 4. An experimental evaluation on real-world data follows in Sect. 5. Finally, we present related work in Sect. 6 and conclude this paper with a short summary and outlook to future work in Sect. 7.

2   Problem Statement

Relational tables on the Web provide structured information on entities and their relationships in the form of attributes. Entities that are instances of the same semantic concept (i.e. have the same type, such as *city* or *country*) are grouped and described by the same attributes. In general, the description of concepts and their relationships in tables can be very versatile and complex. However, most Web table applications, such as entity augmentation [13,14] and fact search [12], focus predominantly on *simple concepts* and *binary relations*.

A simple concept is described by a *single* key column K, which is called the *entity column* and contains names or other identifiers of the entities, and a number of additional attribute columns A_i, which contain property values [10,13]. A binary relation is a relation that holds between two columns. These binary relations can be *Entity-Attribute Binary (EAB)* relations [13], that hold between the key and an attribute of the same concept, for instance the relation *(Country, Area)* in Fig. 1. They can also be *Entity-Entity Binary (EEB)* relations, that hold between two concepts represented by their keys, for instance *(City, Country)*. Based on these assumptions, we can derive the following definitions for single-concept and multi-concept tables:

Definition 1 (Single-concept Relation). *Let R be a relation with attributes $A_1, \ldots, A_n$. Then R corresponds to a* single-concept *relation R_{C_i}, if every $A_i \in R$ is a valid attribute of the same semantic concept C_i and there is a non-transitive functional dependency $K \to A$ (forming a binary relation) between the entity column K and each attribute $A_i \neq K$.*

Definition 2 (Multi-concept Relation). *A relation R corresponds to a* multi-concept *relation, if it can be described as the result of a join (or multiple joins) between several single-concept relations. Multi-concept relations contain multiple entity columns K_i and each column that is not an entity column, is an attribute of exactly one of the concepts defined by the entity columns.*

Following the notion in [11], our overall goal, *semantic normalization* of Web tables is then defined as follows:

Definition 3 (Semantic Normalization). *The objective of* semantic normalization *is the detection and separation of individual semantic concepts C_i contained in a relation R. Multi-concept relations are split into multiple single-concept relations $R_{C_i} \subset R$, one for each concept. Single-concept relations remain unchanged.*

3   Characteristics of Web Tables

As outlined in the previous section, we are interested in semantically meaningful binary relations in tables. Public datasets, such as Web and Open Data tables, pose unique challenges for the identification of these relations. In the traditional DBMS environment, functional dependencies (FDs) are an integral part of schema design to ensure the integrity of the data. They are derived from the application logic or provided by domain experts. For public datasets, we usually do not have access to this information and can only try to infer FDs directly from the data, which leads to both, meaningful and coincidental FDs. *Coincidental functional dependencies* are caused by the distribution of values in the table and do not reflect semantic relationships between attributes. Web tables are very likely to contain coincidental FDs as they are often considerably shorter than to tables in business OLTP/OLAP environments, thus, representing only a small, and not always representative, subset of the attribute domain. Most coincidental FDs are caused by *unique* columns, which are not a key of the table, or *constant* columns.

In contrast, incomplete and anonymized data may cause meaningful functional dependencies to be undetectable. Especially columns containing names, often serving as the key of a concept, are affected by anonymization. In Open Government Data, for instance, personal details such as names and contact details are sometimes replaced with default placeholders such as *N/A*, *redacted* or *undisclosed*. Compared to *NULL* values, these placeholders are not necessarily detected in the data. In addition, we may encounter errors or inconsistencies in the data that violate the functional dependency constraints, such as the following example from our test corpus.

Program Name	Program Address
Beth Emeth Homecare	1080 McDonald Avenue
Beth Emeth Homecare	1080 MacDonald Avenue

Inferring semantic relationships from the header of the tables is equally challenging, due to the varying quality and ambiguity of the column labels. In public tables, we often encounter abbreviations, non-informative labels (e.g. *name* or *value*), long descriptive labels or even missing labels. In addition, inconsistent naming conventions across publishers of Web tables make it difficult to identify the attributes presented in each column as well as the relations between them.

Overall, Web data is very heterogeneous and noisy and it is clear that we will not achieve useful results if we focus only on data correlations or only on schema correlations. Instead, we need to combine evidence from both, the table data as well as the schema to filter out noise and only detect meaningful relations.

4   Semantic Normalization

4.1   Overview

In this section, we describe our normalization approach, including details for each step involved. As outlined previously, to normalize the tables, we require meaningful functional dependencies that reflect the semantics of the data. However, without additional domain knowledge or expert support, we can only extract potential dependencies from the data, which also introduces incorrect dependencies. In order to retrieve only meaningful dependencies, we need to filter the resulting set of candidates. Figure 2 illustrates the process involved in our approach. In the first step (including 1.1 and 1.2) we extract all necessary information from the data to form a dependency graph. After that, we systematically evaluate and filter the dependencies in the graph (2.1 and 2.2). The remaining dependencies then provide the base for the table normalization. Empty columns are excluded from the process.

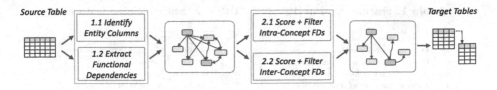

Fig. 2. Overview of semantic normalization process.

4.2   Entity Column Identification

In [10], Venetis et al. use a binary classifier and a set of five features to automatically detect entity columns in single-concept tables, which frequently coincide with the left-most STRING column. The column with the highest score is selected if the classifier identifies more than one candidate per table. We follow a similar classification approach. However, in contrast to single concept tables, multi-concept tables contain multiple entity columns, located at any position in the table. Therefore, we extended the original set of features (see Tables 1) with six additional features (see Table 2).

In most cases, entity columns contain STRING tokens. While Feature 2 already checks for numeric content, we specifically check for STRING content in Feature 6 to distinguish it from other non-numeric tokens, such as DATE tokens. Feature 7 containes the column position divided by the total number of columns in the table. Feature 8 records whether the column has a label, while Feature 9 searches for specific keywords in the label that might indicate an entity column. So far, we check for the most prominent terms *name* and *id*, although further keywords may be added. Finally, Features 10 and 11 look at the correlation between the column in question and its left and right neighbors, using the probability in Eq. 1 as the score. Assuming a reading order from left to right, with attributes of a concept listed *after* the concept's key column, we expect entity columns to have a high correlation to its right neighbor, but a low correlation to its left neighbor. If no neighbor exists, i.e. at the beginning and end of the table, the correlation is set to 0. We tested various state-of-the art classifiers, including SVM and Random Forest. Experimental results are detailed in Sect. 5.

Evaluation on real-world data has shown that in some cases the same concept can be represented by multiple key columns, each of which can serve as the key on their own. In almost all cases, these concepts contained both, a *name* and an *id* column. We cluster these columns and select one representative for the subsequent processing steps. Due to lack of space, we omit the details here.

From the set of entity columns in a table, we select the left-most entity column with the largest number of dependent columns as the *main* entity column, which describes the main topic of the table.

4.3   Extracting Functional Dependencies

As we assume *binary* relations, we are looking for functional dependencies (FD) $X \to A$, where the determinant X is a single column. Furthermore, we require

Table 1. Features used in [10].   **Table 2.** Entity column features.

No	Feature description
1	Fraction of cells with unique content
2	Fraction of cells with numeric content
3	Variance in number of data tokens
4	Average number of data tokens
5	Column index (from the left)

No	Feature description
6	Fraction of cells with String content
7	Relative column position
8	Has column label
9	Label contains *name* or *id*
10	Correlation with left neighbor
11	Correlation with right neighbor

X to be an entity column K. Extracting these dependencies from the table, we receive two types of FDs as depicted in Fig. 3: *inter-concept* FDs $K_i \rightarrow K_j$ between two entity columns, and *intra-concept* FDs $K \rightarrow A_i$ between an entity and an attribute column. The figure also shows that, in addition to meaningful FDs, we also extract several coincidental or transitive FD candidates.

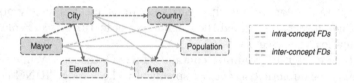

Fig. 3. Dependency graph for example table, including inter-concept and intra-concept dependencies. Meaningful dependencies are marked in blue (Colour figure online).

Normally, we expect functional dependencies to be exact, meaning that the dependency holds for each non-NULL value $x \in X$. However, Web and Open Data tables may contain errors or inconsistencies in the data that violate the constraint. In the literature, several measures have been proposed, that address this issue by using approximate FDs instead of exact FDs [7,8,11]. Here we use the notion introduced as *probabilistic functional dependencies* in [11]. A probabilistic FD is denoted as $X \xrightarrow{p} A$, where p is the probability that $X \rightarrow A$ holds in the data. The probability is calculated as follows, where D_x is the set of distinct values in X and a_x is the value in A that occurs most often with value x:

$$Pr(X \rightarrow A) = \frac{\sum_{x \in D_x} |x, a_x|}{\sum_{x \in D_x} |x|} \tag{1}$$

A threshold θ_p then determines how much noise is tolerated in the data.

4.4   Intra-concept Dependencies

Intra-concept dependencies reflect relationships between an entity column and an attribute column. For each attribute column, we may retrieve several potential entity columns (see Fig. 3). To identify the most likely entity column, we propose a combination of structural and semantic measures to score and rank all candidates and select the one with the highest rank. The structural scores *Column Distance*, *Data Correlation* and *Null Value Distribution* address the column arrangement as well as the distribution of values in the table. In addition, the semantic scores *Common Concept Label* and *Attribute Co-Occurrence* take into consideration the schema of the table and semantic relationships between column labels.

Column Distance/Position. Two different assumptions motivate taking the position of the columns and their distance into consideration. First, we assume that, going from left to right, the concept identifier is mentioned before any related attribute in the table. Therefore, FDs where the determinant X appears before the dependent A in the table should be ranked higher than FDs with the determinant appearing after the dependent. Second, we assume that attributes of the same concept appear in close proximity in the table to indicate their relatedness. Thus, FDs with a smaller distance between determinant and dependent are assigned a higher score. We use the following equation to combine both assumptions into a single score, where θ_d is an adjustable threshold:

$$score_{CD}(A, X) = \begin{cases} 1 - \frac{dist(X,A)}{\theta_d} & , if \; dist(X, A) < \theta_d \\ 0 & , else \end{cases} \tag{2}$$

$$dist(A_i, A_j) = \begin{cases} j - i & , if \; i < j \\ i - 1 & , else \end{cases} \tag{3}$$

Setting θ_d to the number of columns in the table yields a relative score, while using a constant value yields a score independent of the table size.

Data Correlation. We also take the correlation between data values into account, using the notion of *variation of information* (VI) to measure redundancy between an attribute column A and the candidate key column X. Variation of information is defined as follows, where $I(A, X)$ is the mutual information between columns A and X and $H(A)$ and $H(A|X)$ are the entropy and conditional entropy, respectively:

$$VI(A, X) = H(A) + H(X) - 2I(A, X) \tag{4}$$

$$I(A, X) = H(A) - H(A|X)$$

Multi-concept tables may contain redundancy in the data as the result of a $n : 1$ relationship between two concepts (see example in Fig. 1). Similar levels of redundancy between two columns indicate a stronger relationship. Since VI is a distance metric, we use Eq. 5 to calculate the similarity score. To obtain a score in the range [0,1], we use a normalized variant of VI, where $H(A, X)$ is the joint entropy. Note that for the special case of a FD between two constant columns with $H(A) = H(X) = 0$, Eq. 5 is undefined. In this case, we assign a score of 1, as we want this case to be interpreted as any other case where $H(A) = H(X)$.

$$score_{VI}(A, X) = 1 - \frac{VI(A, X)}{H(A, X)} \tag{5}$$

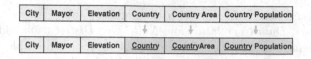

Fig. 4. Common concept labels for the example in Fig. 1

Null Value Distribution. FDs are only defined for instances where the values in both columns X and A are not NULL. However, the distribution of NULL values may help us identify coincidental FDs. For a meaningful FD $X \to A$, we expect key column X to not be NULL for all instances where attribute column A is not NULL. Otherwise, column X may not be the correct key for the attribute. Based on this assumption, we assign the following score in the range [0,1], where N is the number of rows in the table, $x_i \in X$ and $a_i \in A$:

$$score_{NV}(A, X) = \frac{\sum_{n \in N} g(x_n)}{\sum_{n \in N} g(x_n) \cdot g(a_n)} \tag{6}$$

$$g(x) = \begin{cases} 1 & , if\ x \neq NULL \\ 0 & , else \end{cases} \tag{7}$$

Common Concept Labels. Analyzing the characteristics of Web tables, we noticed that in many cases attributes of the same concept shared a common label to indicate their relatedness. Figure 4 shows the header of the example table in Fig. 1 with additional concept labels. Apart from different formatting conventions, concept labels almost always appeared as a prefix to the actual attribute label. To identify such *common concept labels (CCL)*, we first extract all prefixes that are shared by at least two columns in the table, using only the longest common prefix. To ensure valid concept labels, we enforce the following constraints: (1) the concept label must be a noun or compound of nouns, and (2) the remaining attribute label must be a valid noun phrase, i.e. cardinal numbers or single letters are not allowed.

Some tables with multiple nested concepts may even form a hierarchy of CCLs. In this case, a longer common prefix suggests a stronger semantic relatedness and should receive a higher score. Therefore, we use Eq. 8 to score FDs, where $CCL(A, X)$ denotes the common prefix shared by attributes A and X and $maxCCL(A)$ is the longest prefix that column A shares with any other column. We use the word count to measure the length of a prefix.

$$score_{CCL}(A, X) = \frac{|CCL(A, X)|}{|maxCCL(A)|} \tag{8}$$

Attribute Co-Occurrence. Since concept labels are only available for a small subset of columns, we use an additional, more general measure to identify the semantic relatedness between attributes based on the *coherency score* proposed in [3]. Utilizing their *attribute correlation statistics database (AcsDb)*, which is a large collection of unique schemas, including frequency statistics, of tables extracted from the Web, Cafarella et al. compute a score measuring how strongly related the attributes in the table are. For a pair of columns, they calculate the *pointwise mutual information (PMI)* based on the frequency and co-occurrence statistics in AcsDb (see Eq. 9). Attributes that occur together in the same schema more often than they occur individually, are thought to be strongly related.

$$PMI(A_1, A_2) = \log \frac{p(A_1, A_2)}{p(A_1)p(A_2)} \tag{9}$$

$$nPMI(A_1, A_2) = \frac{PMI(A_1, A_2)}{-\log p(A_1, A_2)} \tag{10}$$

We utilized AcsDB to retrieve PMI scores for all column pairs (A, X). To address the ambiguity and sometimes low quality of attribute labels, we apply basic schema normalization techniques [9] as well as synonym expansion to increase the chance of finding corresponding attributes in AcsDb. Using $nPMI$, the normalized variant of PMI, we receive a score between -1 and 1. As we are only interested in positive evidence, we set the score to 0 if $nPMI$ returns a score <0.

$$score_{AC}(A, X) = \begin{cases} nPMI(A, X) & ,if\ nPMI(A, X) > 0 \\ 0 & \text{else} \end{cases} \tag{11}$$

Combining Scores. To score the candidate dependencies we compute the weighted sum of our scores:

$$\begin{aligned} score_{total}(A, X) = {} & \omega_0 + \omega_1 \cdot score_{VI}(A, X) + \omega_2 \cdot score_{CCL}(A, X) \\ & + \omega_3 \cdot score_{AC}(A, X) + \omega_4 \cdot score_{NV}(A, X) \\ & + \omega_5 \cdot score_{CD}(A, X) \end{aligned} \tag{12}$$

From this total score we retrieve the functional dependency with the highest score for each attribute column. Should more than one FD receive the highest score, we choose the main entity column as the most likely entity column (if a FD exists) or resort to the *Column Distance* score to decide. As a result, we always select exactly one entity column as the concept key for each attribute.

4.5 Inter-concept Dependencies

Inter-concept dependencies reflect relationships between two concepts and are represented by dependencies between the entity columns of each concept. Relationships between concepts can be much more complex than dependencies within a concept. Here, we consider 1:1 or n:1 relationships. $n : m$ relationships are not that easily detected in the data and require a more comprehensive semantic analysis. Similar to intra-concept FDs, we score each inter-concept FD using a linear combination of the scores described in Sect. 4.4. The weights ω_i differ for inter-concept FDs, as, for instance, common concept labels are not as prominent here as they are for intra-concept FDs. After scoring each dependency, we select the dependencies with the highest scores, following the steps depicted in Fig. 5. First, we resolve bi-directional dependencies between concepts. Although they may represent a valid 1:1 relationship between the concepts, we need to remove one direction in order to avoid splitting the graph into unconnected subgraphs. Note that the directionality of the relation can be restored later. We keep the dependency with the highest score, unless one of the columns is the main key of the table. In that case we always favor the FD with the main key as the determinant. After resolving bidirectional FDs, some of the columns may still depend on more than one other column. For each of these columns we select the FD with the highest score as described in Sect. 4.4, removing coincidental as well as transitive dependencies in the process. The result is a dependency graph where each column (except the main key) depends only on a single column, resembling a join graph for the concepts in the table. Finally, we group intra- and inter-concept FDs for each entity column to form single-concept tables.

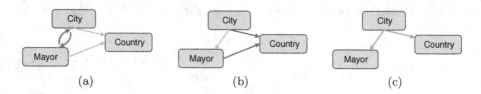

Fig. 5. Selection process for inter-concept dependencies.

5   Experimental Evaluation

Dataset. To evaluate our semantic normalization approach, we selected 100 wide tables from several data sources on the Web, including *Wikipedia, data.gov.uk, NYC Open Data*[1] and *Socrata*[2], to allow for a diverse set of schemata. We manually labeled entity columns and identified concept boundaries for each table. To ensure that our algorithm also works correctly for single-concept tables, we included some in the test corpus. Table 3 shows the main statistics of the corpus.

Table 3. Dataset statistics.

	Min	Avg.	Max
# of rows	2	187.54	2507
# of columns	6	10.24	20
# of empty columns	0	0.22	4
# of entity columns	1	3.92	8

Table 4. Classification accuracy.

Classifier	Accuracy (%)	
	Training	Test
Random forest	**87.1**	**86.6**
SVM (PolyKernel)	74.9	76.6
SVM (RBNFKernel)	85.5	85.3
Logistic regression	74.9	77.1

---

[1] https://nycopendata.socrata.com/.
[2] https://opendata.socrata.com/.

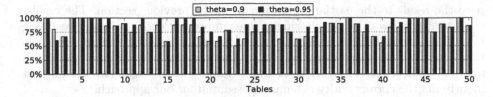

Fig. 6. Evaluation of dependency selection process with respect to θ_p.

Entity Column Detection. For the classification of columns into entity and non-entity columns, we split the dataset into a training and test set, each containing 50 tables, with 518 and 484 columns, respectively. In both sets, the entity/non-entity ratio is roughly 2:3. We tested various classification techniques using the implementations in WEKA [6]. We optimized parameters using 10-fold cross-validation on the training set and then evaluated each classifier on the test set. Table 4 shows the results. Additionally, we evaluated how well our algorithm identified the *main* entity column amongst all entity columns in a table (see Sect. 4.2). Using the hand-labeled entity columns as input, the algorithm selected the correct main entity column for all tables in the test set. However, if the main key is not identified as an entity column by the classifier, our algorithm cannot detect it at a later stage.

Inter-concept and Intra-concept Dependencies. To evaluate the scoring and filtering of inter- and intra-concept dependencies, we used the hand-labeled entity columns as input and compared the results to manually extracted binary relations for each table. For our experiments, we set the parameter $\theta_d = 20$ (Eq. 2), which equals the maximum number of columns in our corpus. We learned the weights ω_i for the scoring function in Eq. 12, using linear regression on the training set, with different weights for intra- and inter-concept FDs. Figure 6 shows the results of the dependency selection step for all tables in the test set. The figure depicts the percent of columns in each table, for which the correct dependency was selected or which was correctly identified as the main key. Empty columns are discarded. Parameter θ_p determines the functional dependencies that are extracted for each table. While a lower value tolerates more noise in the data (see second instance in the figure), it also introduces more incorrect dependencies for many tables. We achieved the best result for $\theta_p \geq 0.95$, which we use in the remaining experiments. While we obtain good results for the majority of tables, there are also some table with an accuracy below 75 %. In most of these tables there is only very little or no evidence for a strong relatedness between columns, with the decision ultimately based solely on the column distance.

Complete Normalization Process. After evaluating each subprocess individually, we now evaluate the complete normalization process. To see how errors in the classification of entity columns affect the overall accuracy, in Fig. 7, we compare the result to the partial evaluation from the previous section. The results show that in a few cases the classifier misses important entity columns, resulting in a significantly lower accuracy for these tables. The entity columns in question contained mostly numeric values, for instance transaction numbers, which are very difficult to tell apart from other numeric attributes. This shows that identifying the correct entity columns is essential for our approach.

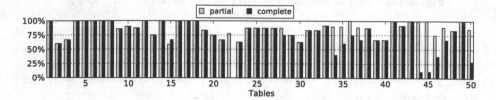

Fig. 7. Evaluation of the complete semantic normalization process ($\theta_p = 0.95$).

Optimization. Some of the tables in our corpus have only very few rows, which makes coincidental dependencies more likely. Also, the smaller the sample size, the harder it is to distinguish between actual and coincidental dependencies. In [11], Wang et al. address this issue by first integrating data from multiple related tables into a single mediated schema, before extracting functional dependencies, thus combining the samples.

We apply a similar optimization approach to our algorithm. However, given the heterogeneity and ambiguity of schemas found in Web tables, we only combine tables that have the *same* schema, including data types, to ensure that the tables have the same semantics. Such tables are quite common, especially on Open Data platforms, where the same statistics are often published each month or for each department.

In our test corpus, we found tables with the same schema for 30 tables. In Fig. 8 we show the results for each of these tables, comparing the accuracy obtained from individual tables to the accuracy obtained from clustering tables with the same schema. The optimization step increases the accuracy for several tables, especially tables 19 and 20. However, there are also three tables where the clustering leads to a slightly lower accuracy. This is the case when a feature, such as redundancy, is significant in a smaller sample, but becomes much less significant in a larger sample.

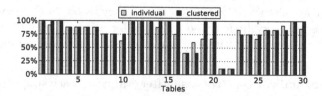

Fig. 8. Evaluation of optimization step ($\theta_p = 0.95$).

6   Related Work

Relational tables on the Web have been studied extensively in recent years, specially regarding their extraction [2,4] and integration [5] as well as their application in search engines [3] and for entity augmentation [13]. Automatically *understanding* Web tables, i.e. identifying the specific semantic concept(s) described in the tables by mapping the columns to concepts, entities and attributes in a knowledge base, has also received some attention. In [12], Wang et al. utilize Probase, a large probabilistic taxonomy, while Venetis et al. use custom *isA* and relationship databases in [10]. The majority of techniques designed to automatically process Web tables is based on the simplifying *single concept assumption*, including [5,10,12–14]. We see semantic normalization as a requisite preprocessing step for these applications in order to process multi-concept Web tables correctly.

Automatic database normalization has also been studied in the literature. However, most approaches, such as [1], require the correct functional dependencies, usually defined manually, as input. However, for public Web tables, we do not have these FDs. In [7], Huhtala et al. propose an efficient algorithm for discovering minimal functional dependencies in relational tables. In [8], Ilyas et al. introduce soft functional dependencies as well as soft keys for columns where the constraint does not necessarily hold for all values. The authors also address the impact of sample size as well as constant columns on the number of functional dependencies. However, the application of join optimization differs significantly from our scenario. The closest to our work is the work on functional dependencies in pay-as-you-go data integration systems by Wang et al. [11]. Here, the authors introduce the notion of semantic normalization. Small Web tables are integrated into a large mediated schema and probabilistic FDs are extracted to transform the mediated schema into 3NF. In contrast to our approach, Wang et al. only rely on correlations in the data to identify functional dependencies. Additionally, our approach does not require the integration of all tables into a large mediated schema, which seems infeasible for a large corpus of millions of Web tables. Instead, we only combine tables with identical schemas.

7   Conclusion and Future Work

In this paper, we present a multi-phase algorithm for the semantic normalization of Web and Open Data tables containing multiple semantic concepts. Identifying concept boundaries within these tables is an important step towards automatic table understanding and for the correct extraction of binary relations as required for entity augmentation or question answering. An evaluation using complex real-world tables shows that our approach can split these multi-concept tables accurately into single-concept tables, given sufficient evidence in the data. The goal for future work is to further improve entity column classification, as it is essential for the semantic normalization of tables.

References

1. Bahmani, A., Naghibzadeh, M., Bahmani, B.: Automatic database normalization and primary key generation. In: Canadian Conference on Electrical and Computer Engineering, CCECE 2008, pp. 000011–000016, May 2008
2. Cafarella, M.J., Halevy, A.Y., Khoussainova, N.: Data integration for the relational web. Proc. VLDB Endow. **2**, 1090–1101 (2009)
3. Cafarella, M.J., Halevy, A.Y., Wang, D.Z., Wu, E., Zhang, Y.: Webtables: exploring the power of tables on the web. Proc. VLDB Endow. **1**(1), 538–549 (2008)
4. Cafarella, M.J., Halevy, A.Y., Zhang, Y., Wang, D.Z., Wu, E.: Uncovering the relational web. In: WebDB (2008)
5. Das Sarma, A., Fang, L., Gupta, N., Halevy, A.Y., Lee, H., Wu, F., Xin, R., Yu, C.: Finding related tables. In: Proceedings of the 2012 ACM SIGMOD International Conference on Management of Data, New York, NY, USA, pp. 817–828 (2012)
6. Hall, M., Frank, E., Holmes, G., Pfahringer, B., Reutemann, P., Witten, I.H.: The weka data mining software: an update. SIGKDD Explor. Newsl. **11**(1), 10–18 (2009)
7. Huhtala, Y., Kärkkäinen, J., Porkka, P., Toivonen, H.: Tane: an efficient algorithm for discovering functional and approximate dependencies. Comput. J. **42**(2), 100–111 (1999)
8. Ilyas, I.F., Markl, V., Haas, P., Brown, P., Aboulnaga, A.: Cords: automatic discovery of correlations and soft functional dependencies. In: Proceedings of the 2004 ACM SIGMOD International Conference on Management of Data, SIGMOD 2004, New York, NY, USA, pp. 647–658. ACM (2004)
9. Sorrentino, S., Bergamaschi, B., Gawinecki, M., Po, L.: Schema normalization for improving schema matching. In: Laender, A.H.F., Castano, S., Dayal, U., Casati, F., de Oliveira, J.P.M. (eds.) ER 2009. LNCS, vol. 5829, pp. 280–293. Springer, Heidelberg (2009)
10. Venetis, P., Halevy, A., Madhavan, J., Paşca, M., Shen, W., Wu, F., Miao, G., Wu, C.: Recovering semantics of tables on the web. Proc. VLDB Endow. **4**(9), 528–538 (2011)
11. Wang, D.Z., Dong, X.L., Sarma, A.D., Franklin, M.J., Halevy, A.Y.: Functional dependency generation and applications in pay-as-you-go data integration systems. In: 12th International Workshop on the Web and Databases, WebDB 2009, Providence, Rhode Island, USA, 28 June 2009
12. Wang, J., Wang, H., Wang, Z., Zhu, K.Q.: Understanding tables on the web. In: Atzeni, P., Cheung, D., Ram, S. (eds.) ER 2012. LNCS, vol. 7532, pp. 141–155. Springer, Heidelberg (2012)
13. Yakout, M., Ganjam, K., Chakrabarti, K., Chaudhuri, S.: Infogather: entity augmentation and attribute discovery by holistic matching with web tables. In: Proceedings of the 2012 ACM SIGMOD International Conference on Management of Data, SIGMOD 2012, New York, NY, USA, pp. 97–108. ACM (2012)
14. Zhang, M., Chakrabarti, K.: Infogather+: Semantic matching and annotation of numeric and time-varying attributes in web tables. In: Proceedings of the 2013 ACM SIGMOD International Conference on Management of Data, SIGMOD 2013, New York, NY, USA, pp. 145–156. ACM (2013)

Toward RDF Normalization

Regina Ticona-Herrera[1,3]([✉]), Joe Tekli[2], Richard Chbeir[1], Sébastien Laborie[1],
Irvin Dongo[1], and Renato Guzman[3]

[1] University of Pau and Adour Countries - LIUPPA, Anglet, France
rchbeir@acm.org, {sebastien.laborie,irvin.dongo}@univ-pau.fr
[2] Lebanese American University, Byblos, Lebanon
joe.tekli@lau.edu.lb
[3] San Pablo Catholic University, Arequipa, Peru
reginapaola.ticonaherrera@univ-pau.fr,
renato.guzman@ucsp.edu.pe

Abstract. Billions of RDF triples are currently available on the Web
through the Linked Open Data cloud (e.g., DBpedia, LinkedGeoData
and New York Times). Governments, universities as well as companies
(e.g., BBC, CNN) are also producing huge collections of RDF triples and
exchanging them through different serialization formats (e.g., RDF/XML,
Turtle, N-Triple, etc.). However, RDF descriptions (i.e., graphs and serializations) are verbose in syntax, often contain redundancies, and could
be generated differently even when describing the same resources, which
would have a negative impact on their processing. Hence, we propose
here an approach to clean and eliminate redundancies from such RDF
descriptions as a means of transforming different descriptions of the same
information into one representation, which can then be tuned, depending on the target application (information retrieval, compression, etc.).
Experimental tests show significant improvements, namely in reducing
RDF description loading time and file size.

Keywords: Rdf graph · Serialization · Redundancies and disparities

1 Introduction

Since several years, the Web has evolved from a Web of linked documents to
a Web of linked data. As a result, a huge amount of RDF statements, in the
form of $< Subject, Predicate, Object >$ triples, are currently available online
through the Linked Open Data cloud thanks to different projects[1](e.g., *DBpedia*,
LinkedGeoData, *New York Times*, etc.). These triples are usually stored in RDF
datasets after being serialized into several machine readable formats such as
RDF/XML, N-Triple, Turtle, N3 or JSON-LD. Therefore, RDF descriptions can
be represented in different ways and formats as shown in Fig. 1.

However, in different scenarios (e.g., collaborative RDF graph generation [6],
automatic RDF serialization [12,16], etc.), RDF descriptions might be verbose
and contain several redundancies in terms of both: the structure of the graph
and/or the serialization format of the resulting RDF-based file. For instance,

---

[1] http://linkedgeodata.org, http://data.nytimes.com/, http://dbpedia.org.

© Springer International Publishing Switzerland 2015
P. Johannesson et al. (Eds.): ER 2015, LNCS 9381, pp. 261–275, 2015.
DOI: 10.1007/978-3-319-25264-3_19

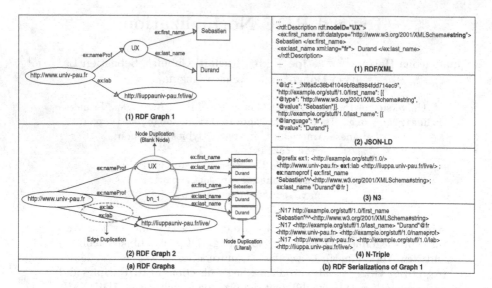

Fig. 1. RDF description (graph and serialization) examples

graphs 1 and 2 in Fig. 1, which have been created by two different users, describe the same RDF information even though they have different structures (e.g., duplication of nodes and edges). Additionally, even more redundancies and disparities[2] will occur when serializing both graphs (e.g., the same RDF resource or blank node can be serialized in different ways, language and data-type declarations can be specified or omitted, namespaces can have different/duplicate prefixes, etc.). Consequently, such RDF redundancies and disparities would naturally have a negative impact on the processing of RDF databases (including storage, querying, mapping, annotating, versioning, etc.).

In this paper, we address the problem of cleaning RDF descriptions by introducing a new method as a means of transforming different RDF descriptions of the same RDF statements into one single (normalized) representation. Our method targets RDF modeling on two levels: (1) the structure/graph (logical) level, by eliminating redundancies in RDF graphs, which is typically useful in graph-based RDF querying, mapping, and versioning applications, and (2) the serialization (physical) level, by eliminating redundancies and disparities in the syntactic structure, and adapting it to the application domain, in order to optimize storage space and loading time.

The rest of this article is organized as follows. Section 2 reviews background and related work in RDF normalization. Section 3 presents motivating examples, highlighting different normalization features left unaddressed by most existing approaches. Our method for normalizing RDF descriptions (graphs and serializations) is detailed in Sect. 4. In Sect. 5, we present our prototype and illustrate experimental results. We finally conclude in Sect. 6.

---

[2] We use disparities to designate different serializations of the same information.

2 Background and Related Work

2.1 Basic Notions

Definition 1 (Statement [st]). *An RDF statement expresses a relationship between two resources, two blank nodes, or one resource and one blank node. It is defined as an atomic structure consisting of a triple with a Subject (s), a Predicate (p) and an Object (o), noted st:$< s, p, o >$, w.r.t. a specific vocabulary V (see Definition 3), where: (a) $s \in U \cup BN$ is the subject to be described (U is a set of Internationalized Resource Identifiers and BN is a set of Blank Nodes), (b) $p \in U$ refers to the properties of the subject, and (c) $o \in U \cup BN \cup L$ is the object (L is a set of Literals)* ♦*

The example presented in Fig. 1.1.a underlines 4 statements with different resources, properties, and blank nodes such as st_1: <http://www.univ-pau.fr, ex:nameProf, UX >, st_2: $< UX, ex:first\_name,$ "Sebastien" ^xsd : string >, st_3: $< UX, ex : last\_name,$ "Durand">, and st_4: <http://www.univ-pau.fr, ex:lab, http://www.univ-pau.fr>.

Definition 2 (RDF Graph [G]). *An RDG graph is a directed labeled graph made of a set of < s,p,o> statements in which each one is represented as a node-edge-node link* ♦

In the remainder of the paper, *"RDF graph"* and *"RDF logical representation"* are used interchangeably.

Definition 3 (RDF Graph Vocabulary [V]). *An RDF Graph Vocabulary is the set of all values occurring in the RDF graph, i.e., $V = U \cup L \cup BN$* ♦

Definition 4 (External Vocabulary [QN]). *An RDF External Vocabulary is a set of QNames[3] (QN) to represent IRI references $\{qn_1, qn_2, \ldots, qn_n\}$. Each qn_i is a tuple $< px_i, ns_i >$ where px_i is a prefix[4] (e.g., foaf, ex, dc,...) and ns_i is a namespace* ♦

For instance $QN = \{(\mathbf{ex}, \text{http://example.org/stuff/1.0}), (\mathbf{mypx}, \text{http://ucsp.edu.pe})\}$, where "ex" is a standard prefix, "mypx" is a local prefix, and http://example.org/stuff/1.0 and http://ucsp.edu.pe/ are namespaces.

Definition 5 (RDF File [F]). *An RDF file is defined as an encoding of either a set of RDF statements or an RDF graph, using a predefined serialization format (e.g., RDF/XML, Turtle, N3, etc.)* ♦

In the remainder of the paper, *RDF file, RDF serialization* and *RDF physical representation* are used interchangeably. Also, the following functions **R, U, L, BN, ST, NS** and **Px** will be used respectively to return all the Resources (IRIs and literals), IRIs, Literals, Blank Nodes, statements, namespaces and prefixes of a graph G or a file F.

---

[3] http://www.w3.org/TR/REC-xml-names/.
[4] Following the W3C Recommendation, we consider that all the prefixes have to be unique for each namespace.

2.2   Related Work

The need for RDF normalization has been identified and discussed in various domains, ranging over knowledge representation, data integration, and service and semantic data mediation. Yet, few existing studies have addressed the issues of logical (graph) and physical (syntax) RDF normalization.

Knowledge Representation, Integration and Semantic Mediation. Various approaches have been developed to normalize knowledge representation in RDF, namely in the bioinformatics domain [1,6,10,11,14]. In [14], the authors provide an approach to map LexGrid [11], a distributed network of lexical resources for storing, representing and querying biomedical ontologies and vocabularies, to various Semantic Web (SW) standards, namely RDF. They introduce the LexRDF project which leverages LexGrid, mapping its concepts and properties to standard - normalized - RDF tagging, thus providing a unified RDF based model (using a common terminology) for both semantic and lexical information describing biomedical data. In a related study [10], the authors introduce the Bio2RDF project, aiming to create a network of coherent linked data across life sciences databases. The authors address URI normalization, as a necessary prerequisite to build an integrated bioinformatics data warehouse on the SW, where resources are assigned URIs normalized around the bio2rdf.org namespace. Several approaches have developed semantic mediators (translators), in order to convert information from one data source to another following the data format which each system understands [7,8]. Most studies consider the original data to be well organized (normalized), thus the resulting RDF data would allegedly follow. Note that in most of these projects, issues of redundancies in RDF logical and syntax representations are mostly left unaddressed.

RDF Graph (Logical) Normalization. In [5], the authors discuss some of the redundancies which can occur in a traditional RDF directed labeled graphs. Namely, an RDF graph edge label (i.e., a predicate) can occur redundantly as the subject or the object of another statement (e.g., $<$ $dbpedia$: $Researcher, dbpedia$: $Workplace, dbpedia$: $University$ $>$ and $<$ $dbpedia$: $Workplace, rdf$: $type, dbpedia$: $Professional$ $>$). Hence, the authors in [5] introduce an RDF graph model as a special bipartite graph where RDF triples are represented as ordered 3-uniform hypergraphs where edge nodes correspond to the $<$ $subject, predicate, object$ $>$ triplet constituents, ordered following the statement's logical triplet ordering. The new model is proven effective in reducing the predicate-node duplication redundancies identified by the authors. In subsequent studies [3,4], the authors address the problem of producing RDF normal forms and evaluating the equivalence among them. The studies in [3,4] specifically target the RDFS vocabulary with a set of reserved words to describe the relationships between resources (e.g., rdfs:type, rdfs:range, rdfs:domain, etc.). They provide a full-fledged theoretical model including notions such as: *RDF lean* (minimal) *graph* as a graph preserving all URIs of its origin graph while having fewer blank nodes, and *RDF normal form* as a minimal and unique representation of an RDF graph, among others. In [9,13], the authors introduce an algorithm allowing to transform an RDF graph into a standard form, arranged in a deterministic way, generating a cryptographically-strong hash identifier for the graph, or digitally signing it. The algorithm takes a JSON-LD input format, and provides an output in N-triple serialization while relabeling certain nodes and erasing certain redundancies. Our approach completes the latter studies by targeting logical (graph) redundancies being out of the scope of [3–5,9,13], namely distinct edge (predicate) duplication, blank node (subject/object) duplication, and combined edge and node duplication.

RDF Syntax (Physical) Normalization. At the physical (syntactic) level, the authors in [16] introduce a method to normalize the serialization of an RDF graph using XML grammar (DTD) definitions. The process consists of two steps: (a) Defining an XML grammar (DTD) to which all generated RDF/XML serializations should comply, (b) Defining SPARQL query statements to query the RDF dataset in order to return results, consisting of serializations compliant with the grammar (DTD) at hand. This is comparable to the concept of semantic mediation using SPARQL queries [8]. Note that the SPARQL statements are automatically generated based on the grammar (DTD).

To sum up, our approach completes and builds on existing methods to normalize RDF information, namely [3–5,9,13,16], by handling logical and physical redundancies and disparities which were (partially) unaddressed in the latter.

3 Motivating Example

We discuss here the motivations of our work, highlighting two different levels: (i) logical redundancies, and (ii) physical redundancies and disparities.

3.1 Logical (Graph) Redundancies

Consider the example given in Fig. 1.a.2. One can easily see several kinds of redundancies:

- **Problem 1 - Edge Duplication:** where identical edges, designating identical RDF predicates, appear more than once,
- **Problem 2 - Node duplication:** where identical nodes, designating identical RDF subjects and/or objects, appear more than once, e.g., in Fig. 1.a.2 highlighting Blank Node duplication and Literal duplication respectively.

3.2 Physical (Serialization) Disparities

Consider now Fig. 2.a which represents a possible serialization of the RDF graph in Fig. 1.a.2 encoded in the RDF/XML format. One can see that several types of redundancies and disparities are introduced: some are inherited from the logical level, while others appear at the physical (serialization) level:

- **Problem 3 - Namespace duplication:** where two different prefixes are used to designate the same namespaces (*ex* and *ex1* in lines 4–5 in Fig. 2.a),

```
1 <?xml version="1.0" encoding="UTF-8" standalone="no"?>          1 <?xml version="1.0" encoding="UTF-8" standalone="no"?>
2 <rdf:RDF xmlns:rdf="http://www.w3.org/1999/02/22-rdf-syntax-ns#"  2 <rdf:RDF xmlns:rdf="http://www.w3.org/1999/02/22-rdf- syntax-ns#"
3 xmlns:dc="http://purl.org/dc/elements/1.1/"                      3 xmlns:dc="http://purl.org/dc/elements/1.1/"
4 xmlns:ex="http://example.org/stuff/1.0/"                         4 xmlns:ex="http://example.org/stuff/1.0/"
5 xmlns:ex1="http://example.org/stuff/1.0/">                       5 xmlns:ex1="http://example.org/stuff/1.0/">
6 <rdf:Description rdf:about="http://www.univ-pau.fr">             6 <rdf:Description rdf:about="http://www.univ-pau.fr">
7   <ex:nameprof rdf:nodeID="UX"/>                                 7   <ex1:nameprof>
8   <ex:nameprof>                                                  8     <rdf:Description>
9     <rdf:Description>                                            9       <ex1:first_name rdf:datatype="http://www.w3.org/2001/XMLSchema#string">
10     <ex1:first_name rdf:datatype="http://www.w3.org/2001/XMLSchema#string">   Sebastien</ex1:first_name>
       Sebastien</ex1:first_name>                                  10      <ex:last_name xml:lang="fr">Durand</ex1:last_name>
11     <ex:last_name xml:lang="fr">Durand</ex:last_name>           11      <ex:last_name xml:lang="en">Durand</ex:last_name>
12     <ex1:last_name xml:lang="fr">Durand</ex1:last_name>         12      <ex1:first_name>Sebastien</ex1:first_name>
13     </rdf:Description>                                          13      <ex:last_name>Durand</ex:last_name>
14   </ex:nameprof>                                                14     </rdf:Description>
15   <ex:lab rdf:resource="http://liuppa.univ-pau.fr/live/"/>     15   </ex1:nameprof>
16   <ex:lab rdf:resource="http://liuppa.univ-pau.fr/live/"/>     16 </rdf:Description>
17 </rdf:Description>                                              17 </rdf:RDF>
18 <rdf:Description rdf:nodeID="UX">
19   <ex:first_name rdf:datatype="http://www.w3.org/2001/XMLSchema#string">
       Sebastien</ex:first_name>
20   <ex:last_name xml:lang="fr">Durand</ex:last_name>
21 </rdf:Description>
22 </rdf:RDF>
```

(a) Of the RDF graph in Fig. 1.a.2   (b) With syntactic disparities

Fig. 2. RDF/XML serializations

- **Problem 4 - Unused namespace:** where the namespaces are declared but never called in the body of the document (*dc* in line 3 in Fig. 2.a).
- **Problem 5 - Node order variation:** i.e., node siblings in the RDF description might be ordered differently when serialized (e.g., nodes in lines 6–16 in Fig. 2.b follow the order of appearance of XML elements, which can be re-ordered differently in another serializations).
- **Problem 6 - Handling typed elements:** objects (literals) elements can be typed or not (lines 9–12 in Fig. 2.b).
- **Problem 7 - Handling language tags:** Distinguishing between identical literals having different language tags (lines 10–11 in Fig. 2.b).

One can clearly realize the compound effect of missing logical and physical normalization when contrasting the serialization in Fig. 1.b.1 with those in Fig. 2, all of which represent the same RDF information (cf. Sect. 1). In the following, we develop our proposal addressing the above problems. Due to space limitations, only RDF/XML format is used in what follows to illustrate RDF serialization results.

4   RDF Normalization Proposal

We first provide a set of definitions, rules, and properties before developing our process.

4.1   Definitions

Definition 6 (Extended Statement [st$^+$]). *An extended statement is a more expressive representation of a statement (st), denoted as:* $st^+ :< s',p',o' >$ *where:*

- $s' :< s, ts >$ *is composed of the subject value (s) and its type (ts $\in \{u, bn\}$)*

– $p' :< p, dt, lng >$ is composed of the predicate value (p), its datatype $dt \in DT^5 \cup \{\perp\}$, and the language tag $lng \in Lang^6 \cup \{\perp\}$. $\perp$ represents a "null" value.
– $o' :< o, to >$ is composed of the object value (o) and its type $to \in \{u, bn, l\}$

The following notation is adopted to represent an extended statement:

$$st^+ :< s_{ts \in \{u,bn\}}, p_{dt}^{lng}, o_{to \in \{u,bn,l\}} \blacklozenge$$

Based on the example of Fig. 1.1.a, st_1 becomes st_1^+: <http://www.univ-pau.fr$_u$, ex:nameProf$_\perp^+$, UX$_{bn}$ >. The function ST^+ will be used in the following to return all the (extended) statements of an RDF Description.

Definition 7 (Extended Statement Equality $[=_{st}]$). *Two extended statements st_i^+ and st_j^+ are equal, noted $st_i^+ =_{st} st_j^+$, if and only if: (1) $st_i^+.s' = st_j^+.s'$, (2) $st_i^+.p' =st_j^+.p'$, and (3) $st_i^+.o' = st_j^+.o'$* ♦

In Fig. 3.a, $st_3^+ =_{st} st_4^+$ since they share the same subject (i.e., u_1), the same predicate (i.e., p_4), and the same object (i.e., u_2).

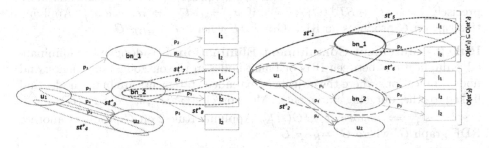

(a) With equal and different extended statements

(b) Of extended statement containment depicting the graph in Fig. 1.a.2

Fig. 3. RDF statement examples

Definition 8 (Outgoings $[O]$). *Given an extended RDF statement st_i^+, the outgoings of st_i^+, noted $O(st_i^+)$, designate the set of extended statements having for subject the object element o_i' of st_i^+:*

$$O(st_i^+ :< s_i', p_i', o_i' >) = \{st_j^+ :< (o_i', p_j', o_j' >, ..., st_n^+ :< o_i', p_n', o_n' >\} \blacklozenge$$

In Fig. 3.b, we identify the following outgoings of st_1^+: $O(st_1^+) = \{st_5^+, st_6^+\}$ $= \{< bn\_1_{bn}, p2_{string}^\perp, l_{1l} >, < bn\_1_{bn}, p3_\perp^{fr}, l_{2l} >\}$

Definition 9 (Statement Containment $[\preceq]$). *An extended statement st_i^+ is said to be contained in another extended statement st_j^+, noted $st_i^+ \preceq st_j^+$, if: (1) $O(st_i^+) \subseteq O(st_j^+)$, (2) both st_i^+ and st_j^+ have the same subject and predicate, and (3) the object type (to) in both statements is a blank node (bn)* ♦

---

[5] DT is a set of datatypes: string, number, date, etc.
[6] Lang is a set of language tags: @fr, @en, etc.

In Fig. 3.b[7], $st_1^+ \preceq st_2^+$ since they share the same subject (i.e., s_1) and the same predicate (i.e., p_1), and have $O(st_1^+) \subseteq O(st_2^+)$ w.r.t. their outgoings.

Definition 10 (RDF Equality $[=_{RDF}]$). *Two RDF graphs G_i and G_j are equal, $G_i =_{RDF} G_j$, if all the statements of G_i occurs in G_j and vice versa. Similarly, two RDF files F_i and F_j are equal, $F_i =_{RDF} F_j$, if: (1) the corresponding graphs are equal, and (2) F_i has the same encoding format (e_i) as F_j in (e_j)* ◆

4.2 Normalization Rules

In this section, we provide a set of rules to resolve the problems listed in Sect. 3.

Solving Logical Redundancies. Given an input RDF graph G, logical redundancies related to node duplication, edge duplication and node/edge duplications (presented in Sect. 3.1) can be eliminated from G by applying the following transformation rules (proofs are provided in [15]):

Rule 1 - Statement Equality Elimination: It is designed to eliminate edge duplications and/or node duplications using equality between statements. More precisely, $\forall st_i^+, st_j^+ \in ST^+(G) \wedge i \neq j$, if $st_i^+ =_{st} st_j^+ \implies$ remove st_j^+. Applying this rule on G produces an RDF Graph G' where $G' =_{RDF} G$.

Rule 2 - Statement Containment Elimination: It is designed to eliminate edge duplications between IRIs or blank nodes in the outgoing statements, and node duplications where the objects of extended statements are blank nodes and linked to other outgoing statements). More precisely: $\forall st_i^+, st_j^+ \in ST^+(G) \wedge i \neq j$, if $st_j^+ \preceq st_i^+ \implies$ remove $st_j^+ \cup O(st_j^+)$. Applying Rule 2 on G produces another RDF graph G' where $G' =_{RDF} G$.

Solving Physical Disparities. Given an input RDF file F, physical disparities related to namespace duplication, unused namespaces, and node order variation (presented in Sect. 3.2) can be eliminated from F by applying the following transformation rules:

Rule 3 - Namespace Duplication Elimination: It is designed to eliminate namespace duplications along with corresponding namespace prefixes such as: $\forall qn_i, qn_j \in QN(F) \wedge i \neq j$, if $qn_i.ns_i = qn_j.ns_j \implies$ remove $qn_j \wedge$ replace $qn_j.px_j$ by $qn_i.px_i$ in (F). Applying Rule 3 on F produces another equal RDF file since duplicated $qn_j.ns_j$ and its corresponding prefix $qn_j.px_j$ have been removed, while replacing $qn_j.px_j$ by $qn_i.px_i$ in the whole file.

Rule 4 - Unused Namespace Elimination: It is designed to eliminate the unused namespaces[8] with their respective prefixes such as: $\forall qn_i.ns_i \in NS^+(F)$, if $qn_i.ns_i \notin NS^+(G) \implies$ remove qn_i. Applying Rule 4 on F produces another equal file where unused $qn_i.ns_i$ and respective prefix px_i have been removed.

---

[7] st_i^+, u_i, p_i, bn_i, and l_i represent corresponding extended statements, IRIs, predicates, blank nodes, and literals.

[8] An unused namespace is a namespace which is mention in the serialization file but which is not use in any of the statements, i.e., it will not appear in the Graph.

Rule 5 - Reordering: It is designed to solve the varying node order problem by imposing a predefined (user-chosen) order on all statements of an RDF File F. More formally: $\forall st_i^+, st_j^+ \in ST^+(F) \implies st_i^+ <_{\tilde{p}} st_j^+$ where $\tilde{p} :< iorder, sortc >$ is a tuple composed of an indexing order "*iorder*" and a sorting criterion "*sortc*". For *iorder*, we follow the six indexing schemes presented in [17] (SPO, SOP, PSO, POS, OSP, OPS) since it is the combination of the three elements of the statement (subject, predicate, object) in an RDF Description. For *sortc*, we adopt *asc* and *des* to represent ascending and descending order respectively. The default value for the parameter $\tilde{p}$ is $< sop, asc >$ representing an ascending order of statements w.r.t. their subjects / objects / predicates (sop). Applying Rule 5 on F produces an equal RDF file where all the statements have been ordered following the (user-chosen) order type parameter $\tilde{p}$.

4.3   Normalization Properties

Our approach verifies the following properties characterizing the quality of the normalization process. This corresponds to the notion of Information Reusability discussed in existing studies[9]. Proofs are provided in [15].

Property 1 (Completeness). *An RDF Description D_i (graph or file) is said to be complete regarding D_j if D_i preserves and does not lose any information in D_j, i.e., each resource, statement and namespace of D_j has a corresponding resource, statement and namespace in D_i* •

Property 2 (Minimality). *An RDF Description D_i is said to be minimal if all the resources, statements and namespaces of D_i are unique (i.e., they do not have duplicates in D_i) and all the namespaces are used* •

Property 3 (Compliance). *An RDF File F_i is said to be compliant if it verifies all the rules of the W3C standard in producing a valid file based on an RDF Graph G_i, i.e., its structure remains compliant with RDF serialization standards (e.g., RDF/XML)* •

Property 4 (Consistency). *An RDF Description D_i is said to be consistent if D_i verifies the completeness, minimality and compliance properties to ensure the data quality of the description* •

4.4   Normalization Process

The inputs of our RDF normalization process are: (a) an RDF graph (logical representation) or an RDF file (physical representation) to be normalized, and (b) user parameters related to the RDF output form and prefix renaming, enabling the user to tune the results according to her requirements (see Fig. 4). In the following, we detail each step of the normalization process.

---

[9] This is comparable to the notion of *map function* in [4] except that the authors do not consider namespaces.

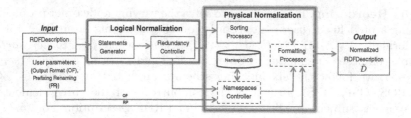

Fig. 4. Architecture of our RDF normalization process

Logical Normalization. Algorithm 1 provides the pseudo-code to remove redundancies from an RDF graph. It starts by erasing the statement redundancies having IRIs or literals. Consequently, it removes the equal statements as well as contained ones with duplicated blank nodes (bn) (and all the outgoings O derived of the bn) following Definition 9.

Algorithm 1. Removing Redundancy

Require: $st^+[]$ //*List of Extended Statements of the RDF Description*
Ensure: $st^+[]$ //*List of Extended Statements without duplication*
1: $N = st^+$.length(); //*Number of Statements in the list*
2: **for** i=1, i $\leq$ N, i++ **do**
3:   **for** j=i+1, j $\leq$ N, j++ **do**
4:     **if** $st^+[i]$.to = "IRI" or $st^+[i]$.to = "Literal" and $st^+[i] =_{st} st^+[j]$ **then**
5:       Remove $st^+[j]$; // *remove statement duplication following Rule 1*
6:     **else**
7:       **if** $st^+[j]$.to = "bn" and $st^+[i]$.to = "bn" and $st^+[i]$.s = $st^+[j]$.s and $st^+[i]$.p = $st^+[j]$.p
        and ($st^+[i] \preceq st^+[j]$ or $st^+[j] \preceq st^+[i]$) **then**
8:         Remove $st^+[j]$ and all its outgoings // *remove blank node duplication - Rule 2*
9:       **else**
10:        **if** $st^+[i]$.o = $st^+[j]$.o **then**
11:         Remove $st^+[j]$ // *remove statement duplication following Rule 1*

Physical Normalization. It is divided into three components based on the types of physical disparities being processed:

1. ***Namespaces Controller (NC):*** it controls namespace duplication by eliminating the redundancies (Rule 3) and the unused namespaces (Rule 4) in the RDF File. The input of this component is the prefix renaming parameter, which allows to customize the renaming of the prefixes while providing a unique way to normalize them.
2. ***Sorting Process (SP):*** Rule 5 establishes the node order variation, to have a unique specification of the statements in the output serialization. It considers the parameter $\tilde{p} :< iorder, sortc >$ given by the user according to the targeted applications.
3. ***Formatting Process (FP):*** it allows to: (a) choose a specific form for the output RDF file, (b) manage the variety of blank node serializations, and (c) manage the datatypes and the languages. Our current solution allows three different output forms (other forms could be devised based on user/application requirements):

Flat: it develops each RDF description one by one as a single declaration, i.e., each subject has one declaration in the file.

Compact: it nests the RDF description, i.e., each description may have another description nested in its declaration.

Full compact: dedicated to RDF/XML format, it nests the RDF description, uses the ENTITY XML construct to reduce space by providing an abbreviation for IRIs[10], reuses the variables in the RDF file, and uses attributes instead of properties for the blank node serialization.

Providing different output types is necessary to satisfy the requirements of different kinds of RDF-based applications (e.g., compact representations are usually of interest to human users when running RDF queries [4], yet less compact/more structured representations could be useful in automated processing, such as in RDF annotation recommendation [12]).

5   Experimental Evaluation

5.1   Experimental Environment

An online prototype system: *RDF2NormRDF*[11], implemented using PHP and Java, was developed to test, evaluate and validate our RDF Normalization approach. It was used to perform a large battery of experiments to evaluate: (i) the effectiveness (in performing normalization), and (ii) the performance (execution time) and storage space of our approach in comparison with its most recent alternatives. To do so, we considered three large datasets: (i) Real_DS made of real RDF files from LinkedGeoData, (ii) Syn_DS1 made of synthetic RDF files generated from the real dataset by including additional random logical redundancies and physical disparities, and (iii) Syn_DS2 consisting of a variation of Syn_DS1 files including more heterogeneity (more duplications and disparities) than Syn_SD1 (cf. Table 1).

Table 1. Features of files in each dataset

Datasets	Features	Size kb	IRIs	BNs	Lit.	St	BN Dup.	Lit. Dup.	St Dup.	Log. Red. %	Ns Dup.	Non-used Ns	Phys. Dis. %
Real_DS	Max	7.2	9	0	40	77	0	34	63	77	7	8	55
	Min	2.8	9	0	7	25	0	1	11	29	1	2	40
	Avg	4.5	9	0	20	45	0	14	30	57	3	5	48
Syn_DS1	Max	4.3	17	12	26	63	3	7	18	69	4	2	78
	Min	0.5	2	1	3	5	0	0	0	0	0	0	0
	Avg	2.2	6	6	14	30	2	5	12	32	2	1	60
Syn_DS2	Max	48.6	47	129	320	784	123	307	753	98	126	3	98
	Min	0.5	2	1	3	5	0	0	0	0	0	0	0
	Avg	7.9	6	19	50	116	16	43	98	64	14	2	70

Experiments were undertaken on an Intel®Core(TM) i7 - 2600 + 3.4 GHz with 8.00 GB RAM, running a MS Windows 7 Professional OS and using a Sun JDK 1.7 programming environment.

---

[10] http://www.w3.org/TR/xml-entity-names/.

[11] Available at http://rdfn.sigappfr.org/.

5.2 Experimental Results

We evaluated the effectiveness and performance of our method, called in this section **R2NR**, in comparison with alternative methods, namely the JSON-LD normalization approach [9,13] and the HDT technique [2].

Table 2. Goals/properties achieved in datasets after normalization processes

Goals/Properties	Real_DS			Syn_DS1			Syn_DS2		
	JSON-LD	HDT	R2NR	JSON-LD	HDT	R2NR	JSON-LD	HDT	R2NR
Solving log. redundancies	57%	57%	57%	5%	32%	32%	12%	64%	64%
Solving phys. disparities	48%	48%	48%	60%	60%	60%	70%	70%	70%
Preserving completeness	True	True	True	True	False	True	True	False	True
Preserving minimality	True	True	True	False	False	True	False	False	True
Preserving compliance	True	True	True	True	True	True	True	True	True
Preserving consistency	True	True	True	False	False	True	False	False	True

1. **Effectiveness:** Results in Table 2 show that our method produces normalized RDF files that fulfill all our normalization properties and goals, in comparison with JSON-LD and HDT which miss certain logical and physical redundancies/disparities. On one hand, the **JSON-LD method** preserves some of the redundancies, i.e., JSON-LD removes only 5 % over a 32 % average of logical redundancies in *Syn_DS1* and 12 % over a 64 % average of logical redundancies in *Syn_DS2*. However, it removes all logical redundancies from the Real_DS (all 57 % of logical redundancies and 48 % of physical disparities were removed). A careful inspection of JSON-LD shows that it preserves blank node duplication and certain literal duplications, which explains the results obtained with the Real_DS (since it does not contain any blank nodes). As a result, the JSON-LD approach does not satisfy the *minimality* and *consistency* properties. On the other hand, the **HDT method** results show, at first glance, that it successfully eliminates all logical redundancies and physical disparities. Nonetheless, a closer look at the results revealed that the HDT technique actually preserves blank node redundancies by assigning them different identifiers and/or representing them as IRIs. Hence, the HDT method actually keeps the logical redundancies (and corresponding physical disparities) and does not consequently satisfy the *completeness*, *minimality* and *consistency* properties.

2. **Performance**: In addition, we verified our approach' performance in terms of loading time and storage space, in comparison with JSON-LD and HDT.

   (a) **Jena loading time:** The complexity of our method comes down to worst case $O(N^2)$ time where N represents the number of RDF statements in the document (cf. [15]). Jena loading time results depicted in Figs. 5 and 6 confirm our approach' polynomial (almost linear) time w.r.t. document size. Also, results in Figs. 5.a and 5.b demonstrate that our method executes faster than JSON-LD's. Note that redundancy reduction using of *JSON-LD* amounts to 5 % on average file size in Syn_DS1, while our method reaches an average 27 % size reduction ratio, which explains the reduction in loading time. In addition, results in Figs. 6.a and 6.b show that our method remains also faster than the HDT method. In fact, as shown in Table 2, the datasets

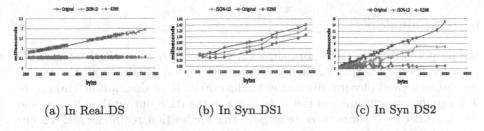

(a) In Real_DS   (b) In Syn_DS1   (c) In Syn DS2

Fig. 5. Average Jena loading time comparison with JSON-LD method

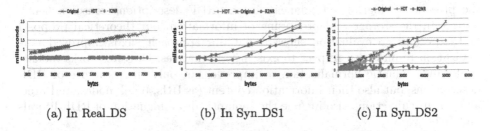

(a) In Real_DS   (b) In Syn_DS1   (c) In Syn_DS2

Fig. 6. Average Jena loading time comparison with HDT method

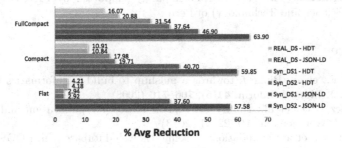

Fig. 7. Average reduction in datasets Syn_DS1 and Syn_DS2 w.r.t. the output format

generated by HDT do not have redundancies and disparities, yet contain a larger number of IRIs with no (zero) BNs. This confirms that HDT is transforming BNs into IRIs, which shows that RDF compression does not always guarantee normalization. Note that we are currently investigating this issue in more details in a dedicated experimental study.

(b) Storage: Neither JSON-LD nor HDT methods provide parameters to customize output format requirements as we do. They work with their predefined outputs, i.e., the JSON-LD method with N-triples, and the HDT method with Bitmap Triples (BT), in comparison with our method which handles the standard formats and thus allows developing different outputs w.r.t the target application. Figure 7 shows that our normalization method improves (reduces) the size of the RDF files in all formats of the datasets processed by JSON-LD and HDT methods.

6   Conclusion

We proposed here an RDF normalization method *RDF2NormRDF* able to: (i) preserve all the information in RDF descriptions, (ii) eliminate all the logical redundancies and physical disparities in the output RDF description, (iii) establish a unique specification of the statements in the RDF output description, and (iv) consider user parameters to handle the application requirements. To our knowledge, this is the first attempt to study and integrate RDF normalization in two aspects: logical redundancies and physical disparities. Understanding that the presence of logical redundancies in the RDF descriptions causes a heavy impact in the storage and processing of information, our theoretical proposal and experimental evaluation showed that our approach yields improved normalization results with respect to existing alternatives. Ongoing works include exploiting semantic normalization to improve, not only the structure of RDF descriptions, but also their information content (as IRIs, blank nodes, and literals). Future directions also include the semantic disambiguation of RDF literals and IRIs.

Acknowledgments. This work has been partly supported by FINCyT (Fund for Innovation, Science and Technology) of Peru.

References

1. Belleau, F., et al.: Bio2rdf: towards a mashup to build bioinformatics knowledge systems. J. Biomed. Inform. **41**(5), 706–716 (2008)
2. Fernández, J.D., et al.: Binary rdf representation for publication and exchange (HDT). J. Web Semant. **19**, 22–41 (2013)
3. Gutierrez, C., et al.: Foundations of semantic web databases. In: PODS 2004, pp. 95–106. ACM (2004)
4. Gutierrez, C., et al.: Foundations of semantic web databases. J. Comput. Syst. Sci. **77**(3), 520–541 (2011)
5. Hayes, J., Gutierrez, C.: Bipartite graphs as intermediate model for RDF. In: McIlraith, S.A., Plexousakis, D., van Harmelen, F. (eds.) ISWC 2004. LNCS, vol. 3298, pp. 47–61. Springer, Heidelberg (2004)
6. Jiang, G., et al.: Using semantic web technology to support ICD-11 textual definitions authoring. J. Biomed. Semant. **4**, 11 (2013)
7. Kerzazi, A., et al.: A model-based mediator system for biological data integration. In: Journes Scientifiques en Bio-Informatique, pp. 70–77 (2007)
8. Kerzazi, A., et al.: A semantic mediation architecture for RDF data integration. In: SWAP, p. 3 (2008)
9. Longley, D.: RDF dataset normalization (2015). http://json-ld.org/spec/latest/rdf-dataset-normalization/
10. Nolin, M.-A., et al.: Building an hiv data mashup using Bio2RDF. Briefings Bioinform. **13**(1), 98–106 (2012)
11. Pathak, J., et al.: Lexgrid: a framework for representing, storing, and querying biomedical terminologies from simple to sublime. J. Am. Med. Inform. Assoc. **16**(3), 305–315 (2009)

12. Salameh, K., Tekli, J., Chbeir, R.: SVG-to-RDF image *Semantization*. In: Traina, A.J.M., Traina Jr., C., Cordeiro, R.L.F. (eds.) SISAP 2014. LNCS, vol. 8821, pp. 214–228. Springer, Heidelberg (2014)

13. Sporny, M., Longley, D.: RDF graph normalization (2013). http://json-ld.org/spec/ED/rdf-graph-normalization/20111016/

14. Tao, C., et al.: A RDF-base normalized model for biomedical lexical grid. In: The 8th International Semantic Web Conference, p. 2 (2009)

15. Ticona-Herrera, R., et al.: Rdf similarity. Technical report (2015). http://rdfn.sigappfr.org/RDFN-TR-15.pdf

16. Vrandecic, D., et al.: RDF syntax normalization using XML validation. In: Proceedings of the SemRUs, p. 11 (2009)

17. Weiss, C., Karras, P., Bernstein, A.: Hexastore: sextuple indexing for semantic web data management. Proc. VLDB Endow. **1**(1), 1008–1019 (2008)

Design Dimensions for Business Process Architecture

Alexei Lapouchnian[1(✉)], Eric Yu[1], and Arnon Sturm[2]

[1] University of Toronto, Toronto, Canada
alexei@cs.toronto.edu, eric.yu@utoronto.ca
[2] Ben-Gurion University of the Negev, Beer-Sheva, Israel
sturm@bgu.ac.il

Abstract. Enterprises employ an array of business processes (BPs) for their operational, supporting, and managerial activities. When BPs are designed to work together to achieve organizational objectives, we refer to them and the relationships among them as the business process architecture (BPA). While substantial efforts have been devoted to designing and analyzing individual BPs, there is little focus on BPAs. As organizations are undergoing changes at many levels, at different times, and at different rates, the BP architect needs to consider how to manage the relationships among multiple BPs. We propose a modeling framework for designing BPAs with a focus on representing and analyzing architectural choices along several design dimensions aimed at achieving various design objectives, such as flexibility, cost, and efficiency.

Keywords: Business process architecture · Goal modeling · Requirements

1 Introduction

Organizations rely on multitude of business processes (BPs) for their everyday functioning as well as for longer term viability and sustainability. These processes need to work together coherently, constituting the *process architecture* of the organization.

BP analysis and design have garnered a lot of attention, unlike the development of business process architectures (BPAs), which is about how BPs should relate to each other to better serve their organization's objectives. Given the vast changes faced by today's enterprises due to business/technological innovations, BPs can no longer be optimized individually. Rather, the onus should be on BPAs, which cannot remain static and need to be re-designed to account for changes inside and outside of organizations.

BPA development involves making trade-offs across processes, particularly regarding how to balance flexibility and agility with other design objectives such as costs, efficiency, and so on. Existing conceptions of BPA (e.g., [3], enterprise architecture (EA) frameworks [13], etc.) mostly see BPAs as a given, or as something to be discovered, with no effort to shape or evolve them. We, however, see the need to explore BPA alternatives and to support the selection among them, while taking complex trade-offs into consideration. An appropriate modeling notation capable of addressing these requirements is needed. This paper proposes temporal and recurrence dimensions that, we

© Springer International Publishing Switzerland 2015
P. Johannesson et al. (Eds.): ER 2015, LNCS 9381, pp. 276–284, 2015.
DOI: 10.1007/978-3-319-25264-3_20

believe, serve as key elements of such a framework. We chose the public transportation domain, which is rich, diverse, and easy to understand to illustrate our approach.

This paper is organized as follows. Section 2 outlines the architectural design space and Sect. 3 discusses two dimensions of that space. Section 4 talks about the analysis of BPAs. Then, Sect. 5 covers the related works and Sect. 6 concludes the paper.

2 Variability Space for Business Process Architectures

In a framework aimed for BPA design, we have to represent a space of BPA alternatives and thus require a modeling notation capable of this and of reasoning about the identified choices. We propose two BPA design dimensions to structure the space of architectural alternatives: (1) *temporal* is about placing a process element (PE) – an activity or a decision – earlier or later relative to other PEs within a BP; (2) *recurrence* is about placing a PE into a BP that is executed more (or less) frequently relative to other BPs. Then, changes to BPAs can be seen as movements along these dimensions.

When determining the positioning of PEs along these dimensions, we look at the trade-off between flexibility and efficiency, a major concern in BPA design. How this tension is resolved is domain- and organization-specific. For a static business domain, BPs in a BPA can be tightly coupled and globally optimized. Instead, in a dynamic domain, BPAs have to provide flexibility to support organizational agility. This flexibility requires that organizations keep their options open. This may incur costs, decrease performance, and increase complexity and hence leads to the above-mentioned trade-off. Thus, the BPA design space model needs to represent where/what options exist and provide ways to analyze the choices w.r.t. important quality objectives (NFRs).

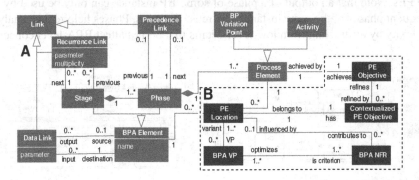

Fig. 1. The metamodel for the approach combines (A) BPA modeling and (B) goal modeling.

These two dimensions were determined based on existing studies (see Sect. 5), our own experience with BPAs, and the analysis of existing BPAs and their potential changes. We are not claiming that no other dimensions exist. In fact, here, we elaborate only on the core portion of our research on designing adaptive BPAs. Further design dimensions for evolving BPAs are proposed in [7, 8]. In addition, ideas on utilizing emergent technologies for system flexibility with the help of BPAs are discussed in [6].

The architectural description we are targeting should outline the major BPA elements and relationships while avoiding over-specification. Figure 1 presents a meta-model for our modeling notation where we combine BP modeling and goal modeling. In Sect. 3, we discuss the modeling of BPAs (the "A" part of the metamodel), while Sect. 4 talks about using a goal-based notation (the "B" part) for comparing them.

3 Dimensions for Designing Process Architectures

3.1 The Temporal Dimension

There may be multiple possible placements for PEs along the temporal dimension in a BP that achieve the same functional objective and comply with the existing functional dependencies, but differ in their non-functional characteristics. These choices need to be resolved by looking at how each variant affects the quality criteria that the organization is interested in. Trip payment options for passengers in Fig. 2A illustrate this: unlike the standard fare charged before a trip, a distance-based fare can be charged after one. Both obtain the payment, but differ in the charged amount, payment fairness, etc.

Definitions. Determining what is better, to charge the customer before a trip, during the trip or after travel depends on one's point of view. Going for payment fairness, we want taxi-like distance-based fares. Here, the system needs a richer context (the traveled distance) only available after travel. So, payment on exit is a better option for fairness. We identify *phases* – portions of a BP such that placing a PE anywhere within them is the same in terms of the functional and non-functional results. Moving PEs *across* phase boundaries may affect action outcomes and decision quality. Unlike most software variability approaches, we analyze in which phase (i.e., when) it is best to place PEs. Note that an output of a phase of some BP instance can only be used by the subsequent phases of the same instance. No reuse happens. Phases help reduce analysis complexity by abstracting from low-level details that do not affect BPA-level concerns.

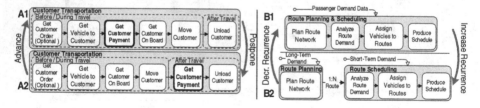

Fig. 2. A: Trip payment options – before (A1) or at (A2) the end of a trip. B: Splitting a single stage into two (B2) or merging two stages into one (B1).

Postponement. We look at postponing or advancing PEs by placing them in later or earlier phases respectively (see Fig. 2A). *Postponement* is a well-known business strategy (e.g., [9]) that delays activities/decisions that require up-to-date information while expecting more precise data to be available at a later point in a BP. This produces better, more context-sensitive outcomes. Instead, *advancement* supports stability/uniformity and is enabled either by PEs that rely on less information and thus have more tolerance for uncertainty or by predicting the missing information. Availability of data required by the postponed PEs and data collection/analysis costs are important concerns here. E.g., for distance-based fares, the system needs to obtain the traveled distance.

3.2   The Recurrence Dimension

Unlike the previous section where PEs were executed for every process instance, the recurrence dimension is about reusing PE outcomes in multiple BP instances: how often should certain decisions or actions be (re)executed and under what conditions?

Definitions. We group PEs with the same execution frequency into process portions called *stages* that contain one or more phases (e.g., in Fig. 2A, Customer Transportation is a stage with two phases). Once a stage executes, its output is available to the subsequent stages (if any) until it is re-executed. In our models, stages are connected with control flow links labeled with "1:N" to indicate the relationship cardinality (see Fig. 2B). A *stage boundary* between two stages points to the two options for placing PEs. Moving a PE across this boundary can lead to a major change in the PE's execution frequency. Figure 2B shows this change in both directions for Route Scheduling. A PE can also be moved across more than one stage. A stage represents a (sub-)process and while the temporal dimension was about intra-process analysis, here the focus is on inter-process relationships – relative rates of execution/change cycles among processes.

Using the Dimension. We identify PEs that can be reused for multiple BP instances (i.e., that are independent of the variations in those instances), at least for a period of time, to arrive at a stage-based configuration. Such PEs form a stage with its output available to be reused by the subsequent stages, thus saving time, money and/or other resources and perhaps affecting other NFRs. E.g., buying a transit pass eliminates the need to pay for each trip separately and hence improves convenience. Another heuristic is to identify PEs with the same execution frequency (e.g., yearly product redesigns accompanied by the product manual updates) or are triggered by the same data-driven trigger (e.g., a passenger demand change forces a bus schedule revision and the subsequent publication of the new schedule). Splitting a stage into two or merging two stages into one (see Fig. 2B) is another type of change in this dimension.

Domain Example. One way to perform public transit route network planning and scheduling is to combine both into a single stage (Fig. 2B1). Whenever routes need to be changed (e.g., due to major demand changes), Route Planning and Scheduling stage is triggered. The stage's data input (the message flow at the top) has all the available passenger demand data. Both route planning and scheduling are bundled together and have the same change cycle, meaning that changing schedules without a route network redesign is not possible. For a predictable constant demand this will work well. However, the rigidity will hurt the company's ability to change its schedules more often than its routes given an evolving passenger demand. As a solution, we can unbundle the two stages as shown in Fig. 2B2, creating Route Planning and Route

Scheduling, which are triggered when long-term or short-term passenger demand changes, respectively. Then, the route network (the outcome of the Route Planning stage) can be reused for multiple Route Scheduling instances (again, note the "1:N" annotation) supporting frequent rescheduling on the same route and thus enabling more flexibility, but likely having higher cost, complexity, etc. Given the choices along the recurrence dimension, an organization needs to pick the one that best fits its needs and preferences, while reflecting its business domain volatility.

Figure 3 presents a BPA fragment for a transit company showing the relationships among stages (we abstract from phases here) and implementing the more flexible options from the previously discussed examples. Data inputs for a stage come from other stages (see Required Capacity) or from the outside – e.g., through monitoring (see Social Network Data). A BPA model gives an overview of the organization's BPs and points to where reuse happens. Figure 3 shows a *possible* BPA configuration. Others are obtained by moving PEs among stages/phases. The starting point for developing a BPA using our notation is an existing BPA in another notation or a collection of BPs. We are working on a goal-driven approach for designing the initial BPA using ideas from [5] that would refine organizational goals into objectives to be achieved by individual PEs.

The next section elaborates on modeling and analysis of BPA configuration choices.

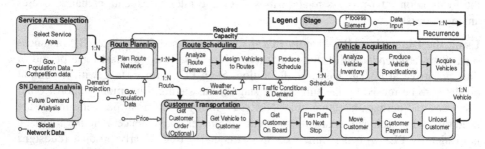

Fig. 3. A fragment of a business process architecture for a public transit company.

4   Analyzing Process Architecture Alternatives

We use a goal model (GM) variant to represent and analyze BPA choices along the two dimensions (see Fig. 1B for its metamodel). GMs are used in Requirements Engineering to capture stakeholder/system objectives. They can represent variability in achieving goals using OR decompositions (exclusive ORs in our case) and capture NFRs to analyze goal refinement choices. We adapted GMs to represent possible PE placements in BPAs and to evaluate them against the relevant NFRs. A GM (see Fig. 4) focuses on a single PE, so multiple GMs will be used to analyze various portions of a BPA.

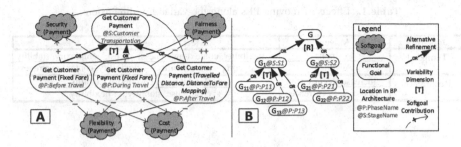

Fig. 4. A: Analyzing the temporal placement of customer payment PE. B: A generic goal model for positioning a PE with the objective G in a BPA.

Identifying BPA Choices. To show how a GM is developed and used, we look at the trip payment options discussed in Sect. 3.1. A GM provides the intentional (vs. the operational) view, has the goal of the PE in question as its root node (see the top Get Customer Payment node in Fig. 4A), and shows how it can be refined when placed into different BPA stages/phases. E.g., Fig. 4A shows the three phases of Customer Transportation stage where Get Customer Payment PE can be placed. The @P: PhaseName and @S:StageName annotations refer to these locations. Each goal refinement looks at the PE placement options along one dimension ([T] for temporal and [R] for recurrence, as in Fig. 4). The refinement choices for the temporal dimension are the stages where a PE can be placed (while respecting the existing functional constraints). The phases within a *previously selected* stage are the choices for the temporal dimension – e.g., in Fig. 4A, customers can pay before (@P:Before Travel), during or after travel. This leads to the common pattern of analysis where [T] follows [R] (see Fig. 4B).

Varying parameters in goal nodes capturing alternative PE placements represent the possibly different implementations of a PE in various places in a BPA. Implementations depend on the available data, etc. E.g., the traveled distance is only known after the trip. Later stages/phases usually have higher contextualization and thus more parameters.

Using NFRs. NFRs for evaluating PE placement choices must be elicited. They are modeled as softgoals and evaluated using *contribution links* (e.g., [5]). The evaluation can be qualitative, with values such as help(+)/hurt(−), make(++)/break(−−) (as in Fig. 4A). A softgoal is *satisficed* if there is more positive than negative evidence for this claim. Given extra information, a more precise evaluation is possible. Softgoal prioritization can help in case of conflicting NFRs that cannot be satisficed at the same time.

Handling Trade-Offs. Moving PEs along the two dimensions affects many NFRs. Some NFRs are domain- and PE-independent. The effects of moving PEs on typical NFRs are shown in Table 1 (cost/flexibility also feature in Fig. 4A). Others are domain- or PE-specific – e.g., security and fairness for fare payment (Fig. 4A). The resulting trade-offs are resolved based on two things: (1) domain dynamics (what changes, how frequently, etc.); (2) the organization's ranking of the above-mentioned quality criteria.

Once the analysis of BPA alternatives for a PE is done, a place (stage/phase) in the BPA is identified for it. It can be seen as the delta between the as-is and to-be BPAs.

Table 1. Effects of moving PEs along the variability dimensions.

Dimension	PE Movement	Effect of Movement on NFRs
Temporal	Postpone	+: flexibility, context-awareness; –: cost, complexity, stability
	Advance	+: cost, complexity, stability; –: flexibility, context-awareness
Recurrence	Increase Recurrence	+: flexibility; –: cost, reuse, stability
	Decrease Recurrence	+: cost, reuse, stability; –: flexibility

Reconfiguring BPAs. Our method's objective is to help organizations find the right balance between flexibility and stability given their preferences/priorities and the level of change in their business domains. Change can have two types of effect on a BPA. Once designed, a BPA has some amount of flexibility to support certain types of domain changes (e.g., varying trip distances are supported by distance-based fare payments). If flexibility in a BPA can accommodate the change (i.e., the changing trip lengths), a BPA does not need a reconfiguration. However, if the assumption that passenger trip lengths are quite different (thus justifying the complexity/cost of distance-based fares) no longer holds, distance-based fares become *too flexible* for the domain and the BPA may need to be changed. Hence, a BPA is modified when the domain dynamics (the rate/range of change) is changes (not when *some* change happens).

5 Related Work

Over the years, BPAs were researched (e.g., [3]) and various sets of relationships among their BPs were identified [2, 3], such as sequence, hierarchy, trigger, etc. In the EA area, BPAs are viewed as BP cooperation and ArchiMate [7] includes causal, mapping, realization, etc. relationships between BPs and services. Other relevant domains are BP variability [1], which focuses on the realization relationships among BPs and the binding time of such processes, and software product lines [10]. The ability to postpone design decisions to the latest feasible point is a major reason to support variability [10]. Advancing decisions in processes improves their efficiency [11]. In goal-based methods for BPs (e.g., [5]), the many ways of achieving business goals are utilized to develop customizable, adaptive, and evolving systems. Methods weaving requirements and BPs have also emerged. E.g., [12] uses NFRs and contexts for runtime BP configuration. In process-aware information systems, Weber et al. [14] identify four dimensions of change and point to the new BP relationships: *creation*, where a BP creates another one, and *recurrence*, where a BP is followed by another BP multiple times.

Overall, existing approaches focus on process-level variability while we look at the BPA-level variability and goal-based trade-off analysis (also missing from most other approaches). Also, for PEs that are decisions themselves, the phases/stages where they can be placed provide options for *when* (temporal dimension) and *how often* (recurrence dimension) to bind them. Hence, we obtain domain-specific variability binding options, which are much richer than those usually discussed in variability research (e.g., [4]).

6 Discussion and Conclusions

A good BPA must reflect its domain. Assumptions about the domain (e.g., its volatility and data availability within it) need to be analyzed to justify the flexibility in the BPA. Such assumptions are not yet captured in our approach. We are working on their formal modeling, which will help specify the conditions that trigger BPA adaptations.

The analysis in Sect. 4 favours local optimization (at the level of individual PEs), which is detrimental to the global optimality. We are developing ways to integrate multiple goal models, each similar to those in Fig. 4, to alleviate this. Also, identification of relevant data and its availability and volatility play a vital role in positioning PEs within a BPA and we are currently working on integrating data into this approach.

We support some model incompleteness to reduce modeling and analysis complexity and to focus on the important domain aspects. GMs only have to be utilized in volatile portions of the domain, where BPA changes might be needed. Some portions of the BPA can be seen as stable and thus not requiring an analysis. Also, most BP modeling details are below our threshold of interest, which is a phase. For analyzing BPA choices, we plan to utilize goal reasoning algorithms already successfully used in a variety of applications (e.g., in [5] for BP configuration).

In summary, we presented an approach for modeling/analysis of BPAs to help organizations be better aligned with their business domain dynamics and their desired level of flexibility. We introduced two BPA design dimensions to identify BPA choices and adopted GMs to analyze these w.r.t. important NFRs. We have identified a research agenda to further refine the approach and are applying the method in several domains that experience significant business/technological volatility in order to evaluate it.

References

1. Chakravarthy, V., Eide, E.: Binding time flexibility for managing variability. In: Proceedings of the MVCDC 2 at OOPLSA 2005, San Diego, CA, USA (2005)
2. Dijkman, R., Vanderfeesten, I., Reijers, H.: Business process architectures: overview, comparison and framework. Enterp. Inf. Syst. doi:10.1080/17517575.2014.928951
3. Dumas, M., La Rosa, M., Mendling, J., Reijers, H.: Fundamentals of Business Process Management, Ch. 2. Springer, Berlin (2013)
4. Galster, M., Weyns, D., Tofan, D., Michalik, B., Avgeriou, P.: Variability in software systems – a systematic literature review. IEEE TSE **40**(3), 282–306 (2014)
5. Lapouchnian, A., Yu, Y., Mylopoulos, J.: Requirements-driven design and configuration management of business processes. In: Alonso, G., Dadam, P., Rosemann, M. (eds.) BPM 2007. LNCS, vol. 4714, pp. 246–261. Springer, Heidelberg (2007)
6. Lapouchnian, A., Yu, E.: Exploiting emergent technologies to create systems that meet shifting expectations. In: Proceedings of ET Track at CASCON 2014, Toronto, Canada (2014)
7. Lapouchnian, A., Yu, E., Sturm, A.: Re-designing process architectures: towards a framework of design dimensions. In: Proceedings of RCIS 2015, Athens, Greece, 13–15 May 2015

8. Lapouchnian, A., Yu, E., Sturm, A.: Re-designing process architectures. Technical report CSRG-625, University of Toronto (2015). ftp://cs.toronto.edu/csrg-technical-reports/625

9. Pagh, J., Cooper, M.: Supply chain postponement and speculation strategies: how to choose the right strategy. J. Bus. Logistics **19**(2), 13–33 (1998)

10. Svahnberg, M., van Gurp, J., Bosch, J.: A taxonomy of variability realization techniques: research articles. Softw. Pract. Experience **35**(8), 705–754 (2005)

11. Subramaniam, S., et al.: Improving process models by discovering decision points. Inf. Syst. **32**(7), 1037–1055 (2007)

12. Santos, E., Pimentel, J., Pereira, T., Oliveira, K., Castro, J.: Business process configuration with NFRs and context-awareness. In: ER@BR (2013)

13. TOGAF Version 9.1. (2011). http://pubs.opengroup.org/architecture/togaf9-doc/arch/

14. Weber, B., Reichert, M., Rinderle-Ma, S.: Change patterns and change support features – enhancing flexibility in process-aware information systems. Data Knowl. Eng. **66**(3), 438–466 (2008)

Interoperability and Integration

A Conceptual Framework for Large-scale Ecosystem Interoperability

Matt Selway, Markus Stumptner[✉], Wolfgang Mayer, Andreas Jordan,
Georg Grossmann, and Michael Schrefl

Advanced Computing Research Centre, University of South Australia,
Adelaide, Australia
{andreas.jordan,matt.selway}@mymail.unisa.edu.au,
{mst,wolfgang.mayer,georg.grossmann}@cs.unisa.edu.au,
schrefl@dke.uni-linz.ac.at

Abstract. One of the most significant challenges in information system design is the constant and increasing need to establish interoperability between heterogeneous software systems at increasing scale. The automated translation of data between the data models and languages used by information ecosystems built around official or de facto standards is best addressed using model-driven engineering techniques, but requires handling both data and multiple levels of metadata within a single model. Standard modelling approaches are generally not built for this, compromising modelling outcomes. We establish the SLICER conceptual framework built on multilevel modelling principles and the differentiation of basic semantic relations that dynamically structure the model and can capture existing multilevel notions. Moreover, it provides a natural propagation of constraints over multiple levels of instantiation.

Keywords: Metamodelling · Conceptual models · Multilevel modelling

1 Introduction

Lack of interoperability between computer systems remains one of the largest challenges of computer science and costs industry tens of billions of dollars each year [1,2]. Standards for data exchange have in general not solved the problem, as standards are not universal or universally applied even within a given industry, leading to *heterogeneous ecosystem* with large groups of software systems built around different standards that must interact to support the entire system lifecycle. We are currently engaged in the "Oil and Gas Interoperability Pilot" (or simply OGI Pilot) that aims for the automated, model-driven transformation of data during the asset lifecycle between two of the major data standards in the Oil & Gas industry ecosystem. The main standards considered by the project are the ISO15926 suite of standards [3] and the MIMOSA OSA-EAI specification [4]. These standards and their corporate use[1] are representative of the

---

[1] Current industrial participants in the OGI Pilot include: Intergraph, Bentley, AVEVA, Worley-Parsons, for ISO15926; IBM, Rockwell Automation, Assetricity for MIMOSA; various Oil & Gas or power companies as potential end users.

© Springer International Publishing Switzerland 2015
P. Johannesson et al. (Eds.): ER 2015, LNCS 9381, pp. 287–301, 2015.
DOI: 10.1007/978-3-319-25264-3_21

interoperability problems faced in many industries today, and the cost of changing a code base and continuous change in the base data make conventionally programmed interoperability solutions too expensive to build and maintain. To enable sensor-to-boardroom reporting, the effort to establish and maintain interoperability solutions must be drastically reduced. This is achieved by developing model transformations based on high level conceptual models.

The core contribution of this paper is threefold: (1) We compare the suitability of different multi-level modelling approaches for the integration of ecosystems in the Oil & Gas industry, (2) provide the SLICER (Specification with Levels based on Instantiation, Categorisation, Extension and Refinement) relationship framework to overcome limitations of existing approaches, and (3) evaluate the framework on an extended version of the comparison criteria from [5].

2 Ecosystem Interoperability

The suite of standard use cases defined by the Open O & M Foundation covers the progress of an engineering part (or plant) through the Oil & Gas information ecosystem from initial specification through design, production, sales, deployment, and maintenance including round-trip information exchange. The data transformations needed for interoperability require complex mappings between models covering different lifecycle phases, at different levels of granularity, and incorporating data and (possibly multiple levels of) metadata within one model.

Notably, different concepts are considered primitive objects at different stages of the lifecycle. For example, during design the specification for a (type of) pump is considered an object that must be manipulated with its own lifecycle (e.g. creation, revision, obsolescence), while during operations the same object is considered a type with respect to the physical pumps that conform to it and have their own lifecycle (e.g. manufacturing, operation, end-of-life). Furthermore, at the business/organisational level, other concepts represent categories that perform cross-classifications of objects at other levels. This leads to an apparent three levels of (application) data: business level, specification level, and physical entity level. To describe these different levels multi-level modelling (MLM) approaches to model-driven engineering seem a natural fit. Ideally, a flexible conceptual framework should represent the entire system lifecycle, in a way that simplifies mapping between disparate models by the interoperability designer.

The ecosystem transformations use a joint metamodel that serves as the common representation of the information transferred across the ecosystem (cf. Fig. 1) and must be able to handle the MLM aspects. As pointed out in [6], such complex domains generally are not dealt with using the classical GAV or LAV (Global/Local As View) querying approach, but require a more general form of mapping describing complex data transformations. Notably, the data integration systems surveyed in [6] generally use languages that do not have MLM or even metamodelling capabilities, and the automated matching capability of the systems listed (e.g., MOMIS, CLIO) is probabilistic. As transformations in the engineering domain must guarantee correctness (e.g., an incorrectly identified part or part type can result in plant failures), heuristic matching cannot

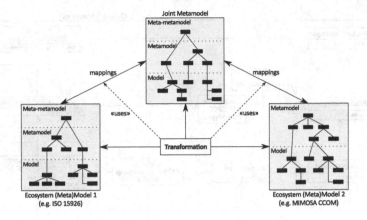

Fig. 1. Ecosystem interoperability through a joint metamodel

replace the need for a succinct and expressive conceptual model for designing the mappings.

A UML-based approach to modelling this situation is shown in Fig. 2, in which ProductCatalogue, ProductCategory, ProductModel, and ProductPhysicalEntity are explicitly modelled as an abstraction hierarchy using aggregation in an attempt to capture the multi-level nature of the domain[2]. Specialisation is used to distinguish the different categories (i.e. business classifications), models (i.e. designs), and physical entities of pumps. Finally, instantiation of singleton classes is used to model the actual catalogue, categories, models, and physical entities.

From both a conceptual modelling and interoperability perspective, this is unsatisfactory: it is heavily redundant [5,7]; the misuse of the aggregation relationship to represent a membership and/or classification relation results in physical entities that are not intuitively instances of their product models; last it creates difficulty in modelling the lifecycles of both design and physical entities as well as the dynamic introduction of new business categories. This directly affects mapping design as the real semantics of the model are hidden in implementation.

3   Multi-Level Modelling Techniques

A number of Multi-level modelling (MLM) techniques have been developed to address the shortcomings of the UML-based model. While they improve on the UML-based model in various respects, no current approach fulfils all of the criteria necessary for ecosystem interoperability (see Sect. 4.1).

Potency-based MLM Techniques [7] were originally introduced for *Deep Instantiation (DI)* to support the transfer of information across more than one level of instantiation. They separate *ontological* instantiation, which are domain

---

[2] While UML 2 supports power types in limited fashion, this example is restricted to more basic UML constructs. Power types are discussed in Sect. 3.

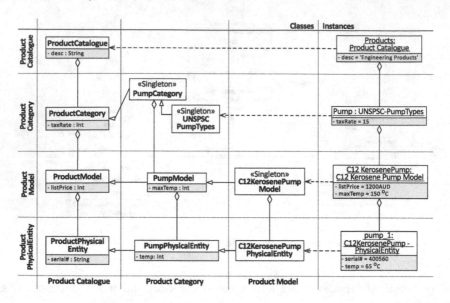

Fig. 2. Product catalogue modelled in plain UML (adapted from [5])

specific instantiation relationships which can be cascaded multiple times, from standard *linguistic* instantiation used in UML meta-modelling. Each model element (e.g. class, object, attribute, or association) has an associated potency value that defines how many times the model element can be ontologically instantiated. That is, a class with potency *2* can have instances of instances, while an attribute with potency *2* can be transferred across two instantiation levels. Modelling elements with potency *0* cannot be further instantiated. For example, the class Pump Model in Fig. 3a defines two attributes: $temp^2$ and $maxTemp^1$. Both attributes have the flat data-type integer as range (representing a value in degrees Celsius); however, $maxTemp^1$ is instantiated only once due to its potency of *1*, while $temp^2$ is instantiated twice and only receives its complete concrete value at the *O0* level, with the assigned value at *O1* interpreted as a default value for its instances.

In a more complete model, the ProductCategory concept would result in an additional level of instantiation. To model the domain appropriately, this approach invariably results in relationships (other than instantiation) crossing level boundaries, which is not allowed under *strict* meta-modelling typically adhered to by potency-based models. For example, if the attributes $temp^2$ and $maxTemp^1$ were modelled as associations to values of the type DegreeCelsius (rather than just integers that must be interpreted as such), no matter at what level DegreeCelsius is placed it would result in an association crossing a level boundary at some point [9]. Moreover, the concept Pump is in violation of strict meta-modelling as it does not (ontologically) instantiate anything from the level above. An alternative would be to remove Pump altogether; however, this would cause a conceptual mismatch with the domain and lead to more complex mappings to models

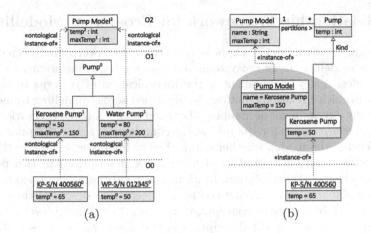

Fig. 3. Extracts of the Deep Instantiation (a) and power type (b) based models of the product catalogue example. Superscripts represent *potency*; the ellipse links the type and instance facets of a concept following the notation of [8].

including such a concept. Finally, DI partially conflates specialisation and instantiation semantics through the use of potency, complicating model transformations.

Power types [10] have also been applied in a multi-level context, e.g., [8]. Basically, for a power type t of a another type u, instances of t must be subtypes of u. Figure 3b shows an example of the Power Type pattern, where Pump Model is the power type for the concept Pump; the concept ProductCategory would be represented as cascading uses of the power type pattern. As discussed in [5], this leads to complex and redundant modelling, which complicates creation and maintenance of interoperability mappings. However, power typing does not conform to strict meta-modelling with non-instantiation relationships crossing level boundaries, which occurs no matter where you try to relocate the concepts. Giving the *partitions* relation instantiation semantics, as argued in [8], to allow it to cross the level boundary leads to the counter-intuitive notion that the type is a partition of itself. Finally, separating the type facet from the object facet leads to redundancy, when the two facets are really the *same object* [10].

M-Objects and M-Relationships. (M standing for multi-level) [11] use a *concretization* relation which stratifies objects and relationships into multiple levels of abstraction within a single single hierarchy. The example situation would be modelled with a top-level concept ProductCatalogue containing the definition of its levels of abstraction: category, model, and physical entity. The lower levels would include the different PumpCategories, PumpModels, and PhysicalPumps, respectively. While the M-Objects technique produces concise models with a minimum number of relations, the concretization relation between two m-objects (or two m-relationships) must be interpreted in a multi-faceted way (since it represents specialisation, instantiation, and aggregation), increasing the difficulty of identifying mappings between models.

4   A Relationship Framework for Ecosystem Modelling

A highly expressive and flexible approach is required to overcome the challenges involved in modelling large ecosystems and supporting transformations across their lifecycles. A key requirement is the identification of *patterns of meaning* from basic primitive relations that can be identified across modelling frameworks and assist the development of mappings between them. A core observation when building transformations for the real world complexity of the OGI pilot is that a higher level in the model, whether ontological or linguistic, expresses the relationship between an entity and its definition (or description) in two possible ways: abstraction and specification. In *abstraction*, entities are grouped together based on similar properties, giving rise to a concept that describes the group—the entities may be groups of concepts, thus permitting a multi-level hierarchy. With *specification*, an explicit description is laid down, and entities (artefacts) are produced that conform to this specification [12,13]. This is a common scenario in the use of information systems and plays a major role in the framework.

The common factor is that a *level of description* (and therefore the consistency constraints for a formalism that expresses these principles) is not purely driven by instantiation. A different level is also established by enriching the vocabulary used to formulate the descriptions. This corresponds to the concept of *extension* in specialisation hierarchies [14]: if a subclass receives additional properties (attributes, associations etc.) then these attributes can be used to impose constraints on its specification and behaviour. Identifying levels based on the basic semantic relationships between entities enables a flexible framework for describing joint metamodels in interoperability scenarios. Moreover, a domain modeller need not utilise the primitive semantic relations directly; rather, they are identifiable in the conceptual modelling framework of a particular model, supporting the development of mappings between models of different frameworks.

In contrast, most of the MLM approaches summarised in Sect. 3 focus on the pre-layering of levels as the main designer input, a corollary from the natural rigidity of the level system in linguistic instantiation as intended by UML. However, it is generally assumed by prior specialized work on relationships such as specialization [14], metaclasses [15], materialization, and aggregation [16] that there can be arbitrary many levels of each. Therefore, interoperability solutions must accommodate mappings to models that are not designed according to the strictness criterion, may cover domains at different levels of granularity, and cannot (from the view of the interoperability designer) be adapted.

To support this we present the SLICER (Specification with Levels based on Instantiation, Categorisation, Extension and Refinement) framework based on a flexible notion of levels as a result of applying specific semantic relationships. In the following description of the relations forming the SLICER framework we refer to Fig. 4, which shows the application of the framework to the product catalogue example in the context of the OGI Pilot. The diagram exemplifies the finer distinctions made by the SLICER framework including: the increase of detail or specificity (from top to bottom) through the addition of

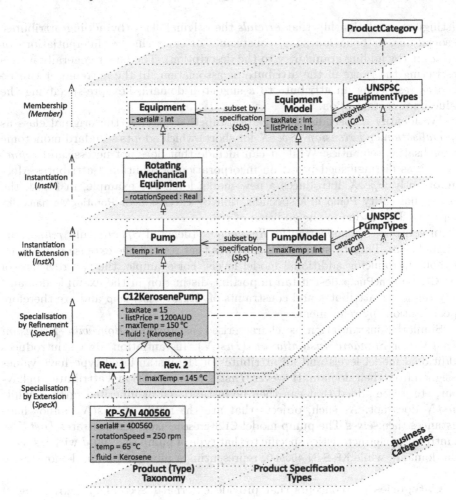

Fig. 4. Product catalogue example modelled using SLICER relationships

characteristics, behaviour, and/or constraints that are modelled; the explicit identification of characteristics for describing models of equipment (i.e. specifications) themselves, not only the physical entities; the second level of classification through the identification of categories and the objects that they categorise; and the orthogonal concerns of the different stakeholders and stages of the lifecycle (shown by separation into 3 dimensions).

Instantiation and Specialisation: Like other frameworks, SLICER uses instantiation and specialisation as the core relations for defining hierarchies of concepts based on increasing specificity (or decreasing abstraction). In contrast to previous MLM techniques, the levels are not strictly specified, but dynamically derived based on finer distinctions of relations between more or less specific types. We characterise specialisation relationships along the lines of [14] to distinguish a relationship that *extends* the original class (by adding attributes, associations, or behaviour, i.e., constraints or state change differentiation) or *refines* it (by adding granularity to the description). The latter generally means restricting the range of the attribute or association. In the extreme, it can be seen as restricting an attribute to a singleton domain, i.e., pre-specifying the value upon instantiation.

We identify a specialisation relationship that extends the original class as *Specialisation by Extension* (*SpecX* for short) which adopts standard monotonic specialisation semantics. While it can include (but does not necessitate) *refinement*, it is distinguished based on incorporation of additional attributes Most importantly, *SpecX* introduces a new model level. For example, in Fig. 4, the conceptual entity Pump is a specialisation by extension of RotatingMechanicalEquipment. This is due to the addition of *temp*.

In contrast, *Specialisation by Refinement* (denoted *SpecR*, only *refines* the original class. Thereby supporting subtypes that restrict the extension of a class without introducing additional model levels. For example, the two revisions of the C12KerosenePump design (an important distinction in the example domain) only refine the attributes and constraints of C12KerosenePump and are therefore specialisations by refinement.

Similarly, instantiation is characterised as *Instantiation with Extension* (*InstX*), or *Standard Instantiation* (*InstN*). Instantiation always introduces additional model levels and all attributes of the instantiated type have values assigned from their domain. However, *InstX* allows additional attributes, behaviour, etc. to be added that can then be instantiated or inherited further, while *InstN* does not. As such, objects that are the result of *InstN* cannot have instances themselves The pump model C12KerosenePump demonstrates *InstX* as it introduces characteristics specific to that type of pump (e.g. *fluid* with a singleton domain), while KP-S/N 400560, representing a physical pump, demonstrates *InstN*.

Categories are concepts that provide external ("secondary") grouping of entities based on some common property and/or explicit enumeration of its members. In the SLICER framework, we explicitly represent categories through two relationships: *Categorisation* (*Cat*) and *Membership* (*Member*). *Categorisation* relates two concepts, one representing a category and the other a type, where the *members* of the category are *instances* of the type. While *Membership* resembles instantiation (the two are mutually exclusive), it does not place any constraints on the assignment of values to attributes (the category and its members can have completely different sets of attributes). However, this does not preclude the specification of membership criteria, or constraints, for allowing or disallowing the possible members of a category. In addition, categories exist on the same level as the type they categorise. This is intuitive from the notion that their membership criteria (if any) are defined based on the type they categorise. Moreover, sub-categories can be specified through *SpecR* only, supporting refinement of the membership constraint. For example, the concepts UNSPSCEquipmentTypes and UNSPSCPumpTypes are a pair of categories indicat-

Table 1. Summary of Relation Properties

	SpecX	SpecR	InstX	InstN	Cat	Member	SbS
Refinement	x	x					
New attributes	x		x				
Must cross 1 level	x		x	x			
Can cross levels		x					
Assign values to attrs.			x	x			
Inst. can have inst.	x	x	x		x	x	x
Cat.-Type Relation					x		
Propagates constraints	x	x	x	x		x	x
Base type attr. access					x	x	x
Used f. Powertype			x				x

ing that their members are equipment models (or specifically pump models for the subcategory) conforming to the UNSPSC standard. The two revisions of C12KerosenePump are both *members* of UNSPSCPumpTypes (and hence are members of UNSPSCEquipmentTypes).

Specifications are the second means of relating an entity to its description, discussed earlier. For this we introduce the *Subset by Specification (SbS)* relation to identify specification types and the parent type of the specification concepts. The specification class (for example EquipmentModel) exists at the same level as the type it refers to as it can define constraints with respect to that type. However, subtypes of the specification type can be defined to reference particular properties; so EquipmentModel can be specialised to refer to properties of Pumps. Together with *InstX*, this relationship can be used to construct the powertype pattern [10].

In the presence of multiple specifications types for a concept, SLICER supports multiple instantiation; that is, an object can be an instance of multiple concepts as long as the concepts are instances of specification types for the same concept. For example, if a second specification type were related to Pump then KP-S/N 400560 could instantiate specifications related to both types. This form of multiple partitioning of a type [10] is not supported in standard meta-modelling or multi-level modelling approaches that allow only single instantiation.

Descriptions and Constraints: The final SLICER component is the explicit handling of object descriptions or constraints through the different relations introduced above. Basically, a description refers only to the attributes specific to its object, can be inherited through specialisation and instances of a type (and members of a category) must satisfy its description (or membership criterion). More important, however, is the handling of complex situations requiring the propagation of constraints across multiple instantiation relations. For example, the specification type PumpModel may define a constraint involving a relation between *maxTemp* (defined by itself) and *temp* (defined by Pump). In this situation the constraint is not applicable to direct instances of PumpModel as they do not assign a value to temp; instead, the constraint applies to KP-S/N 400560 two instantiations away. We handle this in a natural and intuitive way. Basically, we propagate the part of the constraint that is not applicable at a certain level across instantiation relations until it reaches an object for which it can be evaluated.

Due to lack of space to present the full formal framework, Table 1 summarises the properties of the relations described in this section.

4.1   Evaluation and Comparison

The comparison of MLM approaches in [5] provides the criteria for making domain models more concise, flexible, and simple, but does not cover certain aspects important for defining joint metamodels in a an interoperability scenario as in the OGI Pilot[3]. We therefore extend the comparison by additional criteria (#5–7), as summarized in Table 2:

1. ***Compactness*** encompasses *modularity* (all aspects related to a domain concept can be treated as a unit[4]) and *absence of redundancy*. In the ecosystem context, modularity as in [5] is too restrictive as disparate models may group concepts differently. Therefore, we use the *locality* criterion instead (see 5).
2. ***Query Flexibility*** means that queries can be performed to access the model elements at the different levels of abstraction (e.g. querying for all of the product categories, models, physical entities, and their specialisations).
3. ***Heterogeneous Level-Hierarchies*** allow an approach to introduce "new" levels of abstraction for one (sub-)hierarchy without affecting another (e.g. introducing a PumpSeries level of abstraction in between PumpModel and PumpPhysicalEntity without causing changes to any other hierarchy).
4. ***Multiple Relationship-Abstractions*** include classification of relationships, querying and specialisation of relationships (and their classes and metaclasses).
5. ***Locality*** is supported by an approach that can define attributes, relations, and constraints on the model element closest to where they are used (e.g. an attribute most relevant to models or designs themselves should be situated on the concept ProductModel rather than some related concept such as Product).
6. ***Decoupling of Relationship Semantics*** if they have clearly delineated semantics from one another rather than combining the semantics of multiple, commonly understood relations.
   Violating this makes it much harder to perform "*sanity checks* regarding the integrity of metamodelling hierarchies" [17].
7. ***Multiple Categorisation*** considers whether or not the modelling approach can support an element being placed in multiple (disjoint) categories. Since an item could simultaneously belong to multiple categories, it is particularly

---

[3] Except for criterion (2), which is necessary for multiple classification hierarchies.

[4] E.g. Product category, model, and physical entity are grouped together.

Table 2. Summary of comparison (extended and adapted from [5])

Approach/Criterion #	1[a]	2	3	4	5	6	7
UML	$-$	$+$	$\sim/+^{b}$	$-/\sim^{b}$	$+$	$\sim$	$+$
Deep Instantiation [7]	$+$	$+$	$-$	$\sim$	$\sim$	$\sim$	$\sim$
MetaDepth [18]	$+$	$+$	$+$	$\sim$	$\sim$	$\sim$	$\sim$
Dual-Deep Instantiation [9, 19]	$+$	$+$	$+$	$+$	$\sim$	$-$	$-$
Powertypes (Simple) [8]	$-$	$\sim$	$-$	$-/\sim^{b}$	$+$	$\sim$	$\sim$
Powertypes (Ext.)	$-$	$+$	$+$	$-/\sim^{b}$	$+$	$\sim$	$\sim$
Powertype (Onto.) [20]	$-$	$\sim$	$+$	$-$	$+$	$+$	$-$
Materialization [16]	$+$	$\sim$	$-$	$\sim$	$-$	$\sim$	$-$
M-Objects [11]	$+$	$+$	$+$	$+$	$-$	$-$	$-$
SLICER	$+$	$+$	$+$	$\sim$	$+$	$+$	$+$

Legend: (+) Full Support, ($\sim$) Partial Support, ($-$) No Support
[a] redundancy-free only
[b] using OCL

important for business level classification. For example, a pump model being certified by two different standards would place it in the categories for both.

With regard to these criteria, the SLICER approach:

1. allows redundancy-free modelling using a range of relations that include various attribute propagation, inheritance, and assignment semantics;
2. supports query flexibility across relationships and concepts in a domain model;
3. allows heterogeneous level-hierarchies through specific primitive semantic relations and its flexible, dynamic approach to level stratification;
4. supports specialisation and instantiation of domain and range relationships, which constitutes partial support of multiple relationship extractions; however, (meta-)classification of relationships may not be required due to the finer semantic distinctions provided by SLICER;
5. supports locally specified attributes and relations by design;
6. clearly differentiates a number of important, yet distinct, primitive relations that exist between domain concepts; and,
7. supports multiple categorisations through explicit representation of categories for which objects can belong to multiple concurrently.

Mapping Example. Using the semantic distinctions described in the previous section, other models of the ecosystem can be analysed to better identify mappings between them. This requires relating the distinctions made by SLICER to other modelling approaches as without them only partial alignment between models can be found (compare Figs. 3 and 4). For brevity, we only discuss the more interesting case of potency-based models due to the greater mismatch (e.g. Pump is at a different level and the attributes are located on different concepts).

The distinctions between extension, refinement, categories, and specifications that are made in SLICER are apparent in the potency-based model through:

1. An attribute with potency ≥ 2 suggests $InstX$ as the potency indicates that the attribute should be introduced to the concept at the level where its potency $= 1$ (so that it can be given a value at the next instantiation).
2. A subclass that introduces new attributes over its parent class suggests $SpecX$, while a subclass that only refines the domain of an attribute suggests $SpecR$; this is a direct corollary of the same distinction made in SLICER.
3. An object with potency $= 0$ and attributes with only potencies $= 0$ is $InstN$.
4. If all of the instances of an object are specialisations of the same class it suggests an SbS relation between the instantiated class and the specialised class. In particular, if the class has attributes with potency > 0 or the instantiated class has attributes with potency $= 2$ it is most likely the case.
5. Specialised classes that have no attributes with a potency ≥ 1 and that are not identified as related to a specification class (see previous rule) suggests possible use of specialisation for categorisation. The introduction of attributes indicates the boundary between categorisation and categorised object(s).

Applying these rules to the example we can identify $InstN$ by rule (3) between the object at $O0$ on the left and KerosenePump, hence it can be aligned with the bottom level of the joint metamodel At the $O1$ level, KerosenePump instantiates PumpModel from the level above which includes an attribute with potency 2; thereby identifying $InstX$ by rule (1). Moreover, by rule (2) a $SpecX$ relation is identified as KerosenePump adds attributes over the class it specialises (Pump) Finally, all of the instances of PumpModel are subclasses of Pump, matching rule (4) and indicating the relation SbS(PumpModel, Pump). As a result, this pattern can be matched in the joint metamodel. Note that the result would be the same if the attribute $temp^2$ were defined on Pump as $temp^1$. Finally, if the concept Pump were not modelled at all (i.e. it was completely incorporated into PumpModel) then the specialisation to some concept could still be inferred due to the combination of specialisation and instantiation embodied by potency.

Conceptual Complexity and Practical Use. At first glance, the incorporation of the above described relationship patterns would appear to increase the complexity of the modeler's task, by increasing the number of basic relationships considered in a model. However, this merely reflects the difficulty of the task. The core data models of the two standards used for the OGI Pilot experiments involve hundreds of classes with thousands of attributes that can be extended further by individual providers.

First, the new relationships in SLICER were developed by identifying special cases that are known to cause problems in practice, leading to wrapping modelling concerns into generic domain associations, naming conventions, or even both, resulting in highly complex and often inconsistent models [21]. The goal of this approach is to enable development of coherent models in domain areas where persistent and large scale effort has *failed* to produce workable models since the modelling approaches used did not permit expressing the salient features of the domain. This state of affairs is confirmed by the rise of interest in MLM methods even outside the interoperability application domain. Though these methods have been largely developed in the software engineering community and are typically based on UML, they generally revolve around conceptual domain models rather than detailed design models.

A key factor was the identification of the extension and specification relationships as distinguishing factors that needed to be explicitly represented. Separating out these two types of relationships, they can be considered as *meta-properties* that annotate or enrich the families of relationships existing in conventional models. This can be seen as an extension of the classical U, I, and R meta-properties used in OntoClean [22]. In the case of the X (extension) meta-property, this can be partly automatically identified by examining property specifications. In the S (specification) case, it would be used to identify what would before have been a generic domain association, or at best an instance of the generic D (Dependency) meta-property. This indicates an avenue towards automated support for consistent use of these relationships.

Also, given that the need for some sort of MLM capability has been widely recognised, the standard for a complexity comparison would not be conventional modelling methods, but the other MLM methods. Compared to standard potency-based methods (e.g., MetaDepth), a SLICER model employs mostly the standard relationships found in conceptual models, with limited extensions that carry domain semantics that any modeller would be familiar with. In comparison, assigning potency value requires thinking in terms of multiple metalevel instantiations, and potentially restricts instantiations at lower levels. This raises the question whether a potency-based designer can predict the implications of his modelling choice for all later additions to the model. Compared to Deep Instantiation and Dual Deep Instantiation, SLICER supports designer-oriented terminology and a clear separation of semantic concerns. Powertypes are based on implicit constraints that cannot be expressed in standard conceptual models, and require observing implicit dependencies between subtype partitions and powertype membership. The M-Objects and Materialisation relationship both conflate multiple different semantic relations into a single relationship (called concretization in the former and materialisation in the latter). This actually makes the task harder for an analyst since the model does not help him to keep track of the semantic distinctions that gave rise to the different treatment in the domain in the first place.

5   Conclusion

Effective exchange of information about processes and industrial plants, their design, construction, operation, and maintenance requires sophisticated information modelling and exchange mechanisms that enable the transfer of semantically meaningful information between a vast pool of heterogeneous information systems. This need increases with the growing tendency for direct interaction of information systems from the sensor level to corporate boardroom level. One way to address this challenge is to provide more powerful means of information handling, including the definition of proper conceptual models for industry standards and their use in semantic information management and transformation.

In this paper we have described the SLICER framework for large scale ecosystem handling. It is based on the notion that a model-driven framework for creating mappings between models of an ecosystem must identify the underlying semantics of a model based on the relationship between an entity and its description, most clearly embodied by artefacts subject to their specifications. This led to the introduction of a set of primitive relationship types not so far separately identified in the literature. For one, they reflecting the fundamental conceptual modelling distinction between extension and refinement [14], thus allowing to succinctly express distinct semantics encountered in multilevel and multi-classification modelling scenarios. On the other hand, they represent the first time elevation of the *specification* relationship to full "citizenship" status among the relationships in the conceptual model. Together these relationships capture underlying design assumptions encountered in various potency and powertype approaches and enable their separate examination and study. This allows to express the complex mapping of large scale interoperability tasks in a consistent and coherent manner. The identification of extension and specification/description as ontologically relevant meta-properties offers as a natural next step the connection to our work on artifact and specification ontologies [12]. Future work includes the extension of the underlying formal framework and the investigation of the above mentioned connections to formal ontological analysis.

References

1. Young, N., Jones, S.: SmartMarket Report: Interoperability in Construction Industry. McGraw Hill, Technical report (2007)
2. Fiatech. Advancing Interoperability for the Capital Projects Industry: A Vision Paper. Technical report, Fiatech, February 2012
3. ISO. ISO 15926 - Part 2: Data Model (2003)
4. MIMOSA. Open Systems Architecture for Enterprise Application Integration (2014)
5. Neumayr, B., Schrefl, M., Thalheim, B.: Modeling techniques for multi-level abstraction. In: Kaschek, R., Delcambre, L. (eds.) The Evolution of Conceptual Modeling. LNCS, vol. 6520, pp. 68–92. Springer, Heidelberg (2011)
6. Bergamaschi, S., Beneventano, D., Guerra, F., Orsini, M.: Data integration. In: Embley, S.D.W., Thalheim, B. (eds.) Handbook of Conceptual Modeling. Theory, Practice, and Research Challenges, pp. 441–476. Springer, Heidelberg (2011)
7. Atkinson, C., Kühne, T.: The essence of multilevel metamodeling. In: Gogolla, M., Kobryn, C. (eds.) UML 2001. LNCS, vol. 2185, pp. 19–33. Springer, Heidelberg (2001)
8. Gonzalez-Perez, C., Henderson-Sellers, B.: A powertype-based metamodelling framework. Softw. Syst. Model. 5(1), 72–90 (2006)
9. Neumayr, B., Jeusfeld, M.A., Schrefl, M., Schütz, C.: Dual deep instantiation and its ConceptBase implementation. In: Proceeding CAISE 2014, pp. 503–517 (2014)
10. Odell, J.J.: Power types. JOOP 7, 8–12 (1994)
11. Neumayr, B., Grün, K., Schrefl, M.: Multi-level domain modeling with M-objects and M-relationships. In: Proceedings APCCM 2009, pp. 107–116 (2009)

12. Jordan, A., Selway, M., Mayer, W., Grossmann, G., Stumptner, M.: An ontological core for conformance checking in the engineering life-cycle. In: Proceeding Formal Ontology in Information Systems (FOIS 2014). IOS Press (2014)
13. Borgo, S., Franssen, M., Garbacz, P., Kitamura, Y., Mizoguchi, R., Vermaas, P.E.: Technical artifact: an integrated perspective. In: FOMI 2011. IOS Press (2011)
14. Schrefl, M., Stumptner, M.: Behavior consistent specialization of object life cycles. ACM TOSEM **11**(1), 92–148 (2002)
15. Klas, W., Schrefl, M. (eds.): Metaclasses and Their Application. LNCS, vol. 943. Springer, Heidelberg (1995)
16. Goldstein, R.C., Storey, V.C.: Materialization. IEEE Trans. Knowl. Data Eng. **6**(5), 835–842 (1994)
17. Kühne, T.: Contrasting classification with generalisation. In: Proceedings APCCM 2009, pp. 71–78, Australia (2009)
18. de Lara, J., Guerra, E., Cobos, R., Moreno-Llorena, J.: Extending deep meta-modelling for practical model-driven engineering. Comput. **57**(1), 36–58 (2014)
19. Neumayr, B., Schrefl, M.: Abstract vs concrete clabjects in dual deep instantiation. In: Proceeding MULTI 2014 Workshop, pp. 3–12 (2014)
20. Eriksson, O., Henderson-Sellers, B., Ágerfalk, P.J.: Ontological and linguistic meta-modelling revisited: a language use approach. Inf. Softw. Technol. **55**(12), 2099–2124 (2013)
21. Smith, B.: Against idiosyncrasy in ontology development. In: Proceeding Formal Ontology in Information Systems (FOIS 2006), pp. 15–26 (2006)
22. Welty, C.A., Guarino, N.: Supporting ontological analysis of taxonomic relationships. Data Knowl. Eng. **39**(1), 51–74 (2001)

Flexible Data Management across XML and Relational Models: A Semantic Approach

Huayu Wu[1]([⊠]), Tok Wang Ling[2], and Wee Siong Ng[1]

[1] Institute for Infocomm Research, A*STAR, Singapore, Singapore
{huwu,wsng}@i2r.a-star.edu.sg
[2] School of Computing, National University of Singapore, Singapore, Singapore
lingtw@comp.nus.edu.sg

Abstract. Relational model and XML model have their own advantages and disadvantages in data maintenance and sharing. In this paper, we consider a framework that can be used in many applications that need to maintain their potentially large data in relational database and partially publish, exchange and/or utilize the data in different XML formats according to the applications' preferences. The existing relational-to-XML data transformation relies on pre-defined XML view or transformation rules. Thus the output XML data or views are rigid and invariable, and cannot fulfill the requirement of format flexibility in our considered situations. In this paper, we propose a semantic approach that supports the framework under our consideration. We use conceptual models to design both relational data and XML views, and invent algorithms to transform data in one model to the other via conceptual model transformation. We also demonstrate how XPath queries issued to XML views can be translated and processed against a relational database through conceptual models.

Keywords: ORASS · XML · Relational model · Model transformation

1 Introduction

Despite the research efforts and achievements in XML data management, XML is still not yet a successful database model. Most structured data in large amounts are still maintained in relational databases, even though many of them will be published or transferred in XML format. One of the reasons is that the semi-structured XML data itself brings in more difficulties and inefficiency in maintenance and processing than tabular relational data. In fact, most XML documents exchanged on the Internet are not large in size, and most practical XML-enabled database systems (e.g., Oracle DB and IBM DB2) do not use XML as a database model to store large size of data. Those database systems only consider XML as a complex data type embedded in relational tables.

Since relational data and XML data have their own advantages and disadvantages in data maintenance and sharing, modern applications compellingly require a new data management framework, especially when data becomes larger and

© Springer International Publishing Switzerland 2015
P. Johannesson et al. (Eds.): ER 2015, LNCS 9381, pp. 302–316, 2015.
DOI: 10.1007/978-3-319-25264-3_22

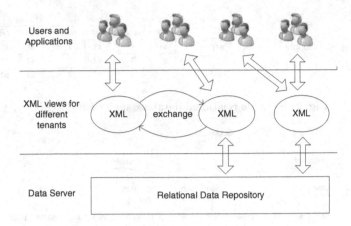

Fig. 1. Data maintenance and consumption across relational and XML models

sharing becomes more important. In this paper we focus on a practical framework in which a large amount of data are maintained in the relational model, while partial data from the relational database are published, exchanged and/or utilized in XML model. Figure 1 describes the framework. We use an example to illustrate how such a framework serves the practical use cases.

In the example, a property agency maintains a relational database server for all housing information in the country. Some agents need to retrieve a part of the data according to locations, to share with other agents who do not have direct access to the server, or share with their customers offline. They may need such data in XML format, as XML data are easily transferrable and interpretable by various applications for better demonstration. However, a copy obtained by an agent may not be up-to-date. When the agent searches the XML data in hand, she may not be able to get the latest information. If the Internet is available, her XML query can be routed to the server and transformed into SQL queries to query the server database to get the latest information, though she cannot update her entire XML data due to security, bandwidth or other reasons.

We can see in this framework, users are physically and probably also logically isolated from underlying relational data repository. They can only directly access different XML views of (part of) the data, which are convenient for sharing but incomplete in content. User queries are subject to XML views, but query evaluation needs to be done in underlying relational database in SQL. To effectively implement such a framework, we need to tackle two main challenges.

Challenge 1. Presentation variance during data publication.

Example 1. Consider a simple relational database designed by ER model, as shown in Fig. 2. This database can be published as different XML views. A regulator (supplier-centric) may be interested in the view in Fig. 3(a) so that he can check all information for an inspected *supplier*; while a broker

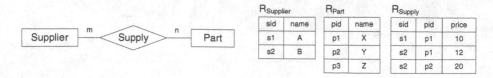

Fig. 2. Example relational database under ER design

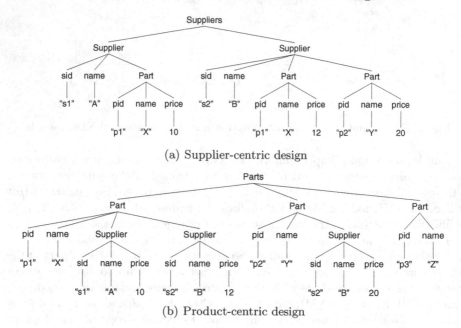

Fig. 3. Two XML views for the database in Fig. 2

(part-centric) may want a view as in Fig. 3(b), in which for a particular *part* he can access all suppliers with different supplying prices.

Publishing relational data in XML format has been studied for several years (detailed reviews can be found in Sect. 2.1). We would like to emphasize that the existing relational-to-XML transformation is restricted by either pre-defined XML view or transformation rules. In other words, given a relational schema, the output XML schema from the existing algorithms is fixed. Worse still, the output XML schema can hardly be re-structured, otherwise the integrity constraints cannot be preserved. As a result, those algorithms cannot be used in our considered framework in which the system should be able to flexibly produce different XML views based on users' preferences.

Challenge 2. Cross-model query interpretation. Queries issued to XML views are in XML query format. It should be well interpreted and translated into SQL to search the relational data store on the server. Due to its complex syntax and the requirement on the knowledge of underlying XML structure, XQuery is not often used in practical application systems. Instead, in such systems queries are specified in either keywords or simple expressions like XPath. In this paper, we do not make query translation between XQuery and SQL, which has been done in many existing works. We consider XPath and keyword query, which are less structured and more challenging to interpret into SQL.

Example 2. Consider a user query asking for the information about part "p1" in the relational database in Fig. 2. If the user receives an XML view as in Fig. 3(a), probably an XPath query is issued as *//Part[pid= "p1"]*. This query will return the whole subtree rooted at a satisfied *Part* node. With the knowledge of the data semantics, we know actually the information about *price* in the subtree does not describe a *part* entity, but a relationship between a *part* and a *supplier*. The answer to this query should ignore the *price*. To find this answer in the underlying relational database, we need this interpretation to compose the SQL query. Also to find part "p3", *//Part[pid= "p3"]* will not return any answer, as no supplier supplies "p3" and thus it is not in the XML view. We need to interpret the query and compose SQL query to search the relational database.

To summarize, we need to well interpret the query and do transformation according to query semantics rather than query structure.

To solve the above challenges, we propose to use semantics-enriched models to describe both relational data and XML views, and make data publishing and query translation go through model mapping. The key rationale is as illustrated above, i.e., both data transformation and query translation need the support from the semantics, rather the structure of data and query. In particular, we use the ER model [1] to design underlying relational database, and the ORA-SS model [13] for XML view creation. With our work, the ER diagram for the underlying relational data will be mapped into one ORA-SS schema diagram, according to our algorithm. Then the ORA-SS schema can be seamlessly transformed into other designs by using the corresponding operators such as *swap*, *drop* and *project*, to satisfy different users' requirements. We use the ORA-SS model because other schemas (e.g., DTD and XMLSchema) cannot capture the data semantics to correctly do the transformation.

When an XML query is issued to one XML view, the query nodes will be mapped to different semantic concepts in ORA-SS. Then these ORA-SS concepts will be mapped to the semantic concepts (e.g., entity or relationship) in ER diagram. The query is interpreted against the ER diagram using the semantic concepts, and then based on the interpretation, SQL query is composed to search the relational database. Searching results will be transformed into XML fragments through the model mapping between ER diagram and ORA-SS diagram again, and returned to the user.

The rest of this paper is organized as follows. In Sect. 2, we revisit background and related work. In Sect. 3, we describe our framework, as well as the techniques used in our framework. We conduct experiment and show the result in Sect. 4. Finally we conclude this paper in Sect. 5.

2   Background and Related Work

2.1   Relational-to-XML Transformation

Publishing relational data in XML format has attracted much research interest. There are different systems [7,17] focusing on defining XML views of relational data, such that user queries can be issued to the XML views. However, those early systems do not consider integrity constraints of the database.

Schema transformation is an essential way to generate XML documents from relational data. [11] proposed a NeT algorithm to derive nested structures from flat relations, and then proposed the CoT transformation to preserve inclusion dependencies. [19] defined a set of rules according to integrity constraints to map from relational schema to XML schema. Similar work was done in [14] as well. [10] targets on level-based functional dependency analysis to capture hierarchical view of relational model to do transformation. There are also works [4] targeting at the efficiency of using SQL engine to publish relational data to XML data.

The common limitation of the existing works is that they only focus on publishing relational data in XML format from the system's point of view. They design rigid algorithms or rules for the transformation, leaving no control on the preference of XML design by users. The output XML schema of these algorithms is fixed, and can hardly be re-structured, otherwise the integrity constraints cannot be preserved. As a result, these works cannot be used for our considered situation, in which relational data should be able to get transformed into different XML views according to users' preferences.

There are also many works on schema mapping between relational data (e.g., [8,15]), and between XML data (e.g., [3]). They do not focus on cross-model mapping. We do not further review them. [16] maps queries between applications and databases. It is neither applicable to our considered case, because those mappings requires users to specify mapping rules to the system, while in our case users may be unaware of the underlying relational database structure.

2.2   XML-to-relational Query Translation

Another related field is to translate XML queries issued to the XML view, into SQL queries to query relational database. In existing work, this query translation is tightly bound to the transformation from XML data into relational data. There are different approaches to decompose XML data into relational tables, based on XML node, XML edge, XML path, and XML DTD and other schemas [9]. For each decomposition approach, there will be an XML-to-relational query translation accordingly. However, these works assume XML databases as initial databases, and leverage RDBMS for data storage and query processing. In our case, we are handling relational data which are viewed as XML documents by users. Moreover, the structure or schema -based query translation may not be able to capture and translate real query semantics, as discussed in Sect. 1.

2.3   ORA-SS Model

ORA-SS model [13] was proposed to model semi-structured data with rich semantics preserved. Conceptually, it is similar to ER model for relational database design, but focuses on preserving such semantic concepts as entity and relationship in XML database design. An ORA-SS schema diagram is a directed graph where each internal rectangle node represents an object class and each leaf circle node represents an attribute type of an object class or a relationship type. Similar to ER diagram, key attributes are marked by solid circles. Relationship type is represented by a label on the edge between two object classes. The label includes the relationship type name, the degree of the relationship and the participation constraints of involved entity classes. Such ORA semantics can be discovered from XML data if they are not available [12]. Functionally, ORA-SS model can capture all semantics reflected in an ER model. The two XML documents shown in Fig. 3 can be modeled by the ORA-SS schema diagrams as shown in Fig. 4.

Besides the capability to capture rich semantics, another reason for which we choose the ORA-SS to model materialized XML views is that an ORA-SS schema diagram can be easily transformed into another equivalent design by the operators of *swap*, *drop* and *project* [2]. Also, the semantics captured by the ORA-SS model and the ER model are quite similar, which will be helpful for the schema mapping under these two models.

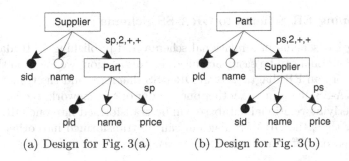

(a) Design for Fig. 3(a)   (b) Design for Fig. 3(b)

Fig. 4. Two example ORA-SS schema diagrams

3   Semantic Framework

We present our approach to manage data across relational and XML models. Different from the existing approaches, we use semantic models (ER model and ORA-SS model) to represent two database schemas and perform schema mapping with semantic models. As such, we can maximally preserve the semantics of data and correctly map two types of data. Figure 5 generally shows how relational data and XML data are inter-transformed under our semantic approach.

We assume that the underlying relational database is designed by ER model. Translating an ER schema diagram into relational schema is maturely studied. In our approach, we are on top of this translation. No matter how relational tables are normalized to reduce redundancy or partially duplicated to improve the performance of frequent queries, we assume the mapping between entity/relationship in the ER diagram and relational tables are consistent. When

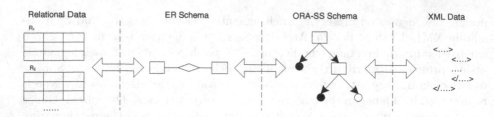

Fig. 5. Transformation from relational data to XML data

the relational database need to be published as one or more XML views, the ER diagram will be wholly or partially mapped to one ORA-SS diagram. Then the positions of the particular entity classes and relationship types in the hierarchy can be swapped based on different users' preferences. Finally, under different ORA-SS schema, different XML views can be materialized and published.

As mentioned in Sect. 1, there are two challenges in our proposed approach, i.e., data transformation and query translation. The rest of the paper will focus on tackling these two challenges.

3.1   Mapping ER Schema to ORA-SS Schema

Translating ER schema into relational schema and publishing XML data based on ORA-SS schema can be easily achieved. In this section, we focus on the transformation from an ER diagram to an ORA-SS diagram, and the transformation from an ORA-SS diagram to another one. Under our framework, the ER diagram for the underlying relational database will be transformed into one ORA-SS diagram. After that, the ORA-SS diagram can be transformed into other ORA-SS designs based on different users' preferences.

ER-to-ORA-SS Transformation. We design a depth-first search (DFS) based algorithm in Algorithm 1 to traverse an ER diagram or a semantically self-contained ER sub-diagram that is interested by an XML view, and create the corresponding ORA-SS schema diagram. The first step of the algorithm, i.e., line 1–4 in Algorithm 1, is to remove cycles in an ER diagram. A cycle is formed in an ER diagram when there are more than one relationship types involving a particular set of entity classes. In this case, we recursively duplicate the object classes, one at a time, in such a cycle, until all cycles are resolved. Example 3 shows an example of removing cycles.

The main recursive DFS function is shown in Function 1. The algorithm starts from any entity class. For simplicity, we only consider the basic form of ER diagram, i.e., with only rectangle and diamond for entity class and relationship type respectively. The basic idea of Function 2 is to align the entities participating each relationship in one path in the ORA-SS diagram. Then the relationship name, degree and participation constraints are marked as the label of the edge connecting the lowest entity, and the relationship attributes are attached to the

---

Algorithm 1. transform(ERD)

Input: an ER diagram ERD
Output: the corresponding ORA-SS diagram
 1: **for all** pair of entity classes *(A, B)* in ERD between which there are more than
    one relationship type involved **do**
 2:     make a copy of B, namely B'
 3:     disconnect one relationship type that links A and B from B, and link it to B'
 4: pick one entity type E_0 to start with
 5: transform(ERD, E_0)

Function 1. transform(ERD, E_0)

 1: mark E_0
 2: create an entity with E_0 in the ORA-SS diagram
 3: set the cursor in the ORA-SS diagram pointing to E_0
 4: **for all** unmarked relationship type neighbor R_i of E_0 **do**
 5:     mark R_i
 6:     **for all** unmarked entity class neighbor E_i of R_i **do**
 7:         **if** E_i also plays a relationship role in ERD **then**
 8:             **for all** unmarked entity class neighbor E_i' of E_i **do**
 9:                 create a down edge from the cursor entity
10:                 transform(ERD, E_i')
11:             create the relationship on the edge pointing to the cursor entity in the ORA-
                SS diagram, which involves all entity neighbors of E_i in the ER diagram
12:         **else**
13:             create a down edge from the cursor entity
14:             transform(ERD, E_i)
15:     create relationship R_i on the edge pointing to the cursor entity in the ORA-SS
        diagram, which involves all entity neighbors of R_i in the ER diagram
16:     **if** R_i also plays an entity role in ERD **then**
17:         let E_c be the cursor entity
18:         **for all** unmarked relationship type neighbor R_j of R_i **do**
19:             mark R_j
20:             **for all** unmarked entity class neighbor E_j of R_j **do**
21:                 create a down edge from the cursor entity
22:                 transform(ERD, E_j)
23:             create relationship R_j on the edge pointing to the cursor entity in the ORA-
                SS diagram, which involves all entity neighbors of both R_i and R_j in the
                ER diagram
24:         reset ORA-SS cursor to E_c
25:     reset ORA-SS cursor to E_0

---

lowest entity.[1]. Line 7–11 and 16–24 deal with a special case of ER diagram, in
which a relationship type also plays an object class role or an object class also
plays a relationship type role. Example 5 illustrates such a case.

---

[1] This step is trivial and omitted in Function 1. Function 2 focuses on the structure
transformation from ER diagram to ORA-SS diagram.

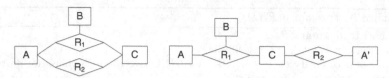

(a) ER diagram with cycle   (b) Resolved ER diagram without cycle

Fig. 6. Cycle removing for an ER diagram

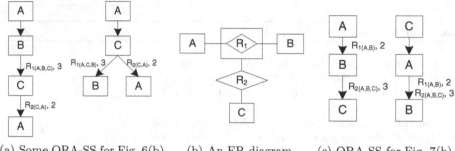

(a) Some ORA-SS for Fig. 6(b)   (b) An ER diagram   (c) ORA-SS for Fig. 7(b)

Fig. 7. ER to ORA-SS transformation example

Example 3. Consider an ER diagram with a cycle, i.e., with more than one relationship types between two entity classes, in Fig. 6(a). By Algorithm 1, either A or C will be duplicated. Suppose we choose A, then the resolved ER diagram without cycle is shown as Fig. 6(b). This process will continue, until no cycle exists in the ER diagram.

Example 4. For the ER diagram (without cycle) in Fig. 6(b), when we start the traversal from the entity class A, we can have two possible ORA-SS diagrams as shown in Fig. 7(a). In particular, after creating the entity class A in ORA-SS diagram and traversing the relationship type R_1, the algorithm can visit either B or C, which results two different ORA-SS diagrams.

Example 5. Consider an ER diagram in Fig. 7(b), in which the relationship typle R_1 also plays an entity role for the relationship type R_2. Based on Function 1, when we start with the object class A, we will have an ORA-SS diagram as shown in the first diagram in Fig. 7(c); when we start with the object class C, we will have the second diagram in Fig. 7c. Note that in an ORA-SS schema diagram, if two relationship types, namely R_1 and R_2, are along the same path, and R_2 contains all object classes that are involved in R_1, then every relationship instance for R_2 contains a relationship instance for R_1, as reflected by Fig. 7(b).

Picking any entity class in an ER diagram to start transformation (line 5 in Algorithm 1) may lead to many data redundancies for the resulting ORA-SS schema. In Example 4, the ER diagram in Fig. 6(b) is transformed into ORA-SS diagrams in Fig. 7(a) by starting the transformation at entity class A. Based on the schemas in Fig. 7(a), an XML instance will contain many redundancies if we assume the participation constraints for R_1 and R_2 are both many-to-many (m:n). We propose two guidelines for generating better ORA-SS diagrams.

Guideline 1. If a binary relationship type between entity classes A and B is one-to-many (1:m), A should appear as the parent of B in the ORA-SS diagram. Similarly, for a ternary relationship type for A, B and C, if the participation constraint is 1:m:n, A should appear as parent/ancestor of B and C.

Guideline 2. An entity class involved in more relationship types should appear as the ancestor of other entity classes involved in less relationship types.

However, to use hierarchical structure to represent relational data can hardly avoid data redundancy (assuming m:n relationship type). Furthermore, in some applications, users may need a design with data redundancy to facilitate their processing. If an ORA-SS diagram requires less data redundancies, the two guidelines can help during choosing the entity class in the ER diagram to start the transformation, or during inter- ORA-SS transformation.

Inter- ORA-SS Transformation. After transforming the ER diagram for the underlying relational database into an ORA-SS diagram, we can make semantically equivalent inter- ORA-SS transformation, based on users' preferences. This operation enables the flexibility and variety of the XML views shown to end customers. Inter- ORA-SS transformation is done by the operators of *swap*, *drop* and *project* over ORA-SS views, as defined in [5]. In this paper, we do not further elaborate it. Interested readers can refer to [5].

3.2 Query Processing

As stated in Sect. 1, XQuery is seldom used by non-IT personnel in business application to search XML data, due to its complexity. Also XQuery-to-SQL is not challenging, as both the languages are structured and intention clear. In this work we consider XPath and keyword query which are easier to compose by users, but vague in conveying real search intentions.

It is easy to understand that keyword queries can be often ambiguous to convey users' search intention. In XPath, a query is composed based on the structure of corresponding XML data. This also limits the expressivity of a query, as illustrated in Example 2. Our framework is good at handling ambiguity during the translation from less (or non-) structured XML query to SQL to search the underlying relational database. Different from the existing query translation approaches, which purely rely on the query structure, we actually capture the data semantics to fully understand the query purpose before searching the database. The query processing in our framework includes four steps:

1. Identify the entity class(es) and/or relationship type(s) in the XML query, according to the ORA-SS model.

2. Mapping the involved entity class(es) and/or relationship type(s) in ORA-SS model to the entity class(es) and/or relationship type(s) in the ER model of the relational database.
3. Interpret the query with ER model, and compose SQL query.
4. Query the relational database, and construct the result in XML format as the user prefers.

We use the same example to go through the query processing under our proposed framework. When a user would like to find all the information about the Part p1, he/she can simply issue the query *//Part[pid="p1"]* to the XML view in Fig. 3(a), without worrying whether the correct information can be return based on the query expression. Then *Part* can be identified as the only entity involved in the query, based on the ORA-SS diagram in Fig. 4(a). From the mapping between the ER diagram in Fig. 2 and the ORA-SS diagram in Fig. 4(a), we understand that the query aims to find information centered at the entity *Part* in the ER diagram. Then we can easily issue a selection to the *Part* table to find the entity's full information, and a join with the relationship table if we also need to return the suppliers that supply this part. Finally, the result can be constructed based on the structure defined in the ORA-SS diagram in Fig. 3(a), and returned to the user. Note that our framework also considers the situation that the query answer cannot be found in XML view, but in the underlying relational database. For example, a query *//Part[pid="p3"]* is issued to the XML view which does not contains part *p3*. This query will actually be processed against the underlying database (not shown) that contains part *p3*.

During query processing, Step 3, i.e., interpreting a query with an ER diagram is the most challenging step. Given an ER diagram and one or several entity classes and relationship types, we need to find a reasonable connection among them, to be the search intention of the query. There are quite a number of heuristics defined in the literatures of XML keyword search, to find search intentions of a set of keywords in a data graph. Our problem is actually simpler than that, i.e., we have already processed query keywords and reduced them into entity and relationship level to search an ER diagram. The existing works can be adapted for our case, but to be self-contained, in this paper we design our version of search heuristics under the context of ER diagram.

Definition 1. (Search Intention) For an ER diagram *ERD* and a set of entity classes and/or relationship types *V* mapped from an XML query, each of which possibly has predicates on its attributes, the search intention of the query is a sub-graph *G* of *ERD* which is formed by (1) the minimum Steiner tree *G'* spanning *V*, and (2) extending *G'* by including all attributes of the entity classes and relationship types in *G'*, and including all participating entity classes and their attributes of the relationship types in *G'*.

Definition 2. (Search Result) The search result of a query is all instances of this query's search intention that satisfy the predicates on the entity classes and relationship types in the search intention.

Definition 3. (Projected Search Result) The search result of a query can be selectively returned to the user, by projecting certain entities, relationships and/or their attributes that are explicitly specified in the XML query.

Based on the defined heuristics, Algorithm 2 shows how we find the query result from underlying relational database. In the algorithm, the set of entity classes of relationship types extracted from XML query will be taken as an input, together with the value predicates associated to each of them, if any. Also, the algorithm considers the output information of the XML query. If the query is in XPath syntax, by the definition of XPath, the output node will be the last element node in the query expression. If the query is a keyword query, the output node will be found by analyzing query keywords, which has been extensively studied in XML keyword search works (e.g., [18]). The algorithm will first find the minimum Steiner tree to span all the input entity class and relationship type nodes in the ER diagram. Then the corresponding relational tables for the nodes in the minimum Steiner tree will be identified. Finally, an SQL query will be composed by joining the relevant tables and applying selection predicates, and outputting desired information.

Algorithm 2. Query Processing

Input: a relational database R, an ER diagram ERD, a set of entity classes and relationship types V, a set of value-based predicates P, a set of entity classes and/or relationship types whose instances will be outputted O

Output: query result set RS

1: find a minimum Steiner tree G spanning V in ERD
2: initiate an empty set of relations T
3: **for all** entity type or relationship class v in V **do**
4:   find the corresponding relation r in R for v
5:   $T = T \bigcup \{r\}$
6: construct SQL query to find RS: $RS = \pi_O(\sigma_P(\bowtie (T)))$

Example 6. Suppose a query to find all parts supplied by the supplier s1 is issued to the example database in Fig. 2. The query can be specified as *//Supplier[sid="s1"]/Part* or *//Part[Supplier/sid="s1"]*, depending on which XML view in Fig. 3 is used. No matter which view is used, the query is mapped by two entity classes, i.e., *Supplier* and *Part*, in which there is a predicate on *Supplier*. The minimum Steiner tree for the two entity classes is exactly the same as the simplified ER diagram shown in Fig. 2. Then the search intention of this query is *Supplier*, *Part*, and their relationship type with attributes. By considering the predicate on *Supplier*, an SQL query can be issued.

Example 7. Consider another example query to find all parts with price of 10. The query can be specified as *//Part[price=10]* or *//Part[.//price=10]* in the two views in Fig. 3. From the corresponding ORA-SS diagrams, we know that *price* is associated to the relationship type between *Part* and *Supplier*, i.e., *Supply*. Then we search the ER diagram with two nodes, *Part* and *Supply*. After finding the minimum Steiner tree, we extend the relationship type *Supply* by including the other participating entity class, i.e., *Supplier* and its attributes, as the search intention of the query. By applying the predicate on *price*, we will have search result, including supplier *s1* supplies part *p1* at the price of 10. The result explains who supply these parts at 10. Only returning part *p1* has price of 10 is meaningless, as *p1* may also have other prices by other suppliers.

Finding the minimum Steiner tree of a set of vertices in a graph is an NP-complete problem. However, since we search in a schema-level ER diagram which is small, the performance will not be an issue. We further validate it by experiment in the next section.

4   Experiment

Model transformation works on schema level, regardless of the size of database. Because the database schema, either the ER or the ORA-SS, is quite small in size, the cost on schema transformation is rather low. Furthermore, schema transformation is a one-time effort, regardless of queries. It will not bring in any impact on query performance. Thus in this section, we focus on query processing.

Our query processing approach involves searching the minimum Steiner tree in an ER diagram. Although we claim that this is not a bottleneck because of the small size of an ER diagram, we further conduct experiment to validate it.

We first adopt the schema for basketball statistic data[2] which follows an ER diagram with 11 nodes. We transform it into an ORA-SS diagram to accept queries. Due to the space limit, as well as our focus on response time rather than semantics, we do not show the exact diagrams. We randomly compose 12 keyword queries (under our framework, XPath and keyword queries are similarly processed) with different number of query nodes and use a PC with a 3.2 GHz CPU and a 8 GB memory to run the queries. Figure 8(a) shows the average response time for queries with different number of nodes by searching the ER diagram. We can see for both cold cache and warm cache cases, the search is quite fast.

Considering more complex databases, we randomly generate a set of ER diagrams with different sizes (from 10 nodes to 100 nodes), and test the searching time with randomized five-node queries. The result is shown in Fig. 8(b). We can see that the searching time does not increase significantly as the complexity of ER diagram increases. In fact there is no obvious difference for a computer to search 10 nodes or 100 nodes. Of course the time will increase if the number of nodes reaches thousands or millions, but the schema complexity of normal business databases will not reach that number.

From many RDBMS performance evaluation reports (e.g., a recent one in [6]), the query processing time for a database in gigabytes can easily reach seconds or minutes. Compared to the result in Fig. 8, we can conclude that the minimum Steiner tree searching in schema level is not a bottleneck for query processing.

---

[2] http://www.databasebasketball.com/.

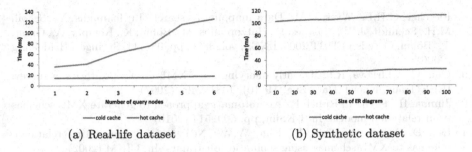

(a) Real-life dataset   (b) Synthetic dataset

Fig. 8. Cost on ER diagram searching

5 Conclusion

In this paper we consider a practical data management framework in which data are maintained in RDBMS and partially published, shared and utilized in XML formats. We highlight the challenges in data transformation and query processing in such a framework that were not addressed by the existing works in schema mapping. In particular, in our considered framework, we required the transformed XML views to be customizable based on users' preferences. We propose a semantics-based approach to resolve these challenges. In particular, we use semantic models, i.e., ER model and ORA-SS model to design relational database and XML views. We design algorithms to achieve model transformation from ER to ORA-SS. We also discuss how to interpret XPath or keyword queries and translated them into SQL queries to query the underlying relational database.

References

1. Chen, P.P.: The ER model: toward a unified view of data. ACM Trans. Database Syst. **1**(1), 9–36 (1976)
2. Chen, Y.B., Ling, T.W., Lee, M., Nakanishi, M., Dobbie, G.: A semantic approach to the design of valid and reversible semistructured views. JCSE **1**(1), 95–123 (2007)
3. Bonifati, A., Chang, E.Q., Ho, T., Lakshmanan, L.V.S., Pottinger, R., Chung, Y.: Schema mapping and query translation in heterogeneous P2P XML databases. VLDB J. **19**(2), 231–256 (2010)
4. Chaudhuri, S., Kaushik, R., Naughton, J.F.: On relational support for XML publishing: beyond sorting and tagging. In: SIGMOD, pp. 611–622 (2003)
5. Chen, Y.B., Ling, T.W., Lee, M., Nakanishi, M., Dobbie, G.: A semantic approach to the design of valid and reversible semistructured views. JCSE **1**(1), 95–123 (2007)
6. Difallah, D.E., Pavlo, A., Curino, C., Cudre-Mauroux, P.: OLTP-Bench: an extensible testbed for benchmarking relational databases. PVLDB **7**(4), 277–288 (2013)
7. Fernandez, M., Kadiyska, Y., Suciu, D., Morishima, A., Tan, W.: SilkRoute: a framework for publishing relational data in XML. ACM Trans. Database Syst. **27**(4), 438–493 (2002)

8. Fletcher, G.H.L., Wyss, C.M.: Data mapping as search. In: Ioannidis, Y., Scholl, M.H., Schmidt, J.W., Matthes, F., Hatzopoulos, M., Böhm, K., Kemper, A., Grust, T., Böhm, C. (eds.) EDBT 2006. LNCS, vol. 3896, pp. 95–111. Springer, Heidelberg (2006)
9. Gou, G., Chirkova, R.: Efficiently querying large XML data repositories: a survey. IEEE Trans. Knowl. Data Eng. **19**(10), 1381–1403 (2007)
10. Jumaa, H., Fayn, J., Rubel, P.: An automatic approach to generate XML schemas from relational models. In: UKSim, pp. 509–514 (2010)
11. Lee, D., Mani, M., Chiu, F., Chu, W.W.: NeT & CoT: translating relational schemas to XML schemas using semantic constraints. In: CIKM (2002)
12. Li, L., Le, T.N., Wu, H., Ling, T.W., Bressan, S.: Discovering semantics from data-centric XML. In: Decker, H., Lhotská, L., Link, S., Basl, J., Tjoa, A.M. (eds.) DEXA 2013, Part I. LNCS, vol. 8055, pp. 88–102. Springer, Heidelberg (2013)
13. Ling, T.W., Lee, M.L., Dobbie, G.: Semistructured Database Design. Springer, US (2005)
14. Liu, C., Vincent, M.W., Liu, J.: Constraint preserving transformation from relational schema to XML schema. World Wide Web **9**(1), 93–110 (2006)
15. Mecca, G., Rull, G., Santoro, D., Teniente, E.: Semantic-Based Mappings. In: ER (2013)
16. Melnik, S., Adya, A., Bernstein, P.A.: Compiling mapplings to bridge applications and databases. In: SIGMOD, pp. 461–472 (2007)
17. Shanmugasundaram, J., Shekita, E., Barr, R., Carey, M., Linday, B., Pirahesh, H., Reinwald, B.: Efficiently publishing relational data as XML documents. VLDB J. **10**(2–3), 133–154 (2001)
18. Wu, H., Bao, Z.: Object-oriented XML keyword search. In: ER (2011)
19. Zhou, R., Liu, C., Li, J.: Holistic constraint-preserving transformation from relational schema into XML schema. In: Haritsa, J.R., Kotagiri, R., Pudi, V. (eds.) DASFAA 2008. LNCS, vol. 4947, pp. 4–18. Springer, Heidelberg (2008)

EMF Views: A View Mechanism for Integrating Heterogeneous Models

Hugo Bruneliere[1]([✉]), Jokin Garcia Perez[1], Manuel Wimmer[2],
and Jordi Cabot[3]

[1] AtlanModTeam, Inria - Mines Nantes - LINA, Nantes, France
{hugo.bruneliere,jokin.garcia-perez}@inria.fr
[2] Business Informatics Group, Vienna University of Technology, Vienna, Austria
wimmer@big.tuwien.ac.at
[3] ICREA - UOC, Barcelona, Spain
jordi.cabot@icrea.cat

Abstract. Modeling complex systems involves dealing with several heterogeneous and interrelated models defined using a variety of languages (UML, ER, BPMN, DSLs, etc.). These models must be frequently combined in different cross-domain perspectives to provide stakeholders the view of the system they need to best perform their tasks. Several model composition approaches have already been proposed addressing this problem. Nevertheless, they present some important limitations concerning efficiency, interoperability and synchronization between the base models and the composed ones. As an alternative we introduce EMF Views, an approach coming with a dedicated language and tooling for defining views on potentially heterogeneous models. Similarly to views in databases, model views are not materialized but instead redirect all model access and manipulation requests to the base models from which they are obtained. This is realized in a transparent way for both the modeler and the other modeling tools using the concerned (meta)models.

Keywords: Modeling · Viewpoint · View · Heterogeneity · Virtualization

1 Introduction

Software systems are becoming increasingly complex, making them more and more difficult to comprehend, develop and maintain. To handle this complexity, they are usually represented by sets of models at different abstraction levels and possibly conforming to different modeling languages [6]. Each one of these languages is specialized to provide the modeling constructs required to deal with a particular concern of the system. Typically, several models must be combined to generate the most adequate perspective of the system for each involved person (depending on his/her role). This is challenging, notably due to the heterogeneity of the models and to the various existing (implicit) relationships between them. Such a common issue is very likely to appear in almost any non-trivial project.

© Springer International Publishing Switzerland 2015
P. Johannesson et al. (Eds.): ER 2015, LNCS 9381, pp. 317–325, 2015.
DOI: 10.1007/978-3-319-25264-3_23

For instance, the TOGAF Enterprise Architecture Platform (TEAP) project[1]
was a joint industrial-academic collaboration to provide support for the governance of enterprise architectures (EAs). Its base platform was SmartEA[2] that
integrates TOGAF[3]. Use case providers wanted to customize SmartEA and also
include both business process information defined using BPMN[4] and requirement specifications defined with ReqIf[5]. Besides integrating BPMN and ReqIf
models as part of SmartEA, they wanted to be able to interconnect these models with the TOGAF ones and to provide partial views on the combined models
depending on given profiles (e.g., with restricted access for security reasons).
Thus, they needed to specify several viewpoints linking the TOGAF, ReqIF and
BPMN metamodels altogether.

To tackle such a challenge, several approaches for model composition have
already been developed so far, e.g., [2,12,15,17,18]. In short, mostly all of them
propose the generation of a completely new composed model automatically populated with elements copied from the set of base models that participate in the
composition. Thus, they present some important limitations in terms of performance (due to the time required to copy model elements into the composed
model), synchronization (due to the lack of change propagation from the contributing models to the generated one, or the other way round), and/or interoperability (the composed model may have a different nature than the contributing
ones and needs to be manipulated using specialized tools).

In this paper we propose EMF Views, an alternative solution based on adapting the concept of database views [1] for models. EMF Views enables the specification of model views grouping elements using concepts coming from different
metamodels. Such views can be adapted to the needs of a specific user. From
a user perspective (and from a modeling tool one as well), the view behaves as
any other regular model. Views are not materialized but computed on-demand,
which brings some important benefits compared to previous approaches. The
expressivity of our view definition language is comparable to *select-project-join*
queries in relational algebra. Our approach has been implemented and made
available as an open source Eclipse component.

The remainder of this paper is structured as follows. Section 2 describes the
related work. Section 3 presents EMF Views while Sect. 4 summarizes its implementation. Finally, Sect. 5 critically discusses EMF Views and outlines our next
steps.

2   State-of-the-Art

Views have a long tradition since the introduction of relational databases, and
have been proposed for object-oriented databases as well [1,23]. View mechanisms have also been discussed in the context of modeling languages such as

---

[1] http://www.teap-project.org.
[2] http://www.obeosmartea.com.
[3] http://www.opengroup.org/togaf.
[4] http://www.bpmn.org.
[5] http://www.omg.org/spec/ReqIF.

UML [4] and ER [9]. In a nutshell, most of these language-specific approaches rely on query languages to define virtual elements in intra-model views. Although this support is useful in a single model context, additional mechanisms are needed to provide inter-model views. In this respect, there are two major research lines: (*i*) multi-viewpoint modeling used to design different viewpoints based on a unified underlying model, and (*ii*) model composition to define correspondences between different models to build a common view.

There are several approaches for multi-viewpoint modeling (cf. [13] for a survey). Atkinson et al. [3] propose Orthographic Software Modeling (OSM), a projective multi-view methodology in which views are dynamically generated via model transformations from a single base model. Cicchetti et al. [10] propose to define viewpoints as subsets of a base metamodel. Integrating/linking different (meta)models does not seem to be a focus in these projective approaches. Burger [7] propose a mechanism to improve views in OSM [3], in which flexible read-only views can be defined at development time. Kramer et al. [19] propose a methodology to create an underlying metamodel from base ones by using structural mappings between metamodel elements. This approach has a mechanism to support extending viewpoints and synchronizing model views with the underlying model. Bork & Karagiannis [5] propose a graphical language to define viewpoints for multi-viewpoint modeling. Finally, Romero et al. [22] propose a complementary approach for specifying and realizing correspondences between viewpoints.

Concerning model composition, there are approaches that can be used to simulate views using different link types between models [12,17,18]. However, they do not explicitly focus on model views but rather provide general capabilities (often referred to as megamodels) needed to reason about connected models. For instance, languages such as the Epsilon Merging Language [18] may help during the composition process, e.g., by facilitating the identification of the elements to merge. Besides these model composition approaches, some dedicated model view approaches are emerging. In [15] the authors introduce, based on Triple Graph Grammars, the possibility to have non-materialized views by extending base metamodels using inheritance. In a successor work [2], they present an approach for materialized views without modifying the base metamodels, but requiring to explicitly populate the views via model transformations. EMF Facet[6] is another approach to define read-only non-materialized views. Finally, Hegedüs et al. [14] present a similar approach to EMF Facet for model interconnection by augmenting base metamodels with derived features.

To summarize, key challenges to be addressed are the following ones:

- **Genericity**: the view mechanism should be applicable for all modeling languages (i.e., to all metamodels and corresponding models);
- **Expressivity**: a *select-project-join*-like support should be at least provided;
- **Non-intrusiveness**: the view mechanism should be applicable without modifying (e.g., internally) the used modeling languages;

---

[6] http://www.eclipse.org/modeling/emft/facet.

- **Interoperability**: a view should be a regular model from user and tool perspectives;
- **Modifiability**: a view should be changeable as a regular model is;
- **Synchronization**: changes in base models should be directly reflected in the views, and vice versa;
- **Scalability**: view creation and manipulation time should be sufficiently limited, as well as corresponding memory usage.

To the best of our knowledge, none of the available approaches fully satisfies all these characteristics. There is always a trade-off between the offered capabilities and some of these properties, such as scalability or synchronization more particularly. Moreover, no approach provides both inter-model view support and the expected expressivity in terms of view definition. EMF Views intends to tackle these challenges as explained in the following.

3   The EMF Views Solution

EMF Views is intended to answer to the concrete need for model views. This section provides an overview of its conceptual framework as well as the related SQL-like DSL it comes with for defining viewpoints.

Before introducing the overall approach itself, we define the terminology used in there. A **viewpoint** is the description of a partitioning and/or restriction of concerns from which systems can be observed. In our modeling context, it consists of a set of concepts coming from one or several metamodels, eventually complemented with some new interconnections between them. A **view** is a representation of a specific system from the perspective of a given viewpoint. In our modeling context, it consists of a set of elements coming from one or several models, eventually complemented with some new interconnections between them. A **virtual model** is a model whose (virtual) elements are just proxies to actual elements contained in other models. The same approach is also applicable at metamodel-level, i.e., a **virtual metamodel**. A **weaving model** is a model that describes links between elements coming from other different models. It conforms to a weaving metamodel that specifies the types of links that can be represented at weaving model-level.

3.1   Conceptual Framework

EMF Views proposes a generic approach allowing to build views on any set of interrelated models that conform to potentially different metamodels. It provides a two-step approach that explicitly separates the specification of viewpoints from the realization and handling of corresponding views. A model view consist of a set of proxy elements, which point to concrete elements from the base models referenced in the view, plus some newly added cross-model relationships between them.

In order to do so, EMF Views relies on a model virtualization approach that is deployed similarly at both metamodel- and model-levels. Thus, such views are

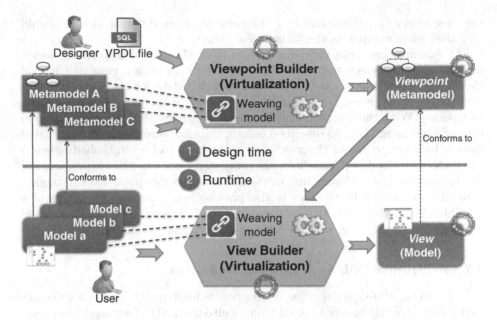

Fig. 1. Overview of the EMF Views approach

actually virtual models that act transparently as regular models via proxies to these interrelated models, but do not duplicate any of the already available data. Each view conforms to a particular viewpoint, which has been previously specified from one or several corresponding metamodels (interconnected together) as a virtual metamodel. Interestingly, the fact that both viewpoints and views are actually virtual (meta)models behaving as normal (meta)models allows for easier viewpoint/view composition. An overview of the EMF Views approach is shown on Fig. 1.

At *design time*, designers may specify a new viewpoint by choosing the concerned metamodel(s), listing the relations she/he wants to represent between them (as well as indicating how to eventually compute them at view-level, see hereafter), and identifying the concepts and properties to be selected. This required information is directly collected from the designer/architect, either manually or using our SQL-like DSL (cf. Sect. 3.2). This input data is stored in a weaving model that is then used by the virtualization mechanism to obtain the actual viewpoint. Therefore, the original metamodel(s) are not modified or polluted by the viewpoint definition. This results in a virtual metamodel, representing the viewpoint, that aggregates several different metamodels according to the given specification.

Similar to the *select-project-join* operations in relational algebra, the viewpoint definition specifies what classes/features from the contributing metamodels should be part of (or, conversely, filtered out from) the view (*projection*), what conditions model elements will need to satisfy in order to appear as a result in the view query (*selection*) and how the elements from different models should be linked when computing the actual view (*join*).

At *runtime*, once the viewpoint is specified, the user can work on querying and handling views that conform to it. To obtain such a view, she/he can choose the set of input models to be used as input data for the view (and that conform to their respective metamodels, themselves used to create the given viewpoint). With those models and the given viewpoint, EMF Views can build the corresponding view. As described before, the view is represented as a virtual model. In order to create the view, new links have to be established between the underlying models. These links are computed from the rules expressing the combination of the corresponding metamodels at the viewpoint-level (though a manual modification by the user is also possible when needed) by means of a matching engine (cf. Sect. 4). The links are stored in a separate weaving model associated with the view, without altering the original models neither.

3.2   A SQL-like DSL for Viewpoint Definition

In order to facilitate the definition of viewpoints more easily, EMF Views comes with a DSL strongly inspired from the very well-known SQL language. The choice of an SQL-like language has been quite natural since SQL has already proven its relevance to deal with similar problems in the database community. As said earlier, it also allows expressing the main needed operations in our model view context, i.e., select, project, and join. Additionally, considering such a widespread language as a base language for our DSL intends to facilitate DSL adoption by potential future users.

Listing 1.1 presents a grammar excerpt of our textual DSL highlighting its four main language features:

- **Create view**: Defines the name of the view(point) as well as the contributing metamodels (and their respective names).
- **Select**: Lists the attributes and relations to be shown (corresponding classes are implicitly selected too). To show all, * is used.
- **From**: Specifies which concepts are going to be linked and provides a name for the new relation (that will be used in the view).
- **Where**: Expresses constraints to filter elements or to define connection between them, using simple mappings or more complex expressions.

Listing 1.1. Partial Grammar of VPDL (ViewPoint Definition Language)

```
Model:"create␣view"viewName=ID"on"metamodel+=Metamodel (","metamodel+=
    ↪Metamodel)* expression+=Expression;
Metamodel: metamodelURL+=EString"as"metamodelName+=MetamodelName;
Expression:"select"select+=Select"from"from+=From"where"(condition+=
    ↪Condition)+;
Select: select+="*"| selectFeature+=SelectFeature (","selectFeature+=
    ↪SelectFeature)*;
SelectFeature: metamodel+=[MetamodelName]"."class+=Class"{"feature+=Feature (","
    ↪feature+=Feature)*"}";
From: join+=Join (","join+=Join)* ;
Join: joinLeft+=JoinPart"join"joinRight+=JoinPart"as"reference+=Reference;
```

```
JoinPart: metamodel+=[MetamodelName]"."class+=Class;
MetamodelName: name=ID; Class: name=ID; Feature: name=ID; Reference: name=ID;
Condition: ECLExpression;
```

To illustrate our language, Listing 1.2 shows a simple example of a viewpoint specification (from the TEAP scenario as introduced in Sect. 1). It selects and aggregates some elements from the original metamodels (*select* part) and establishes new relations between them (*from* and *where* parts).

Listing 1.2. Simple example of viewpoint specification in VPDL

```
create view myEnterpriseArchitectureViewpoint on
      "http://www.obeonetwork.org/dsl/togaf/contentfwk/9.0.0"as TOGAF,
      "http://www.omg.org/spec/BPMN/20100524/MODEL-XMI"as BPMN,
      "http://www.omg.org/spec/ReqIF/20110401/reqif.xsd" as REQIF
select TOGAF.Process{name}, BPMN.Process{processType, processCriticality},
           TOGAF.Requirement{rationale}, REQIF.SpecObject{longName}
from TOGAF.Process join BPMN.Process as detailedProcess,
        TOGAF.Requirement join REQIF.SpecObject as detailedRequirement
where TOGAF.Process.name = BPMN.Process.name and TOGAF.Process.isAutomated = false
        and REQIF.SpecObject.values->exists(v | v.theValue = TOGAF.Requirement.name)
```

4 Eclipse-Based Tooling Support

EMF Views has been implemented on top of Eclipse and its well-known Eclipse Modeling Framework (EMF) providing general model creation and handling capabilities. It mainly consists of four main components which are notably adapting Virtual EMF [11] in a viewpoint/view context.

Viewpoint and View generators combine the two APIs mentioned thereafter and offer Eclipse GUI components, such as viewpoint- and view-specific creation wizards and editors. A *Model View API* (deriving from the EMF model access API) supports virtualization to handle viewpoints and views transparently as any regular EMF models. A *Linking API* is managing inter-model links, and has been connected to the mapping engine of the Epsilon Comparison Language [17] to compute such links at view-level. VPDL (cf. previous section) has been developed from scratch with Xtext[7] and then integrated using model-to-text (in Xtend[8]) and model-to-model transformations (in ATL [16]). For user convenience, VPDL notably comes with a proper editor including syntax highlighting and content-assist displaying the applicable classes, attributes, and references. The open source tool and screencasts are available at GitHub[9].

5 Critical Discussions and Next Steps

EMF Views globally fulfils the expected properties introduced in Sect. 2. *Genericity* is ensured as any existing (meta)models may be considered to build viewpoints and views. The provided DSL and underlying support allow EMF Views

---
[7] https://eclipse.org/Xtext.
[8] http://www.eclipse.org/xtend.
[9] https://github.com/atlanmod/emfviewsSQL.

to offer the expected *expressivity*. *Non-intrusiveness* is naturally achieved since the original metamodels and models are not modified at all. As mentioned earlier, *interoperability* is guaranteed as model views may be used wherever regular models can (mostly in a read-only mode currently). *Modifiability* is currently partially supported, as only changes to attribute values in views are propagated back to the original models. However, since views and actual models do share the same real instances via a proxy mechanism, *synchronization* at view-level is directly obtained.

To start evaluating *scalability*, we performed empirical experiments focusing on time and memory consumption. We compared EMF Views to a simple composition approach based on the execution of ATL model transformations [16] that copy/merge the elements from the contributing models to the composed model (representing the view). This acted as a representative example of typical behavior from existing composition approaches. We found out that our approach is faster at creating views, and shows only a small overhead during their manipulation (mainly due to implemented lazy loading strategy). Regarding memory usage, in our case the required memory is almost equal to the sum of the size of the contributing models since we do not duplicate model elements for the view but use lightweight proxies. In contrast, in traditional composition the fully duplicated composed model has to be kept in memory as well. We believe these results show that EMF Views scales up to large models and related views.

As further work, we plan to enrich our view mechanism with support for a limited set of aggregation operations, focusing on distributive aggregate functions [21] expressed with OCL extensions [8]. This will be linked to addressing the view update challenge [20], as EMF Views is currently limited to individual attribute updates. At tool-level, we plan to offer the option to persist a snapshot of the model view in case users want to export them to other (external) tools. Finally, we would like to improve usability by exploring alternative languages for view definition and manipulation (e.g., graphical ones) beyond our current SQL-based DSL.

Acknowledgment. We thank Juan David Villa Calle and Caue Avila Clasen for their work on past EMF Views versions. This work has been co-funded by the Vienna Business Agency within the COSIMO project (grant number 967327), Christian Doppler Forschungsgesellschaft, and BMWFW, Austria.

References

1. Abiteboul, S., Bonner, A.: Objects and views. SIGMOD Rec. **20**(2), 238–247 (1991)
2. Anjorin, A., Rose, S., Deckwerth, F., Schürr, A.: Efficient model synchronization with view triple graph grammars. In: Cabot, J., Rubin, J. (eds.) ECMFA 2014. LNCS, vol. 8569, pp. 1–17. Springer, Heidelberg (2014)
3. Atkinson, C., Stoll, D., Bostan, P.: Orthographic software modeling: a practical approach to view-based development. In: Maciaszek, L.A., González-Pérez, C., Jablonski, S. (eds.) ENASE 2008/2009. CCIS, vol. 69, pp. 206–219. Springer, Heidelberg (2010)

4. Balsters, H.: Modelling database views with derived classes in the UML/OCL-framework. In: Stevens, P., Whittle, J., Booch, G. (eds.) UML 2003. LNCS, vol. 2863, pp. 295–309. Springer, Heidelberg (2003)
5. Bork, D., Karagiannis, D.: Model-driven development of multi-view modelling tools - the MUVIEMOT approach. In: DATA (2014)
6. Brambilla, M., Cabot, J., Wimmer, M.: Model-Driven Software Engineering in Practice. Synthesis Lectures on Software Engineering. Morgan & Claypool Publishers, San Rafael (2012)
7. Burger, E.: Flexible views for rapid model-driven development. In: VAO Workshop (2013)
8. Cabot, J., Mazón, J.-N., Pardillo, J., Trujillo, J.: Specifying aggregation functions in multidimensional models with OCL. In: Parsons, J., Saeki, M., Shoval, P., Woo, C., Wand, Y. (eds.) ER 2010. LNCS, vol. 6412, pp. 419–432. Springer, Heidelberg (2010)
9. Ceri, S., Fraternali, P., Bongio, A.: Web modeling language (WebML): a modeling language for designing Web sites. Comput. Netw. **33**(1), 137–157 (2000)
10. Cicchetti, A., Ciccozzi, F., Leveque, T.: A hybrid approach for multi-view modeling. ECEASST **50**, 1–12 (2011)
11. Clasen, C., Jouault, F., Cabot, J.: VirtualEMF: a model virtualization tool. In: ER Workshops (2011)
12. Didonet Del Fabro, M., Valduriez, P.: Towards the efficient development of model transformations using model weaving and matching transformations. Softw. Syst. Model. **8**, 305–324 (2009)
13. Goldschmidt, T., Becker, S., Burger, .: Towards a tool-oriented taxonomy of view-based modelling. In: Modellierung (2012)
14. Hegedüs, Á., Horváth, Á., Ráth, I., Varró, D.: Query-driven soft interconnection of EMF models. In: France, R.B., Kazmeier, J., Breu, R., Atkinson, C. (eds.) MODELS 2012. LNCS, vol. 7590, pp. 134–150. Springer, Heidelberg (2012)
15. Jakob, J., Königs, A., Schürr, A.: Non-materialized model view specification with triple graph grammars. In: Corradini, A., Ehrig, H., Montanari, U., Ribeiro, L., Rozenberg, G. (eds.) ICGT 2006. LNCS, vol. 4178, pp. 321–335. Springer, Heidelberg (2006)
16. Jouault, F., Allilaire, F., Bézivin, J., Kurtev, I.: ATL: a model transformation tool. Sci. Comput. Program. **72**(1), 31–39 (2008)
17. Kolovos, D.S.: Establishing correspondences between models with the epsilon comparison language. In: Paige, R.F., Hartman, A., Rensink, A. (eds.) ECMDA-FA 2009. LNCS, vol. 5562, pp. 146–157. Springer, Heidelberg (2009)
18. Kolovos, D.S., Paige, R.F., Polack, F.A.C.: Merging models with the epsilon merging language (EML). In: Wang, J., Whittle, J., Harel, D., Reggio, G. (eds.) MoDELS 2006. LNCS, vol. 4199, pp. 215–229. Springer, Heidelberg (2006)
19. Kramer, M.E., Burger, E., Langhammer, M.: View-centric engineering with synchronized heterogeneous models. In: VAO Workshop (2013)
20. Mayol, E., Teniente, E.: A survey of current methods for integrity constraint maintenance and view updating. In: ER Workshops (1999)
21. Palpanas, T., Sidle, R., Cochrane, R., Pirahesh, H.: Incremental maintenance for non-distributive aggregate functions. In: VLDB (2002)
22. Romero, J.R., Jaen, J.I., Vallecillo, A.: Realizing Correspondences in multi-viewpoint specifications. In: EDOC (2009)
23. Wiederhold, G.: Views, objects, and databases. IEEE Comput. **19**(12), 37–44 (1986)

Collaborative Modeling

Gitana: A SQL-Based Git Repository Inspector

Valerio Cosentino[1]([✉]), Javier Luis Cánovas Izquierdo[1,2], and Jordi Cabot[2,3]

[1] AtlanMod Team, Inria, Mines Nantes, LINA, Nantes, France
valerio.cosentino@inria.fr
[2] UOC, Barcelona, Spain
jcanovasi@uoc.edu
[3] ICREA, Barcelona, Spain
jordi.cabot@icrea.cat

Abstract. Software development projects are notoriously complex and difficult to deal with. Several support tools such as issue tracking, code review and Source Control Management (SCM) systems have been introduced in the past decades to ease development activities. While such tools efficiently track the evolution of a given aspect of the project (e.g., bug reports), they provide just a partial view of the project and often lack of advanced querying mechanisms limiting themselves to command line or simple GUI support. This is particularly true for projects that rely on Git, the most popular SCM system today.

In this paper, we propose a conceptual schema for Git and an approach that, given a Git repository, exports its data to a relational database in order to (1) promote data integration with other existing SCM tools and (2) enable writing queries on Git data using standard SQL syntax. To ensure efficiency, our approach comes with an incremental propagation mechanism that refreshes the database content with the latest modifications. We have implemented our approach in Gitana, an open-source tool available on GitHub.

Keywords: Git · SQL · Conceptual schema

1 Introduction

Software development projects are inherently complex due to the extensive collaboration and creative thinking involved [1]. In the last years, several tools have been created to cope with such complexity by providing specific support for the different development activities. Probably the most relevant are: Source Control Management (SCM) systems to manage code repositories [2,3], storing and tracking the different source code versions; issue tracker systems [4] to provide support on maintenance and evolution activities such as reporting bugs and requesting new features; and code review tools to increase the quality of the final software product [5] by recording the communications between reviewers and code authors. While these tools efficiently track the evolution of a specific aspect of the software project, each provides just a partial view of it and usually

© Springer International Publishing Switzerland 2015
P. Johannesson et al. (Eds.): ER 2015, LNCS 9381, pp. 329–343, 2015.
DOI: 10.1007/978-3-319-25264-3_24

comes with insufficient means (e.g., only command line support or other simple user interfaces) to perform any non-trivial query operation that could shed some light on important aspects of the project status.

This is particularly true for Git [6], that has become the most popular SCM system thanks to superior off-line capabilities, easier branch management and its promotion by most well-known Open Source Software (OSS) forges (e.g., GitHub and BitBucket) [7]. Despite the existence of several project management and monitoring tools built on top of Git, there is still a major lack of data integration efforts between them and they all fall short regarding the possibilities they offer for advanced query functionalities, thus forcing practitioners to resort to learning and using several off-the-shelf tools (e.g., [8,9]), based on predefined queries on individual aspects of the project.

To overcome this situation, in this paper we propose a conceptual schema for Git and an approach that, given a Git repository, exports its data to a relational database derived from our conceptual schema. After the initial creation of the database, an incremental update mechanism will synchronize the database with the repository at any moment by considering only the modifications in the repository that took place after the last update.

Once in our database, we can easily integrate it with data coming from other tools that rely on a database infrastructure (e.g., GHTorrent [10], a scalable and offline mirror of GitHub; and Gerrie [11], a data and information crawler for the code review tool Gerrit). Thus, we are able to offer a shared place to perform cross-cutting analysis of a software development project including for instance the collaboration information (e.g., issues and pull requests) together with the Git (e.g., patches, file renamed) and code review data (e.g., review comments).

Furthermore, our proposed relational database can be exploited using the standard SQL language (and any other database analytical processing tool), thus enabling any practitioner familiar with the SQL language to easily inspect Git repositories according to her specific needs. For this same purpose, we have also predefined several views and stored procedures to simplify the computation of useful metrics and simulation of typical Git commands.

We have implemented our approach in *Gitana*, an open-source tool available on GitHub [12].

The remainder of the paper is organized as follows. Section 2 presents the motivation and the state of the art. Section 3 describes our approach while Sect. 4 discusses the application scenarios. Section 5 reports on the implementation details and evaluation conducted on five projects in GitHub. Finally, Sect. 6 ends the paper with conclusion and future work.

2   Motivation and State of the Art

Git repositories are typically explored via the Git command line. However, this approach is a complex and tedious task which requires developers to have a deep knowledge of Git and shell commands. Moreover, the integration of the queried information with other tools is very limited.

Listing 1.1. Number of deleted files via command line.

```
git log --diff-filter=D
        --summary
| grep 'delete mode'
| wc -l
```

Listing 1.2. Number of modifications on a file (named `FileName`) per user via command line.

```
git log --full-history
        --format=tformat:%a FileName
| gawk
 '{ count[$1]++ } END
  { for(j in count) printf "%s: %s changes\n",
    j, count[j]}'
```

As an example, performing a query to obtain the number of deleted files in a Git repository requires to complement the Git command with shell commands to present the result in a comprehensible way. Listing 1.1 shows a possible way to perform this query. As can be seen, the Git command involves a few advanced parameters and the output must be digested with additional shell commands to filter (i.e., `grep` command) and present the desired information (i.e., `wc` command to count).

When the aim is to perform a more detailed analysis (e.g., where grouping or pattern-matching is required), the definition of queries with the command line can quickly become an almost impossible task for non-experts. For instance, Listing 1.2 shows a query to obtain the number of modifications on a file per user. As can be seen, the query requires using the Git command option `log` together with some advanced parameters and other shell commands (see the use of `gawk` command) to digest, analyze and present the data.

This situation has not been solved by the proliferation of Git-based tools we have witnessed in the last years. This affects not only the day-to-day operations on Git-based projects but also any attempt to analyze Git to unveil interesting facts on the development process, team dynamics, etc., that could be useful to optimize the management and evolution of the project. Or, given the tight relationship between Git and code hosting platforms for open source projects, any effort aimed at mining Git repositories to extract common patterns for OSS development.

Previous works focused on extracting information from SCM systems to perform specific analysis (e.g., bug prediction, logical coupling detection, etc.), in particular most of them ([13–19]) target older SCM systems such as CVS and SVN, while only few focus on Git ([7–9,20,21]). More recent works have the goal of analyzing complete OSS forges such as GitHub or SourceForge ([10,22]). Our approach belongs to the first axis (i.e., it's not linked to a specific code hosting platform) and tries to overcome different limitations of previous works. Such limitations are presented and compared with our proposal below (see Table 1).

Table 1. Comparison with previous works.

	[13, 18]	[15]	[16]	[7, 14, 20]	[17]	[9]	[8, 21]	[19]	Gitana	Git
Generality									x	x
Flexibility									x	
Incrementality									x	x
Exportability			x		x	x			x	
Extensibility	x	x	x		x				x	
Availability		x	x				x	x	x	x

- *Generality.* All previous works target very specific information goals, thus they extract and store only a portion of the SCM data. This partial view of the information makes very difficult to extend and integrate these tools with others targeting a different perspective. Instead, we propose a database which stores all SCM information (coarse and fine-grained).
- *Flexibility.* Many previous works such as [8, 9, 19] do not allow ad-hoc queries, and Git itself makes it possible by only mixing Git commands together with shell scripting, as shown before. On the other hand, the works in [13, 15–18] allow the definition of ad-hoc queries limited to the portion of the SCM information they store. Our approach aims at performing better since it mirrors all information in the Git SCM and stores it in a RDB, thus it is flexible and complete enough to satisfy all user's needs for querying purposes.
- *Incrementality.* None of the previous works provides an incremental propagation mechanism to align the database content with the SCM's latest modifications. Thus, they have to be executed each time the repository changes, which hampers scalability when dealing with large repositories. Our approach includes an incremental propagation mechanism that makes it suitable for both small and large repositories.
- *Exportability.* Works by [16, 17] provide an exporter from the data stored in the database to XML, however the XML structure and level of detail is unclear (e.g., it is not said whether the XML format is a mere representation of each database table). On the other hand, the approach presented in [9] does not rely on a database, but presents the data extracted from the SCM in HTML, XML and plain-text format. Currently, Gitana provides a JSON exporter that restructures the database information to make it available in other technologies beyond SQL.
- *Extensibility.* Previous works such as [13, 15–18] rely on a database that should be modified to add new sources of information, however they do not discuss any extension mechanism. On the contrary, our approach discusses possible integrations with other sources of information such as bug tracking systems and code review tools, as we show in this paper. Such integration can be easily achieved by connecting the concepts of *Developer*, *Commit* or *File* embedded in our model to their equivalent in other sources.

- *Availability.* Only [8,9,15,16,19,21] make their implementations available to download. However only [15,21] have been active in the past years. Gitana has been made available on GitHub at [12].

In short, to the best of our knowledge no previous tool or research effort has proposed a conceptual schema for Git with the intent of (1) promoting data integration with other Git-based tools and (2) enabling advanced query functionalities for Git.

3 Our Approach

We propose a conceptual schema for Git to facilitate data integration between existing Git-based tools and advanced query operations. This schema is materialized as a relational database, for which we have defined an extraction process that populates and keeps it up-to-date. Next we describe the conceptual schema, the corresponding database schema and the extraction process.

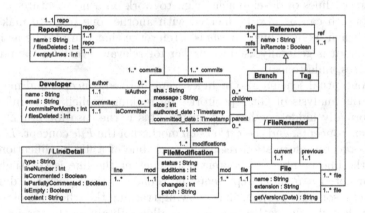

Fig. 1. Git conceptual schema.

3.1 Modeling Git

Git is a decentralized SCM system based on a master-less peer-to-peer replication where any replica of a given project can send or receive any information to or from any other replica. In Git, any developer can either create from scratch her own Git repository or obtain a copy (i.e., clone) of an existing one (i.e., remote). Although Git can theoretically work without a centralized repository, in practice there is usually a central repository that serves as the authoritative copy of the software project, thus it is what everyone fetches from and pushes to.

The structure of a Git *repository* is shown in the conceptual schema of Fig. 1. A non-empty repository is organized following a tree structure where each node is represented by a *commit* to which *references* can be assigned. A commit is uniquely identified (i.e., using SHA) and contains a revision of the files within the repository reflecting the state of the project at a given point in time (i.e., snapshot).

In particular, a commit stores the differences between the files (*Files* and *FileModifications*) that changed between two revisions (the *patch* attribute in *FileModification* stores this raw information). It also contains information (i.e., name and email) regarding the corresponding author and committer (*Developer*), where the former is the one that did the change and the latter is the one that applied the change to the repository. Furthermore, a commit includes a reference to its parent commit(s) (*parent*). Generally, a commit has only one parent, that represents the previous state of the project; however, it can be parent-less (e.g., the commit originates the repository) or have multiple parents (when merging two or more branches).

Commits can be linked to different references, such as *branches* and *tags*. A *branch* represents a line of development that can be local if exists only in the cloned repository or remote if belongs to the remote repository (*inRemote*). The default branch is usually named *master* but new ones can be created to start new separated lines of development (e.g., to work on a new feature or to fix a bug). A branch can be also be merged with another one (e.g., to make a new release or when fixing a bug). A *tag* is a reference that can also be assigned to commits and it is generally used as marker for relevant events in the repository (e.g., releases, milestones).

The conceptual schema also includes some derived concepts and methods to facilitate the analysis of Git repositories by making explicit some information that is normally hidden or hard to query. This is the case for *LineDetail* and *FileRename* concepts, and the *getVersion* method of the *File* concept. *LineDetail* represents precise information regarding each line of a file modification, which includes the line number, its content and whether the line is (partially) commented or not. *FileRename* represents file rename actions occurred during the life-cycle of a software project, thus allowing tracking the whole life of a file in a repository. Finally, the *getVersion* method allows obtaining the version of a file at a particular timestamp.

Additionally, concepts in the schema can be enriched with some calculated metrics, expressed as derived attributes, to analyze the repository. As an example, we have added to the schema some derived attributes like *filesDeleted* and *emptyLines* providing some activity metrics of the repository.

3.2 A Database Schema for Git

The previous conceptual schema is materialized in the relational database schema of Fig. 2. In a nutshell, concepts/attributes in the conceptual schema are mapped into tables/columns in the database schema and associations are mapped into foreign keys (e.g., *author* association in *Commit* concept and *author_id* foreign key in *commit* table) or new tables (e.g., *commit_parent*) depending on the cardinality of the association, following the typical translation strategies. Note that

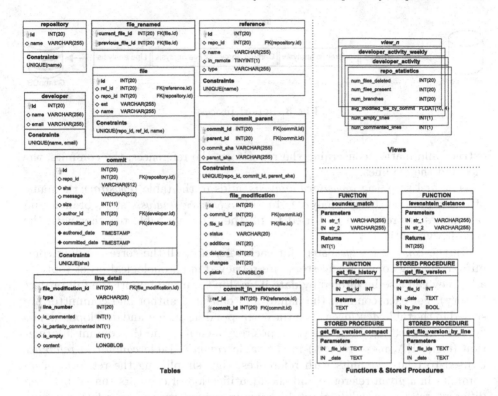

Fig. 2. Database schema for Git.

the *Branch-Tag* taxonomy has been mapped to a new attribute *type* in the table *reference*.

Additionally, several views have been created in the database to calculate the derived attributes in the conceptual schema (e.g., *repo_statistics* or *developer_activity* views). Auxiliary methods have been implemented as either functions or store procedures (see *get_file_version*). Full description of these views and methods can be found in the GitHub repository hosting the tool [12].

3.3   Extraction Process

We have defined an extraction process which interacts with a given Git repository in order to populate a database conforming to the schema previously described. The operations implemented by our approach support both the initial data loading from the SCM system to the database and later incrementally updating it with the latest SCM information. In the remainder of this section both operations are presented.

Initial Import Process. An overview of the process to populate the database is shown in Fig. 3. It is composed of five steps, that respectively analyze and

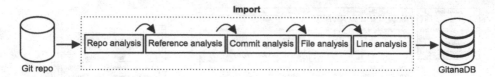

Fig. 3. Import process.

extract information concerning the repository, the references, the commits, the files and the file lines.

The first step, the *repository analysis*, adds to the table *repository* the name of the repository being analyzed. In the *reference analysis* step, branch and tag names are retrieved and used to fill the *reference* table together with the repository identifier.

In the *commit analysis* step, for each reference, all the corresponding commits, including common ancestors, are retrieved in chronological order (optionally, they can also be filtered by date if we don't want to import the full project history). For each commit, the names and emails of author and committer are stored in the table *developer*, then the corresponding author and developer identifiers are inserted together with the repository identifier and the commit information (e.g., SHA, message, etc.) in the table *commit*. Table *commit_in_reference* is used to relate commits with references, thus simplifying the retrieval of all commits in a given reference and the identification of commits shared between different references, while the table *commit_parent* is used to relate commits among each other, storing the relation between a commit and its corresponding parent(s).

In the *file analysis* step, for each commit, the differences (*patch*) between the previous and current versions of the files modified by the commit are retrieved. A patch can concern modifications (i.e., addition, deletion, changes) or renamings (i.e., changing its name or location) of a file. For each file involved in a patch, its name (including its path) and extension as well as the identifiers of the repository and reference the file belongs to are stored in the table *file*. The identifier of the file is used to relate the file with its corresponding modifications and/or to its renamings. In particular, each modification on a file is stored in the table *file_modification*. It contains the identifiers of the related file and commit, the current status of the file at the time of that modification, the content of the patch and the number of additions, deletions and changes. On the other hand, if a file has been renamed, the previous and current identifiers of the files are stored in the table *file_renamed*.

Table *line_detail* is populated during the *line analysis* step. Such a step is in charge of analyzing and extracting the information concerning the individual file lines by using regular expressions on top of the patch information. In particular, the line number, the content of the line and the type of modification (i.e., addition, deletion) are retrieved together with information concerning whether the line is empty, commented or partially commented. The identification of

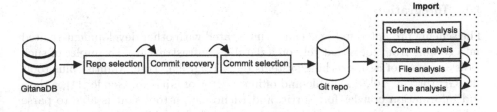

Fig. 4. Update process.

comments is currently able to deal with line and block comments for the following languages: Python, Java, HTML, XML, SQL, JavaScript, C, C++, Scala, PHP, Ruby and Matlab.

Incremental Update Process. The update process, shown in Fig. 4, keeps the information in the SCM system aligned with the one contained in the database. It is composed of three steps that connect the database with the Git repository plus a final *extraction* process to integrate the data not yet in the database.

The first step, *repository selection*, consists of selecting the name of a repository to update and retrieving the corresponding identifier stored in the table *repository*. In the *commit recovery* step, the repository identifier is used to collect the last SHA commit for each reference stored in the database by joining the information contained in the tables *repository, commit, commit_in_reference* and *reference*. Once the pairs {reference name, last SHA commit} are obtained, the *commit selection* step uses them to gather from the Git repository the set of new commits to be added to the database. Optionally, the retrieved commits can be filtered by retaining only those ones created before a certain date. In addition, the Git repository is queried also to collect those references that have no correspondence in the database.

Finally, some steps of the import process, previously presented, are reused to persist the new elements in the database, in particular, the *reference analysis* is started for each new reference found in the Git repository but not in the database, optionally together with a before date; while the *commit analysis* is launched for each set of new commits per reference.

Currently the materialized views in the database are recalculated each time the update process is triggered. Although this solution works properly with small and medium sized repositories, it may be inefficient for large repositories. In this sense, previous research efforts on incremental maintenance of materialized views (e.g., [23–25]) can be used to improve the efficiency of the update process.

4   Application Scenarios

In this section we present some application scenarios for the integration and advanced query functionalities provided by Gitana.

4.1   Integration

The schema presented in Fig. 2 can be integrated with other development-related data coming from tools that rely on a database infrastructure. Examples of such tools are most of issue tracking systems (e.g., BugZilla, Trac, Mantis) plus tools like GHTorrent [10], a scalable and offline mirror of GitHub; Gerrie [11], a data and information crawler for Gerrit, and Bicho [26], a tool that is able to parse different issue tracking systems (e.g., Launchpad, Jira, Allura).

Any of such tools embeds at least one concept that deals with *Commits*, *Files* or *Developers*. By leveraging on such concepts we can connect their database schemas with our Git schema. While the integration for files and commits is straightforward, since it involves a perfect match between file names and SHA identifiers; the integration for developers can only be semi-automatic and require the use of well-known identity matching/entity resolution algorithms (e.g., [27, 28]), as the developer can use different credentials (e.g., login or email) in the different tools she participates to.

As example, we illustrate how our approach can be integrated with GHTorrent and Gerrie. Figure 5 shows an excerpt of the main tables involved in the integration[1]. As can be seen, the concepts of *Developer*, *File* and *Commit* defined in our Git schema are similarly used in GHTorrent (tables *Users* and *Commits*) and Gerrie (tables *Gerrie_file*, *Gerrie_person*).

Integrating with GHTorrent provides a broader view of a GitHub project by combining the collaboration data and the Git SCM data, which GHTorrent does not cover at the moment. Thus, it opens up the possibility of writing queries that rely on both collaboration and code information (e.g., the most conflictive file lines according to the number of pull requests modifying them). Gerrie integration allows extending the analysis of code-reviews with fine-grained information from the Git SCM data (e.g., most influential developers in terms of changes in the files for a particular branch). Note that it is also possible to integrate the three tools, where Gitana may act as pivot representation. This scenario would provide a general view of the collaboration in the development process, for instance, to analyze the most conflictive pull requests (information provided by GHTorrent) in terms of code reviews (information provided by Gerrie).

In these examples, mappings are mostly one-to-one but it could happen that concepts are expressed with a different structure in the tools to be combined. In that case, previous findings on structural conflict resolution in the integration of Entity-Relationship schemas (i.e., [29–31]) should be used to deal with such schematic discrepancies.

4.2   Advanced Query Functionalities

Given the database representation of the SCM system, we can leverage on the plethora of tools and techniques existing in the database realm to perform

---

[1] GHTorrent and Gerrie schemas are available at: http://ghtorrent.org/files/schema.png and http://gerrie.readthedocs.org/en/latest/database/#schema, respectively.

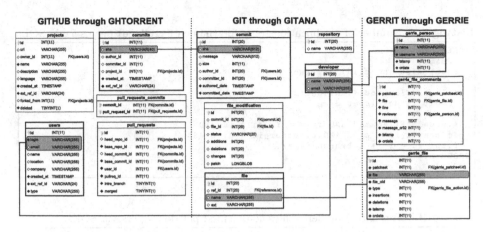

Fig. 5. GHTorrent and Gitana database schema excerpts and the main joining columns.

Listing 1.3. Number of deleted files via SQL.

```
SELECT
    COUNT(DISTINCT fm.file_id)
FROM
    file_modification fm
WHERE
    status ='deleted';
```

advanced queries on the data. In this section we illustrate the advantages of this approach when compared to the traditional command line support.

For this, we show how we can easily rewrite the examples from Sect. 2 in SQL. Listings 1.3 and 1.4 show the SQL queries required to calculate the number of deleted files and the main modifications (i.e., additions, deletions and total) made by each developer in the repository, respectively (as done in Listings 1.1 and 1.2). As can be seen, instead of using a combination of command line and shell commands, with Gitana, pure SQL syntax suffices to get the information.

Additionally, our approach also includes some calculated information not directly available when using the Git command line. This can help developers to uncover valuable information from the repository. For instance, the derived concept *LineDetail* and its corresponding database table *line_detail* may help to discover who are the developers that comment the most the source code. Listing 1.5 shows a SQL query which calculates the number of files including comments per developer. In order to do so without Gitana, developers would need to come up themselves with the regular expressions to analyze the code.

Beyond pure SQL queries, we can also apply on our Git schema any existing ETL (Extract, Transform and Load) and OLAP (On-line Analytical Processing) technologies to perform multidimensional analysis for Git. Such a multidimensional analysis model can be personalized [32] to provide users with appropriate structures allowing them to intuitively analyze and understand the

Listing 1.4. Number of modifications on a file (named `FileName`) per user via SQL.

```
SELECT
  d.name,
  count(distinct c.id) AS changes
FROM
  file f, file_modification fm,
  commit c, developer d
WHERE
  f.name ="FileName" AND f.id = fm.file_id AND
  fm.commit_id = c.id AND c.author_id = d.id AND
  f.ref_id = 1 /* a branch id */
GROUP BY
  d.id;
```

Listing 1.5. Number of files commented per developers via SQL.

```
SELECT
  d.name AS developer,
  count(distinct(fm.file_id)) as num_files
FROM
  line_detail ld, file_modification fm,
  commit c, developer d
WHERE
  ld.file_modification_id = fm.id AND
  fm.commit_id = c.id AND c.author_id = d.id AND
  (ld.is_commented = 1 OR
   ld.is_partially_commented = 1)
GROUP BY
  d.name;
```

SCM information by reporting on current data, looking at historical data and trying to make predictions about future trends. Examples of possible analysis could be the evolution of the number of commits and files per unit of time (e.g., week, month, quarter) as well as the contribution activity of the developers in the project over time.

5   Tool Support and Evaluation

Our method has been implemented in a tool called Gitana, available at [12]. Gitana relies on different technologies. The database import process has been implemented in Python 2.7.6 and rely on version 0.3.1 of GitPython[2], a library to interact with Git repositories. The generated database is by default stored in a MySQL server. The tool is launched via a simple GUI interface shown in Fig. 6.

A JSON export component is also available. The JSON export process is composed of four steps: (1) repository selection; (2) file selection, which collects/ignores the files in the repository fulfilling some conditions (e.g., file extensions, inclusion in specific directories); (3) developer aggregation, which allows

---

[2] https://pypi.python.org/pypi/GitPython.

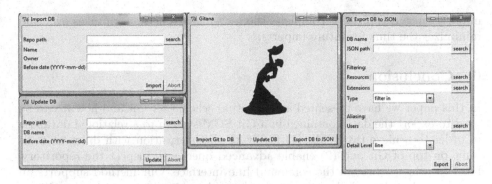

Fig. 6. Gitana GUI interface.

Table 2. Evaluation on 5 open-source projects.

owner-repo	branches	tags	# commits in refs	extraction time
atlanmod-EMFtoCSP	1	-	67	5m37s
atlanmod-gila	2	-	289	8m47s
atlanmod-collaboro	3	3	1083	2h2m33s
octopress-octopress	1	38	5215	5h25m8s
reddit-reddit	2	-	10680	9h39m29s

to merge together different user names corresponding to the same physical person or developers part of the same development sub-team and (4) file export, which collects the information for each file and generates the JSON data. The resulting JSON document follows a tree-based representation of the database where each entry is composed of the overall information (e.g., extension, current status and last modification) of a file, the commits that modified it, the list of changes at file level with the corresponding patches and the list of changes at line level. We believe the JSON support can facilitate the analysis of Git repositories by means of other technologies. For instance we described in [33] a website generator, relying on Gitana, to display the *bus factor*[3] of Git repositories.

Our extraction process has been evaluated on five open-source projects available on GitHub. Beyond validating that the extracted information was correct (either manually or, for some repositories, by discussing it with the project owners), the goal was to check the efficiency of the process. The results of the evaluation are shown in Table 2. We ran the evaluation on a 2.6 GHz Intel Core i7 processor with 8 GB of RAM. Note that even if the database import process can take some time (e.g., 2 hours for 1,000 commits), this only refers to the initial

---

[3] The bus factor of a project is typically defined as the number of key developers who would need to be incapacitated, i.e., hit by a bus, to make the project unable to continue.

import. Once this phase is complete, the incremental mechanism takes over and minimizes the time for future imports.

6   Conclusion

In this paper we have presented a conceptual schema for Git and how it can be used to export the data contained in a Git SCM system to a relational database in order to achieve two goals: (1) facilitate the integration with different tools built on top of Git and (2) enable advanced queries to inspect the repository, hard to achieve through the command line interface. Our method supports an incremental update of the Git data and has been implemented in the Gitana tool, freely available on GitHub at [12].

As further work, we would like to have a deeper integration of all kinds of SCM tools advancing in our idea of having one single central (database-oriented) shared access point for all the project information, enabling lots of interesting cross-cutting queries. Moreover, at the tool level, we would like to speed up the initial import phase by parallelizing the analysis of branches and tags in the repository. We are also interested in making more tunable the output of the JSON exporter, as currently the user is bound to the JSON predefined output structure.

References

1. Cockburn, A., Highsmith, J.: Agile software development: the people factor. Computer **34**(11), 131–133 (2001)
2. Rochkind, M.J.: The source code control system. Trans. Softw. Eng. **4**, 364–370 (1975)
3. O'Sullivan, B.: Making sense of revision-control systems. Commun. ACM **52**(9), 56–62 (2009)
4. Serrano, N., Ciordia, I.: Bugzilla, itracker, and other bug trackers. Software **22**(2), 11–13 (2005)
5. Kemerer, C.F., Paulk, M.C.: The impact of design and code reviews on software quality: an empirical study based on PSP data. Trans. Softw. Eng. **35**(4), 534–550 (2009)
6. Chacon, S., Hamano, J.C.: Pro Git, vol. 288. Apress, Berkeley (2009)
7. Bird, C., Rigby, P.C., Barr, E.T., Hamilton, D.J., German, D.M., Devanbu, P.: The promises and perils of mining Git. In: MSR, pp. 1–10 (2009)
8. Gitstats (2007). http://gitstats.sourceforge.net/
9. GitInspector (2012). https://code.google.com/p/gitinspector/
10. Gousios, G., Spinellis, D.: GHTorrent: Github's data from a firehose. In: MSR, pp. 12–21 (2012)
11. Gerrie (2013). http://gerrie.readthedocs.org/en/latest/index.html
12. Gitana website. https://github.com/SOM-Research/Gitana
13. Fischer, M., Pinzger, M., Gall, H.: Populating a release history database from version control and bug tracking systems. In: ICSM, pp. 23–32 (2003)
14. Zimmermann, T., Weißgerber, P.: Preprocessing CVS data for fine-grained analysis. In: MSR, pp. 2–6 (2004)

15. Robles, G., Koch, S., GonZÁlez-Barahona, J.M., Carlos, J.: Remote Analysis and measurement of libre software systems by means of the CVSAnalY tool. In: RAMSS, pp. 51–55 (2004)

16. Draheim, D., Pekacki, L.: Process-centric analytical processing of version control data. In: IWPSE, pp. 131–136 (2003)

17. Robles, G., González-Barahona, J.M., Ghosh, R.A.: Gluetheos: automating the retrieval and analysis of data from publicly available software repositories. In: MSR, pp. 28–31(2004)

18. Antoniol, G., Di Penta, M., Gall, H., Pinzger, M.: Towards the integration of versioning systems, bug reports and source code meta-models. Electron. Notes Theor. Comput. Sci. **127**(3), 87–99 (2005)

19. Stephany, F., Mens, T., Gîrba, T.: Maispion: a tool for analysing and visualising open source software developer communities. In: ST, pp. 50–57 (2009)

20. Lee, H., Seo, B.K., Seo, E.: A Git source repository analysis tool based on a novel branch-oriented approach. In: ICISA, pp. 1–4 (2013)

21. Dyer, R., Nguyen, H.A., Rajan, H., Nguyen, T.N.: Boa: a language and infrastructure for analyzing ultra-large-scale software repositories. In: ICSE, pp. 422–431 (2013)

22. Williams, J.R., Di Ruscio, D., Matragkas, N., Di Rocco, J., Kolovos, D.S.: Models of OSS project meta-information: a dataset of three forges. In: MSR, pp. 408–411 (2014)

23. Gupta, A., Mumick, I.S., et al.: Maintenance of materialized views: problems, techniques, and applications. IEEE Data Eng. Bull. **18**(2), 3–18 (1995)

24. Staudt, M., Jarke, M.: Incremental maintenance of externally materialized views. VLDB **96**, 3–6 (1996)

25. Ross, K.A., Srivastava, D., Sudarshan, S.: Materialized view maintenance and integrity constraint checking: trading space for time. SIGMOD Rec. **25**, 447–458 (1996)

26. Gonzalez-Barahona, J.M., Izquierdo-Cortazar, D., Robles, G., del Castillo, A.: Analyzing gerrit code review parameters with bicho. ECEASST, vol. 65 (2014)

27. Christen, P.: A comparison of personal name matching: techniques and practical issues. In: ICDM, pp. 290–294 (2006)

28. Goeminne, M., Mens, T.: A comparison of identity merge algorithms for software repositories. Sci. Comput. Program. **78**(8), 971–986 (2013)

29. Lee, M.L., Ling, T.W.: A methodology for structural conflict resolution in the integration of entity-relationship schemas. Knowl. Inf. Syst. **5**(2), 225–247 (2003)

30. Chai, X., Sayyadian, M., Doan, A., Rosenthal, A., Seligman, L.: Analyzing and revising data integration schemas to improve their matchability. VLDB **1**(1), 773–784 (2008)

31. Haas, L.M., Hentschel, M., Kossmann, D., Miller, R.J.: Schema AND data: a holistic approach to mapping, resolution and fusion in information integration. In: Laender, A.H.F., Castano, S., Dayal, U., Casati, F., de Oliveira, J.P.M. (eds.) ER 2009. LNCS, vol. 5829, pp. 27–40. Springer, Heidelberg (2009)

32. Garrigós, I., Pardillo, J., Mazón, J.-N., Trujillo, J.: A conceptual modeling approach for OLAP personalization. In: Laender, A.H.F., Castano, S., Dayal, U., Casati, F., de Oliveira, J.P.M. (eds.) ER 2009. LNCS, vol. 5829, pp. 401–414. Springer, Heidelberg (2009)

33. Cosentino, V., Cánovas Izquierdo, J.L., Cabot, J.: Assessing the bus factor of Git repositories. In: SANER, pp. 499–503 (2015)

Near Real-Time Collaborative Conceptual Modeling on the Web

Michael Derntl[1]([⊠]), Petru Nicolaescu[1], Stephan Erdtmann[1], Ralf Klamma[1], and Matthias Jarke[1,2]

[1] RWTH Aachen University, Lehrstuhl Informatik 5, 52056 Aachen, Germany
{derntl,nicolaescu,erdtmann,klamma,jarke}@dbis.rwth-aachen.de
[2] Fraunhofer FIT, Birlinghoven Castle, 53754 Sankt Augustin, Germany

Abstract. Collaboration during the creation of conceptual models is an integral pillar of design processes in many disciplines. Synchronous collaboration, in particular, has received little attention in the conceptual modeling literature so far. There are many modeling and meta-modeling tools available, however most of these do not support synchronous collaboration, are offered under restrictive licenses, or build on proprietary libraries and technologies. To close this gap, this paper introduces the lightweight meta-modeling framework SyncMeta, which supports near real-time collaborative modeling, meta-modeling and generation of model editors in the Web browser. It employs well-proven Operational Transformation algorithms in a peer-to-peer architecture to resolve conflicts occurring during concurrent user edits. SyncMeta was successfully used to create meta-models of various conceptual modeling languages. An end-user evaluation showed that the editing tools of SyncMeta are considered usable and useful by collaborative modelers.

1 Introduction

In many disciplines modeling is a key activity in defining high-level representations of the target domain [1]. Models obtain meaning through creative social processes, which are driven by the different perspectives of involved stakeholders [2]. Nowadays—as a result of the increased connectedness and the rising need for collaboration in distributed teams—people working together on a software product require technology support for collaboration [3]. Collaborative modeling tools that enable synchronization of concurrent, remote changes are useful to allow distributed teams to perform conceptual modeling tasks independent of their physical location. The idea of synchronous collaboration on documents has been well known since more than half a century ago, when Doug Engelbart presented such features during the "mother of all demos" [4]. However, only recent advances in Web-based information and communication technology—in particular the proliferation of the Web with cheap, ubiquitous, fast, and low-latency broadband offerings—has made the realization of these dreams possible at scale.

There are many Web applications that support near real-time (NRT) collaboration, e.g., in *Google Drive* or *Office 365*. Some of these also offer conceptual

P. Johannesson et al. (Eds.): ER 2015, LNCS 9381, pp. 344–357, 2015.
DOI: 10.1007/978-3-319-25264-3_25

modeling support, e.g., *draw.io*. There are many programming libraries available that allow developers to embed synchronous editing into (Web) applications, the most prominent being the Google Drive Realtime API [5]. However, the main drawback of these offerings is that they are currently only available as black-boxed, proprietary solutions, which offer no means to control or adapt the back-end. This produces lock-in situations that help the vendors, but this inevitably leads to issues and migration effort when APIs are being deprecated, server endpoints relocated, protocols updated, or services simply discontinued.

We therefore advocate solutions that involve open-source software that can be adapted and hosted in-house if needed. In particular, we aim for browser-based conceptual model editors that allow collaboration with unrestricted, concurrent editing, near real-time synchronization of remote edits, awareness of peer actions, and the capability of generating model editors using a collaborative, visual meta-modeling approach. We show in the related work section that there is currently no toolkit that adequately covers this feature spectrum. We therefore propose a framework called SyncMeta, which closes this gap in a lightweight style. In [6], we made initial efforts to outline desirable characteristics and suitable technologies for such a domain-independent meta-modeling framework after experimenting with domain-specific prototypes. SyncMeta goes beyond that by providing means for the collaborative creation of meta-models and the generation of collaborative model editors from the defined meta-models. It also resolves the vendor lock-in situation described above by being released as open source code[1] under a permissive license and being fully based on open source libraries and widely implemented Web protocols. The framework was implemented as a widget-based Web application. This is similar in user experience to advanced desktop based modeling tools and also allows distributing user interface components among screens and even devices [7].

So far, the conceptual modeling literature has emphasized little on the opportunities and challenges of real-time collaboration in the conceptual modeling process. We aim to draw attention to this aspect both from a theoretical and practical perspective. For conceptual modeling practice the framework offers many advantages. As a widget-based app it can be integrated in Web IDEs, which are becoming ever more popular among developers, to support collaborative model-driven development on the Web. Also, as a browser-based collaboration tool it facilitates the communication between modelers and other stakeholders like end-users by removing barriers that typically come with desktop software, like installation and maintenance. It also opens up opportunities for new groups of modelers, like teachers. Synchronous modeling features can be used, for instance, in online teaching (MOOCs, webinars, etc.) for training conceptual modelers.

This paper is structured as follows. In the next section we introduce the theoretical background on conceptual meta-modeling and near real-time collaboration. Sections 3 and 4 describe in detail the fundamental concepts and the implementation details of the SyncMeta framework, respectively. Section 5

---

[1] https://github.com/rwth-acis/syncmeta.

presents a functional and end-user evaluation of usability of SyncMeta's user interface. Section 6 identifies the advancements of the framework over the state of the art, and Sect. 7 concludes the paper and outlines future work threads.

2   Meta-Modeling and Near Real-Time Collaboration

A conceptual model builds an abstraction of the real world, consisting of *objects* and the *relationships* between these objects [8]. Conceptual models are created using a particular *modeling language*, which offers *syntax* and *semantics* definitions. The former constitutes the elements of the language and their syntactical notation, whereas the meaning of these elements is described by the latter. Syntax comprises rules regarding the notation of the elements and relationships in textual or visual form. The modeling language can be defined using another model, the so-called meta-model [9]. In this paper, we focus on graph-based modeling languages, that is, languages whose elements are represented either as nodes or edges. There are two different user interaction paradigms for graph-based diagrams, namely *free-hand editing* and *structured editing* [10]. The former allows the modeler to create and combine model elements without any restrictions, potentially leading to syntactically incorrect models. This is avoided in structured editing by offering only those operations that lead from one valid state to another. Obviously, while free-hand editing provides freedom, structured editing scaffolds the creation of valid models.

A real-time collaborative editing system (RTCES) [11] is an application that enables a group of individuals to work on the same document at the same time independent of their geographical location using the Internet as the communication medium. Building RTCES that "feel" like single-user applications is a major challenge during development, in particular the maintenance of consistency among connected sites [12]. To make the remote users aware of the local changes, the operations are propagated and applied by the remote user agents to their local document replica. During this process the consistency of the local document states can be violated by conflicting user actions.

There are two different approaches to deal with this problem: conflict prevention [13] and conflict resolution [14]. The former approach restricts the user to manipulate only those parts of the document that are not manipulated concurrently by another user. This is the approach implemented, for instance, in MetaEdit+ [15]. The conflict resolution approach, on the other hand, is unrestricted by permitting any action on any portion of the shared document. Here we consider only this kind of unrestricted interaction as NRT collaboration. Such approaches require mechanisms for detecting and resolving conflicts. The most prominent among those is Operational Transformation (OT) [14], which adjusts the parameters of an operation dependent on the conflicting operations' parameters. OT works well with linearly structured objects like character strings. For more complex structures dedicated conflict resolution mechanisms have been developed, such as the one in [16], which introduces a graphical OT algorithm.

To improve the perceived responsiveness of a collaborative editor, changes are immediately applied to the local document state. Concurrently, any change is propagated to the connected sites. This can be done using a centralized or decentralized OT architecture [17]. The former offers a client/server style architecture, where all connected user agents operate through the central server component, which stores the shared document state and handles all conflicts. The obvious disadvantage is that the server constitutes a single-point-of-failure and may become a bottleneck for computation time and network delay. In contrast, peer-to-peer architectures [18] delegate the conflict resolution to the clients.

3 SyncMeta Framework Fundamentals

3.1 Basic Concepts and Roles

An illustration of the concepts and roles in the SyncMeta framework is given in Fig. 1. On the meta-modeling layer the meta-modeler uses the *Meta-Model Editor* to cooperate with remote meta-modelers during the creation of a meta-model, specifically the visual language specification (VLS) which builds the basis for an editor for the specified modeling language. VLS is defined visually using a simple graph-based visual modeling language (VML). At any time an instance of the *Model Editor* can be generated based on the defined VLS. The *Model Editor* instance on the modeling layer allows the creation of a model by collaborating modelers according to the modeling language described by the VLS. The modeling steps performed on the meta-modeling and the modeling layer are similar in nature. On both layers a number of users may collaborate synchronously during the creation of a model based on an underlying VML. On the meta-modeling layer the group of users defines the VLS as a model of a particular VML; on the modeling layer they define a model based on the VML determined by the VLS. The only functional difference between the two layers is the step involving the generation of the *Model Editors* via the meta-modeling layer.

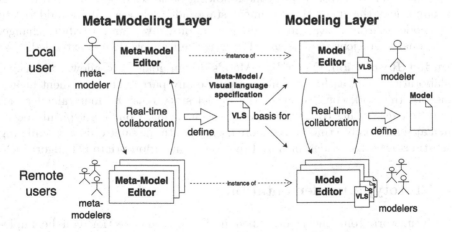

Fig. 1. Concepts and roles in the meta-modeling process.

3.2   Core Components

The main components of the SyncMeta framework are outlined in the following, focusing on components that are necessary for collaborative (meta-) modeling on the Web. Common system components like data storage or user management were omitted here. The prototypical implementation is described in Sect. 4.

Meta-Modeling Hierarchy. The meta-modeling hierarchy defines the abstractions needed to allow meta-modelers define their own graph-based modeling languages. This requires at least three hierarchical tiers: abstracting from the model, the meta-model defines the available model elements, and the meta-meta model (and any meta-models thereof) define the constructs needed and implemented for the graph-based modeling approach.

Model Editor. The model editor is composed of several user interface widgets (see Fig. 2 for a screenshot of the prototype). The *Canvas* widget shows a visual representation of the current state of the model, including mechanisms to manipulate the model (e.g., moving a node, connecting two nodes). The *Palette* widget offers the node and edge types defined in the meta-model. The *Property Editor* widget allows editing of the properties of the elements that are currently selected in the canvas. To facilitate collaborative modeling, an *Awareness* widget is included, which exposes a log of recent modeling actions performed by all collaborators. Finally, an *Export* widget allows exporting the current model in different representations. Additional widgets can be added if needed.

Messaging. Following the OT approach, each action that changes the document state is expressed as an operation. An operation has a particular type and a number of parameters and additional information. SyncMeta requires a messaging infrastructure to exchange state-changing operations among connected user agents. For example, the addition of a node to the canvas would result in an 'add node' operation being sent along with type, position and dimensions of the node as parameters. SyncMeta defines the following basic set of graph-based modeling operations that alter the document state: add/delete attribute, add/delete edge/node to/from canvas, move/resize node on canvas, and attribute change. Additional operations may be added in case the framework gets extended.

Conflict Resolution. One of SyncMeta's design goals is to allow non-locking collaboration—i.e., each user can manipulate any part of the document at any time. To this end, consistency of the model state must be maintained at all connected sites using conflict resolution mechanisms. Conflicts typically occur when at some point in time two or more users update the same object. Conflicting operations need to be identified and resolved using appropriate OT algorithms.

4   Prototype Implementation

The framework components described in the previous section can be implemented using different concrete technologies either from scratch or embedded into existing platforms or applications. Our main non-functional goal was to

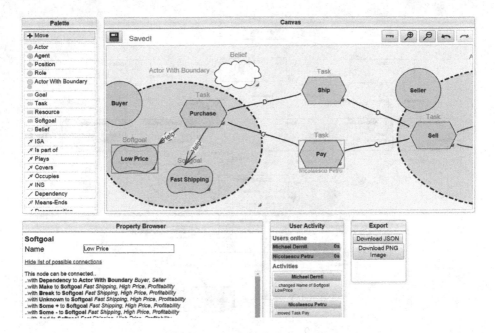

Fig. 2. Screenshot of i^* editor instance

provide an application based on open source libraries that runs without any installation in the Web browser. This section describes a prototypical implementation of the domain-independent meta-modeling framework achieving that.

Software Platform. The user interface was developed using HTML5, CSS3 and JavaScript. As a software platform for the SyncMeta implementation we chose to use the Responsive Open Learning Environments (ROLE) interoperability framework [19], which is implemented in the open source ROLE SDK. This SDK simplifies the development and deployment of Web widget based applications. Despite its name suggesting learning environments as the target applications, the ROLE SDK can actually be used for developing any widget-based application. OpenSocial widgets are embedded in containers called "spaces". Essentially, each space is a widget container that has a unique URL and can be joined by multiple users for collaboration. In SyncMeta a space represents one conceptual model (see Fig. 2 for an example of a space containing an i^* model). Each space offers out-of-the-box features for user management (authentication, authorization, account management) and persistence (data storage API), which greatly simplifies the proof-of-concept prototype development. Additionally this SDK comes with out-of-the-box features for distributing the user interface across various devices [7], which can be useful in many scenarios like demonstrations.

Meta-Modeling Hierarchy. We implemented a basic four-tier meta-modeling hierarchy in the prototype, depicted in Fig. 3. The hierarchy borrows basics from hierarchies of related frameworks like ADOxx [20] and AToM[3] [21] and the one defined by the Meta Object Facility (MOF) [9]. In the hierarchy each

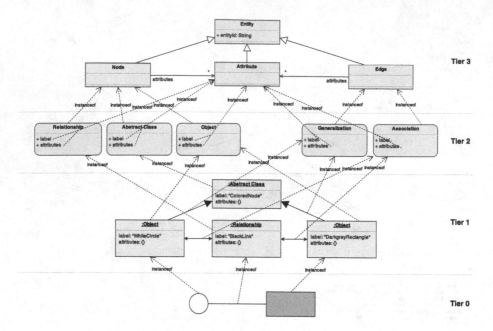

Fig. 3. Simplified meta-modeling hierachy.

tier describes the components of its subordinate tier, i.e., the elements of the subordinate tier are instances of elements contained in the superordinate tier. Concepts on tiers 2 and 3 are implemented in the framework, while models on tier 0 and 1 are user-defined.

Tier 3 ($Meta^3$-Model) defines the basic components of graph-based modeling language in the framework consisting of nodes and edges, each with attributes.

Tier 2 ($Meta^2$-Model) contains the elements of the visual modeling language used in the meta-model editor to define the modeling language an editor should be generated for. In the user interface these elements, along with additional auxiliary ones, are offered in the meta-modeling palette.

Tier 1 (Meta-Model) represents the elements of the meta-model. In Fig. 3 we see two objects on tier 1 inheriting some attributes of a common superclass and an edge linking these two objects. Meta-models are created using the same user interface as the model editors generated from the meta-model. The only difference is the hard-coded $meta^2$-model in the meta-model editor. The visual notation for the meta-models resembles UML class diagrams. Therefore, people who are familiar with meta-modeling concepts and the UML should be fairly comfortable and quick in defining meta-models in SyncMeta. A potentially time consuming task during meta-modeling in SyncMeta is the definition of custom node shapes for elements on tier 0, since those must be defined using the SVG (Scalable Vector Graphics) language.

Tier 0 (Model) hosts the actual model created using a generated model editor instance. In Fig. 3 the model on tier 0 simply consists of two nodes and one edge.

Messaging. The ROLE SDK provides inter-widget communication (IWC) [19] functionality via JavaScript libraries, allowing widgets in a space in the local browser as well as widgets of remote users in the same space to exchange messages using the Extensible Messaging and Presence Protocol (XMPP) [22].

Conflict Resolution. Operational Transformation (OT) is used for conflict resolution during shared editing (see [14] and Sect. 2), in particular the OpenCoWeb JavaScript OT Engine API [23]. The consistency maintenance for NRT collaboration is based on a decentralized peer-to-peer architecture. That is, all operations a client initiates are broadcast to all other clients, which are then responsible for applying the received operations locally after resolving conflicts. The peer OT engines apply the OT algorithms when a conflict is detected and return the new operations to be applied on the local documents. By this, the engine guarantees that after applying the OT operations on all sites, the document states at all connected sites are congruent.

Feature Limitations. Among other limitations, the SyncMeta prototype currently does not support validity checks for models, since the meta-modeling formalism currently offers no way of attaching validation rules or constraints to models and elements. This could however easily be added by extending the meta-meta-model and the model editor implementations, respectively. Due to the widget based design of SyncMeta, the model validation functionality could alternatively also be implemented as a standalone widget that can be selectively added to the modeling space by the user, if needed.

5   Evaluation

Meta-Modeling. To test the functionality of the SyncMeta meta-modeling hierarchy implementation we successfully built meta-models and model editors of diverse conceptual modeling languages that can be visually expressed with a graph-based notation, including Petri nets, UML use case models, ER models, and i^* models [24]. Some of these require non-trivial node shapes and visual layers for aggregate nodes. We also built a fully functional authoring tool for the IMS Learning Design (LD) specification [25] with SyncMeta. This specification does not define a graphical notation, but dozens of concepts and relationships to formally model the design of units of learning [26]. The SyncMeta-based tool is the first real-time collaborative IMS LD authoring application.

End-User Evaluation. To evaluate the usability and usefulness of the NRT model editor with conceptual modelers we performed an end-user study with a model editor generated from the i^* meta-model. i^* models are a well proven means of representing requirements in early system engineering phases [24].

Randomly assigned pairs of modelers with varying degrees of knowledge about i^* modeling were tasked to collaboratively produce an i^* model on seller/buyer relationships in markets (depicted in Fig. 2). In each evaluation session a pair of modelers was placed in a room, each with one computer. The placement of the screens ensured that modelers were not able to observe the screen of the other modeler, but the participants were able and allowed to talk to each other. This setup tries to mimic a common type of virtual collaboration, where collaborators use an audio channel (e.g., Skype). In our view, this setting appeared more realistic than a setting where modelers collaborate without any additional communication, particularly when creating a new model from scratch. Both collaborators were provided with a print-out of the target i^* model to be produced. The participants were asked to create the model from scratch using the editor instances on their computers. When the participants had finished the task, each of them was asked to export the created model, which allowed us to check the local document states on both computers for congruence.

The modelers were also asked to complete an online survey. The survey included 20 questions on usefulness and usability of the editor to be rated on a seven-point Likert scale with options ranging from very bad (1) to very good (7). One question asked for free text feedback. Five additional questions aimed at collecting some background information from participants. These questions asked whether the participant had used any real-time collaborative tool or a graphical editor before. The participants' expertise levels with graphical modeling, conceptual modeling in general, and the i^* modeling language were also measured with seven-point Likert scales.

Participants. The participants were recruited from computer science students and researchers. None of them were involved in the development of SyncMeta. Ten group sessions were conducted with 20 participants in total. Nearly all participants (19 of 20) have used a real-time collaborative tool before, and half of them could rely on experiences with real-time collaborative graphical editors. The participants are using graphical model editors quite frequently ($M = 5.25; SD = 1.68$) and the expertise with conceptual modeling turned out to be moderate ($M = 4.75; SD = 1.55$). The average expertise with i^* was rather mediocre on average ($M = 3.7; SD = 2.36$).

Results. All created pairs of diagrams were fully congruent, indicating no unresolved editing conflicts. Averaging the rating for all usability questions, the participants have evaluated the editor positively ($M = 5.69; SD = 1.17$). The individual ratings per question can be found in Fig. 4. Not surprisingly the question regarding the simplicity of exporting the model achieved the highest rating ($M = 6.4; SD = 1.00$) since one click achieves that. The simplicity of recovering from errors the user made received the lowest average rating of all questions, while still representing a favorable absolute score at $M = 4.95$ ($SD = 1.57$).

Generally the participants seemed to like the user interface ($M = 5.85; SD = .99$) and the usability ($M = 5.7; SD = 1.17$) of the editor. In addition the editor seems to provide the functionality typically expected from a graphical editor ($M = 5.85; SD = 1.31$). The awareness of the actions of the collaborator provided by the activity pane and the highlighting of the node currently modified by the remote user (see Fig. 2, node "Pay" for an example) were also appreciated by the participants. The highlighting in the canvas by displaying the collaborator's name next to the object he or she is currently editing was rated slightly better ($M = 5.65; SD = 1.18$) than the activity log in the awareness widget

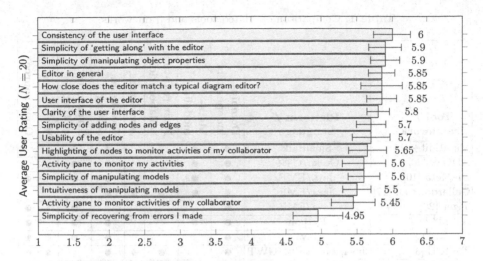

Fig. 4. Survey results of i^* group sessions (quantitative items).

($M = 5.45; SD = 1.36$). This tendency towards the feedback directly on the canvas might be caused by the fact that the user is interrupted in the workflow by switching the focus from the activity pane to the canvas. This might indicate that the users prefer modeling experiences where they can focus on the canvas without any distractions.

To test whether the expertise of the participants had any influence on their ratings of the editor usability a Pearson correlation analysis was performed. No significant correlation could be found at a confidence level of 95 %, indicating expertise does not seem to influence the appreciation of certain modeling features.

To obtain qualitative feedback, one open question asked the participants to name some positive and negative aspects of the tool. Seven participants pointed out that they like the tool in general and the user experience it provides. The intuitiveness and simplicity of the user interface were mentioned positively five times. The negative aspect emphasized most frequently ($n = 8$) was the lack of a standard drag-and-drop functionality (i.e., click–drag–release) to add nodes from the palette directly onto the canvas. In the widget-based environment this had to be implemented as a click–point–click sequence instead, which appeared awkward to some users. In addition, some minor problems of the prototype with adding nodes using drag-and-drop ($n = 4$), the selection of edges ($n = 3$), the navigation on the canvas ($n = 3$) or a delayed synchronization ($n = 2$) were mentioned by at least two participants each.

6   Related Approaches and Toolkits

As evident from Table 1 there are several existing tools that address functional and non-functional features similar to the SyncMeta framework. We limited the

Table 1. Comparison of related tools and frameworks.

Tool / Framework	Platform / Language	Domain-independent	Meta-modeling	Near real-time collaboration	Awareness	Free-hand editing	Structured editing	Web-based	Open source
MetaEdit+ [15]	Smalltalk	●	●			●	●		
ADOxx [20]	Gecko/C++	●	●			●	●		
DiaMeta [10]	Java/EMF	●	●			●	●		●
Gallardo et al. [27]	Java/EMF	●	●	●	●	●			
Tiger [28]	Java/EMF	●					●		●
AToM³ [21]	Python	●	●			●	●		●
GenGED [29]	Java	●				●	●		●
TWICE [3]	Java/JavaScript/GWT	●		●	●	●		●	●
SyncMeta	JavaScript/ROLE SDK	●	●	●	●	●		●	●

table to domain-independent tools that support at least four of the requirements reflected in the columns. Comparing the available tools, one can observe that most tools lack the NRT collaboration functionality and put more emphasis on the implementation of application specific features like code generation in software engineering or simulation of business process models. Different to SyncMeta, most surveyed tools require installation and maintenance effort.

MetaEdit+ [15] is a toolkit for the definition of a modeling language and the generation of related editors for these languages. Domain-specific languages can be defined as meta-models in a graphical manner without any programming skills. A generated editor allows code and documentation generation based on the created model or the simulation of the models behavior. It uses free-hand as well as structured editing as user interaction approach. The tool is not NRT collaborative according to our understanding, since it locks portions of the model being concurrently edited by other modelers.

Another meta-modeling platform is ADOxx [20], which allows the development of domain specific visual modeling languages and the generation of a modeling tool for them. While MetaEdit+ is focusing more on software engineering, the original background of the ADOxx platform is in business process simulation and evaluation. The platform supports free-hand and structured editing.

DiaMeta [10] is a platform-independent Java based framework for the generation of generic editors for visual languages. It uses meta-models for the specification of the diagram language, while the abstract syntax of the model is defined as UML class diagram. The framework builds on the Eclipse Modeling Framework (EMF). As a further tool built upon the EMF and its modeling architecture Ecore, Gallardo et al. [27] propose a method for the development of domain-independent collaborative modeling tools.

In contrast to the aforementioned projects the Tiger Project [28] follows a different approach in the specification of the visual language. It provides a graphical

tool to specify the language by syntax grammars. It follows the structured editing approach for user interaction, deriving further operations to be performed
from graph rules based on the syntax grammar.

The meta-modeling framework AToM^3 [21] puts the focus on the simulation
of the created model. Furthermore, it enables the modeling of different parts of a
system using different formalisms and the conversion between models of different
formalisms. It uses graph grammars for the definition of models and formalisms.

GenGED [29] allows the definition of a visual language in a graphical manner
to generate a graphical editor for that language, which features the free-hand as
well as the structured editing mode. It also is based on grammar rules for the
language specification, and it allows the simulation of the created models.

The Toolkit for Web-based Interactive Collaborative Environments (TWI-
CE) [3] is implemented with Google Web Toolkit. Similar to SyncMeta, TWICE
is committed to the use of common Web technologies, and the source code is
released under a permissive license. Differently to SyncMeta it focuses on the
collaboration of users in a co-located setting, that is, all users manipulate a
model collaboratively in a room on a shared screen while using different devices.

The comparison in Table 1 reveals that the most distinguishing characteristics of SyncMeta are the unrestricted, near real-time collaboration features, the
support of modelers through awareness of collaborators' activities, and that the
framework runs in the Web browser without any need for additional software.

7   Conclusion

In this paper, we presented the near real-time collaborative modeling framework
SyncMeta. It is a domain-independent, Web-based conceptual modeling toolkit
which allows the construction of arbitrary graph-based modeling languages and
the generation of model editors for these meta-models. While in this regard
it is similar to several existing tools, distinguishing functional characteristics
of SyncMeta are the unrestricted, near real-time collaboration support during
meta-modeling and modeling, and the support of modelers through awareness
of collaborators' activities. From a non-functional perspective the distinguishing
characteristics are that the framework is fully open source, and that it runs in
the Web browser without the need for installing additional software.

There are offerings with similar collaboration features as SyncMeta. It was
argued that the Web environments by Google, Microsoft and similar companies are offered as closed-source, cloud-hosted black boxes, with no guarantee of
sustained operation, and other lock-in measures like the restriction to vendor-
specific user accounts. SyncMeta in contrast is not only open source, it also
relies exclusively on open, widely implemented protocols and libraries and can
be hosted in-house. Its modular architecture allows replacing particular technologies with others (e.g., a different algorithm for conflict resolution, or a different
messaging protocol) without affecting the overall functionality.

The end-user evaluation has shown that SyncMeta is considered as a useful
and usable framework for conceptual modeling. Particularly through its openness

it can be embedded in a broader, collaborative systems engineering workflow by extending it, e.g., with model management features or code generation.

Further work on SyncMeta proceeds along several threads. First, we are working on the Yjs library [30], which overcomes the drawbacks of Operational Transformation, particularly when representing complex data types and scaling the processing of operations to high numbers of concurrent users and edits. This will also require further usability and functionality evaluations with larger numbers of collaborators. Second, we are working on improving the conceptual modeling capabilities by introducing view abstractions in the modeling canvas. This will allow custom views—i.e., diagram canvases with custom palettes, filters and transformations—on a model. Third, we are working on an intelligent agent that computes suggestions of modeling actions to collaborating modelers based on a formal model of the intended modeling process for a modeling language. This shall help preventing conflicts and facilitating an efficient near real-time collaboration experience, in particular with a large number of collaborators.

Acknowledgments. This research was supported by the European Commission through the METIS project (grant no. 531262-LLP-2012-ES-KA3-KA3MP).

References

1. Krogstie, J., Opdahl, A.L., Brinkkemper, S. (eds.): Conceptual Modelling in Information Systems Engineering. Springer, Heidelberg (2007)
2. Dittrich, Y., Floyd, C., Klischewski, R.: Social Thinking-Software Practice. MIT Press, Cambridge (2002)
3. Schmid, O., Lisowska Masson, A., Hirsbrunner, B.: Real-time collaboration through web applications: an introduction to the toolkit for web-based interactive collaborative environments (TWICE). Pers. Ubiquit. Comput. **18**(5), 1201–1211 (2014)
4. Levy, S.: Insanely Great: The Life and Times of Macintosh, the Computer that Changed Everything. Penguin Books, New York (1994)
5. Google: Google Drive Realtime API (2014). https://developers.google.com/drive/realtime/
6. Derntl, M., Erdtmann, S., Nicolaescu, P., Klamma, R., Jarke, M.: Echtzeitmetamodellierung im Web-Browser. In: Modellierung 2014. LNI, vol. 225, pp. 65–80 (2014)
7. Kovachev, D., Renzel, D., Nicolaescu, P., Koren, I., Klamma, R.: DireWolf framework for widget-based distributed user interfaces. J. Web Eng. **13**(3&4), 203–222 (2014)
8. Olivé, A.: Conceptual Modeling of Information Systems. Springer, Heidelberg (2007)
9. Atkinson, C., Kuhne, T.: Model-driven development: a metamodeling foundation. IEEE Softw. **20**(5), 36–41 (2003)
10. Minas, M.: Generating meta-model-based freehand editors. In: Electronic Communications of the EASST 1 (2007)
11. Ellis, C.A., Gibbs, S.J.: Concurrency control in groupware systems. ACM SIGMOD Rec. **18**(2), 399–407 (1989)
12. Sun, C., Ellis, C.: Operational transformation in real-time group editors: issues, algorithms, and achievements. In: Procceeding of the 1998 ACM Conference on Computer Supported Cooperative Work (CSCW 1998), pp. 59–68. ACM (1998)

13. Xue, L., Zhang, K., Sun, C.: Conflict control locking in distributed cooperative graphics editors. In: Proceeding of the 1st International Conference on Web Information Systems Engineering (WISE), pp. 401–408. IEEE (2000)
14. Sun, C., Jia, X., Zhang, Y., Yang, Y., Chen, D.: Achieving convergence, causality preservation, and intention preservation in real-time cooperative editing systems. ACM Trans. Comput.-Hum. Interact. **5**(1), 63–108 (1998)
15. Tolvanen, J.P., Pohjonen, R., Kelly, S.: Advanced tooling for domain-specific modeling: MetaEdit+. In Sprinkle, J., Gray, J., Rossi, M., Tolvanen, J.P. (eds.) The 7th OOPSLA Workshop on Domain-Specific Modeling (2007)
16. Fatima, Z., Agarwal, A., Gupta, G., Sharma, M.: Group editer using graphical operational transformation. In: Proceeding of the 5th National Conference on Computing For Nation Development (INDIACom 2011) (2011)
17. Greenberg, S., Marwood, D.: Real time groupware as a distributed system: concurrency control and its effect on the interface. In: Proceeding 1994 ACM Conference on Computer Supported Cooperative Work (CSCW 1994), pp. 207–217. ACM (1994)
18. Sun, C., Jia, X., Yang, Y., Zhang, Y.: REDUCE: a prototypical cooperative editing system. In: Proceeding of the 7th International Conference on Human Computer Interaction, pp. 89–92 (1997)
19. Govaerts, S., et al.: Towards responsive open learning environments: the ROLE interoperability framework. In: Kloos, C.D., Gillet, D., García, R.M.C., Wild, F., Wolpers, M. (eds.) EC-TEL 2011. LNCS, vol. 6964, pp. 125–138. Springer, Heidelberg (2011)
20. Fill, H.G., Karagiannis, D.: On the conceptualisation of modelling methods using the ADOxx meta modelling platform. Enterp. Modell. Inf.n Syst. Architect. **8**(1), 4–25 (2013)
21. De Lara, J., Vangheluwe, H., Alfonseca, M.: Meta-modelling and graph grammars for multi-paradigm modelling in AToM³. Softw. Syst. Model. **3**(3), 194–209 (2004)
22. Saint-Andre, P.: Extensible Messaging and Presence Protocol (XMPP): Core (2011). http://tools.ietf.org/html/rfc6120
23. OpenCoWeb: Open Cooperative Web Framework 1.0. http://opencoweb.org/ ocwdocs, code at https://github.com/opencoweb/coweb-jsoe
24. Yu, E.: Towards modelling and reasoning support for early-phase requirements engineering. In: Proceeding of the 3rd IEEE International Symposium on Requirements Engineering (RE 1997), pp. 226–235. IEEE (1997)
25. IMS Global: Learning Design Specification 1.0 (2003). http://www.imsglobal.org/ learningdesign/
26. Koper, R., Olivier, B.: Representing the learning design of units of learning. Educ. Technol. Soc. **7**(3), 97–111 (2004)
27. Gallardo, J., Bravo, C., Redondo, M.A.: A model-driven development method for collaborative modeling tools. J. Netw. Comput. Appl. **35**(3), 1086–1105 (2012)
28. Ehrig, K., Ermel, C., Hänsgen, S., Taentzer, G.: Generation of visual editors as eclipse plug-ins. In: Proceedings of the 20th IEEE/ACM International Conference on Automated Software Engineering, pp. 134–143. ACM (2005)
29. Bardohl, R., Ermel, C., Weinhold, I.: GenGED – a visual definition tool for visual modeling environments. In: Pfaltz, J.L., Nagl, M., Böhlen, B. (eds.) AGTIVE 2003. LNCS, vol. 3062, pp. 413–419. Springer, Heidelberg (2004)
30. Nicolaescu, P., Jahns, K., Derntl, M., Klamma, R.: Yjs: a framework for near realtime P2P shared editing on arbitrary data types. In: Cimiano, P., Frasincar, F., Houben, G.-J., Schwabe, D. (eds.) ICWE 2015. LNCS, vol. 9114, pp. 675–678. Springer, Heidelberg (2015)

Dynamic Capabilities for Sustainable Enterprise IT – A Modeling Framework

Mohammad Hossein Danesh[1(✉)], Pericles Loucopoulos[2], and Eric Yu[1,3]

[1] Department of Computer Science, University of Toronto, Toronto, Canada
{danesh, eric}@cs.toronto.edu
[2] Manchester Business School, The University of Manchester, Manchester, UK
pericles.loucopoulos@mbs.ac.uk
[3] Faculty of Information, University of Toronto, Toronto, Canada

Abstract. A key consideration of researchers and practitioners alike in the field of information systems engineering is the co-development of information systems and business structures and processes that are in alignment, that this alignment reflects the challenges presented by the business ecologies and that the developed systems are sustainable through appropriate responses to pressures for their evolution. These challenges inevitably need to be addressed through development schemes that recognize the intertwining of information systems, business strategy and their ecosystems. The paper presents the conceptual modeling foundations of such a scheme providing a detailed exposition of the issues and solutions for sustainable systems in which *Capability* plays an integrative role using examples from an industrial-size application. The contribution of the paper is on its proposition of conceptual modeling techniques that are applicable to both business strategies and information systems development.

Keywords: Capability modeling · Enterprise engineering · Sustainability

1 Introduction

As enterprises compete in fast-paced changing ecosystems they need to constantly adapt their service/product offerings to gain and sustain competitiveness. There is an intrinsic relationship between an enterprise, its ecosystems, and its information technology (IT) systems to the extent that changes in one affect the others. This intertwining [1] is central to achieving sustainability for IT systems [2, 3]. Sustainability has been a major concern in strategic management [4] and more recently in Information Systems Engineering (ISE) due to adaptation requirements and the aforementioned intertwining relationship [5, 6].

The approach proposed in this paper is motivated by long-standing research in strategic management, particularly in research that deals with change and the creation of sustainable advantage. Such research has shown that there is sustainable advantage

---

Authors are listed alphabetically to reflect equal contribution.

© Springer International Publishing Switzerland 2015
P. Johannesson et al. (Eds.): ER 2015, LNCS 9381, pp. 358–366, 2015.
DOI: 10.1007/978-3-319-25264-3_26

in possessing, building and protecting valuable, rare and inimitable competencies, in accordance to environmental changes [4]. These meta-level abilities that allow the continuous integration and reconfiguration of an organization's resource base and processes are referred to as Dynamic Capabilities [4].

In this context Enterprise Capabilities are defined as an organization's ability to appropriately assemble, adapt, integrate and deploy valued resources, differentiated skills and organizational routines, usually in combination or co-presence [4, 7].

Building on the notion of capability, this paper presents a conceptual framework upon which the concept of "capability" is operationalized, and presents a set of analytical processes that exploits this notion for designing sustainable IT capabilities and services. The concepts being proposed in the paper are exemplified through examples from a pilot project involving a world-leading digital enterprise.

2 Existing Works on the Notion of Enterprise Capability

In the various strands of research within strategic management one can distinguish between two prevailing views namely those of the Resource Based View (RBV) and the Dynamic Capability View (DCV). In RBV, researchers focus their attention on identifying possession of valuable, rare, inimitable and non-substitutable resources of enterprise as a source of sustainable advantage [4, 8]. In contrast, researchers in DCV focus on the dynamic aspect of enterprise and propose (i) sensing mechanisms that identify dynamic and changing requirements within the enterprise ecosystem, (ii) promotes shared vision and adoption of appropriate business models to seize opportunities, and (iii) reconfigures the resource base through collaborative and complementary capabilities to transform the enterprise into a new desirable state [4].

In the field of ISE the notion of capability has been considered as a means of dealing with agility, flexibility, and business/IT alignment [4]. ISE researchers and practitioners argue that capability as the fundamental abstraction concept focuses on stable business components and that business capability modeling and SOA complement each other thus facilitating the alignment between technical and business architecture [9, 10]. Nevertheless, the use of capability within the ISE field is still in its infancy and the body of knowledge is still fragmented and indeterminate [11]. There are open issues on the role of capability in Business Process Modeling (BPM) [11–13], in SOA [14, 15], and in Enterprise Architecture (EA) [16, 17]. However the definition and relationships of capabilities to other artifacts used in the design process is not investigated thoroughly which can results in misuse [18].

3 'MariServ': An Example Use Case

This section provides a brief description of an industrial use case that is used in the remainder of the paper to demonstrate concepts and processes of the proposed approach. This use case concerns a company, hitherto referred to as MariServ that provides IT services to shipping companies from the construction of a ship, to its lifetime of chartering, to its eventual decommissioning. Typically, these services include commercial operations, financial management, optimal routing, communication support, fleet performance management, social networking for cooperating shipping companies, and regulatory compliance.

Focusing on regulatory compliance, the capability of MariServ is limited to a particular standard consisting of rules referring to security issues. It lacks a wide and general compliance monitoring system concerning all of the different types of maritime regulations that apply on approximately 15,000 ports worldwide as well as in international waters. In this case study, the notion of capabilities is used to depict the "to-be" requirements of MariServ and its relations to other capabilities, goals and objectives, business processes and operational resources, services, and organizational structure and actors. The compliance capability of MariServ has the potential for application in wider spectrum of its business ecology.

4   The Conceptual Framework for Capability-Centered Modeling

This paper proposes the notion of capability as an *integrative* conceptual representation that can relate the ecosystem and changes within the context of an enterprise to operational and service implementations while describing strategic objectives and social settings. Figure 1 presents this role in specifying and integrating different viewpoints within an enterprise. Each view has been studied in the literature and there exist conceptual models that represent an individual view. However, the interrelation and alignment of artifacts among different views has received less attention.

Each one of these views and their interrelationships are described, through the prism of capability, in the remainder of this section. This is done in the form of presenting graphically a meta-model for each view together with an instantiation of selected concepts of a meta-model from the MariServ case (distinguished by "e.g." in each diagram) and followed by a narrative explanation. It should be noted that these meta-models (in Figs. 2, 3, and 4) are sections of a single integrated meta-model but presented separately for reasons of readability.

The Teleological and Social Views: The sub meta-model shown in Fig. 2 elaborates on the concepts in the teleological and social views. Research in conceptual modeling

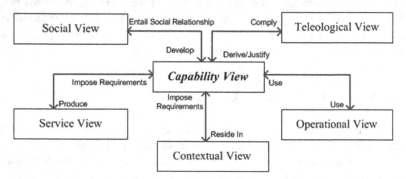

Fig. 1. Conceptual framework – overview of the centrality of capability

has dealt with the social view by modeling actors, roles and business partners and teleological view by modeling goals and business rules [19, 20].

We posit that representing capabilities as the conduit in which strategic objectives and social collaborations unite allows evolutionary decision making. This is supported by studies in strategic management that identify a social and teleological perspective when building capabilities caused by deliberate learning processes that individuals with different skillsets participate in as part of a team [21]. Furthermore, enterprise capabilities develop an identity over time as a result of gradual learning which makes them "path dependent" [7]. By relating *capabilities* and *social aspects* and their relations to both organizational actors and the strategic objectives of the enterprises one can identify and plan for barriers to change and development [7].

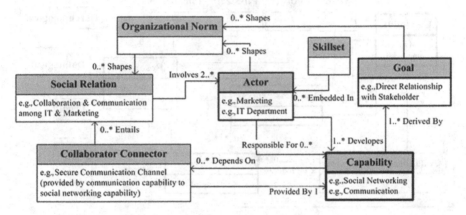

Fig. 2. Concepts of social and teleological view

MariServ Examples. By analyzing relations among capabilities and strategic objectives, MariServ realizes it can outsource its capability to update and maintain back-end software without compromises. On the other hand by analyzing the importance of *Direct Relationship with Stakeholders*, MariServ identifies the need to attain a *Social Networking* capability.

In MariServ, the *Social Networking* capability is dependent on a *Secure Communication Channel* and the *Marketing* capability of MariServ to operate. The social setting responsible for each of the mentioned dependencies can pose resistance towards the implementation of the new capability. In this case, one should consider facilitating collaboration and communication among IT and marketing departments within MariServ as presented in the examples of Fig. 2.

Operational and Service Views: The operational view (which deals with concepts such as processes, resources and transactions) and the service view (that focuses on business and IT service modeling and alignment) have been investigated thoroughly in combination or individually. Figure 3 represents the sub meta-model describing concepts from the operation and service views.

By modeling the relations among technical and operational alternatives and capabilities, one can reason about how different viewpoints and interpretations of multifaceted functional and non-functional requirements are satisfied using capability and social views. This requirement is amplified as studies find decentralization and complementarities as factors that boost innovative product/service offering [4]. Modeling relations among capability, operational and service views is proposed to address such requirements by expressing orchestration choices and providing means to describe the effects of alternatives on strategic objectives. Furthermore, through the relations among capability and social views, one can depict kinds of collaborations required among actors to achieve such complementary relations.

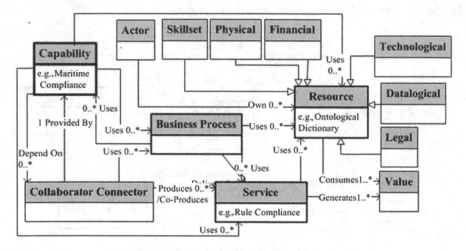

Fig. 3. Concepts of operational and service views

The operational view complements the social views by linking alternative implementations to the rightful stakeholders and their intentions to enable decision making regarding both technical and social aspects of the relationship. The capability view allows linkage and integrated reasoning among the concepts in different views.

MariServ Examples. The *Ontological Dictionary* (a resource) chosen to interpret rules and the depth of its coverage (a quality attribute of the resource) can affect other capabilities and their owners such as MariServ's legal team due to the dependencies among the capabilities. By capturing such relationships one can identify technical, strategic and social consequences of a change. At MariServ considering the input of the legal team when making decisions regarding *Ontological Dictionary* can result in a *Maritime Compliance* capability that is aligned with legal competencies and enables effective implementation of the *Rule Compliance* service.

Contextual View: Context in ISE refers to situational cognition of an IS and its specification and modeling is used to enable system and service adaptation [6]. However, in this framework we go beyond the boundaries of an IS and deal with the organizational setting and the ecosystem in which the IS resides in.

Identifying the value of business processes and capabilities is vital when making strategic decisions [17]. However, since neither generate value directly it is difficult to assign values and far more difficult to understand how stakeholders benefit from the generated value. By modeling the relationships among capabilities, business processes and services, one can analyze the path in which the value is generated. The relations among these artifacts and the ecosystem they reside in are provided in Fig. 4.

MariServ Examples. *Maritime Compliance* capability cannot generate value unless it provides *Rule Compliance* and consequent services that allow automated identification of inconsistencies at the right time and in the right place. One needs to answer questions regarding how a service generated by capability produces revenue whether directly or indirectly e.g. through improving efficiency such as the *Maritime Compliance*, to evaluate capabilities and services.

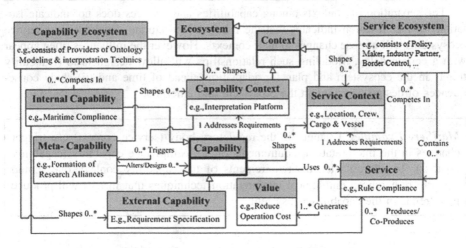

Fig. 4. Concepts of the contextual view

By tracing the value to the social view through capabilities, one can get a sense on how the value is appropriated to different stakeholders. The relationship among generated value and the teleological view can analyze the outcome of strategic objectives and enable decision making on investment choices. Capabilities and services reside in their corresponding ecosystems and hence need to change at different pace and in response to different requirements. Therefore in Fig. 4 two separate entities represent service and capability ecosystems.

> *MariServ Examples.* The *Maritime Compliance* capability relies on ontological interpretation of rules and regulations; hence the capability ecosystem deals with O*ntology Modeling and Interpretation Technics*. On the other hand the *Rule Compliance* service resulted from the capability enables reporting inconsistencies according to ports and regulations, hence the service ecosystem deals with *Policy Makers*, *Industry Partners*, and *Boarder Control* among others. While changes in the capability ecosystem relate to techniques and technologies, changes in the service ecosystem deal with regulatory bodies and policy-making procedures.

The context of capabilities and service are shaped by their ecosystems, hence changes in the ecosystem will impact their situational conditions.

> *MariServ Examples.* The *Rule Compliance* service should adapt its consistency reporting to its situational conditions, i.e., *Local legislation*, *Crew medical state*, *Cargo state*, and *Vessel emission*. To enable such adaptation the capability should be flexible towards change i.e., the *Interpretation Platform* should accommodate changes in business and regulatory dictionaries.

The separation of contexts among capabilities and services does not indicate isolation; in fact research indicates that the relationships among capability and service ecosystems will impose changes to their contexts. However the influences often appear with a certain delay. Modeling such relationships will allow designers to (i) identify trends in the ecosystem and plan for adaptation ahead of time and (ii) study consequences of design decisions in the ecosystem and enterprise.

> *MariServ Examples.* Changes in the regulatory body (service context) can trigger changes in the interpretation requirements of the compliance capability (capability context). To address the new requirements of the capability context, one should perform research on ontological interpretation techniques that in turn will produce new trends in the capability ecosystem.

5 Using the Meta-Model to Design Capabilities and Services

The meta-model presented in Sect. 4 provides an overview of the concepts that we propose in order to have a fully integrated capability-centered approach to sustainable enterprise IT. To answer how these concepts may be used we present a process that depicts three ways of exploiting the capability-centered models namely *descriptive*, *relational* and *evaluative* models in a synergetic manner.

The descriptive models will allow answering questions regarding (i) functional and non-functional requirements of capabilities, processes, services and the supporting organizational structure, (ii) strategic objectives and investment profile of the enterprise and capabilities contribution in satisfying them, (iii) abstraction levels particularly meta-capabilities and meta-processes that trigger and execute change within the enterprise, and (iv) orchestration alternatives and deployment configurations [22]. The relational models enable answering questions regarding (i) economic benefits of services, capabilities and business processes, (ii) collaboration requirements of actors within and outside the boundaries of the enterprise, and (iii) the influences that the alternatives might have on the contextual variables and situations. The evaluative models answer questions regarding (i) the causes and effects among attributes from capability and service context and their interplay with the ecosystem, (ii) the short-term and long-term impacts of a decision on competitive positioning of the enterprise in the ecosystem by analyzing the structure of capabilities, services and their contexts, and (iii) the consequences and impacts of a change in different views.

6 Discussion

In fast paced environments where continuous adaptation and realignment of IT services is required, the challenge in identifying the correct evolutionary requirements is amplified. This paper has argued that there is a need for the development of an ISE methodology and support software tools for the design of services that meet the challenges of alignment, agility and sustainability in relation to requirements that arise as a result of changes in the enterprise domain. To this end the paper proposes an interdisciplinary approach exploring insights from strategic management, systems thinking and information systems engineering.

The integrative factor in the proposed work is the notion of 'enterprise capability', which represents the confluence of research from strategic management and ISE. The conceptual models and the analytical processes presented herein represent the genesis of a *capability-centric* development paradigm.

Acknowledgement. The work of P. Loucopoulos is partly supported by grant NPRP 7-662-2-247. The work of E. Yu and M.H. Danesh is partly supported by Natural Sciences and Engineering Research Council of Canada

References

1. Jarke, M., Loucopoulos, P., Lyytinen, K., Mylopoulos, J., Robinson, W.: The brave new world of design requirements. Inf. Syst. **36**, 992–1008 (2011)
2. Bleistein, S.J., Cox, K., Verner, J., Phalp, K.T.: B-SCP: a requirements analysis framework for validating strategic alignment of organisational IT based on strategy, context and process. Inf. Softw. Technol. **46**, 846–868 (2006)
3. Sousa, H.P., do Prado Leite, J.C.S.: Modeling organizational alignment. In: Yu, E., Dobbie, G., Jarke, M., Purao, S. (eds.) ER 2014. LNCS, vol. 8824, pp. 407–414. Springer, Heidelberg (2014)
4. Teece, D.J.: Explicating dynamic capabilities: the nature and microfoundations of (sustainable) enterprise performance. Strateg. Manag. J. **28**, 1319–1350 (2007)

5. Ulrich, W., Rosen, M.: The business capability map: the "Rosetta Stone" of business/IT alignment. Enterp. Archit. **14** (2014)
6. Bērziša, S., Bravos, G., Gonzalez, T., Czubayko, U., España, S., Grabis, J., Henkel, M., Jokste, L., Kampars, J., Koc, H., Kuhr, J.-C., Llorca, C., Loucopoulos, P., Juanes, R., Pastor, O., Sandkuhl, K., Simic, H., Stirna, J., Zdravkovic, J.: Capability driven development: an approach to designing digital enterprises. Bus. Inf. Syst. Eng. **57**, 15–25 (2015)
7. Leonard-Barton, D.: Core capabilities and core rigidities: a paradox in managing new product development. Strateg. Manag. J. **13**, 111–125 (1992)
8. Barney, J.: Firm resources and sustained competitive advantage. J. Manag. **17**, 99–120 (1991)
9. Cook, D.: Business-capability mapping: staying ahead of the Joneses, MSDN library (2007). http://msdn.microsoft.com/en-us/library/bb402954.aspx
10. Greski, L.: Business capability modeling: theory & practice. Architecture & Governance (2014)
11. Harmon, P.: Capabilities and processes. BPTrends (2011)
12. McDonald, M.P.: Capability is more powerful than process (2009). http://blogs.gartner.com/mark_mcdonald/2009/07/02/capability-is-more-powerful-than-process
13. Rosen, M.: Business processes start with capabilities. BP Trends (2010) http://www.bptrends.com/publicationfiles/12-07-10-COL-BPM%20&%20SOA–BusProcesses%20begin%20with%20Capabilities%201003%20v01–Rosen.pdf
14. Frey, F.J., Hentrich, C., Zdun, U.: Capability-based service identification in service-oriented legacy modernization. In: Proceedings of the 17th European Conference on Pattern Languages of Programs (EuroPLoP), Kloster Irsee, Germany (2013)
15. Homann, U.: A business-oriented foundation for service orientation. Microsoft Developer Network (2006)
16. Wittle, R.: Examining capabilities as architecture. BPTrends (2013)
17. Rosen, M.: Business architecture: are capabilities architecture? BPTrends (2013)
18. Azevedo, C.L.B., Iacob, M.-E., Almeida, J.P.A., van Sinderen, M., Ferreira Pires, L., Guizzardi, G.: An ontology-based well-founded proposal for modeling resources and capabilities in ArchiMate. In: 2013 17th IEEE International Enterprise Distributed Object Computing Conference (EDOC), pp. 39–48 (2013)
19. Kavakli, E.V., Loucopoulos, P.: Focus issue on legacy information systems and business process engineering: modelling of organisational change using the EKD framework. Commun. AIS **2** (1999)
20. Yu, E.S.: Social modeling and *i**. In: Borgida, A.T., Chaudhri, V.K., Giorgini, P., Yu, E.S. (eds.) Conceptual Modeling: Foundations and Applications. LNCS, vol. 5600, pp. 99–121. Springer, Heidelberg (2009)
21. Zollo, M., Winter, S.G.: Deliberate learning and the evolution of dynamic capabilities. Organ. Sci. **13**, 339–351 (2002)
22. Danesh, M.H., Yu, E.: Modeling enterprise capabilities with i*: reasoning on alternatives. In: Iliadis, L., Papazoglou, M., Pohl, K. (eds.) CAiSE Workshops 2014. LNBIP, vol. 178, pp. 112–123. Springer, Heidelberg (2014)

Variability and Uncertainty Modeling

A Working Model for Uncertain Data with Lineage

Liang Wang[1], Liwei Wang[2(✉)], and Zhiyong Peng[1(✉)]

[1] Computer School, Wuhan University, Wuhan China
{nywl,peng}@whu.edu.cn
[2] International School of Software, Wuhan University, Wuhan China
liwei.wang@whu.edu.cn

Abstract. Lineage is important in uncertain data management since it can be used for finding out which part of data contributes to a result and computing the probability of the result. Nonetheless, the existing works consider an uncertain tuple as a set of tuples that can be stored in a relational table. Lineage can derive each tuple in the table, with which one can only find out the tuples rather than specific attributes that contribute to the result. If uncertain tuples have multiple uncertain attributes, for a result tuple with low probability, users cannot know which attribute is the main cause of it. In this paper, we propose an approach to model uncertain data. Compared with the alternative way based on the relational model, our model achieves a low maintenance cost and avoids a large number of redundant storage and join operations. Based on our model, some operations are defined for querying data, generating lineage and computing probability of results. Then we discuss how to correctly compute probability with lineage and an algorithm is proposed to transform lineage for correct probability computation.

Keywords: Uncertain data · Data modeling · Lineage · Probability computation

1 Introduction

Probabilistic databases have received considerable interest in recent years, due to their relevance to many applications like data cleaning and integration, information extraction, scientific and sensor data management, and others [1]. Each uncertain value in probabilistic databases has an existence probability to claim its confidence. This uncertainty may propagate during data operations and cause results with low confidence. Hence, data lineage was proposed to identify the derivation of a data item for better confidence understanding [2].

Generally, lineage is defined according to data model and operations on data. Traditional uncertain data management adopted a two-layer approach to manage uncertain data: an underlying logical model and a working model. The logical model [2, 3] represents uncertain data with *probabilistic or-set-? tables model* [4] to simultaneously represent tuple-level and attribute-level uncertainty. The working model is defined based on relational model and an *uncertain tuple* in probabilistic or-set-? tables model was stored as a set of tuples. It models uncertainties by assigning each tuple a probability value to assert its existence probability and we call such tuples *possible tuples*.

© Springer International Publishing Switzerland 2015
P. Johannesson et al. (Eds.): ER 2015, LNCS 9381, pp. 369–383, 2015.
DOI: 10.1007/978-3-319-25264-3_27

This working model works well when each tuple has only one uncertain attribute. In this situation, every possible tuple is identified by an *ID*, through which we can locate a specific uncertain attribute. However, if tuples contain multiple uncertain attributes, the working model seems to be unsatisfactory. We illustrate the problem in Example 1.

Example 1. Figure 1(a) exemplifies *probabilistic or-set-? tables model*. For a location, multiple sensors monitor its temperature and humidity. Since different sensors may obtain different data for the same location, the attributes *temperature* and *humidity* are uncertain in the table *Sensor*. An uncertain attribute of a tuple has multiple values and each value has a probability to claim its confidence. The attribute *t_p* is the existence probability of the tuple that at least one of the sensors monitoring the same location works well. We assume that all the probabilities of uncertain attributes have been obtained and the attributes *temperature* and *humidity* are independent. Figure 1(b) shows how to store the uncertain relation *Sensor* in relational databases. The uncertain tuple with *ID* = 1 in Fig. 1(a) has been mapped to four possible tuples identified by attributes *ID* and *ID'*. In Fig. 1(c), for the result tuple (l1, t1, h2) with the demonstrated query, its lineage can be denoted as "Sensor.1.2", which means this result is produced by a possible tuple in *Sensor* whose *ID* is 1 and *ID'* is 2.

Sensor(ID,location,temperature,humidity)				
ID	location	temperature	humidity	t_p
1	l1	(t1,0.6)l(t2,0.4)	(h1,0.8)l(h2,0.2)	0.9

(a) The logical model

Sensor(ID,ID',location,temperature,humidity)					
ID	ID'	location	temperature	humidity	probability
1	1	l1	t1	h1	0.432
1	2	l1	t1	h2	0.108
1	3	l1	t2	h1	0.288
1	4	l1	t2	h2	0.072

(b) The working model

Sensor1(location,temperature,humidity)				Query:
location	temperature	humidity	probability	SELECT location, temperature, humidity
l1	t1	h1	0.432	FROM Sensor WHERE temperature='t1';
l1	t1	h2	0.108	The lineage of the result tuple (l1, t1, h2): Sensor.1.2

(c) The result tuple of a query and its lineage

Sensor_1(ID,location,t_p)			Sensor_2(a_ID,ID,temperature,p)				Sensor_3(a_ID,ID,humidity,p)			
ID	location	t_p	a_ID	ID	temperature	p	a_ID	ID	humidity	p
1	l1	0.9	1	1	t1	0.6	3	1	h1	0.8
			2	1	t2	0.4	4	1	h2	0.2

(d) A naïve solution for deriving uncertain attributes

Fig. 1. Traditional uncertain data management

In this example, if the probability of a result tuple is low, it means the tuple has a low confidence. The lineage of the result tuple (l1, t1, h2) can only claim that the result is generated by the possible tuple (1, 2, l1, t1, h2). If users want to know which element (temperature or humidity) is the main cause of the low confidence for this result tuple, the ideal answer is "the value h2 of the attribute *humidity* in the uncertain tuple with *ID* = 1 leads to the low confidence". However, the existing method can only return a possible tuple as an answer, since the working model of existing methods have transformed attribute-level uncertainty into tuple-level uncertainty. That is, such lineage can be seen as tuple-level lineage and is not adequate in some situations.

To handle attribute-level uncertainty, we need to identify each value of the uncertain attributes in a tuple. When executing a query, it needs to simultaneously obtain identifications of attribute values and produce attribute-level lineages for result tuples. It seems to be reasonable to store attribute values and their identifications together. An alternative way treats each attribute value of an uncertain tuple as a tuple associated with identification and probability value. It is shown in Fig. 1(d) and through join operations we can get possible tuples. Hence it reduces to tuple-level lineage and can be solved by the existing methods. Unfortunately, if we want to obtain a possible tuple, we have to execute the join operation multiple times, which is a time consuming operation. To accelerate query, we can create a materialized view for possible tuples. But the maintenance cost for it is huge. Besides, with the number of uncertain attributes in the same uncertain tuple increasing, the storage space for possible tuples exponentially increases.

In this paper, we adopt Object Deputy Model (ODM) [18] to solve the mentioned problems. ODM has been proved to be more efficient in storage and bilateral links between objects achieve low query and maintenance costs. The properties are natural supports for the solution of storage and computational inefficiency caused by the relational database. Further, due to the fact that ODM is designed for certain data, we extend it to the Probabilistic Object Deputy Model (PODM) to represent data uncertainty. Then, we define a comprehensive set of operations that can get query results while generating data lineages. At last, based on the lineage, we define a probability computation operation for results and propose an algorithm to transform lineage for correct probability computation.

The remaining sections are organized as follows. We review the related work in Sect. 2 and introduce our working model for uncertain data management in Sect. 3, following which we elaborate operations on uncertain data and lineage generated by different operations in Sect. 4. For the probability computation operation, we discuss how to guarantee correctness in Sect. 5 before concluding the paper in Sect. 6.

2   Related Works

In [5], the lineage of an output record is defined as identifying a subset of input records relevant to the output record. In [6], three kinds of lineage (also is called provenance) in databases is discussed. "Why" provenance gives the reason the data was generated, say, in the form of a proof tree that locates source data items contributing to its creation. "How" provenance further gives some information about how an output tuple is derived according to the query. "Where" provenance provides an enumeration of the source data items that were actually copied over or transformed into create this data item. The "where" provenance describes the relationships between input and output attributes in tuples, while "why" and "how" provenance describes the relationships between input and output tuples. All above works mainly focus on lineages in certain data. In uncertain data management, most existing works focus on three aspects about lineage: representation, storage and probability computation.

Trio [2] is the first database simultaneously considering uncertainty and lineage. It assumes all base tuples are independent. In [7], for correlated base tuples, it also utilizes lineages to record how to produce result tuples from base tuples. Additionally, it utilizes a forest of junction trees to represent correlation between tuples. Our method, which is different from them, combines the "where" and "how" provenances. We can provide not only the description the relationships between input and output attributes in tuples, but also information about how an output tuple is derived according to the query.

Since lineages may be much larger than data itself, reference [8] proposes to use approximate lineage, which is a much smaller formula keeping track of only the most important derivations. In [9], it tries to find the core of lineage, namely the part of lineage that appears in the computation of every query equivalent to the given one. This information can be identified and compressed previously without considering the query plan.

When utilizing lineages to compute probabilities of result tuples, how to transform lineages to the *read-once* form to guarantee correctness is exploited in [10]. However, some queries (e.g. aggregate query) lead the complex of probability computation to #P-hard. Reference [11] utilizes previously computed query results to reduce the complex from #P-complete to PTIME. For aggregate queries, references [12] and [13] discuss how to efficiently compute probability in specific situations. In [14], efficiently aggregate query answering, partly according to lineage, on uncertain data with cardinality constraints is proposed. Without exact probability computation, reference [15] incrementally refines lower and upper bounds on the probability of the formulas until the desired absolute or relative error guarantee is reached. However, all these works on uncertain data only consider tuple-level lineage. In [16] and [17], how to represent tuple and attribute uncertainty in probabilistic databases is discussed. However, they compute probabilities of result tuples during the query evaluation and lineage is not considered. To guarantee the correctness of probability computation, they restrict the query plan selection and the most efficient query plan may not be selected.

3 Data Modeling

An object may play different roles in different scenarios and belong to different classes. ODM considers different roles of an object as deputy objects and they represent different aspects of the object. Schemas of objects and deputy objects are respectively defined by classes and *deputy classes*. For a deputy class, classes used for creating it are called *source classes* of it. Each deputy object in the deputy class corresponds to at least one *source object* in the source class. The concept of roles [19] is different from the one of views. Besides inherited attributes and methods, roles usually require some additional attributes and methods. So deputy objects can be extended with additional attributes and methods and they can play roles instead of their objects. Through operations including SELECT, PROJECT, EXTEND, JOIN, UNION and GROUPING, we can get a deputy class from one or more classes. For a deputy class, operations on source classes are called the *deputy rule*. Correspondence between an object and its deputy object is represented by a bilateral link. It is different from relational model, because tuples in different relational tables/views often have no links. Compared with relational model, ODM has three advantages:

- First, it avoids join operations by creating deputy classes and querying on them.
- Second, when updates occur on an object, all its deputy objects will be updated through bilateral links, by which can achieve a lower update cost than that on materialized views.
- Third, an inherited attribute in a deputy class just stores a link rather than a value, which can avoid a large number of redundant storage.

In the following, we extend ODM to PODM and give the definition of probabilistic object, probabilistic class, probabilistic deputy object and probabilistic deputy class in PODM. Then we introduce how to transform an uncertain table in probabilistic or-set-? tables model into a set of probabilistic classes and a probabilistic deputy class.

Definition 1 (Probabilistic Object). A probabilistic object is defined as $o = <ID, \{a_j\},$ $p, \{m_k\}>$, where ID and p respectively are the identification and probability of o, $\{a_j\}$ represents a set of attributes of o, and $\{m_k\}$ represents a set of methods including reading and writing attribute values of o.

Definition 2 (Probabilistic Class). A probabilistic class is defined as $C = <\{o_i\}, ID,$ $\{T_j:a_j\}, double:p, \{m_k\}>$, where $\{o_i\}$ represents a set of probabilistic objects, $\{T_j:a_j\}$ represents attributes of probabilistic objects defined by users in C (T_j and a_j respectively are attribute type and name), ID and $double:p$ ($double$ and p respectively are attribute type and name) are pre-defined attributes respectively representing the identification and probability, and $\{m_k\}$ represents methods of probabilistic objects in C.

Definition 3 (Probabilistic Deputy Object). A probabilistic deputy object is defined as $o^d = <ID, \{a_j^d\}, p, \{m_k\}>$, where ID and p respectively are the identification and probability of o^d, $\{a_j^d\}$ represents a set of inherited attributes of o^d, and $\{m_k\}$ represents a set of methods including mapping from a probabilistic object to the probabilistic deputy object o^d and its inversion mapping, reading and writing attribute values of o^d.

Definition 4 (Probabilistic Deputy Class). A probabilistic deputy class is defined as $C^d = <\{o_i^d|o_i^d \to o_i^s, sp(o_i^s)==true\}, ID, \{T_j^d:a_j^d\}, double:p, \{m_k\}>$, where $o_i^d \to o_i^s$ represents that o_i^d is the probabilistic deputy object of o_i^s, sp represents the selection condition that o_i^s should satisfies, ID and $double:p$ are pre-defined attributes respectively presenting the identification and probability, $\{T_j^d : a_j^d\}$ represents a set of attributes inherited from a probabilistic class C^s, and $\{m_k\}$ represents methods of probabilistic deputy objects in C^d.

To detail operations in methods, the following symbols are introduced first.

- $R \Rightarrow T$: R invokes T.
- $\uparrow exp$: The result of an expression exp is returned.
- $o.a$: the value of the attribute a of the probabilistic object o
- $o.a := v$: Attribute a of probabilistic object o is updated with v.
- $f_{a \rightarrow a'}$: A function to convert attribute value from a to a'.
- $mapping(o)$: A function to map from the probabilistic object o to its probabilistic deputy object(s).
- $mapping^{-1}(o^d)$: A function to map from the probabilistic deputy object o^d to its probabilistic source object(s).

When reading/writing attributes not inherited, the methods are defined as follows.
$read(o, a)$

$$read(o, a) \Rightarrow \uparrow o.a, write(o, a, v) \Rightarrow o.a := v$$

$o.a, write(o, a, v)$

$$read\left(o_i^d, a_i^d\right) \Rightarrow \uparrow f_{a_j \rightarrow a_j^d}\left(read\left(o, a_j\right)\right), write\left(o_i^d, a_j^d, v_j^d\right) \Rightarrow write(o, a_j,$$
$$f_{a_j^d \rightarrow a_j}(v_j^d))$$

$o.a := v$

When reading/writing inherited attributes, the methods are defined as follows.
$read(o_i^d, a_j^d) \Rightarrow \uparrow f_{a_j \rightarrow a_j^d}(read(o, a_j)), write(o_i^d, a_j^d, v_j^d) \Rightarrow write(o, a_j, f_{a_j^d \rightarrow a_j}(v_j^d))$

When mapping from a probabilistic object o to its probabilistic deputy objects $\{o_1^d, o_2^d, ..., o_n^d\}$, the method is defined as

$$mapping(o) \Rightarrow \uparrow \{o_1^d, o_2^d, ..., o_n^d\}$$

For any o_i^d in $\{o_1^d, o_2^d, ..., o_n^d\}$, if it is generated according to o and another probabilistic object o', the inversion mapping is defined as

$$mapping^{-1}(o_i^d) \Rightarrow \uparrow \{o, o'\}.$$

An uncertain table R in probabilistic or-set-? tables model consists of three parts: certain attribute, uncertain attribute and tuple probability. All the certain attributes are represented by a probabilistic class C in PODM. The probability of a certain attribute value is always 1 and its existence probability equals to that of the uncertain tuple containing it. Hence tuple probability can be stored in C. Since the ID of each uncertain tuple corresponds to certain one tuple probability, the ID of probabilistic objects in C can be seen as the ID of uncertain tuples. In Fig. 2, uncertain table R contains certain attributes a_1, a_2 and they are represented by the probabilistic class C. IDs and probabilities of uncertain tuples in R are respectively stored in ID and p of probabilistic objects in C. Besides, each uncertain attribute in R is represented by a probabilistic class in PODM. Since an uncertain attribute of an uncertain tuple in U has multiple values, each of these values is stored as different probabilistic objects in the same probabilistic class.

$R(a_1,a_2,a_3,a_4)$

ID	a_1	a_2	a_3	a_4	t_p
1	1	3	(1,0.8)I(7,0.2)	(5,0.7)I(9,0.3)	0.8
2	5	2	(1,0.6)I(7,0.4)	(5,0.9)I(3,0.1)	0.9

(a) The probabilistic or-set-? tables model (underlying logical model)

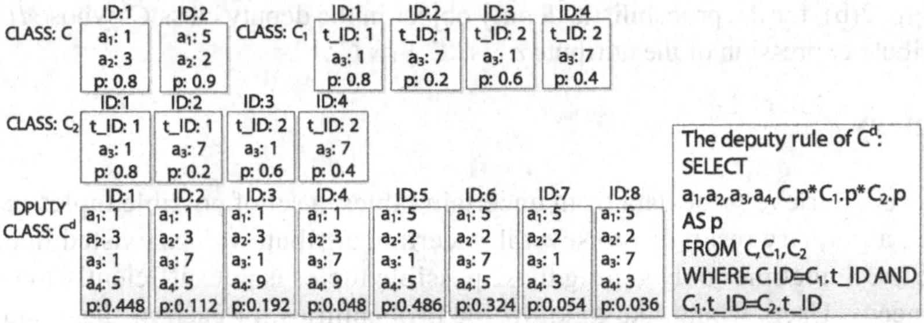

CLASS: C

	ID:1	ID:2
	a_1: 1	a_1: 5
	a_2: 3	a_2: 2
	p: 0.8	p: 0.9

CLASS: C_1

	ID:1	ID:2	ID:3	ID:4
	t_ID: 1	t_ID: 1	t_ID: 2	t_ID: 2
	a_3: 1	a_3: 7	a_3: 1	a_3: 7
	p: 0.8	p: 0.2	p: 0.6	p: 0.4

CLASS: C_2

	ID:1	ID:2	ID:3	ID:4
	t_ID: 1	t_ID: 1	t_ID: 2	t_ID: 2
	a_3: 1	a_3: 7	a_3: 1	a_3: 7
	p: 0.8	p: 0.2	p: 0.6	p: 0.4

DPUTY CLASS: C^d

	ID:1	ID:2	ID:3	ID:4	ID:5	ID:6	ID:7	ID:8
	a_1: 1	a_1: 1	a_1: 1	a_1: 1	a_1: 5	a_1: 5	a_1: 5	a_1: 5
	a_2: 3	a_2: 3	a_2: 3	a_2: 3	a_2: 2	a_2: 2	a_2: 2	a_2: 2
	a_3: 1	a_3: 7	a_3: 1	a_3: 7	a_3: 1	a_3: 7	a_3: 1	a_3: 7
	a_4: 5	a_4: 5	a_4: 9	a_4: 9	a_4: 5	a_4: 5	a_4: 3	a_4: 3
	p:0.448	p:0.112	p:0.192	p:0.048	p:0.486	p:0.324	p:0.054	p:0.036

The deputy rule of C^d:
SELECT
$a_1,a_2,a_3,a_4,C.p*C_1.p*C_2.p$
AS p
FROM C,C_1,C_2
WHERE $C.ID=C_1.t\_ID$ AND
$C_1.t\_ID=C_2.t\_ID$

(b) The probabilistic object deputy model (working model)

Fig. 2. Two-layer approach to manage uncertain data

In Fig. 2, uncertain attributes a_3 and a_4 are respectively represented by the probabilistic classes C_1 and C_2.

Traditional methods store uncertain tuples as a set of possible tuples, which makes queries on uncertain tables possible. Through creating the probabilistic deputy class, each probabilistic deputy object in it represents one possible tuple. In this way, we can retrieve information from uncertain tables by operations on the probabilistic deputy class. Hence the uncertain table R is stored in PODM as a set containing several probabilistic classes and a probabilistic deputy class:

$$\{C, \{C_i|C_i.t\_ID == C.ID\}, C^d\}$$

where all certain attributes are represented by the probabilistic class C, each uncertain attribute is represented by a probabilistic class in $\{C_i\}$ and each possible tuple is represented by a probabilistic deputy object in C^d. For each probabilistic class in $\{C_i\}$, it satisfies the constraint "$C_i.t\_ID == C.ID$". It means that a probabilistic object in C_i has an attribute t_ID whose value equals to certain one ID in C.

When we retrieve information from C^d, inherited attribute values can be obtained through the operations $mapping^{-1}(o^d)$ and $f_{a_j \to a_j^d}(read(o, a_j))$. Note that a view in a relational database cannot instead of a probabilistic deputy class, because ID in a probabilistic deputy class is not inherited from a source probabilistic class.

If users want to know how the uncertainty of a result generated, it should return not only uncertain tuples but also specific uncertain attributes in logical level. For a probabilistic deputy object in C^d, we can get corresponding uncertain tuple ID from the class C. Since the ID in C_i can identify one specific uncertain attribute value in an uncertain tuple, we combine IDs in C and C_i to locate an attribute value in R and define it as the *attribute expression*:

$$a\_\lambda = className\,(t\_ID)\,.t\_ID \mapsto className\,(a\_ID)\,.a\_ID$$

where t_ID and a_ID respectively are identifications of an uncertain tuple and an uncertain attribute value, and for a probabilistic object identification, the function *className()* returns the name of the probabilistic class containing this probabilistic object.

In Fig. 2(b), for the probabilistic deputy object in the deputy class C^d whose *ID* is 1, the attribute expression of the attribute a_3 is "$C.1 \mapsto C_1.1$".

4 Query

When users retrieve some data from uncertain tables, a set of possible tuples may be returned as a query result due to several uncertain attribute values existed in uncertain tables. Sometimes only getting these possible tuples is not sufficiently to satisfy users' needs. Users would like to know the probabilities for each of them and how they were generated from uncertain tables. As we have mentioned in Sect. 3, we have constructed a working model to convert the query on uncertain tables to probabilistic deputy classes, so any query result is generated by one or more probabilistic deputy objects. In the meanwhile, since each value of inherited attributes in a probabilistic deputy class can be located by attribute expressions, in order to reflect where every attribute value of the query result comes from, we give a definition of lineage for each result.

Definition 5 (Lineage). A lineage of a result t is defined as $\lambda(t) = (a\_\lambda) \ominus \lambda'$, where a_λ is an attribute expression, $\ominus$ is the disjunctive/conjunctive operation and λ' is null or a lineage. Part of it containing attribute expression(s) (and disjunctive/conjunctive operation(s)) is called a *sub-lineage*.

In the following, we define operations on uncertain tables and the lineage of a result t obtained by them.

- Selection

$$\mathbf{Select}\,(R, sp) = \{\mathbf{Select}\,(C^d, sp)\,, \lambda\,(t) = a\_\lambda_{R1} \wedge \ldots \wedge a\_\lambda_{Rn}\}$$

where sp represents the selection condition, $a\_\lambda_{R1}, \ldots, a\_\lambda_{Rn}$ are attribute expressions of the inherited attributes in C^d corresponding to all the uncertain attributes in R. For uncertain table R, the selection operation on it actually executes on C^d.

For example, assume that the selection operation on the uncertain table $R(a_1, a_2, a_3, a_4)$ in Fig. 2(a) is $\sigma_{a_1>3\,AND\,a_2=2\,AND\,a_3<8\,AND\,a_4<5}(R)$. For the result "(5, 2, 1, 3)", according to $mapping^{-1}(o^d)$, we find all source probabilistic objects of the probabilistic deputy object with $ID = 7$. Then we find a_1, a_2 belonging to C, a_3 belonging to C_1 and a_4 belonging to C_2 according to the deputy rule through $f_{a_j \to a_j^d}(read(o, a_j))$. So we can determine that an attribute corresponds to which probabilistic object and obtain its probabilistic class name and *ID*. The lineage of the result is $(C.2 \mapsto C_1.3) \wedge (C.2 \mapsto C_2.4)$.

- Projection

$$\mathbf{Project}\,(R, a_j) = \{\mathbf{Project}(C^d, a_j^d), \lambda\,(t) = a\_\lambda_{\mathrm{Rj}}\}$$

where a_λ_{R1} is the attribute expression of the projected uncertain attribute in R.

For example, assume that the projection operation on the uncertain table $R(a_1, a_2, a_3, a_4)$ in Fig. 2(a) is $\pi_{a_3}(R)$. The lineage of the result "(9)" is $C.1 \mapsto C_1.2$.

- Join

$$\mathbf{Join}\,(\{R, S\}, \{(a_j, b_{j,})\}) = \{\mathbf{Join}(\{C_1^d, C_2^d\}, \{(a_j^d, b_{j'}^d)\}), \lambda\,(t) = ((a\_\lambda_{\mathrm{R1}}) \wedge \ldots \wedge (a\_\lambda_{\mathrm{Rn}})) \wedge$$
$$((a\_\lambda_{\mathrm{S1}}) \wedge \ldots \wedge (a\_\lambda_{\mathrm{Sn}}))\}$$

where $\{R, S\}$ are two uncertain tables for joining and $(a_j, b_{j'})$ is a pair of join attributes.

For example, assume that the join operation on the uncertain tables $R(a_1, a_2, a_3, a_4)$ in Fig. 2(a) and $S(a_3, a_4, a_5, a_6)$ in Fig. 3(a) is $R \bowtie S$. The lineage of the result "(5, 2, 7, 5, 3, 7)" is $((C.2 \mapsto C_1.4) \wedge (C.2 \mapsto C_2.3)) \wedge ((D.1 \mapsto D_1.1) \wedge (D.1 \mapsto D_2.2))$.

- Union

$$\mathbf{Union}\,(\{R_i\}) = \{\mathbf{Union}(\{C_i^d\}), \lambda(t) = a\_\lambda_{R_i 1} \wedge \ldots \wedge a\_\lambda_{R_i n}\}$$

where $\{R_i\}$ contains a set of uncertain tables, $a\_\lambda_{R_i 1}, \ldots, a\_\lambda_{R_i n}$ are attribute expressions of all uncertain attributes belonging to the probabilistic class C_i^d. Without duplicate elimination, each result corresponds to only one probabilistic deputy object.

| ID | $S(a_3, a_4, a_5, a_6)$ | | | | |
	a_3	a_4	a_5	a_6	t_p
1	(7,0.3)|(4,0.7)	(7,0.6)|(5,0.4)	3	7	0.8

(a) The probabilistic or-set-? tables model (underlying logical model)

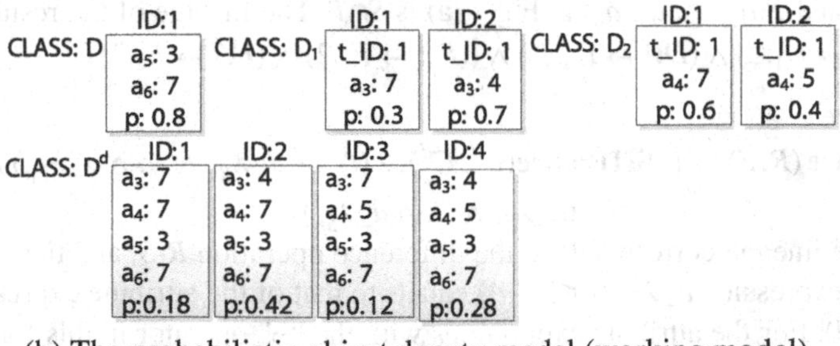

(b) The probabilistic object deputy model (working model)

Fig. 3. An example of uncertain data representation

T(a_3,a_4,a_5,a_6)					
ID	a_3	a_4	a_5	a_6	t_p
1	(6,0.3)|(4,0.7)	(8,0.6)|(5,0.4)	3	7	0.9

(a) The probabilistic or-set-? tables model (underlying logical model)

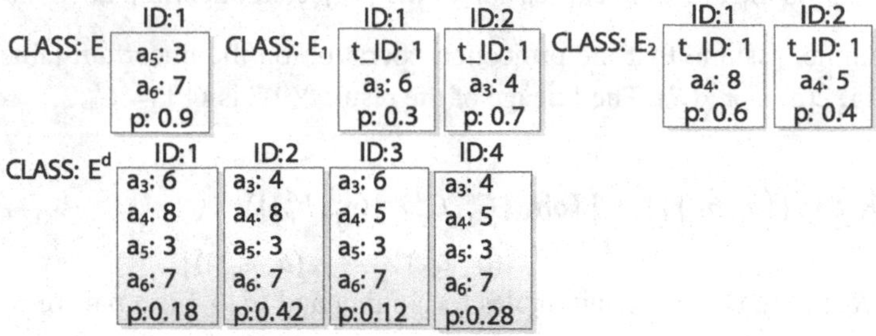

CLASS: E

ID:1
a_5: 3
a_6: 7
p: 0.9

CLASS: E_1

ID:1	ID:2
t_ID: 1	t_ID: 1
a_3: 6	a_3: 4
p: 0.3	p: 0.7

CLASS: E_2

ID:1	ID:2
t_ID: 1	t_ID: 1
a_4: 8	a_4: 5
p: 0.6	p: 0.4

CLASS: E^d

ID:1	ID:2	ID:3	ID:4
a_3: 6	a_3: 4	a_3: 6	a_3: 4
a_4: 8	a_4: 8	a_4: 5	a_4: 5
a_5: 3	a_5: 3	a_5: 3	a_5: 3
a_6: 7	a_6: 7	a_6: 7	a_6: 7
p:0.18	p:0.42	p:0.12	p:0.28

(b) The probabilistic object deputy model (working model)

Fig. 4. An example of uncertain data representation

For example, assume that the union operation on the uncertain tables $S(a_3, a_4, a_5, a_6)$ in Fig. 3(a) and $T(a_3, a_4, a_5, a_6)$ in Fig. 4(a) is $S \cup T$. The lineages of the result tuples "(7, 7, 3, 7)" and "(6, 8, 3, 7)" respectively are $(D.1 \mapsto D_1.1) \wedge (D.1 \mapsto D_2.1)$ and $(E.1 \mapsto E_1.1) \wedge (E.1 \mapsto E_2.1)$.

- Intersection
  Intersect$(R, S) = \{$**Intersect**$(C_1^d, C_2^d), \lambda(t) = (a\_\lambda_{R1} \wedge \ldots \wedge a\_\lambda_{Rn}) \wedge (a\_\lambda_{S1} \wedge \ldots \wedge a\_\lambda_{Sn})\}$
  where the value of the attribute expression a_λ_{Rk} ($k \in [1, n]$) equals to that of the attribute expression a_λ_{Sk} ($k \in$ For example, assume that the union operation[1, n]).

For example, assume that the union operation on the uncertain tables $S(a_3, a_4, a_5, a_6)$ in Fig. 3(a) and $T(a_3, a_4, a_5, a_6)$ in Fig. 4(a) is $S \cap T$. The lineage of the result "(4, 5, 3, 7)" is $((D.1 \mapsto D_1.2) \wedge (D.1 \mapsto D_2.2)) \wedge ((E.1 \mapsto E_1.2) \wedge (E.1 \mapsto E_2.2))$.

- Difference
  Difference $(R, S) = \{$**Difference**$(C_1^d, C_2^d), \lambda(t) = (a\_\lambda_{R1} \wedge \ldots \wedge a\_\lambda_{Rn}) \wedge (\neg$
  $$(a\_\lambda_{S1} \wedge \ldots \wedge a\_\lambda_{Sn}))\}$$
  where the lineage corresponds to the difference operation R-S, and the value of the attribute expression a_λ_{Rk} ($k \in [1, n]$) equals to that of the attribute expression a_λ_{Sk} ($k \in [1, n]$). For the attribute expressions with the "$\neg$" operator in this formula, they just used for probability computation rather than data derivation.

For example, assume that the union operation on the uncertain tables $S(a_3, a_4, a_5, a_6)$ in Fig. 3(a) and $T(a_3, a_4, a_5, a_6)$ in Fig. 4(a) is S-T. The lineage of the result "(4, 5, 3, 7)" is $((D.1 \mapsto D_1.2) \wedge (D.1 \mapsto D_2.2)) \wedge (\neg((E.1 \mapsto E_1.2) \wedge \ldots \wedge (E.1 \mapsto E_2.2)))$.

- Duplicate Elimination

  Eliminate$(R, a_i) = \{$**Eliminate**$(C^d, a_i), \lambda(t) = a\_\lambda_{Ri}^1 \vee \ldots \vee a\_\lambda_{Ri}^n\}$

  where $a\_\lambda_{Ri}^1, \ldots, a\_\lambda_{Ri}^n$ are the attribute expressions of the projected attribute a_i in R and their corresponding attribute values are equality.

  For example, assume that the projection operation with duplicate elimination on the uncertain table $R(a_1, a_2, a_3, a_4)$ in Fig. 2(a) is $\delta(\pi_{a_3}(R))$. The lineage of the result "(1)" is $(C.1 \mapsto C_1.1) \vee (C.1 \mapsto C_1.3)$.

- Grouping and Aggregation

  Agg$(R, a_i, op(a_j), sp) = \{$**Agg**$(C^d, a_i, op(a_j), sp), \lambda(t) = ((a\_\lambda_{Ri}^1 \wedge a\_\lambda_{Rj}^1) \wedge \ldots \wedge (a\_\lambda_{Ri}^p$

  $\wedge a\_\lambda_{Rj}^p)) \wedge ((\neg a\_\lambda_{Ri}^{p+1}) \wedge \ldots \wedge (\neg a\_\lambda_{Ri}^q))\}$

  where a_i is the grouping attribute, op is an aggregation operation and sp represents the selection condition that aggregated results should satisfy. In this formula, except for the possible tuples that have a contribution to the result tuple, uncertain tuples containing possible tuples that have the same value on the attribute a_i but are non-existent are also recorded. Since any two possible tuples of an uncertain cannot simultaneously contribute to the same result, in a lineage, if any one possible tuple occurs, others belonging to the same uncertain tuple are ignored in it.

  For example, for the uncertain table $R(a_1, a_2, a_3, a_4)$ in Fig. 2(a), assume a grouping operation on the attribute a_3 and a *max* operation on the attribute a_4. The lineage of the result "(1, 3)" is $((C.2 \mapsto C_1.3) \wedge (C.2 \mapsto C_2.4)) \wedge (\neg(C.1 \mapsto C_1.1))$. It means that the uncertain tuple with $ID = 1$ in R is either non-existent or existent with the value of attribute a_3 being 7. In this group, it contains only one possible tuple.

5 Probability Computation

After we obtain a lineage of a result tuple, it can be used for probability computation through the **Pro** operation defined as follows.

$$\mathbf{Pro}(t, \lambda)$$

where t is the result and λ is its lineage.

In this paper, we focus on how to correctly compute probabilities of results by lineages under PODM, so we propose a simple method whose complexity is #P-hard. Next, we first introduce the data structure of lineages and then detail how to compute probabilities of results under objects independence assumption.

Each lineage is stored in tree structure in this paper. In the lineage tree, it contains two kinds of nodes. One is the operator node, which represents the operation "$\wedge$" or "$\vee$". One is the data node, which represents an attribute expression. In a data node, it has an attribute to claim whether it is existent. It is used for representing "$\neg$". Assume that a join and a duplicate elimination operations on the uncertain tables $R(a_1, a_2, a_3, a_4)$ in Fig. 2(a) and

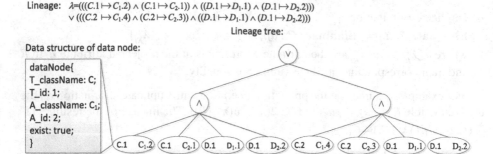

Fig. 5. The lineage tree of the result "(7, 5)"

$S(a_3, a_4, a_5, a_6)$ in Fig. 3(a) is $\delta(\pi_{a_3,a_4}(R\bowtie S))$. The lineage tree of the result "(7, 5)" is shown in Fig. 5. For the data node, we also show its data structure in the figure.

For the operator "$\wedge$", assume we want to compute the probability of an expression "$a\wedge b$". It represents "$Pro(a)*Pro(b)$" in the assumption that all the probabilistic object are independent, where $Pro(a)$ is the probability of the attribute value corresponding to a. For the operator "$\vee$", assume we want to compute the probability of an expression "$a\vee b$". It represents "$1-(1-Pro(a))*(1-Pro(b))$" in the assumption that all the probabilistic objects are independent. As for an attribute expression, it can locate specific probabilistic objects in probabilistic classes and obtain their probabilities. Hence we can utilize these information for probability computation.

However, directly utilizing a lineage to compute the probability may cause some errors. For tuple-level lineage, each possible tuple is annotated with an identification and an identification occurring less than twice in the same lineage guarantees to correctly compute the probability [1, 10]. It is also suitable for attribute-level lineage. For attribute-level lineage, two situations will cause incorrect probability computation. The first situation is that different attribute expressions correspond to the same uncertain tuple and the second situation is that an attribute expression occurs more than one time in a lineage.

We still use the above example to illustrate it. For the result "(7, 5)", the lineage is $((((C.1 \mapsto C_1.2) \wedge (C.1 \mapsto C_2.1)) \wedge ((D.1 \mapsto D_1.1) \wedge (D.1 \mapsto D_2.2))) \vee (((C.2 \mapsto C_1.4) \wedge (C.2 \mapsto C_2.3)) \wedge ((D.1 \mapsto D_1.1) \wedge (D.1 \mapsto D_2.2)))$. For the part "$(C.2 \mapsto C_1.4) \wedge (C.2 \mapsto C_2.3)$", it means that the uncertain tuple whose ID is 2 in R is existent and the values of attributes a_3 and a_4 are respectively 7 and 5. So the existence probability is $Pro(C.2) * Pro(C_1.4) * Pro(C_2.3)$. Based on above analysis, we can transform the lineage to $(C.1 \mapsto (C_1.2 \wedge C_2.1) \wedge D.1 \mapsto (D_1.1 \wedge D_2.2)) \vee (C.2 \mapsto (C_1.4 \wedge C_2.3) \wedge D.1 \mapsto (D_1.1 \wedge D_2.2))$ for probability computation. We denote the transformed lineage as $(\lambda_1 \wedge \lambda_2) \vee (\lambda_3 \wedge \lambda_2)$ for simplification. If we directly utilize it for probability computation, it is computed as follows.

$$1 - (1 - Pro(\lambda_1) * Pro(\lambda_2)) * (1 - Pro(\lambda_3) * Pro(\lambda_2)) = Pro(\lambda_1) * Pro(\lambda_2) +$$
$$Pro(\lambda_3) * Pro(\lambda_2) - (Pro(\lambda_1) * Pro(\lambda_2)) * (Pro(\lambda_3) * Pro(\lambda_2))$$

From above formula, we can find that $Pro(\lambda_2)$ is multiplied twice in the part "$(Pro(\lambda_1) * Pro(\lambda_2)) * (Pro(\lambda_3) * Pro(\lambda_2))$". As this part represents the co-occurrence probability of λ_1, λ_2 and λ_3, $Pro(\lambda_2)$ in it should occur once. Based on above analysis, we transform $(\lambda_1 \wedge \lambda_2) \vee (\lambda_3 \wedge \lambda_2)$ to $(\lambda_1 \vee \lambda_3) \wedge \lambda_2$ for eliminating duplicates.

For a query without self-join operations, based on above analysis, in order to correctly compute probability of a result, we propose an algorithm to transform its lineage as follows.

```
Algorithm 1. exchange_lineage(λ).
Input: Original lineage expression λ;
Ouput: Transformed lineage expression λ'.
1:  t=disjunction_extract(λ);
2:  V=split(t);
3:  W={∅};
4:  for every λᵢ in V do
5:      if not atom(λᵢ) then
6:          V'=split(λᵢ);
7:          for every λᵢ' in V' do
8:              if not atom(λᵢ') then
9:                  λᵢ'=exchange_lineage(λᵢ');
10:         if(root(λᵢ)=='∧')
11:             V'=transform(V');
12:         W=add(V');
13:     else W=add(λᵢ);
14:  λ'=readonce(W);
15: return compose(λ',exchange_lineage(λ-λ'));
```

In Algorithm 1, the function $disjunction\_extract(\lambda)$ finds a disjunctive operator in the lineage tree in the depth first way and returns the sub-tree whose root is this disjunctive operator. The function $split(t)$ obtains all sub-trees whose roots are children of the root of t. The function $atom(\lambda_i)$ determines whether λ_i contains only one node. If it is not, we obtain all sub-trees whose roots are children of the root of λ_i. For all sub-trees contain more than one node, we utilize the function $exchange\_lineage(\lambda'_i)$ to transform them. The function $transform(V')$ finds all sub-trees that represent attribute expressions belonging to the same uncertain tuple and combines them to one node. The combination method remains the part before " $\mapsto$ " unchanged and generates a conjunction for the part after " $\mapsto$ ". As a disjunctive lineage is a disjunction of one or more conjunctions of sub-lineages, the function $readonce(W)$ extracts a common sub-lineage from two or more of them and constructs a conjunctive formula for the common sub-lineage and the remaining part. We have used an example above Algorithm 1 to illustrate the methods realized by $transform(V')$ and $readonce(W)$.

6 Conclusion

In this paper, we propose PODM to manage uncertain data with lineage. In this model, all certain attributes are represented by a probabilistic class and each uncertain attribute is respectively represented by a probabilistic class. In the probabilistic deputy class created according to these probabilistic classes, each probabilistic deputy object represents a possible tuple in the original uncertain table. With bilateral links between probabilistic objects and probabilistic deputy objects, it achieves a low maintenance cost and operations on probabilistic deputy objects can obtain inherited attribute values without join operations. Since inherited attributes in the probabilistic deputy class never store attribute values, it avoids redundant data storage. Based on this model, for an uncertain tuple with multiple uncertain attributes, we define a new lineage that can derive a specific attribute from the result. We define several operations on PODM for querying data and generating lineage. We also discuss how to guarantee probability computation correctness with lineage and propose an algorithm to transform a lineage for correctly computing probability.

Acknowledgements. This work was partially supported by the NNSFC under grant No. 61232002 and No. 61202033, NSFHP under grant No. 2011CDB448, Ph.D. SFWU under grant No. 2012211020207.

References

1. Das Sarma, A., Theobald, M., Widom, J.: Exploiting lineage for confidence computation in uncertain and probabilistic databases. In: ICDE, pp. 1023–1032 (2008)
2. Benjelloun, O., Sarma, A.D., Halevy, A., Theobald, M., Widom, J.: Databases with uncertainty and lineage. VLDB J. **17**(2), 243–264 (2008)
3. Das Sarma, A., Theobald, M., Widom, J.: Working models for uncertain data. In: ICDE (2006)
4. Green, T.J., Tannen, V.: Models for incomplete and probabilistic information. In: Grust, T., Höpfner, H., Illarramendi, A., Jablonski, S., Fischer, F., Müller, S., Patranjan, P.-L., Sattler, K.-U., Spiliopoulou, M., Wijsen, J. (eds.) EDBT 2006. LNCS, vol. 4254, pp. 278–296. Springer, Heidelberg (2006)
5. Cui, Y., Widom, J., Wiener, J.L.: Tracing the lineage of view data in a warehousing environment. ACM Trans. Database Syst. **25**(2), 179–227 (2000)
6. Buneman, P., Tan, W.C.: Provenance in databases. In: SIGMOD, pp: 1171–1173 (2007)
7. Kanagal, B., Deshpande, A.: Lineage processing over correlated probabilistic databases. In: SIGMOD, pp. 675–686 (2010)
8. Ré, C., Suciu, D.: Approximate lineage for probabilistic databases. In: VLDB, pp. 797–808 (2008)
9. Amsterdamer, Y., Deutch, D., Milo, T., Tannen, V.: On provenance minimization. ACM Trans. Database Syst. **37**(4), 3123 (2011)
10. Sen, P., Deshpande, A., Getoor, L.: Read-once functions and query evaluation in probabilistic databases. In: VLDB, pp: 1068–1079 (2010)
11. Dalvi, N., Ré, C., Suciu, D.: Queries and materialized views on probabilistic databases. J. Comput. Syst. Sci. **77**(3), 473–490 (2011)

12. Akbarinia, R., Valduriez, P., Verger, G.: Efficient evaluation of sum queries over probabilistic data. IEEE Trans. Knowl. Data Eng. **25**(4), 764–775 (2013)
13. Murthy, R., Ikeda, R., Widom, J.: Making aggregation work in uncertain and probabilistic databases. IEEE Trans. Knowl. Data Eng. **23**(8), 1261–1273 (2011)
14. Cormode, G., Srivastava, D., Shen, E., Yu, T.: Aggregate query answering on possibilistic data with cardinality constraints. In: ICDE (2012)
15. Fink, R., Huang, J., Olteanu, D.: Anytime approximation in probabilistic databases. VLDB J. **22**(6), 823–848 (2013)
16. Sen, P., Deshpande, A., Getoor, L.: Representing tuple and attribute uncertainty in probabilistic databases. In: ICDM Workshops, pp. 507–512 (2007)
17. Singh, S., Mayfield, C., Shah, R., Prabhakar, S., Hambrusch, S., Neville, J., Cheng, R.: Database support for probabilistic attributes and tuples. In: ICDE, pp. 1053–1061 (2008)
18. Peng, Z.Y., Kambayashi, Y.: Deputy mechanisms for object-oriented databases. In: ICDE, pp. 333–340 (1995)
19. Bachman, C.W., Daya, M.: The role concept in data models. In: VLDB, pp. 464–476 (1977)

Capturing Variability in Adaptation Spaces:
A Three-Peaks Approach

Konstantinos Angelopoulos[1]([⊠]), Vítor E. Silva Souza[2], and John Mylopoulos[1]

[1] University of Trento, Trento, Italy
angelopoulos@disi.unitn.it
[2] Federal University of Espírito Santo, Vitória, Brazil

Abstract. Variability is essential for adaptive software systems, because it captures the space of alternative adaptations a system is capable of when it needs to adapt. In this work, we propose to capture variability for an adaptation space in terms of a three dimensional model. The first dimension captures requirements through goals and reflects all possible ways of achieving these goals. The second dimension captures supported variations of a system's architectural structure, modeled in terms of connectors and components. The third dimension describes supported system behaviors, by modeling possible sequences for goal fulfillment and task execution. Of course, the three dimensions of a variability model are inter-twined as choices made with respect to one dimension have impact on the other two. Therefore, we propose an incremental design methodology for variability models that keeps the three dimensions aligned and consistent. We illustrate our proposal with a case study involving the meeting scheduling system exemplar.

Keywords: Variability · Three-peaks · Requirements · Architecture behaviors

1   Introduction

Adaptive software systems are expected to cope with uncertain environments where requirements cease to be fulfilled now and then. When this happens, an adaptive system needs to reconfigure itself to a different configuration one or more times until fulfillment of its requirements is restored. Possible reconfigurations define an *adaptation space* whose dimensions and size determine the degree of adaptivity of the system-at-hand. Much of the literature on adaptive software systems in the past 15 years has focused on variability that is grounded on requirements or architectures [3,5,8,11,14,17,21]. In this respect, such proposals are limited with respect to adaptivity in that they are blind to dimensions of adaptation other than the native one.

The one and only objective of this paper is to go beyond this one-dimensional view of adaptation spaces by defining adaptation spaces that accommodate three complementary dimensions. The first dimension captures variability in fulfilling

© Springer International Publishing Switzerland 2015
P. Johannesson et al. (Eds.): ER 2015, LNCS 9381, pp. 384–398, 2015.
DOI: 10.1007/978-3-319-25264-3_28

requirements and represents variability in the problem part of the adaptation space. The other two dimensions capture variability with respect to behavior and architecture. These dimensions capture variability in the solution space of the system-to-be, representing how, by whom and in what sequence requirements are to be fulfilled. Together, the three dimensions constitute the adaptation space where an adaptive system searches for alternative reconfigurations.

More specifically, we propose a parametrized model for adaptation spaces that is constituted by a requirements, an architectural and a behavioral dimension. Moreover, the paper proposes a technique for building such models by adopting a three-peaks approach where an adaptation space is defined iteratively by introducing some requirements, deciding on their architectural and behavioral dimensions, and then going back and introducing more requirements, including ones that are determined by architectural and behavioral decisions, extending [23]. The requirements dimension is captured by extended goal models in the spirit of [19].

The rest of the paper is structured as follows. Section 2 presents the research baseline of this work. Section 3 proposes a notation for modeling behavioral and structural variability. In Sect. 4 the three-peaks process is introduced, and it is evaluated in Sect. 5 with an extended version of the meeting scheduling exemplar. Finally, in Sect. 6 we discuss and compare related work, while Sect. 7 concludes.

2   Preliminaries, with Motivating Examples

This section presents the research baseline for this paper.

Goal Models. Following our previous work [1], we use goal models to represent requirements. Figure 1 shows the goal model for a *Meeting Scheduler* system, used as a case study in this paper. The main requirement for the system, *Schedule Meeting*, is represented by the top goal *G0: Schedule Meeting*. AND/OR refinements with traditional Boolean semantics, allow us to refine this goal into finer levels of granularity, down to simple *tasks* that can be operationalized by a component of the system-to-be. Alternatively, *domain assumptions* indicate properties that must hold for the system to work, such as the availability of local and hotel rooms [10].

Even though OR refinements indicate that one of the lower level goals needs to be implemented to satisfy the parent goal, for an adaptive system it is useful to implement all alternatives because this allows multiple reconfigurations during adaptation. Hence, some (in our example, all) OR refinements can be marked as *variation points* (see labels *VP1–VP3* in Fig. 1). In this case, all tasks associated with each variation must be implemented and the system can switch from one configuration to another during adaptation [18], as long as it adheres to its *behavior model* (discussed next).

Awareness Requirements (*AwReqs*) [19] impose constraints on the failure of other requirements, and trigger adaptation. They serve as requirements for the monitoring component of the adaptive system's feedback loop. The degree of failure of other requirements is measured by variables named indicators. For

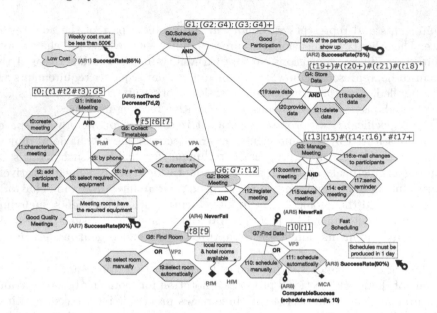

Fig. 1. Goal model for the meeting scheduler case study with flow expressions.

example, $AR4$ prescribes that $G6$ should never fail, whereas $AR3$ indicates that 90 % of the time *schedules are produced within one day*. Therefore, the restrictions of the associated monitored indicators are $I_4 = 100\%$ and $I_3 \geq 90\%$. If the indicators stray this range, then the associated *AwReq* fails.

Another source of variability along the requirements dimension consists of *control variables*. These represent the amount of resources and effort allocated for the system-to-be while it fulfills its requirements. For instance, *FhM* represents *from how many* participants the system should collect time tables before goal $G5$ is considered satisfied (a percentage value). *MCA* is another control variable that represents the *maximum conflicts allowed* for the timeslot chosen for the meeting and participant time tables. *RfM* is yet another, representing how many local (on the premises) rooms have been allocated for meetings, while, *HfM* represents how many hotel rooms are reserved for meetings, and finally *VPA* indicates whether the system has authorization to access personal time tables.

Control variables and variation points, hereafter *requirement control parameters* (*ReqCPs*), can be adjusted at runtime by the adaptation mechanism, to fix failing *AwReqs*. The qualitative relation between *AwReqs* and parameters is captured through a systematic process called *system identification*. During this process the domain expert captures the positive or negative influence that a parameter change can have on an *AwReq*. More specifically, the differential relationship $\Delta(I_2/MCA) < 0$ means that by increasing MCA by one unit the success rate of $AR2$ will decrease. Similarly, $\Delta(I_5/MCA) > 0$ means that by increasing MCA the success rate of $AR5$ will increase. Differential relations are symmetric with respect to increases/decreases, meaning that if MCA is decreased the success rate of $AR5$ will also decrease.

Behavioral Models. We represent the behavior of the system using *flow expressions* [7,16] as attachments to each goal (in Fig. 1, goals *G0–G7*). These are extended regular expressions that describe the flow of system behavior, with each atomic component of allowed sequences of fulfillment of sub-goals that lead to the fulfillment of a parent goal.

The operators ; (sequential), | (alternative), opt() (optional), * (zero or more), + (one or more), # (shuffle) allow us to specify sequences of system actions that constitute a valid behavior. Shuffle specifies that its operands are to be fulfilled concurrently. For example, *G0 # G1* means that goals *G0* and *G1* are to be fulfilled in parallel. Of course, each of these goals has its own flow expression to describe in what order its own subgoals and tasks are to be fulfilled/executed.

Some of the operators presented above introduce behavioral variability that cannot be captured with goal models. For example, the meeting scheduler may send a number of reminders to participants before a meeting (*NoR*), thereby lowering the chances any participant will forget, and increasing chances that *Good Participation* will succeed. To this end, behavioral variability introduced behavioral control parameters.

Software Architecture. The software architecture of a system constitutes a high-level representation of its structure. It depicts how system components are interconnected and what properties they have. The software architecture is highly coupled to the requirements of a system since the latter prescribes what needs to be achieved and *why*, while the former describes *how* fulfillment is achieved. Architectures are described in terms of the concepts of components and connectors. A component constitutes the basic building block of architecture and is responsible for carrying out operations towards the fulfillment of goals. Components can be software, hardware components or human actors that interact with the system through an interface. Components interact with each other within an architecture using communication links, named connectors. For our purposes, we use class diagrams to represent an architecture [9], where classes model components, while associations model connectors. For example, the class diagram in Fig. 2 shows the architecture of the meeting scheduler system.

Variability is captured in architectural models in terms of alternative components that can fulfill the same goal, but with different qualitative properties (e.g., better performance but lesser usability). Variability here can also be introduced by having a number of component instances participating in the runtime

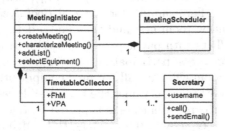

Fig. 2. Architectural diagram for the meeting scheduler

architecture. For instance, the meeting scheduler may have an additional component to what is shown on Fig. 2 that takes in meeting scheduling requests and distributes them among one or more servers each of which consists of the architecture shown in Fig. 2. This kind of variability is exploited by the RAINBOW framework [8].

3   Capturing and Exploring Variability

In the previous section we motivated the need to introduce variability along all three dimensions – goals, behaviors, architecture – to ensure that the system has a large space of adaptation options in trying to cope with one or more requirements failures. In this section, we continue the line of research reported in [18] by demonstrating how to elicit and capture behavioral and architectural *control parameters* (*CPs*) and their impact on system requirements.

Variability in Behavior. The semantics of AND/OR refinements are clear at design time: If goal G is AND/OR refined into sub-goals $G1, ..., Gn$, then the functionality of the system-to-be needs to include functions that fulfill all/at least one of $G1, ..., Gn$.

Behavior talks about the allowable sequences of fulfillment of $G1, ...Gn$ at runtime. Each sequence needs to include one or more of $G1, ..., Gn$, but not all. So, it can be the case that for an AND-refinement we have sequences that fulfill only some of $G1, ..., Gn$ and for OR refinements we have sequences that fulfill all of $G1, ..., Gn$. For example, the goal *Manage Meeting*: although all of its tasks must be implemented, *confirm meeting* and *cancel meeting* are actually conflicting and their use cannot coincide in the same execution sequence. Therefore, the | operator indicates that only one of the two is allowed for any one execution of the system as shown in Fig. 1.

The ; operator is useful when modeling the behavior of an AND-refinement and prescribes the order in which sub-goals/tasks must be fulfilled. It is common practice in software design to impose only one possible order, thereby limiting the reconfiguration capabilities of the system-to-be. In our framework, the designer is encouraged to select multiple alternative behaviors for fulfilling a goal. Accordingly, we introduce *behavioral control parameters* (*BCPs*) that are assigned to the goal's behavior and whose possible values are all the allowed sequences. A *BCP* is defined as $(|[parameter\ name]alt_1\ ...\ alt_n)$, using infix notation for the alternative operator. For example, for the goal *Book Meeting* if the meeting organizers select a meeting room first and then find a date, participation might be low because of conflicts with participant time tables. If they select the date first and the room afterwards, participation may improve but it is not guaranteed that the selected room will have all required equipment. A *BCP* defined by $(|[BCP1]\ G6; G7\ G7; G6)$ takes as values the two possible sequences $G6; G7$ and $G7; G6$. It's impact on the requirements is captured by the differential relations $\Delta(I_2/BCP1)[G6; G7 \rightarrow G7; G6] > 0$ and $\Delta(I_7/BCP1)[G6; G7 \rightarrow G7; G6] < 0$ while the new behavior for the goal *Book Meeting* is depicted in Fig. 3a.

Another variability factor of system behavior is related to the multiplicity of the fulfillments of a goal or a task. When there is the option for the system to fulfill multiple times a goal or a task, the designer must consider the impact of this variability on *AwReqs*. For example, the task *t17 send reminder* is performed by the system–to-be and can be executed multiple times if the goal *Good Participation* is failing. To this end, as depicted in Fig. 3c, we substitute when needed the operators * and + with a *BCP* (in this case named *NoR*) and based on a differential relation such as $\Delta(I_2/NoR) > 0$ the adaptation mechanism can adjust its value when *Good Participation* is failing. The range of values of NoR varies from one to five executions of task *t17*.

The repetitive execution of a task or fulfillment of a goal raises the issue of time synchronization. In the previous example, if $NoR = 3$ and all the reminders are sent one after the other within seconds, the outcome is likely to be an unhappy one. Hence, we introduce a *behavioral function wait()* that takes as argument a *BCP* with a range of values related to time units, in this case days. This function is part of the behavioral model as shown in Fig. 3c and *BCP3* is defined as (| [*BCP3*] 1*day* 2*days* 3*days*).

Next, we revisit OR-refinements in order to extract additional variability. The traditional perception of these refinements at runtime is that the satisfaction of any subgoal would lead to the satisfaction of the parent goal. Therefore, the *ReqCPs* associated to an OR-refinement have as candidate value one of the subgoals. The system-to-be though may require in certain occasions the fulfillment of all the subgoals to guarantee the satisfaction of the parent goal. For example, scheduling a meeting requires the fulfillment of the goal *G5: Collect Timetables* that can be achieved by either contacting the participants *by phone*, *by e-mail* or collecting them *automatically* from a common system calendar. However, when one or more of the invited participants do not use the system calendar the third option could harm *AR2*, since these participants will not receive any invitation for the meeting. Dealing with such a situation requires the utilization of all the alternatives under the OR-refinement. This means that participants who do not have an account for using the system's calendar and therefore their timetables must be collected either *by phone* or *by e-mail* while the timetables of the remaining participants can be collected automatically by the system. To capture this additional variability a new *BCP* is introduced defined as (| [*BCP2*] *VP1* t5#t7 t6#t7), as depicted in Fig. 3b.

Variability in Architecture. We consider next the third peak, architecture, looking for opportunities to introduce variability. In order to be fulfilled, each goal or task must be assigned to at least one component[1]. For this peak, as we mentioned before, there are two sources of variability. The first is related to each component's multiplicity. Certain components may be instantiated multiple times for requirements to be fulfilled. For example, as shown in Fig. 2, an instance of the component *TimetableCollector* can be associated with multiple instances of the component *Secretary*. The number of instances of the latter, is an adjustable variable that affects the operational cost of the meeting scheduling

---

[1] Each component must be able to satisfy on its own the assigned goal.

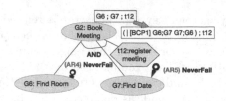

(a) BCP from AND-refinement

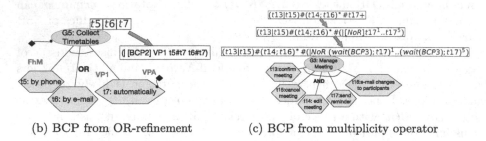

(b) BCP from OR-refinement   (c) BCP from multiplicity operator

Fig. 3. Behavioral control parameters (BCPs) elicitation

process ($AR1$), but also how fast the meetings are scheduled ($AR3$). We refer to such variables as *architectural control parameters* ($ACPs$) following the same definition construct as $BCPs$. In this case we introduce the number of secretaries NoS parameter defined as (| [NoS] 1..5) that will substitute the abstract multiplicity notation, representing explicitly the presence of a new configuration point. The impact of this ACP on the requirements is captured by the differential relations $\Delta(I_1/NoS) < 0$ and $\Delta(I_3/NoS) > 0$.

The second source of architectural variability is related to the selection among multiple candidate components that are assigned with the same goal/task. For the goal *Find Room* we have two candidate software components that are both part of the system and can be used interchangeably. The first component finds the cheapest room reducing the overall cost of the meetings, but does not guarantee that all the required equipment will be present, while the other one finds the best equipped room but might exceed the budget available for scheduling meetings. These two components can be used either interchangeably or concurrently. The concurrent use of both components allows the users select which result is more suitable for them. In specific occasions such as low budget periods, the system may switch to the exclusive use of the component that provides the best price. Therefore, as shown in Fig. 4b we add to the architecture model an ACP named $ACP1$ with candidate values all the possible uses of these components, with the following definition (| [$ACP1$] *BestEquipRoomFinder* *BestPriceRoomFinder* *BestEquipRoomFinder#BestPriceRoomFinder*). The shuffle operator indicates concurrent use of the operand components.

Variability in the Environment. In all previous cases we have seen that factors from the system's environment affect adaptation decisions. Accordingly, our variability model needs to capture variability in the environment and its impact on requirements. Towards this end, we introduce a domain model, as

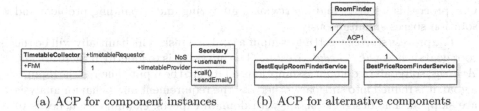

(a) ACP for component instances   (b) ACP for alternative components

Fig. 4. Architectural control parameters (ACPs) elicitation

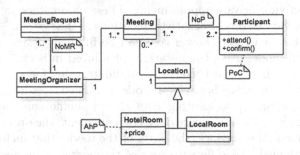

Fig. 5. Domain model for the meeting scheduler environment

shown in Fig. 5. Environmental variability is captured here with a new type of parameter named *environmental parameter* (*EP*).

An *EP* can indicate the number of instances of a domain entity and therefore its multiplicity in the domain model. The difference from architectural multiplicity is that in the case of the environment the adaptation mechanism has no control on the value of the *EPs*. For instance, there is no control on the *number of meeting requests* (*NoMR*) the meeting organizers are sending, neither the *number of participants* (*NoP*) attending a meeting, and therefore these are represented as *EPs*.

The attributes and the operations of domain entities constitute another source for environmental variability. For example, participants may confirm their participation, but in the end not attend a meeting. The *EP percentage of consistency* (*PoC*) captures this, while the *average hotel price* captures the current average cost for reserving a hotel room for meetings.

EPs influence the *AwReqs* in the same manner as *CPs*. However, the adaptation mechanism can only monitor them, identifying undesired situations and change *CPs* to compensate for changes. For example, when the *PoC* is decreased because participants tend to forget meetings they are supposed to attend and the participation is harmed according to the differential relation $\Delta(I_2/PoC) < 0$, then the adaptation mechanism can increase the *NoR* to compensate.

4   A Three-Peaks Modeling Process

The modeling process for three-peaks models is depicted in Fig. 6. It guides the elicitation of all elements of a three-peaks model, including control parameters.

Our process is iterative and intertwined, analyzing and expanding problem and solution spaces simultaneously.

The process starts by getting as input a goal or a task, which initially will be the root goal such as *G0: Schedule Meeting*. The next step is to identify if there are any *AwReqs*, softgoals or domain assumptions related to the input. Then, if the input is a goal, it is refined into subgoals, otherwise the requirement and behavior analysis are skipped. The designers, along with domain experts, examine what needs to be fulfilled and how, starting from eliciting parameters required for that inserted goal to be satisfied, such as how many conflicts are allowed before finding a date or if the system can view private appointments. These parameters are *ReqCPs* and their values may vary during alternative executions of the system.

Continuing the analysis of how a goal can be fulfilled, designers provide an initial behavioral model using the notation introduced in Sect. 2. If the goal is OR-refined then each subgoal becomes a candidate value for a *ReqCP* such as $V1 - V3$ in Fig. 1. Then, the behavioral model is refined by adding a BCP with range of values according to existing *ReqCP* and shuffle combinations of the refinements as in Fig. 3b. In the case of AND-refinement, the order in which the operands of sequential behaviors (the parts of the model that include only the ; operator) is examined. If a different order of the operands implies influence to different *AwReqs* a new *BCP* is introduced with range of values, all the potential orders. Concluding this iteration of behavior analysis, the process examines every ∗ and + operators in order to substitute them with a *BCP*, if needed, as described in Sect. 3. In that case also the *wait(BCP)* function with its own *BCP* is added. The last step leads to a new refinement of the goal since a *wait* task is added as a refinement of the examined goal.

Moving to the architecture peak, designers associate the input goal or task to one or more components of the architecture. This determines who is responsible for the satisfaction of the goal or task. When more than one component is assigned, an *ACP* is added and can be tuned by the adaptation mechanism at runtime in order to activate the most suitable component or a combination of them for fixing failing requirements. Next, if the new component can be instantiated multiple times at runtime and this number has impact on the *AwReqs* while under the control of the adaptation mechanism, the associated multiplicity is substituted with an *ACP*. Then, the assigned components get as attributes the *ReqCPs* and *BCPs* of the goal, as they must be aware of what behavior must follow and what are the values of these parameters. Once the previously elicited variability has been embedded in the assigned component, the designers of the architecture, provided that the goal is fully or partially operationalized, add the tasks produced by the refinement as operations. Finally, the designers may provide additional attributes and operations to the component of more technical nature that are not related neither to requirements nor to behavior. For every new attribute or operation, the process must investigate whether there is need of adding new requirements. If the initial input was a goal then it is refined again, but in case of a task then it is the parent goal that must be processed again.

The last step of the process inspects if the current set of configurations is able to guarantee the satisfaction of all the *AwReqs* related to the investigated goal

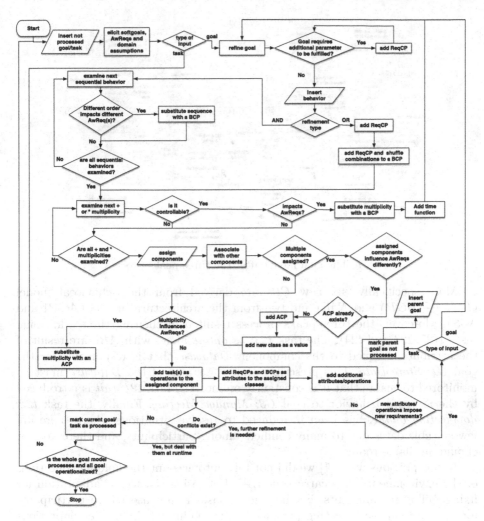

Fig. 6. The three-peaks process as a flowchart

under any possible environmental condition. In case there are situations where the system is not able to guarantee success of all the related *AwReqs*, then two actions can be taken: (a) perform further refinements, finding new *CPs*, goals or tasks; or (b) deal with conflicting requirements, using the conflict resolution mechanism of our previous work [1]. When all goals and tasks are processed and every goal is operationalized, the process terminates.

5   Evaluation

Following the three-peaks process presented in the previous section we produce a goal model with annotated behavior (Fig. 7) and an architectural model (Fig. 8) which includes several additional parameters over what was presented in Sect. 2.

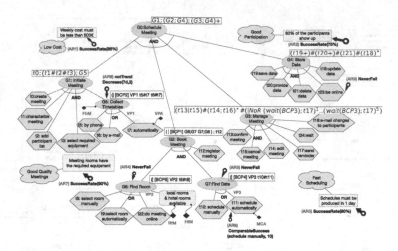

Fig. 7. The goal model after the three-peaks process

More specifically, six new *CPs* are derived from the behavioral model (*BCP*1 − *BCP*5 and *NoR*) and two from the architectural model (*ACP*1 and *NoS*). Moreover, the three-peaks process resulted in eliciting three additional tasks (*t22*, *t23* and *t24*). The task *t23: be online* along with *AR9* are result of the attribute assigned to the component *Database* that is responsible for the goal *G4: Store Data*. This prescribes that the status of the *Database* must be monitored to ensure that it is constantly online. The task *t24:wait* is introduced by the assigned behavior to goal *G3: Manage Meeting*. Finally, the task *t22: do meeting online*, has been introduced to resolve situations where there are few suitable dates due to many conflicts among participants, and there are not enough available rooms.

In our previous work [1] we did not take into account the holding conditions of the environment. This carries the risk of choosing the wrong adaptation for fixing failing requirements. For instance, consider the case where participants forget to attend their meetings, resulting in the failure of *AR2*. The adaptations offered by the original goal model (not generated by the three-peaks process) for fixing *AR2* are either to start viewing private appointments of participants by setting *VPA* to true or decreasing *MCA* allowing fewer conflicts. Neither one anticipates the real cause of the failure. The three-peaks model though offers the parameter *NoR* that increases the number of reminders thereby tackling the source of the problem, and capable of increasing the success rate of *AR4*.

Another case where requirements-only variability proves to be insufficient concerns the room selection by the meeting organizers before or after finding a date for their meeting. Each of the alternative orders works well in different contexts. Selecting room first guarantees good quality meetings, since the meeting organizers select a room that can provide all the required equipment, assuming that the invited participants are available the same date the room is available, otherwise the success rate of *AR5* is at risk. On the other hand, when meeting organizers select date first, it is more likely that they will find a date convenient

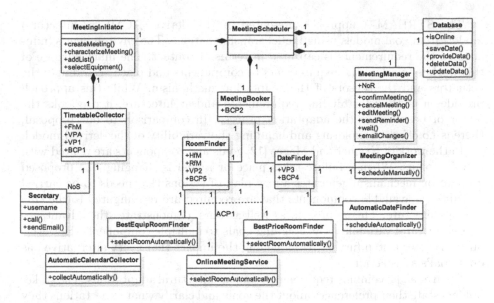

Fig. 8. The architecture model after the three-peaks process

to most invited participants, but a sufficiently equipped room might not be found in periods with high workload for the meeting scheduler, decreasing the success rate of $AR7$. Using behavioral variability, an adaptive meeting scheduler executes the order that complies with the existing context by tuning $BCP1$. Moreover, to maintain the equilibrium between the success rate of $AR1$ and $AR7$ when the system selects rooms automatically the system can use either a component that finds the cheapest room available or the best equipped respectively, exploiting architectural variability and more particularly $ACP1$.

The previous failure scenarios show that the high variability models of the three-peaks process can handle better changes in the system's environment where the requirements-only model would provide ineffective adaptations. A limitation of our approach is that dependencies among CPs are not captured. For instance, it makes sense for MCA to be changed only if the value of $VP3$ is set first to "schedule automatically". In order to alleviate this obstacle, we are planning to extend our notation in order to capture this kind of constraint. Another limitation on the scalability of our proposal is that for every variable introduced into the model, its impact on all $AwReqs$ must be examined.

6 Related Work

Having as a starting point a goal model, Yu et al. [22] propose heuristics to derive other models such as feature models, statecharts and component-connector models. Their purpose is to express the same level of variability in different dimensions of the system. On the other hand, our approach intends to capture the interaction of the system's dimensions and incrementally elicit additional variability along each one of them.

The STREAM-A approach presented in [15] derives ACME architectural models from goal models using model transformations. The environment's influence on the requirements is captured in terms of context. The main purpose of this work is to relate the requirements to components and place accordingly the actuators and the sensors of the adaptation mechanism. While this approach provides a method for binding requirements and architecture, it overlooks the factor of behavior in the adaptation process. In comparison to our proposal, there is no effort of exploring and handling the variability of the derived model.

In the work of Kramer and Magee [12] goals and components are related with reactive plans. When a failure takes place or a goal is changing, the proposed adaptation mechanism generates a new plan of actions that needs to be carried out and the available components that are required are reconfigured to the current architecture. This approach, as well as ours, demonstrates the advantages of architectural variability, by assigning goals to multiple components. However, our work goes a step further by modelling the impact that every alternative has on a goal's satisfaction.

Chen et al. [4] combine requirements and architectural adaptations. The stakeholders state their preferences about the goals and clarify what expectations they have for their satisfaction. Then the adaptation framework monitors the system and if expectations are not met it attempts to find a different architectural solution. In case the problem is still not solved, a new specification is derived from the goal model attempting a reconfiguration at requirements level. This approach doesn't take into account the variability of the behavior which we express by flow expressions. Furthermore, the priority is given to architectural alternatives while in our work all the dimensions are candidates for offering solutions.

Other approaches that derive architectures from goal models such as the work of Chung et al. [6] and Lamsweerde [20], offer systematic methods for relating goals, behaviors and architectures. Even though there is no notion of variability for adaptation purposes within these approaches, they can by exploited to elaborate the production of the three-peaks model as proposed here.

In [13], Lapouchnian et al. describe how to derive high variability business process models from goal models and how softgoals can guide the reconfiguration of a business process. Our approach follows the same line of research, adding a richer notation and capturing behavioral variability that goes beyond the OR-refinements of the goal models. Moreover, we introduce environmental variables that can influence the satisfaction of our goals and drive the reconfiguration process.

Finally another variability management approach that is related to our work is Dynamic Software Product Lines (DSPLs). In [2] DSPLs are applied to service-based system for adapting at runtime to the user's requirements by adding and removing features and reconfiguring the business process in order to support these changes. Our work examines architectural variability at a component level, where different components can satisfy the same goal(s) but may influence different indicators. The reason is that the components share common features, while they differ to some others. Therefore, DSPLs could be complementary to our work to express at a deeper level of detail our architectural variability.

7   Conclusions and Future Work

We propose a systematic process for extracting incrementally variability from goal models. The source of variability lies in the three peaks of a software system: requirements, behavior and architecture. We investigate how variability can be elicited along each peak, introducing behavioral and architecture control parameters and how to model environmental variability. We also present a three-peaks process to derive incrementally high variability requirements, behavioral and architecture models. Finally, we have evaluated our models through execution scenarios of the meeting scheduler exemplar, showing that offering adaptations along three peaks enables the system to handle more failures.

Our future research includes the implementation of a tool that will support the design of our three-peaks models, as well as further evaluation with real case studies. We also plan to experiment with a simulation of the meeting-scheduler exemplar using the three-peaks model extending our previous work [1] on handling multiple failures by exploring the quantitative form the differential relations and applying control theoretical adaptation techniques for multivariable system with multiple inputs and multiple outputs.

Acknowledgment. This work has been supported by the ERC advanced grant 267856 "Lucretius: Foundations for Software Evolution" (April 2011 – March 2016, http:// www.lucretius.eu) and Brazilian research agencies CAPES and CNPq (process numbers 402991/2012-5, 485368/2013-7 and 461777/2014-2).

References

1. Angelopoulos, K., Souza, V.E.S., Mylopoulos, J.: Dealing with multiple failures in zanshin: a control-theoretic approach. In: Proceedings of the 9th International Symposium on Software Engineering for Adaptive and Self-Managing Systems, SEAMS 2014, pp. 165–174. ACM, New York (2014)
2. Baresi, L., Guinea, S., Pasquale, L.: Service-oriented dynamic software product lines. Computer **45**(10), 42–48 (2012)
3. Baresi, L., Pasquale, L., Spoletini, P.: Fuzzy goals for requirements-driven adaptation. In: Proceedings of the 18th IEEE International Requirements Engineering Conference, pp. 125–134. IEEE (2010)
4. Chen, B., Peng, X., Yu, Y., Nuseibeh, B., Zhao, W.: Self-adaptation through incremental generative model transformations at runtime. In: Proceedings of the 36th International Conference on Software Engineering, ICSE 2014, pp. 676–687. ACM, New York (2014)
5. Cheng, B.H.C., Sawyer, P., Bencomo, N., Whittle, J.: A goal-based modeling approach to develop requirements of an adaptive system with environmental uncertainty. In: Schürr, A., Selic, B. (eds.) MODELS 2009. LNCS, vol. 5795, pp. 468–483. Springer, Heidelberg (2009)
6. Chung, L., Supakkul, S., Subramanian, N., Garrido, J., Noguera, M., Hurtado, M., Rodríguez, M., Benghazi, K.: Goal-oriented software architecting. In: Avgeriou, P., Grundy, J., Hall, J.G., Lago, P., Mistrík, I. (eds.) Relating Software Requirements and Architectures, pp. 91–109. Springer, Heidelberg (2011)

7. Dalpiaz, F., Borgida, A., Horkoff, J., Mylopoulos, J.: Runtime goal models. In: Proceedings of the IEEE 7th International Conference on Research Challenges in Information Science, pp. 1–11. IEEE (2013)
8. Garlan, D., Cheng, S.-W., Huang, A.-C., Schmerl, B., Steenkiste, P.: Rainbow: architecture-based self-adaptation with reusable infrastructure. Computer **37**(10), 46–54 (2004)
9. Ivers, J., Clements, P., Garlan, D., Nord, R., Schmerl, B., Silva, J.R.: Documenting component and connector views with uml 2.0. Technical report, DTIC Document (2004)
10. Jureta, I., Mylopoulos, J., Faulkner, S.: Revisiting the core ontology and problem in requirements engineering. In: Proceedings of the 16th IEEE International Requirements Engineering Conference, pp. 71–80. IEEE (2008)
11. Kramer, J., Magee, J.: Self-managed systems: an architectural challenge. In: Future of Software Engineering (FOSE 2007), pp. 259–268. IEEE (2007)
12. Kramer, J., Magee, J.: A rigorous architectural approach to adaptive software engineering. J. Comput. Sci. Technol. **24**(2), 183–188 (2009)
13. Lapouchnian, A., Yu, Y., Mylopoulos, J.: Requirements-driven design and configuration management of business processes. In: Alonso, G., Dadam, P., Rosemann, M. (eds.) BPM 2007. LNCS, vol. 4714, pp. 246–261. Springer, Heidelberg (2007)
14. Oreizy, P., et al.: An architecture-based approach to self-adaptive software. IEEE Intell. Syst. **14**(3), 54–62 (1999)
15. Pimentel, J., Lucena, M., Castro, J., Silva, C., Santos, E., Alencar, F.: Deriving software architectural models from requirements models for adaptive systems: the stream-a approach. Requirements Eng. **17**(4), 259–281 (2012)
16. Pimentel, J., Castro, J., Mylopoulos, J., Angelopoulos, K., Souza, V.E.S.: From requirements to statecharts via design refinement. In: Proceedings of the 29th Annual ACM Symposium on Applied Computing, SAC 2014, pp. 995–1000. ACM, New York (2014)
17. Souza, V.E.S.: Requirements-based software system adaptation. Ph.D. thesis, University of Trento, Italy (2012)
18. Silva Souza, V.E., Lapouchnian, A., Mylopoulos, J.: System identification for adaptive software systems: a requirements engineering perspective. In: Jeusfeld, M., Delcambre, L., Ling, T.-W. (eds.) ER 2011. LNCS, vol. 6998, pp. 346–361. Springer, Heidelberg (2011)
19. Souza, V.E.S., Lapouchnian, A., Robinson, W.N., Mylopoulos, J.: Awareness requirements. In: de Lemos, R., Giese, H., Müller, H.A., Shaw, M. (eds.) Software Engineering for Self-Adaptive Systems. LNCS, vol. 7475, pp. 133–161. Springer, Heidelberg (2013)
20. van Lamsweerde, A.: Requirements Engineering: From System Goals to UML Models to Software Specifications. Wiley, Hoboken (2009)
21. Whittle, J., Sawyer, P., Bencomo, N., Cheng, B.H.C., Bruel, J.-M.: Relax: incorporating uncertainty into the specification of self-adaptive systems. In: Proceedings of the 2009 17th IEEE International Requirements Engineering Conference, RE, RE 2009, pp. 79–88. IEEE Computer Society, Washington (2009)
22. Yu, Y., Lapouchnian, A., Liaskos, S., Mylopoulos, J., Leite, J.C.S.P.: From goals to high-variability software design. In: An, A., Matwin, S., Raś, Z.W., Ślęzak, D. (eds.) Foundations of Intelligent Systems. LNCS (LNAI), vol. 4994, pp. 1–16. Springer, Heidelberg (2008)
23. Nuseibeh, B.: Weaving together requirements and architectures. IEEE Comput. **34**(3), 115–117 (2001). http://www.computer.org/csdl/mags/co/2001/03/r3115-abs.html

Taming Software Variability: Ontological Foundations of Variability Mechanisms

Iris Reinhartz-Berger[1](✉), Anna Zamansky[1], and Yair Wand[2]

[1] Department of Information Systems, University of Haifa, Haifa, Israel
{iris,annazam}@is.haifa.ac.il
[2] Sauder School of Business,
University of British Columbia, Vancouver, Canada
yair.wand@ubc.ca

Abstract. Variability mechanisms are techniques applied to adapt software product line (SPL) artifacts to the context of particular products, promoting systematic reuse of those artifacts. Despite the large variety of mechanisms reported in the literature, a catalog of variability mechanisms is built ad-hoc and lacks systematization. In this paper we propose an ontologically-grounded theoretical framework for mathematically characterizing well-known variability mechanisms based on analysis of software behavior. We distinguish between variability in the *product dimension*, which refers to differences in the *sets* of product's behaviors, and variability in the *element dimension*, which focuses on differences in the *particular* behaviors.

Keywords: Software product line engineering · Variability analysis · Variability mechanisms · Systematic reuse

1 Introduction

Software Product Line Engineering (SPLE) supports developing families of software products [6, 10]. It further promotes systematic reuse of development artifacts using *variability mechanisms*, which are techniques applied to adapt software product line (SPL) artifacts to the context of particular products. Different catalogs of variability mechanisms have been suggested over the years, e.g., [1, 2]. Examples of common mechanisms are: (1) *configuration*, in which SPL artifacts are selected to be included in the product; (2) *parameterization*, in which values for specific products are assigned to parameters defined in the SPL artifacts; (3) *inheritance*, in which SPL artifacts are specialized in the context of a particular product; and (4) *extension*, in which the product can add behavior or functionality to that proposed by the SPL.

Usually, a catalog of variability mechanisms is built ad-hoc and concentrates on design and code artifacts. In order to tame variability mechanisms, i.e., better understand their nature at different development stages and define them systematically, we suggest a formal framework for representing properties of variability mechanisms. The framework is based on the ontological model of Bunge [3, 4] and defines SPL and software products as things exhibiting behavior. The framework further identifies the relationships among SPL and software products enabling mathematical definition of well-known variability mechanisms.

© Springer International Publishing Switzerland 2015
P. Johannesson et al. (Eds.): ER 2015, LNCS 9381, pp. 399–406, 2015.
DOI: 10.1007/978-3-319-25264-3_29

In the rest of the paper Sect. 2 reviews related work, Sect. 3 elaborates the framework and its theoretical foundations, Sect. 4 verifies the suitability of the framework to identify well-known variability mechanisms, and Sect. 5 summaries the work and presents future research directions.

2   Related Work

While some previous work has studied how to represent variability in SPL, e.g., [13, 14], we focus here only on studies that discuss variability mechanisms. In [8] seven variability mechanisms are described, including inheritance, uses (i.e., reuse of abstract functionality), extension, parameterization, configuration, generation, and template instantiation. As well, [2] also refers to inheritance, parameters, templates, and generators, all in the context of architecture design. It further refers to component substitution, which supports selecting from existing variants and inserting into artifacts. The study in [1] lists 13 implementation approaches and techniques for coding variability, including inheritance and parameterization. It also provides specific, code- related mechanisms, e.g., static libraries, dynamic link libraries, and aspect-oriented programming. The study in [17] distinguishes between four different approaches to model variability, depending on the flexibility allowed in SPL: parameterization, information hiding, inheritance, and variation points.

Despite the existence of many variability mechanisms, the catalogs and categorization frameworks of these mechanisms are built ad-hoc and lack structured development. Furthermore, the process of selecting variability mechanisms is often done based on preferences and earlier experience. However, associated costs and impacts are not always fully analyzed and understood [5]. In this work, we suggest an ontology-based framework in which software products and SPLs are considered things exhibiting behavior. The relationships between those things form the ground for representing variability mechanisms in a formal and precise way.

3   The Framework and its Theoretical Foundations

To formalize variability mechanisms, we adapt Bunge's work [3, 4], which emphasizes the notion of systems, defines behavior in a well-formalized way through the concepts of states and events, and has been adapted to conceptual modeling of information systems [15, 16]. The elementary unit in Bunge's ontology is a *thing*, which possesses properties, known to humans via attributes. Software products and SPLs are things whose behavior is of interest.

3.1   Things and Their Behavior

In Bunge's ontological model, things are represented by states and behavior, manifested by state variables x_i, each of which has an associated set of values $Range(x_i)$.

Definition 1 (States and State Families). Let $SV = \{x_1,...,x_n\}$ be a set of state variables associated with a thing. A *state* s for SV is an assignment of values to the state variables of SV, such that for each i: $s(x_i) \in Range(x_i)$. We denote s by $(v_1, v_2, ...)$, where $v_i \in Range(x_i)$ is the value assigned to x_i in s. The set of all states for SV is denoted by STATES(SV). A *state family* is a set of states $S \subseteq STATES(SV)$.

Consider for example a library management system, and suppose that SV contains *mode* and *status*, where: Range(mode) = {regular, event}; Range(status) = {open, closed}. A library can be open or closed. If open, it can be for an event or regularly (for the public to borrow books). We can define the states: s′ = (regular, open), s″ = (event, open). S = {s ∈STATES(SV) | s(status) = open} is thus a state family containing s′ and s″ in which the library is open.

The above example shows how state families can be described using logical conditions on some of the state variables $x_1,...,x_n$. We shall refer to $\{x_1,...,x_n\}$ as the set of *relevant state variables* for S. In our example {status} is the set of relevant state variables for S. Note that by restricting our view only to the relevant state variables, and defining appropriate mapping of their values, a state family can be viewed as a single state. In our example, S can be represented also as a ("single") state with respect to SV′ = {status}. Thus, we use the term state to refer also to state families.

States of a thing are changes due to internal and external interactions. According to Bunge, an external event is a change in the state of a thing as a result of an action of another thing. We abstract here from internal interactions by assuming a given set of external events that trigger behavior. Thus a behavior can be perceived as a transition from an initial state to a final state due to a sequence of external events.

Definition 2 (Behavior). A **behavior** is a triplet b = (S₁, <e>, S*), where S₁ and S* are states (termed initial and final, respectively) and <e> is a sequence of external events. The set of relevant state variables of the behavior b includes the sets of relevant state variables of S₁ and S*.

For example, consider the behavior of opening the library. It transitions from the initial state in which status = closed to a final state in which status = open due to an external event of librarian arriving. Here the initial and final states are actually state families with respect to SV = {mode, status}.

Definition 3 (Thing). A thing has the form T = (SV,E,B), where SV is a set of relevant state variables, E is a set of relevant external events, and B is a set of behaviors for SV and E.

As noted, we consider software products and SPL as things. The former are fully specified descriptions and thus are executable. The latter are partially specified descriptions. SPL artifacts are created and maintained to increase reuse and consequently decrease cost in the development of similar software products. This is primarily achieved by applying variability mechanisms. Therefore, understanding the nature of the exact relationship between the descriptions of software products and SPLs is the key to our ontological interpretation of variability mechanisms.

3.2 Relations Between Software Products and SPLs

Software products are obtained by introducing modifications to the set of SPL behavior or "concretizing" individual product behaviors. Accordingly, we distinguish between the *product dimension* and the *element dimension*.

Product Dimension. In the product dimension we examine the relationship between the whole set of behaviors of the software product (B_P) and the set of behaviors of the SPL (B_{SPL}). We distinguish between two possible cases:

- $B_P \subseteq B_{SPL}$: corresponds to configuration in which some SPL's behaviors are optional and thus a specific software product does not necessarily need to exhibit them.
- $B_{SPL} \subset B_P$: corresponds to extension in which the software product adds behaviors over the SPL's behaviors without "violating" existing SPL's behaviors.

Consider two behaviors of a library SPL: B_{SPL} = {regular-opening, opening-for-an-event}. In one case, a product can be obtained from SPL by configuration, exhibiting only regular-opening. In a second case, another product can be achieved by extension, exhibiting in addition to the SPL behaviors the product-specific behavior – opening-for-staff. A third product can exhibit both regular-opening and opening-for-staff, applying a combination of configuration and extension.

Element Dimension. In the element dimension we examine the relationship between a single behavior of a software product, b_P, and the corresponding (single) behavior of the SPL – b_{SPL}. Our analysis is based on the idea that the software product's behavior "concretizes" the SPL's behavior. While more details may be added, they cannot "violate" the SPL behavior. Such concretization, however, allows for variation in the use of state variables and their values. This is formalized by the notion of mappings.

Definition 4 (Mappings). Let b_{SPL} and b_P be behaviors of an SPL and a software product with relevant state variables SV_{SPL} and SV_P, respectively. A *state variable mapping* is a function S: $SV_{SPL} \rightarrow \wp(SV_P)$. Suppose that $x \in SV_{SPL}$ and $S(x) = \{y_1,...,y_n\}$ where $y_1,...,y_n \in SV_P$. A *value mapping* V_S maps each element in Range(x) to (one or more) elements of Range(y_1) × ...Range(y_n).

As an example, consider SV_{SPL} = {library_status}, where Range(library_status) = {open, closed}. Let SV_P = {back_door_status, front_door_status}, where Range(back_door_status) = Range(front_door_status) = {on, off}. We can define a state mapping S(library_status) = {back_door_status, front_door_status}. Then we may define a value mapping V_S as follows: V_S(open) = {(on, on)} and V_S (closed) = {(on, off), (off, on), (off, off)} (the library is open if and only if both its doors are on). The state variable mapping and value mapping induce in a natural way a mapping from states of SPL to sets of software products' states. We will denote this mapping by $M_{S,V}$. In our running example, the SPL state in which the library status is closed is mapped to the set of states: {(back_door_status = on, front_door_status = off); (back_door_status = off, front_door_status = on); (back_door_status = off, front_door_status = off)}.

The SPL state in which the library status is open is mapped to a set containing only one state: $\{(back\_door\_status = on, front\_door\_status = on)\}$.

We now turn to define *concretization* in which the product behaviors must be a faithful reflection of the SPL behaviors. We treat the sequence of external events as triggers of the corresponding behaviors. Concretization is only defined with respect to the initial and final states.

Definition 5 (Concretization). Let $b_{SPL} = (S_{1SPL}, <e_{SPL}>, S^*_{SPL})$ and $b_P = (S_{1P}, <e_P>, S^*_P)$ be behaviors of an SPL and a software product, respectively. Let SV_{SPL} and SV_P be sets of relevant state variables of the SPL and the software product, respectively, and suppose that there exist a state mapping $M_{S,V}$ from $S_{1SPL} \cup S^*_{SPL}$ to $\wp(S_{1P} \cup S^*_P)$ as defined above. We say that b_P is a *concretization* of b_{SPL} (with respect to $M_{S,V}$) if the following conditions are fulfilled: (1) $S_{1P} \subseteq M_{S,V}(S_{1SPL})$; (2) $S^*_P \subseteq M_{S,V}(S^*_{SPL})$.

For example, the SPL behavior for opening the library can be formalized as $b_{SPL} = (S_{1SPL}, <e_{SPL}>, S^*_{SPL})$, where $S_{1SPL} = \{s \in STATES(SV_{SPL}) \mid status = closed\}$, $S^*_{SPL} = \{s \in STATES(SV_{SPL}) \mid status = open\}$ and $<e_{SPL}> = <librarian-arrived>$. Consider the two following products' behaviors:

- $b_{P1} = (S_{1P1}, <e_{P1}>, S^*_{P1})$, where $S_{1P1} = \{s \in STATES(SV_{P1}) \mid back\_door\_status = off \ OR \ front\_door\_status = off\}$ and $S^*_{P1} = \{s \in STATES(SV_{P1}) \mid back\_door\_status = on \ AND \ front\_door\_status = on\}$; $<e_{P1}> = <librarian-arrived>$
- $b_{P2} = (S_{1P2}, <e_{P2}>, S^*_{P2})$, where $S_{1P2} = S_{1P1}$, and $S^*_{P2} = \{s \in STATES(SV_{P2}) \mid back\_door\_status = on \ AND \ front\_door\_status = off\}$; $<e_{P2}> = <librarian-arrived>$.

It can be checked that only b_{P1} is a concretization of b_{SPL}.

3.3   Mappings' Properties

We now examine the different ways in which concretizations may be achieved by taking a closer look at the properties of $M_{S,V}$ ingredients: the state variable mapping S and the value mapping V_S. Recall that S and V_S are functions of type F: $X \rightarrow \wp(Y)$, where X and Y are state variables or values in their range, respectively. We consider the following properties:

1. Total – for each $x \in X$ there is $y \in \wp(Y)$ such that $F(x) = y$.
2. Onto – for each $y \in Y$ there is $x \in X$ such that $y \in F(x)$.
3. Multiplicity of mapping –
   (a) For each $x \in X$, if $|F(x)| = 1$ then the multiplicity of mapping is *1-to-1*, where $|F(x)|$ is the size of the set $F(x)$. If in addition $F(x) = \{x\}$, then F is *identity*.
   (b) If there is $x \in X$ such that $|F(x)| > 1$ then the multiplicity of mapping is *1-to-m*.
   (c) If there is $y \in Y$ such that $|\{x \mid y \in F(x)\}| > 1$ then the multiplicity of mapping is *m-to-1*.

These cases are shown in Table 1, along with examples in the state variables and values levels.

Table 1. Properties of mappings

Properties	Case type	Example at the state variables levels	Example at the values levels
Not total	Optionality	Library Status, Door Sign → Door Status	Library Status (Open, Closing, Closed) → Door Status (Open, Closed)
Not onto	Expansion	Library Status → Door Status, Librarian Status	Library Status (Open, Closed) → Door Status (Open, Closing, Closed)
1-to-1 (identity)	Use-as-is	Library Status → Library Status	Library Status (Open, Closed) → Library Status (Open, Closed)
1-to-1 (non-identity)	Renaming	Library Status → Door Status	Library Status (Open, Close) → Door Status (Unlocked, Locked)
1-to-m	Refinement	Library Status → Front Door Status, Back Door Status	Library Status (Closed) → Door Status (Key Locked, Alarm Locked)
m-to-1	Consolidation	Library Status, Door Sign → Door Status	Library Status (On Holiday, On Maintenance) → Door Status (Closed)

(Rectangles and ellipses represent state variables of SPL and software products, respectively)

4 Characterizing Variability Mechanisms

The presented framework provides a formal way of capturing variability mechanisms. Some of these mechanisms are applied in the product dimension by adding to or removing from the *set* of SPL behaviors. The extension mechanism in [8] and the configuration and configurator mechanisms in [2, 8], respectively, are applied in the product dimension. Other variability mechanisms are applied in the element dimension by concretizing a *certain* SPL *behavior*. Table 2 presents the characterizations of some variability mechanisms, discussed in [2, 8], by considering different forms of concretization in the element dimension. Particularly, it shows for each mechanism the corresponding S and V_S properties and demonstrates via examples from the library domain.

Table 2. The Element Dimension – Variability Mechanisms and their S and V_S Properties

Variability mechanism	Explanations	Properties	Examples (Relevant state variables and their values)
Inheritance	Behavior specialization is achieved by introducing new state variables	S: 1-to-1 (identity), not onto V_S: 1-to-1	b_{SPL}: $library\ is\ closed \xrightarrow{librarian\ arrives} library\ is\ open$ (LibraryStatus:{open, closed}) b_P: $library\ is\ closed\ and\ it\ is\ an\ holiday \xrightarrow{librarian\ arrives} library\ is\ open\ for\ an\ event$ (LibraryStatus:{open, closed}; TimePeriod:{holiday, ...}; Opening Mode: {regular, event})
Parameters	Parameters are bound to actual values without changing other state variables or their values	S: 1-to-1 (identity) V_S: 1-to-1 (identity), not total	b_{SPL}: $n\ doors\ are\ closed \xrightarrow{librarian\ arrives} n\ doors\ are\ open$ (DoorsStatus:{open, closed}; n: {1, 2, 3, 4}) b_P: $2\ doors\ are\ closed \xrightarrow{librarian\ arrives} 2\ doors\ are\ open$ (DoorsStatus:{open, closed}; n: {2})
Template instantiation/Templates	Product-specific parts in a generic component are adapted	S: 1-to-1 (not-identity) V_S: 1-to-1	b_{SPL}: $entrance\ is\ closed \xrightarrow{librarian\ arrives} entrance\ is\ open$ (EntranceStatus:{open, closed}) b_P: $door\ is\ locked \xrightarrow{librarian\ arrives} door\ is\ unlocked$ (DoorStatus:{locked, unlocked})
Generation/Generator	Specialization and addition are needed to realize the specification	S: 1-to-m, not onto V_S: 1-to-m, not onto	b_{SPL}: $entrance\ is\ closed \xrightarrow{librarian\ arrives} entrance\ is\ open$ (EntranceStatus:{open, closed}) b_P: $door\ is\ locked\ and\ windows\ are\ closed\ and\ alarm\ is\ on \xrightarrow[arrives]{librarian} door\ is\ unlocked\ and\ windows\ are\ open\ and\ alarm\ is\ off$ (DoorStatus:{locked, unlocked}; WindowStatus:{locked, unlocked}; AlarmStatus:{on, off})
Component substitution	Component substitution corresponds to optionality (some variants are selected; others – not)	S: 1-to-1 (identity), not total V_S: 1-to-1 (identity)	b_{SPL}: $door\ is\ closed\ or\ sign - closed\ is\ on \xrightarrow{librarian\ arrives} door\ is\ open\ or\ sign - closed\ is\ off$ (DoorStatus:{open, closed}; SignClosed:{on, off}) b_P: $door\ is\ closed \xrightarrow{librarian\ arrives} door\ is\ open$ (DoorStatus:{open, closed})

5   Summary and Future Work

We have presented an ontological framework for analyzing variability mechanisms, in which relations between SPL and software products can be formally captured. This allows for a mathematical characterization of a variety of well-known variability mechanisms which can be applied at different development stages. The proposed framework is grounded in Bunge's ontology: both SPLs and software products are viewed as things which have certain sets of behaviors. Variability mechanisms can then be applied in two dimensions: (i) the product dimension, modifying the set of SPL behaviors, capturing such mechanisms as configuration and extension; (ii) the element dimension, concretizing a certain SPL behavior, capturing such mechanisms as inheritance, parameterization, template instantiation, and generation.

Future research will evaluate the framework with respect to different catalogs of variability mechanisms; utilize the framework for automatic analysis of variability in different SPL-software product settings [11]; and provide guidelines for selecting and utilizing variability mechanisms.

References

1. Anastasopoules, M., Gacek, C.: Implementing product line variabilities. In: Proceedings of the 2001 Symposium on Software Reusability (SSR 2001), pp. 109–117 (2001)
2. Bass, L., Clements, P., Kazman, R.: Software Architecture in Practice. SEI Series in Software Engineering, 3rd edn. Pearson, New York (2012)
3. Bunge, M.: Treatise on Basic Philosophy, Ontology I: The Furniture of the World, vol. 3. Reidel, Boston (1977)
4. Bunge, M.: Treatise on Basic Philosophy, Ontology II: A World of Systems, vol. 4. Reidel, Boston (1979)
5. Clements, P.: Managing variability for software product lines: working with variability mechanisms. In: 10th International Software Product Line Conference, pp. 207–208 (2006)
6. Clements, P., Northrop, L.: Software Product Lines: Practices and Patterns. Addison-Wesley, Reading (2001)
7. Gildea, D., Jurafsky, D.: Automatic labeling of semantic roles. Comput. Linguistics 28(3), 245–288 (2002)
8. Jacobson, I., Griss, M., Jonsson, P.: Software Reuse: Architecture, Process, and Organization for Business Success. Addison-Wesley Longman, Reading (1997)
9. Morton, T., Kottmann, J., Baldridge, J., Bierner, G.: Opennlp: A java-based NLP toolkit (2005). http://opennlp.sourceforge.net
10. Pohl, K., Böckle, G., van der Linden, F.: Software Product-line Engineering: Foundations, Principles, and Techniques. Springer, Berlin (2005)
11. Reinhartz-Berger, I., Itzik, N., Wand, Y.: Analyzing variability of software product lines using semantic and ontological considerations. In: Jarke, M., Mylopoulos, J., Quix, C., Rolland, C., Manolopoulos, Y., Mouratidis, H., Horkoff, J. (eds.) CAiSE 2014. LNCS, vol. 8484, pp. 150–164. Springer, Heidelberg (2014)
12. Reinhartz-Berger, I., Sturm, A., Clark, T., Cohen, S., Bettin, J.: Domain Engineering: Product Lines, Languages, and Conceptual Models. Springer, Berlin (2013)
13. Sinnema, M., Deelstra, S.: Classifying variability modeling techniques. Inf. Softw. Technol. 49(7), 717–739 (2007)
14. Svahnberg, M., Van Gurp, J., Bosch, J.: A taxonomy of variability realization techniques. Softw. Pract. Experience 35(8), 705–754 (2005)
15. Wand, Y., Weber, R.: On the deep structure of information systems. J. Inf. Syst. 5(3), 203–223 (1995)
16. Wand, Y., Weber, R.: An ontological model of an information system. IEEE Trans. Softw. Eng. 16, 1282–1292 (1990)
17. Webber, D., Gomaa, H.: Modeling variability in software product lines with the variation point model. Sci. Comput. Program. 53(3), 305–331 (2004)

Modeling and Visualization of User Generated Content

Personalized Knowledge Visualization in Twitter

Chen Liu[1], Dongxiang Zhang[2], and Yueguo Chen[3](✉)

[1] Rakuten Institute of Technology, Tokyo, Japan
chen.liu@mail.rakuten.com
[2] National University of Singapore, Singapore, Singapore
zhangdo@comp.nus.edu.sg
[3] Renmin University of China, Beijing, China
chenyueguo@ruc.edu.cn

Abstract. In recent years, Twitter has been playing an essential role in broadcasting real-time news and useful information. Many commercial media companies have used Twitter as an effective online tool to broadcast breaking news. Consequently, the tweets are organized in chronological order. In this paper, we investigate the tweet exhibition problem from a new perspective. We treat Twitter as an important source of knowledge gaining and exploit its semantic attributes. In particular, we focus on extracting and distilling knowledge from tweets and proposing a new organization and visualization tool to exhibit the tweets in a semantic manner. We conducted our experiments on Amazon Turk and the user feedback from crowd demonstrated the effectiveness of our new Twitter visualization tool.

Keywords: Semantic tweets · Wikipedia · Personalization · Visualization

1 Introduction

In recent years, the microblogging service, Twitter, has attracted hundreds of millions users and brought in great impact to our social life. As stated in [16], the majority (over 85 %) of trending topics in Twitter are headline news or persistent news in nature. It provides an open platform that enables ordinary people to conveniently publish information, e.g., updated statuses, re-post tweets, to each other. Even the largest commercial media companies in the world, such as CNN and BBC, have opened their broadcasting channels in Twitter for a real-time spread of breaking news.

As a conventional visualization manner, the tweets are presented to users in chronological order. However, such organization brings unfriendly user experience when information is overloaded and many important or interested tweets can get missed. To address the problem, some researchers proposed tweet ranking to sort tweets by their relevance with user interests [4]. The more relevant of a tweet to a user's profile, the higher position it is displayed in the tweet stream to reduce the probability of missing. However, such a ranking is based on the

© Springer International Publishing Switzerland 2015
P. Johannesson et al. (Eds.): ER 2015, LNCS 9381, pp. 409–423, 2015.
DOI: 10.1007/978-3-319-25264-3_30

simple `tf-idf` model and the semantic in tweets is not captured. In addition, as pointed out in [11], these similarity methods may not work well due to the noisy and short characteristics of tweets.

In this paper, we propose Twitter+, a novel solution to Twitter visualization. Our goal is to provide an alternative solution to information overloading by allowing users to browse the tweets in a semantic manner. We will elaborate two main challenges that we meet in building such a new visualization tool:

1. *Knowledge Organization:* Since tweets are short and noisy in essence, how to effectively group the tweets based on their semantic meaning and assign them with readable and meaningful labels to facilitate user browsing is the first challenge.
2. *Knowledge Personalization:* In order to provide a personalized view, the system needs to take into account user interest and provides a personalized ranking mechanism that works well in the semantic domain.

To address above issues, we use Wikipedia as the knowledge base and build a knowledge graph by extracting the hierarchical structure inside. In this way, we can add the semantic dimension to the Twitter stream. In particular, each tweet will be mapped into the nodes of the knowledge graph. Then a number of trees are built by connecting the mapped nodes. Each subtree denotes a semantic category. These subtrees are further ranked based on a tree kernel model so that the most relevant tweets fitting for user interests will be displayed on the top. To sum up, the contributions of the paper include:

1. We add a knowledge dimension to the tweet stream so that users can find useful or interesting tweets conveniently.
2. We propose a hierarchical manner to organize tweets according to structural information in Wikipedia and devise a kernel based method to measure similarities between tweet sets for personalized ranking.
3. We implement a new tweet visualization tool that is complementary to existing Twitter interface.

2   Related Work

There have been tremendous research efforts on better exploration of Twitter. Most of the works are focused on the social and media perspectives, such as topic detection, personal recommendation, sentiment analysis, spam analysis and event detection. In the following, we will address the most related works with our paper, including topic detection, personal recommendation and visualization.

2.1   Topic Detection

Topic is a key factor to organize tweets in a meaningful way. In order to discover topics from tweets, Naaman et al. [19] characterized contents on Twitter via a manual coding of tweets into various categories, e.g., from "information sharing" to "self promotion". To achieve better results, Ramage et al. [20] characterized users and tweets using labeled LDA, which is a semi-supervised learning model, to map tweets into multiple dimensions, including substance, style, status, and social characteristics of posts. Furthermore, Bernstein et al. [1] utilized search engines to enrich tweets with more relevant keywords in order to capture the topics of tweet stream. Our system differs from above works in that it is not necessary to pre-define a set of topics. With a knowledge base, our system is more general and shows better coverage.

Recently, Meij et al. [11] proposed the problem to link each tweet with related Wikipedia articles. They built plenty of features for each tweet and use machine learning methods to determine the most proper concepts which are used to describe the tweets. Their results yielded that the bag of concepts representation assists in understanding the text more precisely. Although we also need to extract semantic from tweets, this is only a preliminary step in our system. We have more contributions in how to organize and visualize the knowledge after the semantic mining.

2.2 Personal Recommendation

As the tweet stream floods in, users are faced with a "needle in a haystack" challenge when they wish to read the news that are most interesting to them. To solve this information overloading problem, one solution is to provide a personal ranking mechanism so that interesting tweets will be displayed first. For example, Guy et al. [12] re-ranked the stream based on three dimensions in the user profile: people, terms, and places. Their results indicated that building a profile based on the user's stream data can be effective for the personalization task. Chen et al. [5] studied URL recommendation on Twitter as a means to better guide user attention. They evaluated several algorithms through a controlled field study and concluded that both topic relevance and social voting are useful for recommendations. Duan et al. [10] extracted a set of features from Twitter feeds, such as content relevancy and account authority, then they employed a machine-learning strategy to determine the ranking of feeds.

2.3 Tweet Visualization Tools

The other solution to the information overloading is designing a more meaningful interface to explore tweets. For example, Dork et al. [9] designed an evolving, interactive and multi-faceted visualization tool to browse the events discussed in Twitter. Similarly, Vox Civitas [8] is a visual analytic tool to help journalists extract the most valuable news from large-scale social media content. Recently, TwitInfo [17] was proposed to generate an event summary dashboard, which identifies event peak and provides a focus+context visualization of long running events. The above works perform data mining from the whole tweet corpus. There are other works targeted at the personal tweets generated from the following people. For example, Eddi [1] grouped tweets from a user's tweet stream based on topics mined explicitly or implicitly and allowed users to browse the tweets first with the topics they were most interested in.

3   Problem Statement

To investigate the semantic attributes of tweets, our method is to map the tweets to the well-organized concepts in a knowledge base. We select Wikipedia as our knowledge base for two reasons. First, the articles in Wikipedia are of high quality. Ambiguity (single term with multiple different meanings) and synonym (multiple terms have similar meaning) are well solved in Wikipedia. For example, given a word "Apple" in a tweet, from the disambiguation page of "Apple"[1], we can obtain all the possible meaning of this term, including companies, films, music, etc. Then we can choose the most appropriate one depending on the context. In Wikipedia, each article describes a unique topic and is associated with a concept as the semantic attribute. In this paper, we use the article title, which is a succinct and well-formed phrase, to represent the concept. The second reason to use Wikipedia is that there is ontology embedded in the article to capture relationships between different concepts. One of such examples is category, which provides a hierarchy structure to organize the semantic. In the following, we use C to denote the concept set and $\mathbb{C}$ to represent the category set. For any concept c from C, we can find at least a category $\mathsf{c} \in \mathbb{C}$ such that c belongs to category c, denoted by $c \rightarrowtail \mathsf{c}$. Thereafter, we can define the knowledge graph based on the concept and category.

Definition 1. *Knowledge Graph*
A knowledge graph organizes the concepts and categories from Wikipedia in a directed acyclic graph $G = (V, E)$ with a root node R, where $V = \{v | v \in C \text{ or } v \in \mathbb{C}\}$ and for any edge $(u, v) \in E$, we have $u \rightarrowtail v$. The root node R in Wikipedia is `Category:Fundamental`.

Given the knowledge graph, we can first clear the noisy terms and then associate the remaining meaningful terms to Wikipedia concepts. Let $T(u)$ denote the tweet stream generated by the followings of user u. Since one tweet can be assigned to multiple concepts and there could be thousands of tweets in $T(u)$, a great number of concepts may be extracted from $T(u)$. How to effectively and efficiently organize these concepts is a challenging problem. In this paper, we define it as a knowledge organization problem which connects the related concepts in a number of knowledge subtrees derived from the knowledge graph. A knowledge tree is a subtree of the knowledge graph and its leaf nodes are required to be concept nodes. The knowledge organization problem can be formally defined as follows:

Definition 2. *Knowledge Organization*
Given a collection of weighted concept nodes $C' \subset C$ extracted from $T(u)$, the knowledge organization problem finds k sub-trees $\mathbb{S} = \{S_1, S_2, ..., S_k\}$ covering all the nodes in C' and with the minimum weight under a certain weighting function.

---

[1] http://en.wikipedia.org/wiki/Apple_%28disambiguation%29.

Note that the knowledge organization is similar to a clustering problem, which returns a set of clusters or knowledge trees. To support personalized recommendation, we assign a relevance score to associate each knowledge tree with respect to the user. Hence, the knowledge personalization problem is essentially a ranking problem to order the resulted knowledge trees so that the most relevant subtrees are displayed on the top to avoid missing by the user.

Definition 3. *Knowledge Personalization*
Given a collection of knowledge subtrees $\mathbb{S} = \{S_1, S_2, ..., S_k\}$ *and a user profile* u, *the knowledge personalization corresponds to a ranking function* $F(S_i, u) \rightarrow R$ *such that* $R_{S_i} > R_{S_j}$ *if* S_i *is more interesting to the user than* S_j.

4   Knowledge Organization

Given a tweet stream of a user, we first use existing techniques [6] to extract concepts by using Wikipedia as the knowledge base. When a set of concepts are extracted from the tweet stream, how to connect and organize them is a challenging problem. our purpose here is to connect similar concepts together, so that users can quickly find the inter-connections between tweets. Many algorithms have been developed to solve this issue, but for longer documents. For example, in [13], they compute the similarities between concepts and then apply traditional hierarchical clustering methods to group them. However, we argue that there are two drawbacks of the traditional methods. First, it is time consuming to compute pair-wise similarities between concepts considering their large volume. Second, it still fails to reveal the inner connections between concepts, such as why they are connected.

In this section, we propose to organize the concepts based on their shared parent categories. This is because the concepts are essentially leaf nodes in the knowledge graph. Two concepts will be connected to the same category if they are semantic related. The category could be used to explain why they are connected to some extent. For example, "iPhone" and "Macbook Pro" will be connected to the same category "Apple product". Given the concepts extracted from a tweet stream, a number of various organizations could be set up according to their different parent categories. Then, based on each parent category, we can obtain a subtree structure from the knowledge graph to denote the corresponding concept organization. From now on, we define each organization as a subtree and provide its formal definition as follows:

Definition 4. *Subtree S of tweet stream T*
Given a tweet stream T *and the concepts* $C(T)$ *extracted from* T, *a subtree* $S = (r, V, E)$ *could be built from the knowledge graph.* V *represents the concept nodes* C' *where* $C' \subseteq C(T)$ *and root* r *of* S *denotes a common parent category of* C'. *We call the root category as the label of the subtree. The edges in* E *are derived from the links between* r *and* V *in the knowledge graph.*

We further define the covering ability of subtree S, denoted by $Cover(S, C(T))$. Here we do not require a subtree to cover all the concepts in

$C(T)$, but to be meaningful, we require it contains at least two concepts. During the subtree generation process, if a category node A contains a set of concept nodes and is the unique child concept of B, we consider A a better result than B because it is more compact. Then the subtree built based on B could be pruned. This property is important to reduce the optimal subtree search space in the following section.

Since the concepts could be covered by different categories, then a number of subtrees could be built. To find the best set of subtrees covering all the mapped concept nodes, we need to assign a weight to a subtree as the evaluation of its quality. Our subtree weighting function takes into account the following properties, such as concept relevance or subtree representability. Besides, several other factors will affect the tree weight as well, such as the tree height or the root category generality. Note that the smaller the weight, the better the result.

- **Concept Relevancy.** When tweet stream is mapped to a set of concepts, some of the concepts may be more important than others. We assign a weight to determine the relevance score. Since the weight of subtree is an aggregation of concept relevance, we assign lower weight to a subtree if it contains concepts which are more relevant to the stream.
- **Subtree Representability.** Since a concept could belong to several subtrees, we need to measure the representability of the subtree to the concept. In another word, we prefer the subtree which can better represent corresponding concepts.

4.1   Subtree Weight

The weighting function for a subtree S with respect to the tweet stream T, denoted by $\mathcal{W}(S, T)$, needs to take into account the tree size and the concept weight it covers. We first identify the weight function $\mathcal{W}(c, T)$ of a concept node c. Given a tweet in the stream, as mentioned before, we extract keyphrases and detect related concept nodes using the association between keyphrase and concept, which is built offline. Each keyphrase p has a probability $P(c|p)$ to be associated with c. We utilize the concept frequency CF to measure the importance of a concept. If a concept is mapped with multiple tweets, we consider it more important. Finally, we can get

$$\mathcal{W}(c, T) = \sum_{p \in T} P(c|p) * CF(c).$$

To measure the relevance $\mathcal{R}(c, S)$ between concepts and its covering subtree, we identify the relevancy in two aspects. First, we consider similarity among the concepts covered by the tree. If the concepts are semantically close to each other, we consider the tree as coherent. Second, the tree height. According to the hierarchy of knowledge graph, the larger distance between the tree root to its covering concept, the less representative it is. Formally, we define $\mathcal{R}(c, S)$ as follows:

$$\mathcal{R}(c, S) = \frac{\sum_{c_i \in S.V \,\wedge\, c \neq c_i} Sim(c, c_i)}{h(S) * |S.V|}$$

We employ the method in [18] to compute $Sim(c, c_i)$. The higher value means the two concepts are more relevant. $h(S)$ is the tree height which is the longest path in the tree and $|S|$ represents the number of concept nodes contained in S.

Another concern is that a more general category will naturally connect to more concepts in the knowledge graph. For example, in the extreme case, the root of knowledge graph will connect to all the concepts in Wikipedia. However, the tree formed based on the root category is useless to the users. Therefore, we need to normalize the weighting function to reduce the side effect. Given a subtree S, the normalization factor $\sigma(S)$ is defined as $\sigma(S) = \frac{d(S)}{l(S)}$, where $d(S)$ represent the depth of the subtree root node in the whole knowledge graph. A small value of $d(S)$ means the subtree is close to the root node in the knowledge graph, i.e., the category is more general. $l(S)$ is the outdegree of the root node in the knowledge graph, a generative category will be with a larger outdegree, such as "United States of America". In summary, the final weighting function $\mathcal{W}(S, T)$ for a subtree S can be defined as

$$\mathcal{W}(S, T) = \frac{1}{\sigma(S) \cdot \sum_{c \in S.V} \mathcal{W}(c, T) \cdot \mathcal{R}(c, S)}$$

4.2   Optimal Subtree Selection

Given the weighting function of a subtree, our goal is to pick m subtrees so that they cover all the required concept nodes and the total weighting is smallest. We formalize the problem as follows:

Definition 5. *Optimal Subtree Selection*
Given a set of concepts C and their covering subtrees $\mathbf{S}$, find $\mathbf{S}' \subseteq \mathbf{S}$, so that $Cover(\mathbf{S}', C) = C$, and the weight $\sum_{s \in \mathbf{S}'} W(s)$ is minimized.

Our goal is to find a set of subtrees from them which cover all the concept nodes and are assumed as the best organization solution. In this case, $\{D, E\}$ is not a candidate because the set does not cover C_1 to C_5. We prove that the above problem is NP-hard.

Proposition 1. *The Optimal Subtree Selection problem is NP-hard.*

Proof. This proposition could be easily proved by a reduction from the weighted set cover (WSC) problem [15]. We reduce an instance of WSC problem to an instance of the Subtree selection problem as follows: for every element e in the WSC problem, we create a concept node c. For every subset g of the WSC instance, we create a subtree S which covers the corresponding concepts nodes. The weight of S equals to that of g. Given this mapping, it is easy to show that there exists a solution to the WSC problem which minimizes the weights if and only if there exists a solution to the subtree selection problem.

4.3   Subtree Selection Solution

In this section, we adapt a greedy algorithm [7] to solve the optimal subtree selection problem with theoretical guarantee. This algorithm will not only select

---

Algorithm 1. Greedy Algorithm for Subtree Selection Problem

Input: tweet stream T, Extracted Concept $C(T)$, Subtree Set S
Output: A subset of S which minimize the weight function
1 **begin**
2     $U = C(P)$, $R = \emptyset$;
3   **while** $U! = \emptyset$ **do**
4   let S_i be the subtree that has the minimum weight $w_(S_i)/|S_i \cap U|$
5        $R = R \cup S_i$;
6        $U = U \backslash S_i$;
7   return R;

---

the subtree set but also order them. We illustrate the algorithm in Algorithm 1. Given all the possible subtrees, we first order them by the score function below.

$$Score(S) = \frac{W(S,T)}{S \cap U} \qquad (1)$$

U is the set of elements that aren't covered as yet. According to the function, a subtree with a higher weight and covering more concepts will be selected out first. Both of the two criteria make sense when we rank and present the subtrees to users. Another implicit advantage of the formula is that it considers the redundancy between subtrees as well. This property embeds in the formula $S \cap U$ as at each step, it favors the subset which contains more new concepts compared to those already being covered.

5 Knowledge Personalization

Given a set of subtrees returned by the greedy algorithm, a naive solution is to display them ranked by their selection order. However, such an arrangement doesnot consider any influence from the targeting user. In this section, we consider how to re-order the subtrees when personalization is involved.

Generally speaking, a subtree will be assigned a higher score if it is more matching with the personal interests. In this paper, we use the tweets composed or re-tweeted by the user as the source to identify his personal interests, which shows more personal preference than the tweets coming from his followers. We adopt the technique presented above, including knowledge extraction and knowledge organization, to transform the personal tweet stream into subtrees. The problem becomes identifying the relevance score between a subtree and another set of personal subtrees. The best subtree is considered the most relevant to personal interests.

There exist several different ways to compute the similarity between tree-structure items. For example, we can decompose subtrees into a set of nodes and utilize Jaccard's distance to measure the difference. This method is efficient but sacrifices the result quality. First, it only considers the exact match of nodes, but ignores the textual similarity between them. Second, it fails to measure the structural connections due to decomposition. Another way is to adopt tree edit distance [2]. This metric is defined as the minimal number of edit operations to transform one tree to another. However, computing tree edit distance suffers an expensive computational cost. To avoid above problems, we propose a tree kernel method to solve the problem. Compared to above methods, our kernel based approach can be computed in linear time and capture both structural and textual similarity.

Tree kernels have been originally designed based on the idea of efficiently counting the number of tree fragments that are common to both argument trees. A larger function value means a more similar result. Different fragments of trees could be defined according to various situations, such as subtrees or subset trees [21]. In our case, our tree fragments are path based. It is formally defined as a path which starts from the root and ends at any internal node.

$$F(S) = \forall_{n \in S} Path(S.root \leftarrow n) \tag{2}$$

There are three reasons that encourage us to process in this way:

1. Since our tree is generated from Wikipedia, each path is enriched with semantic meanings. Generally, two trees are semantically similar if they share a common path. Therefore, there is no need to require a matching between subtrees as in subset tree splitting.
2. The path based fragments provide more chances to find a match between trees. It is equivalent as achieving "Query Expansion" effects in finding similarities between user and a subtree.
3. Intuitively, the match between longer paths is more important than the short ones in finding similarities. This idea could be naturally expressed in our method as the longer path, the more fragments derived from it.

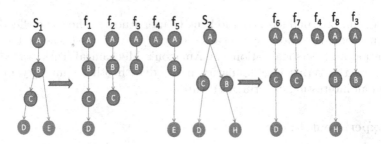

Fig. 1. Example of fragment generation

We give an example to illustrate fragments generated for two trees in Fig. 1. In the original tree kernel, it takes each fragment equally and only considers the structures. For example, the kernel score $\kappa(S_1, S_2) = 2 * 1 + 2 * 2 = 6$ only because they share the substructures f_3, f_4. In our case, we hope to take the node contents of the fragment into consideration as well when comparing them. Since nodes in fragments contain category or concept information, we could evaluate the content importance of two fragments from two subtrees that have the same structure. Let f_a be a fragment in S_a and f_b for S_b, we represent each of them as a vector $v = (w_1, w_2, ..., w_t)$ with each dimension corresponding to a node. Each element in v is evaluated by the metric introduced in Sect. 4. Then for two fragments, we obtain:

$$\kappa_f(f_a, f_b) = v_a \cdot v_b \tag{3}$$

where v_a and v_b are weighted vectors for f_a and f_b separately. $v_a \cdot v_b$ is the dot product.

Consequently, the tree kernel between two subtrees S_a and S_b is defined as follows:

$$\kappa(S_a, S_b) = \sum_{n_a \in S_a} \sum_{n_b \in S_b} \Delta(n_a, n_b) \tag{4}$$

whereby

$$\Delta(n_a, n_b) = \sum_{i=1}^{|F(S_a) \bigcup F(S_b)|} I_i(n_a) I_i(n_b) \kappa_f(f_i^a, f_i^b)$$

and where $I_i(n)$ is an indicator function which determines whether fragment f_i is rooted in node n.

Given the tree kernel, the relevancy between a subtree S and user u is defined as:

$$Rel(S, u) = \kappa(S, \mathbb{S}(u)) \tag{5}$$

$\mathbb{S}(u)$ is the subtree set generated from the user's personal tweets. Finally, the subtrees generated from last sections will be re-ordered based on Eq. 5 and returned to the user.

6   Experimental Study

In this section, we conduct a comprehensive user study to show the effectiveness of our methodology. More specifically, we test the effectiveness of knowledge organization and personalization via Amazon's Mechanical Turk. In addition, we compare the kernel tree similarity method with other similarity measures based on an interesting experiment setup.

6.1   Experimental Setup

To make the experiment results fair enough, we hire workers from Amazon's Mechanical Turk[2] to evaluate our performance. To guarantee the quality, we perform a qualification test on the workers before they join our evaluation. The qualification requires a worker must be active, such as publishing more than 50 tweets and with at least 10 followings. The statistics of the works are listed in Table 1 and we can see that each worker has shared plenty of personal tweets and following tweets so that we can obtain more accurate feedback for evaluation.

---

[2] https://www.mturk.com/.

Wikipedia. The Wikipedia data used in this paper is obtained from [18]. It was released in 22nd, July, 2011 and processed by the Wikipedia Miner tool. The dataset contains $739,980$ categories and $3,573,789$ concepts in total. There are $80,381,903$ links in the knowledge graph.

Table 1. Statistics of worker

workers	50
Avg. followings	283
Avg. personal tweets	572
Avg. following tweets	2451

Table 2. Comparison of different methods

Method	MAP	Precision@10	MRR
$tf - idf$	0.569	0.351	0.609
$Yahoo!$	0.5723	0.382	0.597
$Subtree$	0.761	0.558	0.801

6.2 System Interface

Figure 2 shows our new interface using tweets from "Alon Y. Halevy" as an exmaple. Alon Y. Halevy is a famous researcher with the interests on data integration, personal information management, etc. We organize his tweet stream in several layers and each layer represents a hierarchy structure of the knowledge. The topic labels on the left are more general than those on the right. Users can quickly identify the related tweets by clicking the node label and the relevant tweets will pop up. We can see that "distributed file system" or "data management" are two major topics matching his research interests. It is worth noting that we not only illustrate the tweets that are related but also how they are related. This makes Twitter+ different from the conventional topic extraction works such as [1] and serve as a complementary interface to existing Twitter system.

6.3 Effectiveness of Knowledge Organization

To measure the effectiveness of our knowledge organization methodology, we compare our subtree generation method with two conventional solutions: labels ranked by tf-idf and labels summarized using $Yahoo!API$[3]. The evaluation metrics include *Mean Average Precision (MAP)*, *Precision@10* and *Mean reciprocal rank (MRR)*. *MAP* rewards approaches returning relevant results earlier, and also emphases the rank in returned lists. *Precision@n* is the fraction of the top-n results retrieved that are relevant. *MRR* gives us an idea of how far we must look down in the ranked list to find a relevant result. Due to the high quality of Wikipedia ontology, the results in Table 2 show that our method is significantly better than the other two methods based on the feedback of Amazon Turk workers.

We further provide an illustrative example in Table 3 where top-5 subtrees labels associated with celebrities of five Twitter accounts who are influential in different domains. As shown in the Table, the majority of the results are quite reasonable. This verifies that our method is capable of categorizing tweets

---

[3] http://developer.yahoo.com/search/content/V1/termExtraction.html.

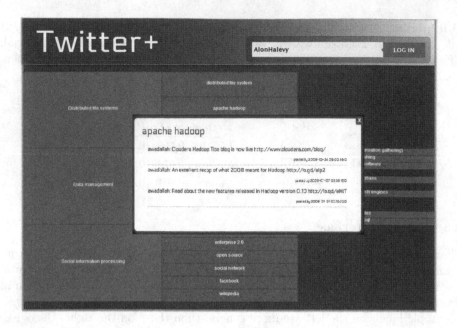

Fig. 2. Knowledge visualization of Alon Y. Halevy's tweets

into meaningful subtrees. For example, in Bill Gates's tweets, we successfully capture that currently he cares more about "philanthropy". This also illustrates that users' interests can indeed be reflected from their tweets.

6.4   Effectiveness of Kernel Similarity

Despite the lack of ground truth, we measure the effectiveness of kernel similarity based on the intuition that an effective similarity measure should be able to rank his followings higher than those randomly chosen users. For each worker, we compute the similarity between his tweets and each of his followings'. These are used as positive test examples and their relevancy scores are set to be 5. As negative test examples, we select the equivalent number users that the worker does not follow and process by the same procedure. Their relevancy is set as 1. Then different similarity functions are compared by the extent to which they rank the positive test users over the negative users. Specifically, we use Normalized Discounted Cumulative Gain ($NDCG$) [14] as the measurement metric. This metric takes both relevancy and ranking order into consideration and is widely used to measure the result quality. Our comparison methods include (1) *Jaccard Similarity*, builds the same model as BOW but uses Jaccard similarity as the measurement. (2) *Bag of Words*(BOW), models a user by his tweets to a $tf - idf$ weighted vector and utilizes cosine similarity. and (3) *Bag of Concepts and Categories*($BOCC$), similar to BOW except that the elements in the vector are concepts and their connected categories as in the subtrees instead of words. Table 4 shows the comparison results across the four

Table 3. Example labels for celebrities

BillGates	BarackObama	Oprah	WayneRooney	TechCrunch
Member states of the United Nations	Presidents of the United States	Oprah Winfrey	Premier League players	Social networking services
Global health	Member states of the United Nations	Member states of the United Nations	Premier League clubs	Web 2.0
Economics	Barack Obama	Cities in Texas terminology	Association football	Venture capital firms
Infectious diseases	States of the United States	English-language television series	Biology	Member states of the United Nations
Public health	Taxation	Grammy Award winners	Member states of the United Nations	Software companies of the United States

methods. Two conclusions are drawn from the results. First, Wikipedia semantics can improve the similarity computation between short text, such as tweets, since both of the two bottom methods outperform the upper two. We believe the reason lying in that *BOCC* representation reduces the ambiguity problem in tweets. This problem is more serious in short text than the normal documents as there is less context information to help understand the tweet topics. Second, structural information in Wikipedia is valuable to further improve the performance. Compared to the third method, our kernel method achieves a 17.8 % increase of the performance.

Table 4. Similarity function comparison

Method	NDCG Score
Jaccard	0.408
BOW	0.423
BOCC	0.470
Subtree	0.554

Table 5. Different list comparison

Method	NDCG Score
Timeline	0.303
BOW	0.324
BOCC	0.346
Subtree	0.383

6.5 Effectiveness of Knowledge Personalization

The personalization is essentially a re-ranking step to emphasize those more relevant to user's personal interests. To evaluate whether our method yields a gain, we need to transform our subtree results into a plain list of tweets since in current Twitter or many other interfaces, the tweets are shown in a list. We make them comparable by leveraging the evaluation strategy in search result clustering area. According to the method in [3], our list is produced by choosing the tweets from the top ranked subtree and going down until we reach a pre-set number of the total unique tweets. Such a subtree linearization would preserve the order in which subtrees are presented but just expand their content

In this experiment, we compare our personalization method with the other three tweet lists. One is simply ordered by the post time of the tweets and the

other two are ordered by cosine similarities between the two tweet streams or
the streams mapped to Wikipedia. We did not compare with Jaccard similarity
measure as it shows worse performance than the cosine similarity in the above
experiment.

1. *Timeline*, orders the unread tweets by their received time. This is provided
   by existing Twitter system.
2. *BOW*, models tweets in `tf-idf` weighted vectors and orders them by cosine
   similarity with users' personal tweets.
3. *BOCC*, replaces the elements in above vectors to concepts and categories
   mapped by tweets.

In our AMT experiment setup, we only consider top 30 tweets in each ranked
list. Each worker is asked to evaluate the relevancy of the ordered tweets according to his interests. To evaluate the rank lists, we employ the $NDCG$ score as
in previous experiments. In Table 5, we compare the average NDCG scores of
the ranked lists generated by different methods. As shown, the score for "Timeline" is relatively low. It indicates that current ordering of tweet stream indeed
has problems and cannot rank the user's interested tweets in the higher positions. However, our *Subtree* method achieves the best performance and shows a
significant improvement.

7   Conclusion

In this paper, we present a new tweet visualization tool as a complement to existing Twitter interface to ease the problem of information overloading. Based on
the hierarchical ontology of Wikipedia, we build a knowledge graph and derive
a number of subtrees to automatically categorize the tweet stream. In addition,
to further facilitate user reading, the subtrees are ordered based on their relevancy with the user's preference and labeled by explicit terms. We conducted
a comprehensive user-study on our methodologies, including the subtree label
quality, kernel method effectiveness, subtree relevancy and achieved promising
feedbacks.

Acknowledgement. Dongxiang Zhang is supported by the NExT Search Centre
(grant R-252-300-001-490), supported by the Singapore National Research Foundation
under its International Research Centre @ Singapore Funding Initiative and administered by the IDM Programme Office. Yueguo Chen is partially supported by National
863 High-tech Program (Grant No. 2014AA015204), and the National Science Foundation of China under grant No. 61472426.

References

1. Bernstein, M., et al.: Eddi: interactive topic-based browsing of social status
   streams. In: UIST, pp. 303–312. ACM (2010)

2. Bille, P.: A survey on tree edit distance and related problems. Theoret. Comput. Sci. **337**, 217–239 (2005)
3. Carpineto et al.: A survey of web clustering engines. ACM Comput. Surv. 41(3), 17:1–17:38, July 2009. ISSN 0360–0300
4. Chen, J., et al.: Make new friends, but keep the old: recommending people on social networking sites. In: CHI, pp. 201–210 (2009)
5. Chen, J., et al.: Short and tweet: experiments on recommending content from information streams. In: CHI, pp. 1185–1194. ACM (2010)
6. Chen, L., et al.: Integrating web 2.0 resources by wikipedia. In: MM, pp. 707–710. ACM (2010)
7. Chvátal, V.: A greedy heuristic for the set-covering problem. Math. Oper. Res. 4(3), 233–235 (1979)
8. Diakopoulos, N., et al.: Diamonds in the rough: social media visual analytics for journalistic inquiry. In: IEEE VAST, pp. 115–122 (2010)
9. Dörk, M., et al.: A visual backchannel for large-scale events. IEEE Trans. Vis. Comput. Graph **16**(6), 1129–1138 (2010)
10. Duan, Y., et al.: An empirical study on learning to rank of tweets. In: COLING, pp. 295–303. ACL (2010)
11. Edgar, M., et al.: Adding semantics to microblog posts. In: WSDM, pp. 563–572. ACM (2012)
12. Guy, I., et al.: Personalized activity streams: sifting through the "river of news". In: RecSys, pp. 181–188. ACM (2011)
13. Hu, X., et al.: Exploiting wikipedia as external knowledge for document clustering. In: KDD, pp. 389–396. ACM (2009)
14. Järvelin, K., et al.: Cumulated gain-based evaluation of IR techniques. ACM Trans. Inf. Syst. **20**(4), 422–446 (2002)
15. Khuller, S., et al.: A primal-dual parallel approximation technique applied to weighted set and vertex covers. J. Algorithms **17**, 280–289 (1994)
16. Kwak, H., et al.: What is twitter, a social network or a news media? In: WWW, pp. 591–600. ACM (2010)
17. Marcus, A., et al.: Twitinfo: aggregating and visualizing microblogs for event exploration. In: CHI, pp. 227–236. ACM (2011)
18. Milne, D., et al.: Learning to link with wikipedia. In: CIKM, pp. 509–518. ACM (2008)
19. Mor, N., et al.: Is it really about me? Message content in social awareness streams. In: CSCW, pp. 189–192. ACM (2010)
20. Ramage, D., et al.: Characterizing microblogs with topic models. In: ICWSM, AAAI (2010)
21. Vishwanathan, S.V.N., et al.: Fast kernels for string and tree matching. In: NIPS, pp. 569–576 (2002)

A Multi-dimensional Approach
to Crowd-Consensus Modeling and Evaluation

Silvana Castano, Alfio Ferrara, and Stefano Montanelli[✉]

Università degli Studi di Milano, DI - Via Comelico 39, 20135 Milano, Italy
{silvana.castano,alfio.ferrara,stefano.montanelli}@unimi.it

Abstract. In this paper, we propose a multi-dimensional approach to support modeling and consensus management in collective crowdsourcing applications/problems. We define the notion of crowd-consensus, and, for each dimension of analysis, we set pre-defined dimensional levels capturing the different variabilities characterizing the crowd-consensus along that dimension in different applications/problems. The design of a crowdsourcing task for a given target problem requires to characterize the task with respect to each dimension following a pattern-based approach.

Keywords: Crowd-consensus modeling · Crowd-consensus evaluation · Pattern-based task design

1   Introduction

Crowdsourcing solutions are getting more and more attention in the recent literature about social computing and collective intelligence elicitation [14]. The most widely accepted definition of crowdsourcing is given in [1], where crowdsourcing is defined as "a type of participative on-line activity in which an individual, [...], proposes to a group of individuals of varying knowledge, heterogeneity, and number, via a flexible open call, the voluntary undertaking of a task". Several applications have been proposed in many different contexts for crowdsourcing the execution of complex, time-consuming activities where the use of automatic procedures is not completely effective, such as for example collaborative filtering [12] and web-resource tagging [8]. In a usual crowdsourcing approach, a same task is assigned to (even high) number of workers each one providing her/his own answer, and the task result must be derived by assessing the level of agreement between the different answers and by deciding if a consensus has been reached. In this context, two levels of support tools are required. A first support is at conceptual level, where methodological approaches are required to formalize all the requirements and factors that influence crowd consensus evaluation, in order to correctly design and implement the corresponding task to be delivered to crowd workers for execution. A second support is at implementation level, where mechanisms are required for assessing the level of answer agreement and decide if the consensus has been reached for a given task and for deriving the final task result

P. Johannesson et al. (Eds.): ER 2015, LNCS 9381, pp. 424–431, 2015.
DOI: 10.1007/978-3-319-25264-3_31

out of answers with agreement. A number of consensus management mechanisms have been proposed, but little attention has been devoted to consensus management at conceptual level.

1.1   Related Work

The notion of consensus is frequently employed in crowdsourcing applications/systems as a solution for quality assessment of task results, either based on an explicit or implicit notion of worker agreement. Usually, consensus is described as an emerging property that could only be obtained by combining together multiple contributions provided by workers [2,3]. On the opposite, in [7], the authors claim that a key problem in quality assessment of crowdsourcing results is how to select the most appropriate worker contribution among the set of answers provided by a group of workers rather than to combine provided contributions. Examples in this direction are discussed in [6,13].

Other work pertinent to our paper are focused on the problem of classifying crowdsourcing approaches/systems. In [5], the focus of the classification is on the outcome produced by the involved workers. A different point-of-view is presented in [15], where the proposed crowdsourcing categories are organized in a taxonomy for providing a basic classification of applications, algorithms, performances, and datasets. In [11], the classification criteria are about technical aspects of quality assessment, like for example task design and data storage. Four different kinds of crowdsourcing systems are designed in [9] according to the type of job/task assigned to workers. Finally, a classification of crowdsourcing approaches based on the process organization is provided in [10].

Original Contribution of the Paper. In this paper, we focus on the conceptual level, and we propose five "dimensions of analysis" to support task modeling and consensus management in collective crowdsourcing applications/problems. With respect to classifications of crowdsourcing approaches/systems previously described, the consensus dimensions we propose are focused on classifying the different levels of variability that influence task modeling in consensus-based crowdsourcing problems. Moreover, we support the setup of the most suitable task configuration by fixing an appropriate dimensional level for each consensus dimension. To this end, four reference patterns to be used as starting point for task design are also provided, which have been derived from classification of common crowdsourcing problems and applications previously discussed.

2   Task Modeling and Crowd-Consensus Evaluation

Conventional crowdsourcing approaches are based on a *marketplace organization* where a *requester* (e.g., an individual, an organization) submits a number of *tasks* (i.e., atomic units of work) to distributed *crowd workers* in exchange for a *revenue*. Tasks are usually assigned in a "one-to-many" fashion, in that each task a set of different workers receive a same task to execute, and consensus

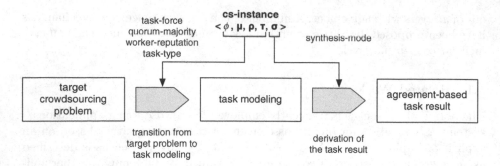

Fig. 1. Dimension-based task modeling and consensus evaluation

management capabilities are required to derive the final task result on the basis of the level of agreement among the, possibly different, answers autonomously provided by each worker [7]. In this context, we introduce the notion of crowd-consensus as follows.

Crowd-consensus is the outcome of a decision-making process for evaluating the degree with which the workers $W = \{w_1, \ldots, w_k\}$ involved in the execution of a task T agree on a specific answer $\bar{A}$ to be given as task result based on the answers $A = \{a_1, \ldots, a_k\}$ supplied by each individual worker.

A task T is defined as *successfully completed* when a crowd-consensus is reached among the workers in W, and it is *unsuccessfully completed* otherwise.

Requirements and conditions under which worker agreement and crowd-consensus are built and evaluated can vary from one crowdsourcing scenario to another. We argue that different *dimensions* influence the consensus evaluation, namely *task-force ϕ, quorum-majority μ, worker-reputation ρ, task-type τ,* and *result-synthesis σ,* each one characterized by different levels of variability (see Sect. 3). We call **cs-instance** $= \langle \phi, \mu, \rho, \tau, \sigma \rangle$ the dimensional 5-tuple modeling the task-instance to be executed by crowd workers, where each dimension of cs-consensus is bound to a specific dimensional level. According to Fig. 1, a cs-instance specification covers two stages of crowdsourcing problem modeling. First, a cs-instance enforces the *transition from target problem to task modeling* stage by properly configuring the relevant task features through the task-force, quorum-majority, worker-reputation, and task-type dimensions, respectively. Second, the cs-instance supports the *derivation of the task result* stage by setting the modality for evaluating the agreement on the worker answers and for deriving the final task result through the synthesis-mode dimension.

3   Crowd-Consensus Modeling Dimensions

Each dimension of crowd-consensus is characterized by different levels of variability expressing different requirements of the target problem to be solved.

Variability of the Task-Force (ϕ). This dimension deals with the number of workers $|W| = k$ to be involved in the execution of a certain task. In a task-force, each worker autonomously executes a received task and independently produces the answer according to her/his personal problem-understanding and expertise. A task-force can be considered as a notion of *implicit* and *dynamic* group of workers, in that (i) a worker $w \in W$ is not aware of other workers being involved in the execution of a certain task, and (ii) the composition of the set W changes from one task to another to avoid mutual and history-based influence among workers. The task-force dimension has an impact on the overall expenses required for the completion of the crowdsourcing activities, since each worker in a task-force has to be paid/rewarded for each executed task. We distinguish two possible levels of variability along this dimension: *broad* and *narrow*[1].

- *Broad* denotes a task-force with large/wide number of workers needed (i.e., $k > th_k$). The choice of a broad task-force is suitable when the collection of a large number of different task answers (i.e., worker opinions) is an important requirement or has a higher priority than the expense limitation.
- *Narrow* denotes a task-force with a small/limited number of workers involved (i.e., $2 \le k \le th_k$). The choice of a narrow task-force is suitable when the variability of possible answers is limited and/or when the expense limitation has a higher priority than the collection of a large number of worker opinions.

Variability of the Quorum-Majority (μ). This dimension is based on the notion of *quorum q* and it captures the different majority constraints to be applied for deciding whether a crowd-consensus is reached within the task-force W. The quorum q is a setup-time parameter and it represents the minimum percentage of workers in W that must agree on the task result $\bar{A}$ for reaching the crowd-consensus. We distinguish two possible levels of variability along this dimension: *qualified* and *non-qualified*.

- *Qualified* denotes a strong majority with a quorum value $0.5 < q \le 1$. A qualified quorum-majority means a *strong* constraint for the formation of crowd-consensus, and thus a lower likelihood that an agreement emerges among the workers in W (especially with a broad task-force). The choice of adopting a qualified quorum-majority is suitable when the priority of the requester is to ensure that a task is successfully completed only if the result is shared/supported by a large number of workers in the task-force W.
- *Non-qualified* denotes a weak majority with a quorum value $0 < q \le 0.5$. A non-qualified quorum-majority means a *weak* constraint for the formation of crowd-consensus, and thus a higher likelihood that an agreement emerges among the workers in W. The choice of adopting a non-qualified quorum-majority is suitable when the priority of the requester is to successfully complete as much as tasks as possible, though only a small crowd-consensus emerges within the workers in W.

---

[1] For implementation, a threshold th_k is specified to set the maximum number of workers that can belong to a narrow task-force.

Variability of the Worker-Reputation (ρ). This dimension captures the levels of *worker trustworthiness* t and expresses the reliability of a worker in executing tasks. At the beginning of the crowdsourcing activities, a worker w has an initial trustworthiness value $t_w = t^0 \in [0, 1]$ which is periodically updated (e.g., after a certain bulk of task executions) to capture the capability of the worker w to successfully complete tasks, and thus to participate in crowd-consensus[2]. We distinguish two possible levels of variability along this dimension: *novice* and *expert*.

- *Novice* denotes a worker reputation characterized by a low trustworthiness $t \in [t_{min}, t^0]$. A worker is novice if her/his trustworthiness is lower than its initial value meaning that it is decreased during the crowdsourcing activities. For a novice worker w, a *minimum-trustworthiness* value t_{min} is also defined to specify a lower bound on t_w under which the worker w is banned from the crowdsourcing activities and it is excluded from task assignments.
- *Expert* denotes a worker reputation characterized by a high trustworthiness $t \in [t^0, 1]$. A worker is expert if her/his trustworthiness is increased during the crowdsourcing activities with respect to the initial trustworthiness value t^0.

Variability of the Task-Type (τ). This dimension considers the nature of the request associated with the task T, based on the kind of worker contribution that is required for accomplishing the task. We distinguish two possible levels of variability along this dimension: *proposition* and *choice*.

- *Proposition* denotes a task request where the worker has to formulate (i.e., propose) an answer from scratch as result of task execution. In a proposition task, the expected answer can be any kind of worker-generated content, like for example a free text/numeric answer as well as a drawing or another visual/multimedia artifact. This task type enables the worker to express her/his creativity thus enforcing the use of crowdsourcing as a solution for knowledge creation and autonomous formation of the so-called wisdom of the crowd.
- *Choice* denotes a task request where the worker has to select her/his answer among a set of predefined alternatives. In a basic situation, the set of predefined alternatives are manually defined by the requester, and/or they are built through (semi-)automatic techniques (e.g., [5]). For more complex problems, we can envisage that the predefined alternatives are crowd-generated through the preliminary execution of proposition tasks (see Sect. 5). In the choice task, the worker contribution in terms of creativity and degree of freedom is quite limited since the opportunity to express a personal answer and/or point-of-view is not allowed. In other words, it is not possible for a worker to express her/his disagreement with all the predefined alternatives by proposing a new different answer.

---

[2] Techniques for progressively updating the worker trustworthiness are out of the scope of this paper. A possible solution based on reinforcement learning is discussed in [4].

Variability of the Synthesis-Mode (σ). This dimension captures the mechanisms used for determining the task result $\bar{A}$ based on the different task answers returned by the workers in the task-force for the assigned task. We distinguish two possible levels of variability along this dimension: *equivalence* and *statistics*.

- *Equivalence* denotes a synthesis-mode based on counting the number of "equivalent" worker answers to determine the task result $\bar{A}$. Given the set of answers A, the task result $\bar{A}$ is the answer with the highest frequency in A. Similarity evaluation techniques can be employed to determine whether two worker answers can be considered as equivalent, besides those identical [4].
- *Statistics* denotes a synthesis-mode based on calculating the task result $\bar{A}$ through statistical inference on the set of worker answers in A. The goal is to determine the task result $\bar{A}$ by considering the distribution of answers in A. This is done by calculating and/or combining one or more statistical indicators, like for example arithmetic mean, variance, or deviation [6].

The synthesis-mode, either equivalence or statistics, can be calculated through (i) a *nominal mechanism* where each answer in A has the same weight in determining $\bar{A}$, and ii) a *weighted mechanism* where each answer in A can have a different weight according to a given criterion (e.g., the worker reputation).

4 Pattern-Based Task Specification

To help the requester in specifying task instances, we provide four pre-defined *reference patterns* based on existing classifications of common problems/approaches in the crowdsourcing literature [5,9–11]. Each pattern codifies the best/recommended practice for consensus management by fixing the level/value of one or more cs-consensus dimensions (see Table 1).

Table 1. Reference patterns and recommended dimensional levels in the cs-instance

Reference pattern	Recommended dimensional levels in the cs-instance				
	ϕ dimension	μ dimension	ρ dimension	τ dimension	σ dimension
Crowd-creation	Broad	μ	Novice	Proposition	σ
Crowd-rating	Narrow	Qualified	ρ	Choice	σ
Crowd-ranking	Broad	Non-qualified	ρ	Choice	σ
Crowd-fact	ϕ	Qualified	Expert	τ	σ

Crowd-creation pattern ⟨broad, μ, novice, proposition, σ⟩. This pattern captures consensus features of crowdsourcing applications requiring the crowd to create/generate new contents based on their own feeling and creativity. Examples of crowd-creation applications are social bookmarking [8], story-telling (e.g., YoCrowd - http://www.yocrowd.com/), and product design (e.g., Design-Crowd - http://designcrowd.com/). We capture these features by fixing task-force ϕ=broad, worker-reputation ρ=novice, and task-type τ=proposition. The choice of task-type proposition as well as task-force broad are due to the nature of crowd-creation applications to encourage workers to express and share their personal point-of-view, so that a large number of different and (possibly) original contributions are collected. The choice of worker-reputation novice is motivated by the consideration that an original contribution in response to a proposition task can be generated by a worker without imposing a specific degree of trustworthiness.

Crowd-rating pattern ⟨narrow, qualified, ρ, choice, σ⟩. This pattern captures consensus features of crowdsourcing applications requiring the crowd to give opinions and rate on ideas, products, or services, as well as to parse, evaluate, and/or filter a given set of information with the aim at selecting the top-preferred one (e.g., HeyCrowd - http://heycrowd.com/). We capture these features by fixing task-force ϕ=narrow, quorum-majority μ=qualified, and task-type τ=choice. The task-type choice is due to the fact that a predefined set of answer options is always available in these kind of applications. In addition, the quorum-majority qualified ensures that the task result has been determined on the basis of a strong agreement, while the task-force narrow is set to foster the agreement formation, and thus to enforce the successful task completion.

Crowd-ranking pattern ⟨broad, non-qualified, ρ, choice, σ⟩. This pattern captures consensus features of crowdsourcing applications requiring the crowd to sort a number of available items according to a given ranking criterion (e.g., the relevance of an item with respect to a certain topic of interest). An example of crowd-ranking application is Rankr [12]. We capture these features by fixing task-force ϕ=broad, quorum-majority μ=non-qualified, and task-type τ=choice. The task-type choice is fixed since the items to rank are known. The task-force broad as well as the quorum-majority non-qualified are recommended to survey the opinion of a sufficiently large number of workers and to properly rank in the task result all the answers that reached the quorum.

Crowd-fact pattern ⟨ϕ, qualified, expert, τ, σ⟩. This pattern captures consensus features of crowdsourcing applications requiring the crowd to resolve the so-called *factual questions*, namely questions for which a right answer exists and the crowd has the goal to discover it. An example of crowd-fact application are web-sites for community-based question answering (e.g., Yahoo! Answers - http://answers.yahoo.com/). We capture these features by fixing quorum-majority μ=qualified and worker-reputation ρ=expert. The quorum-majority qualified is chosen to limit the likelihood that a fortuitous agreement among few workers is reached on a wrong answer. The choice of worker-reputation expert is based on the idea that experts have higher likelihood than novices to give the right answer.

5   Concluding Remarks

In this paper, we presented a multi-dimensional approach for task modeling and crowd-consensus evaluation. For evaluation purposes, we employed our web-based crowdsourcing system called Argo which implements consensus and trustworthiness management techniques. Case-study and experimental results are available on the Argo website[3].

In future research work, we plan to investigate modeling aspects of articulated crowdsourcing processes where different cs-instances are generated and combined to address the requirements of complex crowdsourcing applications.

References

1. Arolas, E.E., de Guevara, F.G.L.: Towards an integrated crowdsourcing definition. J. Inf. Sci. **38**(2), 189–200 (2012)
2. Barowy, D.W., Curtsinger, C., Berger, E.D., McGregor, A.: AutoMan: a platform for integrating human-based and digital computation. In: Proceedings of the 27th Annual ACM SIGPLAN OOPSLA Conference, Tucson, AZ, USA (2012)
3. Bozzon, A., Brambilla, M., Ceri, S., Mauri, A.: Reactive crowdsourcing. In: Proceedings of the 22nd International WWW Conference, Rio de Janeiro, Brazil (2013)
4. Castano, S., Ferrara, A., Genta, L., Montanelli, S.: Combining crowd consensus and user trustworthiness for managing collective tasks. Future Gener. Comput. Syst. (to appear, 2015)
5. Chiu, C.M., Liang, T.P., Turban, E.: What can crowdsourcing do for decision support? Decis. Support Syst. **65**, 40–49 (2014)
6. Demartini, G., Difallah, D.E., Cudré-Mauroux, P.: ZenCrowd: leveraging probabilistic reasoning and crowdsourcing techniques for large-scale entity linking. In: Proceedings of the 21st WWW Conference, Lyon, France (2012)
7. Doan, A., Ramakrishnan, R., Halevy, A.Y.: Crowdsourcing systems on the world-wide web. Commun. ACM **54**(4), 86–96 (2011)
8. Finin, T., et al.: Annotating named entities in Twitter data with crowdsourcing. In: Proceedings of the NAACL HLT Workshop on Creating Speech and Language Data with Amazon's Mechanical Turk. Los Angeles, CA, USA (2010)
9. Geiger, D., Rosemann, M., Fielt, E.: Crowdsourcing information systems a systems theory perspective. In: Proceedings of the 22nd Australasian Conference on Information Systems, Sydney, Australia (2011)
10. Geiger, D., Seedorf, S., Schulze, T., Nickerson, R.C., Schader, M.: Managing the crowd: towards a taxonomy of crowdsourcing processes. In: Proceedings of the Americas Conference on Information Systems (AMCIS), Detroit, MI, USA (2011)
11. Hoßfeld, T., et al.: Survey of web-based crowdsourcing frameworks for subjective quality assessment. In: Proceedings of the 16th IEEE International Workshop on Multimedia Signal Processing (MMSP), Jakarta, Indonesia (2014)
12. Luon, Y., Aperjis, C., Huberman, B.A.: Rankr: a mobile system for crowdsourcing opinions. In: Proceedings of the 3rd International MobiCASE Conference, Los Angeles, CA, USA (2012)
13. McCann, R., Shen, W., Doan, A.: Matching schemas in online communities: a Web 2.0 approach. In: Proceedings of the 24th International Conference on Data Engineering (ICDE 2008), Cancún, Mexico (2008)
14. Scekic, O., Truong, H.L., Dustdar, S.: Incentives and rewarding in social computing. Commun. ACM **56**(6), 72–82 (2013)
15. Yuen, M.C., King, I., Leung, K.S.: A survey of crowdsourcing systems. In: Proceedings of the 3rd IEEE International PASSAT/SocialCom Conference, Boston, MA, USA (2011)

---

[3] http://islab.di.unimi.it/LiquidCrowd/casestudy.php.

Principles for Modeling User-Generated Content

Roman Lukyanenko[1(✉)] and Jeffrey Parsons[2]

[1] College of Business, Florida International University, Miami, USA
roman.lukyanenko@fiu.edu
[2] Faculty of Business Administration, Memorial University of Newfoundland, St. John's, Canada
jeffreyp@mun.ca

Abstract. The increasing reliance of organizations on externally produced information, such as online user-generated content (UGC), challenges the common assumption of representation by abstraction in conceptual modeling research and practice. This paper evaluates these assumptions in the context of online citizen science that relies on UGC to collect data from ordinary people to support scientific research. Using a theoretical approach based in philosophy and psychology, we propose alternative principles for modeling UGC.

Keywords: Conceptual modeling · User generated content · Citizen science

1 Introduction

As traditionally, information systems (IS) were developed and primarily used within organizational boundaries [1, 2]; consequently, conceptual modeling research and practice have typically assumed that models are developed in the context of a well-defined and stable internal organizational setting. Novel forms of information production are emerging that challenge the assumption of internally-focused conceptual modeling. In contrast to information produced by employees or others closely associated with an organization, digital information is increasingly being created by members of the general public who often are casual content contributors (the crowd) - giving rise to the proliferation of *user-generated content* (UGC). Major sources of UGC are social media [3] and crowdsourcing [4], and can take diverse forms such as comments, tags, product reviews, videos, maps [4–8].

Of special interest in this paper is *structured* UGC (rather than less-structured forms, such as forums, blogs, or tweets). Structured user-generated information has advantages for rapid analysis and aggregation. For example, Cornell University launched eBird (www.ebird.com) to collect amateur bird sightings to support its research [8]. Data collection in this project involves populating pre-specified fields (e.g., indicating what biological species the bird is) to generate data in a form immediately useful and amenable to scientific analysis. As data in this project is intended to support scientific research, a key problem is creating structures (e.g., selecting classes) that would be both useful for project sponsors – biology experts and usable by potentially non-expert volunteers.

While incorporating UGC within organizational decision-making processes can be beneficial, these environments pose fundamental conceptual modeling challenges.

P. Johannesson et al. (Eds.): ER 2015, LNCS 9381, pp. 432–440, 2015.
DOI: 10.1007/978-3-319-25264-3_32

In the next section we evaluate major tenets of conceptual modeling research in the context of UGC. Considering the limitations of traditional approaches, we propose alternative principles for modeling UGC.

2 Challenges of Collecting User-Generated Content

Traditionally, conceptual models are used to capture information requirements during the earliest stages of IS development to guide the design and maintenance of the resulting IS [9]. Typically, information requirements are provided by the eventual consumers of data (e.g., managers, employees who require data to perform some tasks). Consequently, the *conceptual models reflect the intended uses of data, which are assumed to be established in advance and stable over time* [10, 11].

Organizational settings with innate governance structures made it possible to *reconcile any conflicting perspectives and promote common understanding on how to collect and interpret data*. Indeed, often a final conceptual model represents a global, integrated view but often does not represent the view of any individual user [12].

The fundamental *approach to conveying domain semantics in a unified conceptual model is representation by abstraction* [13, 14]. Abstraction makes it possible to deliberately ignore the many individual differences among phenomena and represent only relevant information (where consumers of data determine what is relevant based on known uses of data). Abstraction lies at the heart of popular conceptual modeling grammars. For example, a typical script made using the entity-relationship (ER) or Unified Modeling Language (UML) grammars depicts classes, attributes of classes, and relationships between classes. Classes (e.g., student, chair) abstract from differences among instances (e.g., particular students or specific chairs).

Representation by abstraction is assumed to enable complete and accurate representation of relevant domain semantics [15]. *Close contact with users (typical to traditional organizational settings) makes it feasible to capture all relevant perspectives.* Even after IS deployment users are often trained to provide data in the desired format. As Lee and Strong [16] contend, "[a]t minimum, data collectors must know what, how, and why to collect the data".

In contrast to more traditional settings where information creation is (assumed to be) well understood and controlled, in UGC projects there are typically no constraints on who can contribute information. Indeed, engaging broad and diverse audiences is frequently their raison d'être. When developing conceptual models for UGC, some requirements may originate from system owners or sponsors - a relatively well understood group - but the actual information is contributed by distributed heterogeneous users. Many such users lack domain expertise (e.g., product taxonomy or deep medical knowledge) and have unique views or conceptualizations that may be incongruent with those of project sponsors and other users [17]. While traditional systems represented a "consensus view" among various parties, the diverse and often unpredictable user views online make it infeasible to reach such consensus.

Without the ability to reach every potential contributor online, it is impossible to construct an accurate and complete representation of modeled domains. Indeed, the concept of completeness breaks down in this setting, as it assumes all relevant information and potential uses can be determined in advance. A conceptual model representing a domain as perceived by some users may marginalize, bias, or exclude possibly valuable conceptualizations of other users [10]. For example, to maximize scientific utility of the data obtained, projects like eBird typically adopt a scientific taxonomy (e.g., requiring online contributors to identify biological species). One potential negative consequence of this is low quality (e.g., accuracy, completeness) of information stored in the IS as non-experts may struggle to select the correct class or even decide to abandon data entry entirely [18]. Another consequence is lower engagement with systems that do not adequately represent the perspectives of particular users [19].

Presently, no established principles for modeling UGC exist [20], but some question whether traditional conceptual modeling in these new contexts is harmful and call for new approaches [10]. This work aims to answer this call by formulating design principles for modeling UGC.

3   Principles for Modeling UGC

As argued above, in UGC applications it may not be possible to reach a shared understanding among parties and, at the same time, create a data collection environment that supports the discovery and capture of unanticipated phenomena. This limits the applicability of traditional abstraction-based conceptual modeling approaches to UGC.

Abstraction-based conceptual models depict stylized [21] - generalized and simplified - representations of actual complex user experiences and beliefs about the structure of the world. Conceptual modeling grammars based on representation by abstraction assume that models elicited from different users will be similar enough to permit the creation of a unified view. UGC, however, brings with it an expectation that each user has a unique conceptualization of the phenomena in the domain. Thus, constructing a unified global schema may be infeasible.

Focusing users on any one view biases UGC projects to the view of some contributors and may preclude other views from being represented. Ontologically, it has been posited that the world is made of unique objects that humans perceive as stimuli [22, 23]. Humans create abstractions, such as classes, to capture some equivalence among objects for some purpose [24, 25]. Psychology research contends that prior experience, domain expertise, conceptualization, and ad hoc utility result in different abstractions of the same domain between contributors and for the same contributor over time [24, 26]. For example, a citizen scientist may create a class of "oiled birds" to refer to distinct objects (birds) that are covered in oil; this class helps the citizen scientist to communicate vital cues about a potential environmental disaster. The same objects seen a few days earlier could have been classified as "things to photograph" by a group of tourists or "cormorants" by scientists. Modeling using particular "privileged" classes (e.g., cormorant) promotes some uses, possibly at the expense of others. Chosen classes reflect the way humans reason about and understand reality. For example, preferences (e.g., political, cultural, ethic, gender) have been showed to impact the content of Wikipedia (e.g., [27]) and, consequently, its more structured companion, DBpedia.

To support these requirements, the foundation of conceptual modeling for UGC should rely on principles that are to the extent possible, invariant across people, rather than conditioned upon pre-determined abstractions and assuming specific uses. This leads to the formulation of the first principle:

Principle 1. **Modeling UGC should be based on minimal-abstraction, user- and use-invariant representations.**

This principle departs from traditional conceptual modeling driven by abstractions [13]. As class-based abstractions naturally vary across people and driven by predefined uses, they do not satisfy the first principle. To derive user and use-invariant structures, we turn to philosophy (specifically, ontology) that studies what exists in the world independent of human observers. We adopt the ontology of Mario Bunge [22] to reason about reality. Bunge's ontology has been popular in conceptual modeling research as it maps well to IS constructs and has been able to explain and predict a variety of information systems phenomena (for a meta-analysis, see [28]).

Bunge [22] postulates that the world consists of "things" (which can also be thought as instances, objects, or entities). We apply the notion of instances to things in the physical, social and mental worlds. Examples of instances include specific objects that can be sensed in the physical world (e.g., this chair, Herbert Simon) as well any mental objects (e.g., specific promise, Anna Karenina).[1]

Following Bunge, we argue that the instance is an elementary constituent of reality. Ontologically, the existence of an instance does not depend upon abstractions and hence is largely observer-independent. As a consequence, the objective of modeling is to represent potentially relevant instances as fully and faithfully as possible. This leads to the formulation of the second principle:

Principle 2. **The primary construct in modeling UGC should be the instance.**

According to Bunge, every instance possesses properties. Properties are always attached to instances and cannot exist without them; the materiality of properties directly derives from materiality of things. Every instance is unique in some way as different instances fail to share some of their properties.

According to Bunge [22], people are unable to observe properties directly, and perceive them instead as attributes, where several attributes can refer to the same property. The existence of an attribute does not imply that a particular property exists (e.g., the attribute name may be an abstraction of an undifferentiated bundle of properties).

Attributes are basic abstractions of reality insofar as any attribute (e.g., color red, roughness of texture, height of a building) is a generalization formed by humans and can be used to form higher-level abstractions (e.g., things with similar attributes can be grouped into classes). Thus, the third principle is:

---

[1] Notably, while Bunge's ontology is focused on the material world, ontologies with explicit social focus (e.g., Searle [31]) uphold the fundamental role of instances in shaping reality.

Principle 3. **Attributes should be used to represent properties of instances.**

According to Bunge, people use classes (termed "natural kinds" in Bunge) to group instances with common attributes. Classification (a major form of abstraction) allows humans to ignore irrelevant (for the purpose of classification) attributes of instances, thereby gaining cognitive economy and ability to infer unobservable attributes of instances. For example, by stating something is a bird, speakers can save the effort to communicate attributes assumed to be true of birds (e.g., has wings, probably can fly).

Using classes improves communication efficiency and lessens the effort of having to provide an exhaustive list of attributes of an instance. Classes are also intuitive when reasoning about instances. It is unnatural for users to refer to instance x in terms of its attributes alone. It is likely that users refer to x using some class (e.g., dog, employee, bank, account). Finally, knowing what classes users assign to instances reveals any biases in the kinds of attributes users attach to instances. The classes known to a person influence human perception (as illustrated by stereotype effects); knowing the classes users attach to instances, therefore, illuminates gaps and biases in the provided attributes. Classes become a convenient and natural mechanism by which users can reason about instances and describe their properties of interest. They also help a model reader understand the attributes provided. Therefore classes are conceptualized as constructs that can be attached to instances.

Principle 4. **A class can be attached to an instance to represent a set of attributes assumed to be true of other instances described by the same class.**

Finally, as classes can be expressed as bundles of attributes, online contributors should be encouraged, but not required, to provide the definitions (e.g., the kinds of attributes assumed true of the class instances) of the classes used. This leads to the following principle.

Principle 5. **Classes may be defined explicitly (e.g., in terms of attributes), if contributors wish to do so.**

Following from the above principles, modeling UGC is based on representing particular instances via attributes and classes as perceived by particular users at certain moments in time. In contrast to representation by abstraction, the principles proposed assume representational uniqueness, where every representation of an instance may be distinct from every other (i.e., expressed using different attributes and classes), including representations of the same instance by the same user at different times. At the same time, representational uniqueness does not imply that every stored representation be unique, as two different users may independently provide the same set of attributes and classes for the same instance; however in UGC environments shared classes and attributes are difficult to determine in advance.

As a consequence of representational uniqueness, capturing abstractions a priori no longer becomes necessary. This deviates fundamentally from traditional conceptual modeling that seeks discovery and representation of domain specific class-based abstractions that capture commonalities among instances. While the constructs used in the principles above are present in traditional grammars – the notion of instances (entities), classes (entity types), attributes exist in ERD, UML and other grammars, the way they are put together differs from traditional grammars sharply.

Representational uniqueness leads to IS development without relying on abstraction-driven grammars. Returning to the eBird project, where currently data collection is driven by forcing participants to classify an observed instance into a (potentially unfamiliar) class, the proposed principles would lead to a fundamentally different implementation. In particular, eBird would need to first switch from the current relational data model to a more schema-less one. A model should be capable of supporting the representational uniqueness assumption and allow instance attributes to be stored independent of existing abstractions. Once the database is chosen, the instructions in the data collection interface would call for observations of individual entities in the domain (here, specific birds). In accordance with the principles above, the instances could be described using any number of attributes and classes.

Under the proposed principles, crowd volunteers are able to provide information according to their own conceptualization of reality without having to conform to a particular structure. This data can then be stored in the schema-less database (e.g., instance-based [29]). As the data collection interface and the underlying schema no longer forces or even suggests a particular set of classes, observations of instances can be stored without any classification labels. If a volunteer fails to classify a particular instances to the level of specificity required by project's data consumers (e.g., scientists) – as would be common given low level of domain expertise among non-expert crowds [4, 10, 19] – the project can potentially turn to other users asking for help in classifying based on the attributes provided. In addition or as an alternative, machine learning algorithms can be employed for automatic classification. The implementation of the proposed principles on projects like eBird has the potential to increase both the quality of citizen science data [4] and participation rates [5]. Unlike UGC projects that implement traditional approaches to modeling and assume a basic level of expertise from citizen scientists, the principles above allow for the full spectrum of contributors to participate. We believe, the proposed principles represent a realistic compromise in UGC. Non-experts do not always know the phenomenon that was observed. It is more realistic to expect a volunteer to report some features of unknown things than to expect a precise classification and identification.

4   Implications and Future Work

With the growing importance of UGC [3, 4, 7, 30], as exemplified by the case of online citizen science, a pressing question is how to perform conceptual modeling in such environments. To address the emerging challenges, we propose a set of conceptual modeling principles intended to guide development of UGC projects. The principles are intended to support ever-increasing organizational efforts to harness information outside organizational boundaries. This is a cost-effective model of information production, but realizing its full potential remains difficult due to a misalignment between prevailing approaches to modeling and the characteristics of UGC environments. We expect the proposed principles to support creation and use of these types of IS.

While we motivated this work by the challenges inherent in such domain as crowd-sourcing and social media, it may also apply to traditional, corporate settings. The commonly assumed prerequisite of a shared domain understanding is considerably relaxed in the proposed principles. This makes them also applicable to heterogeneous corporate environments, especially given that even in traditional environments disagreements among parties may exist. Hence, using the proposed principles enables capturing individual views within companies and may prove beneficial for some corporate initiatives such as grassroots sense making, increased organizational agility and fostering divergent thinking and creativity.

In implementing the proposed principles, we hope that companies begin to harness collective intelligence of online crowds for internal analysis and decision making. Among other uses, contributions of crowds expand an organization's "sensor" network, bringing in large amounts of data from diverse audiences. Among other things, the principles proposed appear to meet two seemingly mutually-exclusive goals: leveraging crowds to satisfy organizational information needs and harnessing creativity and unanticipated insights of the crowds. We argue both can be achieved as long as fundamental assumptions about modeling change. By focusing on instances (rather than class-based abstractions), crowd contributors with different levels of domain expertise and motivation can contribute data.

With this research, we contribute to conceptual modeling theory. By proposing principles for modeling UGC, we identify a new perspective in conceptual modeling research - representational uniqueness based on concrete instances. This constitutes a significant change to the way conceptual modeling is normally understood and used; including its role in IS development, and functions performed by analysts and users.

A promising question for future research is whether modeling using the proposed principles can be further enhanced with the help of conceptual modeling scripts. In the example of a hypothetically-redesigned eBird (above), the analysis phase proceeds without relying on modeling scripts. A question remains whether development can be made more effective with the use of modeling scripts. The principles proposed here may guide development of conceptual modeling grammars - or rules and constructs that analysts can use to create "instance-based" conceptual modeling scripts.

References

1. Fry, J.P., Sibley, E.H.: Evolution of data-base management systems. ACM Comput. Surv. **8**, 7–42 (1976)
2. Zuboff, S.: In the Age of the Smart Machine: The Future of Work and Power. Basic Books, New York (1988)
3. Susarla, A., Oh, J., Tan, Y.: Social networks and the diffusion of user-generated content: evidence from YouTube. Inf. Syst. Res. **23**, 23–41 (2012)
4. Lukyanenko, R., Parsons, J., Wiersma, Y.F.: The IQ of the crowd: understanding and improving information quality in structured user-generated content. Inf. Syst. Res. **25**(4), 669–689 (2014)
5. Lukyanenko, R., Parsons, J., Wiersma, Y.: The impact of conceptual modeling on dataset completeness: a field experiment. In: ICIS 2014 (2014)

6. Krumm, J., Davies, N., Narayanaswami, C.: User-generated content. IEEE Pervasive Comput. **7**, 10–11 (2008)
7. Louv, R., Dickinson, J.L., Bonney, R.: Citizen Science: Public Participation in Environmental Research. Cornell University Press, Ithaca (2012)
8. Hochachka, W.M., Fink, D., Hutchinson, R.A., Sheldon, D., Wong, W., Kelling, S.: Data-intensive science applied. Trends Ecol. Evol. **27**, 130–137 (2012)
9. Wand, Y., Weber, R.: Research commentary: information systems and conceptual modeling - a research agenda. Inf. Syst. Res. **13**, 363–376 (2002)
10. Lukyanenko, R., Parsons, J.: Is traditional conceptual modeling becoming obsolete? In: Ng, W., Storey, V.C., Trujillo, J.C. (eds.) ER 2013. LNCS, vol. 8217, pp. 1–14. Springer, Heidelberg (2013)
11. Chen, P.P.: Suggested Research Directions for a New Frontier – Active Conceptual Modeling. In: Embley, D.W., Olivé, A., Ram, S. (eds.) ER 2006. LNCS, vol. 4215, pp. 1–4. Springer, Heidelberg (2006)
12. Parsons, J.: Effects of local versus global schema diagrams on verification and communication in conceptual data modeling. J. Manage. Inf. Syst. **19**, 155–184 (2003)
13. Mylopoulos, J.: Information modeling in the time of the revolution. Inf. Syst. **23**, 127–155 (1998)
14. Smith, J.M., Smith, D.C.P.: Database abstractions: aggregation and generalization. ACM Trans. Database Syst. **2**, 105–133 (1977)
15. Olivé, A.: Conceptual Modeling of Information Systems. Springer, Berlin (2007)
16. Lee, Y.W., Strong, D.M.: Knowing-why about data processes and data quality. J. Manage. Inf. Syst. **20**, 13–39 (2003)
17. Erickson, L., Petrick, I., Trauth, E.: Hanging with the right crowd: matching crowdsourcing need to crowd characteristics. In: AMCIS 2012 Proceedings, pp. 1–9 (2012)
18. Lukyanenko, R., Parsons, J.: Rethinking data quality as an outcome of conceptual modeling choices. In: 16th International Conference on Information Quality, pp. 1–16 (2011)
19. Parsons, J., Lukyanenko, R., Wiersma, Y.: Easier citizen science is better. Nature **471**, 37 (2011)
20. Lukyanenko, R., Parsons, J.: Conceptual modeling principles for crowdsourcing. In: International Workshop on Multimodal Crowd Sensing, pp. 3–6 (2012)
21. Kaldor, N.: Capital accumulation and economic growth. In: Lutz, F.A., Hague, D.C. (eds.) The Theory of Capital, pp. 177–222. Macmillan, London (1961)
22. Bunge, M.: Treatise on Basic Philosophy: Ontology I: The Furniture of the World. Reidel, Boston (1977)
23. Rosch, E.: Principles of categorization. In: Rosch, E., Lloyd, B. (eds.) Cognition and Categorization, pp. 27–48. Wiley, Hoboken (1978)
24. Murphy, G.L.: The Big Book of Concepts. MIT Press, Cambridge (2004)
25. Smith, E.E., Medin, D.L.: Categories and Concepts. Harvard University Press, Cambridge (1981)
26. McCloskey, M., Glucksberg, S.: Natural categories: well defined or fuzzy sets? Mem. Cogn. **6**, 462–472 (1978)
27. Bilic, P., Bulian, L.. Lost in translation: contexts, computing, disputing on Wikipedia. In: iConference (2014)
28. Saghafi, A., Wand, Y.: Conceptual models? A meta-analysis of empirical work. In: Hawaii International Conference on System Sciences (2014)
29. Parsons, J., Wand, Y.: Emancipating instances from the tyranny of classes in information modeling. ACM Trans. Database Syst. **25**, 228–268 (2000)

30. Lukyanenko, R., Parsons, J.: Information quality research challenge: adapting information quality principles to user-generated content. J. Data Inf. Qual. (JDIQ) **6**(1), 3 (2015)
31. Searle, J.R.: Intentionality: An essay in the Philosophy of Mind. Cambridge University Press, Cambridge (1983)

Ranking Friendly Result Composition
for XML Keyword Search

Ziyang Liu[1], Yichuang Cai[2], Yi Shan[3](✉), and Yi Chen[4]

[1] LinkedIn, Mountain View, CA, USA
ziliu@linkedin.com
[2] Microsoft, Redmond, WA, USA
yica@microsoft.com
[3] School of Computing, Informatics, and Decision Systems Engineering,
Arizona State University, Tempe, AZ, USA
yshan1@asu.edu
[4] School of Management, New Jersey Institute of Technology, Newark, NJ, USA
yi.chen@njit.edu

Abstract. This paper addresses an open problem of keyword search in
XML trees: given relevant matches to keywords, how to compose query
results properly so that they can be effectively ranked and easily understood by users. The approaches adopted in the literature are oblivious to
user search intention, making ranking schemes ineffective on such results.
Intuitively, each query has a search target and each result should contain
exactly one instance of the search target along with its evidence about its
relevance to the query. In this paper, we design algorithms that compose
atomic and intact query results driven by users' search targets. To infer
search targets, we analyze return specifications in the query, the modifying relationship among keyword matches and the entities involved in the
search. Experimental evaluationsvalidate the effectiveness and efficiency
of our approach.

Keywords: Keyword search · XML Tree · Search intent

1 Introduction

Keyword search provides a simple and friendly mechanism to access the information in XML documents for users who do not know structured query languages,
or for the applications in which data schemas are too complex or fast-changing.

To generate results for XML keyword search, we need to (1) Identify relevant
matches to input keywords. (2) Compose query results based on the relevant
matches. (3) Rank the results according to their relevance to input keywords.

Much research has been performed to address the challenges in the first and
third steps. For identifying relevant matches, existing approaches use *Variants*
of *Lowest Common Ancestors* to connect keyword matches, referred as VLCA in
this paper [1–9]. For example, consider a query "*Arizona, state*" on the XML tree
in Fig. 1. In Fig. 1, we assign integer IDs to some nodes to facilitate illustration.

© Springer International Publishing Switzerland 2015
P. Johannesson et al. (Eds.): ER 2015, LNCS 9381, pp. 441–449, 2015.
DOI: 10.1007/978-3-319-25264-3_33

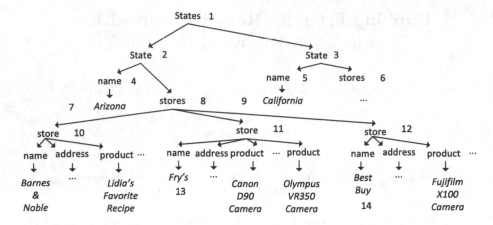

Fig. 1. Sample XML tree about stores

Although there are many matches to *state*, only the one (with ID 2) that corresponds to the *Arizona* match is considered relevant. For result ranking, a variety of ranking factors were proposed [2], including result size, keyword proximity, number of keyword matches in the result, as well as IR-style ranking functions.

However, little study has been done on the second problem: composing query results. In text search, a retrieval unit is a *document* in the repository, and thus the problem of result composition is inapplicable. In contrast, XML keyword search engines returns subtrees in the XML data tree in order to provide finer-grained query results and to better satisfy the users' needs. Given relevant keyword matches, how to dynamically extract a right subtree from the original data tree as a query result is not trivial. At the same time, the way of composing results in XML keyword search has crucial effects on result ranking and user search experience, as to be shown in the following examples.

Example 1. A user seeking the stores selling cameras in Arizona would issue a query "*Arizona, camera, store*". Consider the XML tree in Fig. 1 as a fragment of the XML data.

One popular query result composition methods is named as *Subtree Result* in this paper, adopted in many existing work [4,10,11]. Subtree Result defines a query result as a tree rooted at a VLCA node consisting of *all* relevant matches that are descendants of this VLCA node and the paths connecting them.

For this query, Subtree Result only returns one result: the tree rooted at a *state* node (0.0) that contains the match to *Arizona* and *all* the matches to *store* and *camera*, as shown in Fig. 2(a).

Such a result is not informative to the user. As we would imagine, there are hundreds of stores in Arizona that sells cameras. Instead of checking all of them, a user would want to find the top ranked stores to visit, where the ranking might consider store reputation (mimic to page-rank), the number of cameras sold (related to TFIDF ranking), location (related to local search), among other factors. However, with all the camera stores in Arizona in a single result, none of the existing ranking methods can rank the stores. With such a query result, a user would have to spend prohibitive amount of efforts to check the information of every store, design a ranking method herself to rank the stores in order to decide which few of them to visit.

Besides Subtree Return, the other commonly used result composition method is called *Pattern Match*, used in [2,3]. Pattern Match defines a query result as a tree rooted at a VLCA node consisting of *exactly one relevant match to each query keyword* and the paths connecting them.

For the query "Arizona, camera, store" on the XML tree in Fig. 1, some sample results generated by Pattern Match are shown in Fig. 2(b). As we can see, each result has information of exactly one camera.

Such results can be annoying. The data may have thousands of cameras, sold by hundreds of stores. The query user looks for *stores* that sells cameras, not cameras. It's much more desirable to have all distinct stores ranked automatically, rather than having the user to read tens of cameras sold by the same store and to manually rank the stores in order to find the top ranked ones.

In contrast to the two existing approaches, this paper presents techniques that compose query results based on the inferred search semantics. For the same query "Arizona, camera, store", we identify that the user is looking for camera stores in Arizona, and compose query results such that each result contains information of one distinct store, along with the related matches to *Arizona* and *camera* as evidence of its relevance, such as the two query results shown in Fig. 2(c). In this way, when results are properly ranked, a user obtains the top ranked stores.

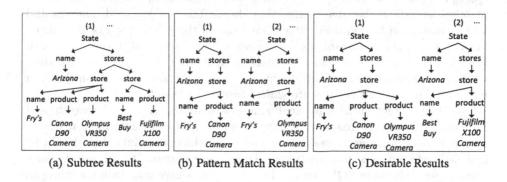

(a) Subtree Results   (b) Pattern Match Results   (c) Desirable Results

Fig. 2. Results of query "Arizona, camera, store"

Intuitively, each keyword search has a goal, either a real world entity or relationship among entities, as observed in [12]. We use term *search target* to refer to the information that the user is looking for in a query, and *target instance* to denote each instance of the search target in the data. Each desirable query result should have *exactly one target instance* along with all associated evidence that shows the relevance to the user query, so that top-k ranked results corresponds to top-k ranked search target instances.

Based on this intuition, we propose a novel technique to automatically compose atomic and intact query results for XML keyword searches. Unlike the existing approaches, which are oblivious to users' search intentions, the proposed query result composition is *driven by the user search target* and hence is ranking friendly. Experimental evaluations validate the effectiveness and efficiency of our approach.

2   Target Driven Query Result Composition

Users who issue queries often desire the information of one or a set of entities that satisfy certain conditions. We name such entities as *target entities*. In this section we first introduce the data model and then discuss how to automatically infer target entities from user query and data. Then we discuss how to compose meaningful query results based on relevant matches and inferred target entities. We propose strategies to compose query result centred around the inferred search target.

2.1   Identifying Target Entities

We first introduce the data model of an XML tree. We consider a node in the XML data tree has one of three categories: *entities* that represent real world objects, *attributes* that describe the corresponding entities, and *connection nodes* which connect other nodes but do not have much meaning. Our system adopts the heuristics developed in [10] to infer node categories. For example, in Fig. 1, *state* is entity and *name* is its attribute. *stores* is a connection node. A *Keyword Query* is a set of words. A *Query Result* is a tree that consists of a set of relevant matches as well as the edges connecting them. Relevant keyword matches and VLCA nodes can be identified by applying one of the existing works introduced in Sect. 1. An entity instance is a *relevant entity instance* if it is on the path from a VLCA node to a relevant match in XML tree. The types of such data nodes are *relevant entities*. Consider query "Arizona, camera, store" on the data in Fig. 1. The relevant entities are *state*, *store* and *product*.

We infer target entities by analyzing the matches to input keywords and the XML data structure. Often a query has two parts: the information that a user is looking for, referred as *return node*, and the constraints that should be satisfied, referred as *search predicates*. These are analogous to the *select* clause versus *where* clause in SQL queries. For example, a user may look for stores in Arizona and issue a query "Arizona, store". In this query, store is a return node, and Arizona is a search predicate.

To automatically detect return nodes and search predicates from a user query, we observe that if an entity or attribute node name is specified in a query without information about its associated attribute values, then likely this node name represents a return node, and its attribute values are what the user is looking for. On the other hand, an attribute value node (e.g., *Arizona*), or a pair of attribute name and value (e.g., *state, Arizona*) is likely to be a search predicate.

In a query Q, we consider a keyword k to be a *return node* if one of the following conditions holds, otherwise is a search predicate. (1) k matches an entity e, and there is no keyword k' that matches its attribute or attribute value. (2) k matches a connection or attribute node u, and there is no keyword k' matching a node v, such that u is an ancestor of v.

Next we can infer target entities from return nodes. If a return node is an entity node, then it is a target entity. If a return node is an attribute (e.g., *address*), then the associated entity (e.g., *store*) is considered as a target entity. Otherwise, a return node is a connection node (e.g., *stores* in sample XML document), then its nearest descendant entities are considered as target entities (e.g., *store*).

However, not all user queries provide hints for return nodes. In case the query keywords do not contain return nodes, we exam all the relevant entities and the relationships between search predicates and theses entities to identify target entities. We have two observations. First, a user may use the attribute values to modify an entity (e.g. find stores that are named Fry's), or attribute values of a related entity to modify an entity (e.g. find stores that are in the state of Arizona). Second, each search predicate keyword shall modify the instances of the target entity. In other words, removing any keyword from the query will return different target instances. For example, users will not use query "Arizona, Store" to search for states that are named as "Arizona" and that have a store, because every state has stores, and keyword "store" does not modify instances of state. We propose *modifier* to represent a keyword match of search predicate and *modifying relationship* to represent the relationship between the match and an entity. Based on observation 1, we define modifier to be any attribute value or name-value pair A connected to entity E. Furthermore, based on observation 2 A is a modifier if there is at least one instance of E in the XML tree which does not have A connected. Otherwise A cannot be used to further describe E. So E is a target entity when all search predicates can modify it. Due to limited space, we refer readers to [13] for details.

When there is only a single target entity, it is called the *center entity* of the query result which is the search target of the keyword query. When there are no target entities found, we consider all relevant entities as target entities. When there are multiple target entities found, besides these entities we further consider the relationships between them as search targets.

2.2 Composing a Query Result

As illustrated by examples in Sect. 1, results of XML keyword search should be *atomic*, i.e., consist of a single target instance; and *intact*, i.e., contain the whole target instance together with all its supporting information. For example, in Example 1, each query result should correspond to a distinct store. Atomicity enables the ranking method to rank the target instances and show the top-k most relevant ones to the user. Also, each query result should have all supporting information, all cameras sold by the store. With intactness, a ranking method can have the whole view of each target instance to give a fair ranking. A keyword match of k is supporting information of a target instance e if there are no other keyword matches of k with closer relationship to e [13].

When a query has multiple target entities, we observe that atomicity and intactness may not be simultaneously achievable. Consider Fig. 1 for example. If both *state* and *store* are known as target entities, then all *store* nodes in the subtree of a *state* node constitute the supporting information of the *state* node. According to intactness, they should all be included in the result that contains this *state* node. However, according to atomicity, only one *store* node can be included in one result. In this case, since atomicity and intactness are not both achievable, we choose to achieve intactness using subtree result. The reason is that subtree result can be achieved much more efficiently and scalably than pattern match.

3   Algorithms

In this section, we present the indexes and algorithms to efficiently identify target entities and generate meaningful query results.

3.1   Indexes

A core operation in the query composition is to identify target entities. There are three indexes we use: *Label Index*, *Node Index*, and *Modifier Index*. Label Index is an inverted index, which retrieves a list of data nodes given a keyword.

As discussed in Sect. 2, we have different strategies for determining target entities in two different situations. If there are return nodes specified in input keywords, target entities are the entities associated with the return nodes. Otherwise, if return nodes are not specified, we check the modifying relationship between each attribute value that matches a keyword and each relevant entity involved in the query.

For the first case, we need to quickly determine a return node's associated entity. To support this, we build a node index for all entity and attribute nodes. For an entity node, the entry includes all its attribute values. For an attribute node, the entry includes its value and its parent entity.

For the second case, we need to quickly determine whether a predicate is a modifier of an entity type. To support it, we build a Modifier index that records the modifying relationship between attribute values (more accurately, attribute name value pairs) and entities. In this index, each attribute value has an associated list that records the entities that are *not* modified by this attribute value. Since the entities that are modified by an attribute value are far more than those that are not modified by it, we record the negative cases.

3.2   Generating Query Results

Now we present our algorithms, which takes relevant matches to a keyword query (which are obtained by adopting one of the existing approaches [2–4,14]) and indexes as input, and composes meaningful query results.

Based on Sect. 2, first we need to identify target entities based on return nodes and search predicates. Second, we use target entities to construct query results which are atomic and intact.

For a keyword search, first we use the label index to retrieve all nodes matched by a query keyword. Second, we compute VLCA nodes, each of which is the root of one or more query result trees, for which purpose we use the algorithm proposed in XKSearch [4]. Third, we find relevant entity instances by checking whether they are on the paths from a VLCA to a relevant match (Sect. 2.1). Fourth, using node index, we can determine the node category of each keyword match and then infer the return nodes and search predicates of the query. To achieve that, we first retrieve the entry for each keyword. If it is matched by an entity or attribute name, we further exam whether other keywords match a value of an attribute of the entity or a value of the attribute. If a keyword is matched by an entity or attribute name but no other keywords match its value, then this keyword is a return node, otherwise search predicate. If return nodes are present, we can find the corresponding target entity instances by accessing the node index as well. Otherwise, we find the modifying relationship and consequently the target entity instances using the modifier index by checking whether each search predicate modifies the target entity instance or not. Next, if a center entity exists, then each relevant entity instance that is an instance of the center entity leads to a query result. For each such entity instance e, a query result is generated consisting of e and its supporting information (Sect. 2.2), together with their connections. Such a result is atomic and intact. If there is no center entity, it takes the default mode, which requires the result to be *intact*. To achieve that, it generates query results by returning the relevant matches in the subtrees rooted at VLCA nodes.

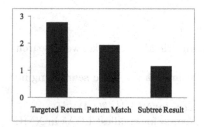
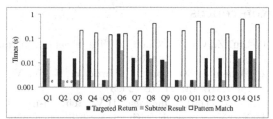

Fig. 3. User scores of query results   **Fig. 4.** Processing time on Baseball data

4   Experiments

In this section we present experimental study of our approach for composing query results, *Targeted Return*.

We have tested one data set: Baseball [15] with 15 real user queries.[1] We compare Targeted Return with Subtree Result and Pattern Match. For each

---

[1] https://db.tt/omJcnxdX.

query, the users were given the results generated by the three approaches with shuffled order, and were asked to give an overall satisfaction score based on their impression of each query result of scale [1, 3]: 3 means I have no trouble finding what I am looking for; 2 means I need efforts to extract the desired information from the results; 1 means I cannot find what I am looking for without re-organizing the results. The average scores of the three approaches of all 15 test queries given by the user is shown in Fig. 3. Targeted Return got the best score of 2.76, followed by Pattern Match 1.92 and Subtree Result 1.15. This indicates that the organization of query results by Targeted Return is closest to users' expectations. Response time of all three systems over the 15 queries are shown in Fig. 4. *e* indicates that the system Pattern Match fails to return any query results because of its data size limit. As we can see, Targeted Return achieves comparable processing time with other systems.

5   Conclusions

Our approach of composing query results is driven by search targets, and produces atomic and intact results. To identify user search targets, we infer return nodes for keyword searches and the modifying relationship among attribute values and entities in the data. Then we determine the target entities and center entity for a keyword search, based on which query results are composed. Experimental evaluation has shown the effectiveness and efficiency of our approach.

Acknowledgements. This work is partially supported by NSF CAREER Award IIS-0845647, Google Cloud Service Award and the Leir Charitable Foundations.

References

1. Liu, Z., Cai, Y., Chen, Y.: TargetSearch: a ranking friendly XML keyword search engine. In: ICDE (2010)
2. Cohen, S., Mamou, J., Kanza, Y., Sagiv, Y.: XSEarch: a semantic search engine for XML. In: VLDB (2003)
3. Li, Y., Yu, C., Jagadish, H.V.: Schema-free XQuery. In: VLDB (2004)
4. Xu, Y., Papakonstantinou, Y.: Efficient keyword search for smallest LCAs in XML databases. In: SIGMOD (2005)
5. Zhou, J., Bao, Z., Wang, W., Zhao, J., Meng, X.: Efficient query processing for XML keyword queries based on the IDList index. VLDB J. **23**(1), 25–50 (2014)
6. Zeng, Y., Bao, Z., Ling, T.W., Li, G.: Removing the mismatch headache in XML Keyword search. In: SIGIR (2013)
7. Le, T.N., Bao, Z., Ling, T.W.: Schema-independence in XML keyword search. In: Yu, E., Dobbie, G., Jarke, M., Purao, S. (eds.) ER 2014. LNCS, vol. 8824, pp. 71–85. Springer, Heidelberg (2014)
8. Zeng, Y., Bao, Z., Ling, T.W., Li, G.: Efficient XML keyword search: from graph model to tree model. In: Decker, H., Lhotská, L., Link, S., Basl, J., Tjoa, A.M. (eds.) DEXA 2013, Part I. LNCS, vol. 8055, pp. 25–39. Springer, Heidelberg (2013)

9. Shi, J., Lu, H., Lu, J., Liao, C.: A skylining approach to optimize influence and cost in location selection. In: Bhowmick, S.S., Dyreson, C.E., Jensen, C.S., Lee, M.L., Muliantara, A., Thalheim, B. (eds.) DASFAA 2014, Part II. LNCS, vol. 8422, pp. 61–76. Springer, Heidelberg (2014)

10. Liu, Z., Chen, Y.: Identifying meaningful return information for XML keyword search. In: SIGMOD (2007)

11. Lin, R.-R., Chang, Y.-H., Chao, K.-M.: Improving the performance of identifying contributors for XML keyword search. SIGMOD Rec. **40**(1), 5–10 (2011)

12. Cheng, T., Chang, K.C.-C.: Entity search engine: towards agile best-effort information integration over the Web. In: CIDR (2007)

13. Liu, Z., Cai, Y., Chen, Y.: Ranking friendly result composition for XML keyword search, ASUCIDSE-2015-001. Technical report, Arizona State University (2015)

14. Li, G., Feng, J., Wang, J., Zhou, L.: Effective keyword search for valuable LCAs over XML documents. In: CIKM (2007)

15. Baseball   Dataset.   http://www.ibiblio.org/xml/books/biblegold/examples/ baseball

Schema Discovery and Evolution

How is Life for a Table in an Evolving Relational Schema? Birth, Death and Everything in Between

Panos Vassiliadis[1]($\boxtimes$), Apostolos V. Zarras[1], and Ioannis Skoulis[2]

[1] Department of Computer Science and Engineering,
University of Ioannina, Ioannina, Greece
{pvassil,zarras}@cs.uoi.gr
[2] Opera Software, Oslo, Norway
giskou@gmail.com

Abstract. In this paper, we study the version history of eight databases that are part of larger open source projects, and report on our observations on how evolution-related properties, like the possibility of deletion, or the amount of updates that a table undergoes, are related to observable table properties like the number of attributes or the time of birth of a table. Our findings indicate that (i) most tables live quiet lives; (ii) few top-changers adhere to a profile of long duration, early birth, medium schema size at birth; (iii) tables with large schemata or long duration are quite unlikely to be removed, and, (iv) early periods of the database life demonstrate a higher level of evolutionary activity compared to later ones.

Keywords: Schema evolution · Patterns of change · Design for evolution

1 Introduction

Databases evolve over time, both in terms of their contents, and, most importantly, in terms of their schema. Schema evolution affects all the applications surrounding a database, as changes in the database schema can turn the surrounding code to be syntactically or semantically invalid, resulting in runtime crashes, missing or incorrect data. Therefore, the understanding of the mechanics of schema evolution and the extraction of patterns and commonalities that govern this process is of great importance, as we can prepare in time for future maintenance actions and reduce both effort and costs.

However, in sharp contrast to traditional software, *the study of the mechanics of schema evolution is quite immature, as it includes only few case studies, each with limited insights into the topic* [1,4,5,7,8,10]. The reason is quite simple, as the research community would find it very hard to obtain access to monitor database schemata for an in-depth study over a significant period of time.

I. Skoulis—work done while in the Univ. Ioannina.

© Springer International Publishing Switzerland 2015
P. Johannesson et al. (Eds.): ER 2015, LNCS 9381, pp. 453–466, 2015.
DOI: 10.1007/978-3-319-25264-3_34

Fortunately, the emergence of open-source repositories (svn/sourceforge/github) allows us to overcome this limitation, as we can gain access to the entire history of data-intensive project files, including the versions of the files with the database schema definition.

In this context, we embarked in the adventure of uncovering the internal mechanics of schema evolution, after having collected and analyzed a large number of database histories of open-source software projects. In [9], we have reported on our findings for the compatibility of database schema evolution with Lehman's laws and show that whereas the essence of Lehman's laws holds, the specific mechanics have important differences when it comes to schema evolution. In this paper, we come with a different contribution, that zooms into the details of the evolution of individual tables rather than entire relational database schemata and explore *how evolution-related properties, like the possibility of deletion, life duration, or the amount of updates that a table undergoes, are related to observable table properties like the number of attributes or the version of birth of a table.* Our guiding research questions and their answers follow.

Which tables are the ones that attract updates? Clearly, low-change tables dominate the landscape. In fact, when we studied the relation of life duration and amount of updates, we observed the *inverse Γ pattern* which states that updates are not proportional to longevity: with the exception of few long-lived, "active" tables, all other types of tables come with an amount of updates that is less than expected. These active tables with the larger amount of updates are long lived, frequently come from the first version of the database and, unexpectedly, they are not necessarily the ones with the largest schema size, but typically start as medium sized (although some of them, exactly due to their updates, eventually obtain a large number of attributes).

Which tables eventually survive and which ones are deleted? Our study has identified that there exist *"families"* of tables whose survival or removal is related to their characteristics.

- *Wide survivors.* The relation of schema size with duration revealed another interesting pattern, which we call *the Γ pattern*: "thin" tables, with small schema sizes, can have arbitrary durations, whereas "wide" tables, with larger schema sizes, last long.
- *Entry level removals.* The population of deleted tables is mostly located in a cluster of newly born, quickly removed, and with few or no updates; at the same time, we observe very low numbers of removed tables with medium or long durations. The schema size of deleted tables is typically small, too.
- *Old timers: time is on my side.* It is quite rare to see tables being removed at old age; although each data set comes with a few such cases, typically, the area of high duration is overwhelmingly inhabited by survivors!

We attribute the above phenomena to the "dependency magnet" nature of tables: the more dependent applications can be to their underlying tables, the less the chance of removal is. Large numbers of attributes, or large number of queries developed over time make both wide and old tables unattractive for removal.

Roadmap. The rest of this paper is structured as follows. In Sect. 2 we discuss related work. In Sect. 3 we present the experimental setup of our study. We present our findings in Sects. 4 and 5. In Sect. 6, we discuss threats to validity. In Sect. 7, we conclude with a discussion on the practical implications of our findings and open issues. For more material, including links to software and data, we indicate the web page of this paper (http://www.cs.uoi.gr/~pvassil/publications/2015_ER/) to the reader.

2   Related Work

The first empirical study on the evolution of database schemas has been performed in the 90's [8]. The author monitored the database of a health management system for a period of 18 months. The results of the study showed that all the tables of the schema were affected and the schema had a 139 % increase in size for tables and 274 % for attributes. The author further observed a significant impact of the changes to the related queries. The second empirical study has been performed much later [1]. In this study, the authors analyzed the database back-end of MediaWiki, the software that powers Wikipedia. The results showed a 100 % increase in schema size. However, 45 % of changes did not affect the information capacity of the schema (but were rather index adjustments, documentation, etc.). The authors further provided a statistical study of lifetimes, change breakdown and version commits. This line of research was based on PRISM (recently re-engineered to PRISM++ [2]), a change management tool. Certain efforts investigated the evolution of databases, along with the applications that depend on them. In [10], the authors focused on 4 case studies. They performed a change frequency and timing analysis, which showed that the database schemas tend to stabilize over time. In [5], two case studies revealed that schemas and source code do not always evolve in sync. In [7], the authors investigated ten case studies. The results indicated that database schemas evolve frequently during the application lifecycle, with schema changes causing a significant amount of code level modifications.

Recently, we studied the applicability of Lehman's laws of software evolution in the case of database schemas [9]. In this study we found evidence that schemas grow so as to satisfy new requirements. However, this growth does not evolve linearly or monotonically, but with periods of calmness and short periods of focused maintenance activity. After a sufficient amount of changes, the schema size reaches a more mature level of stability.

Whereas previous work has mostly focused on the macroscopic study of the entire database schema, in this paper, we zoom into the lives of tables. We study the relationship of table duration, survival and update activity with table characteristics like schema size, time of birth etc. To the best of our knowledge, this is the first time that such findings appear in the related literature.

3   Data Compilation and History Extraction

In this section, we briefly present the eight datasets that we collected, sanitized, and studied using our open source SQL diff tool, Hecate. Those datasets include

Dataset	Versions	Lifetime	Tables @Start	Tables @End	Attributes @Start	Attributes @End	% commits with change
ATLAS Trigger	84	2 Y, 7 M, 2 D	56	73	709	858	82%
BioSQL	46	10 Y, 6 M, 19 D	21	28	74	129	63%
Coppermine	117	8 Y, 6 M, 2 D	8	22	87	169	50%
Ensembl	528	13 Y, 3 M, 15 D	17	75	75	486	60%
MediaWiki	322	8 Y, 10 M, 6 D	17	50	100	318	59%
OpenCart	164	4 Y, 4 M, 3 D	46	114	292	731	47%
phpBB	133	6 Y, 7 M, 10 D	61	65	611	565	82%
TYPO3	97	8 Y, 11 M, 0 D	10	23	122	414	76%

Fig. 1. The datasets we have used in our study [9]

Content Management Systems (CMS's), Web Stores, along with Medical and Scientific storages (see Fig. 1).

In a nutshell, for each dataset we gathered as many schema versions (DDL files) as we could from their public source code repositories (cvs, svn, git). We have targeted only changes at the database part of the project as they were integrated in the trunk of the project. The files were collected during June 2013. For all of the projects, we focused on their release for MySQL (except ATLAS Trigger, available only for Oracle). The files were then processed by our tool, Hecate, that allows the accurate detection of (a) updates at the attribute-level, and specifically, attributes inserted, deleted, having an changed data type, or participation in a changed primary key, and, (b) changes at the table-level, with tables inserted and deleted, in a fully automated way. For lack of space, we refer the reader to [9] and its supporting web page.

4   Table Schema Size, Duration and Updates

In this section, we study the relations between the tables' schema size, duration and updates.

Table schema size and duration. We have analyzed the lifetime duration of all the tables in all the studied datasets and we have observed that there is an interesting relation of durability with table schema size.

We computed the durations (in number of versions) for each table in each dataset. Then, again for each table, we produced a normalized measure of duration, by dividing the duration of the table by the duration of its database (more accurately, for the number of versions that we have monitored). This results in

Tables...	Range	#Tables	Pct.
Short lived	< 0.33	302	41.94%
medium duration	0.33 - 0.77	149	20.69%
Long lived	> 0.77	269	37.36%
Long but not full dur.	*(0.77 – 1.0)*	*81*	*11.25%*
from v0 to v.last	*1.0*	*188*	*26.11%*

Fig. 2. Distribution of tables with respect to their normalized duration

all tables having a normalized duration in the range of (0 ... 1]. Then, in an attempt to simplify intuition, we decided to classify tables in three categories: (i) short-lived (including both the ones who were removed from the system at some point and the ones whose short duration is due to their late appearance), (ii) tables of medium duration and (iii) long lived tables. To avoid manually specifying the limits for each of these categories, we used a k-means clustering [3] that, based on the normalized duration of the tables, split the data set in three clusters, and specifically at the values 0.33 and 0.77. The details of the breakdown appear in Fig. 2, where we devote one line per category, along with the breakdown of the long-lived category in two subclasses: (a) tables that live long, but not throughout the entire lifetime of the database and (b) tables that live from the very first till the last version that we have monitored (and thus, have a normalized duration equal to 1.0).

One of the fascinating revelations of this measurement was that there is a 26.11 % fraction of tables that appeared in the beginning of the database and survived until the end. In fact, if a table is long-lived, there is a 70 % chance (188 over 269 occasions) that it has appeared in the beginning of the database.

Figure 3 gives the statistical distribution of the average table schema size (i.e., the average number of attributes a table has throughout its lifetime). Typically, the average table schema size has a very small deviation, as tables do not change largely. Thus, the average size is pretty close to the respective max and min size for the studied tables. *On average, half of the tables (approx. 47 %) are small tables with less than 5 attributes. The tables with 5 to 10 attributes are approximately one third of the tables' population and the wide tables with more than 10 attributes are approximately 17 % of the tables.* The first category has quite a few outliers, both high and low (and a large standard deviation of 20 %), whereas the two others do not really oscillate too much (with standard deviations of 14 and 13 %, respectively). Interestingly, the datasets with less evolutionary activity (atlas, typo3 and biosql) are the ones concentrating outlier values.

The first column of Fig. 4 gives the relation of a table's schema size (measured as the number of attributes of a table's schema at its birth and depicted in the x-axis of each of the scatterplots) with duration (y-axis in the scatterplots), in three characteristic data sets, Atlas, Coppermine and Mediawiki (for lack of space, it is impossible to show all datasets). We observe a phenomenon that we call Γ pattern: *tables with small schema sizes can have arbitrary durations, whereas tables with larger schema sizes last long.*

Pct of tables with num. of attributes ...

	≤5	5-10	>10
atlas	10,23%	68,18%	21,59%
biosql	75,56%	24,44%	0,00%
coppermine	52,17%	30,43%	17,39%
ensembl	54,84%	38,06%	7,10%
mediawiki	61,97%	19,72%	18,31%
phpbb	40,00%	44,29%	15,71%
typo3	21,88%	31,25%	46,88%
opencart	57,20%	33,05%	9,75%
Average	46,73%	36,18%	17,09%

Fig. 3. Distribution of tables with respect to their average schema size

Plainly put, the facts that the first column of Fig. 4 vividly demonstrates are as follows. There is a large majority of tables, in all datasets, whose size is between 0–10 attributes. We have visualized data points with transparency in the figure; therefore, areas of intense color mean that they are overpopulated with data points that fall on the same x, y coordinates. Many of the data sets are limited to really small sizes: biosql tables do not exceed 10 attributes, ensembl tables do not exceed 20 attributes and atlas, mediawiki and coppermine have one or two tables "wider" than 20 attributes (btw., observe the outlier table of 266 attributes in atlas, too). Due to their large percentage, small tables dominate the area near the beginning of the axes in the figure. What is interesting however, is that *their small size does not determine their duration*: this is depicted as a distribution of data points in parallel with the y-axis for all the small sizes of a table's schema. On the other hand, we observe that *whenever a table exceeds the critical value of 10 attributes in its schema, its chances of surviving are high. In fact, we observe that in most cases the "wide" tables are created early on and are not deleted afterwards.* We conjecture that the explanation for the "if you're wide, you survive" part of the Γ pattern is due to the impact a table deletion has: the wider a table, the higher a chance it acts as a fact table, frequently accessed from the queries in the applications surrounding the database. Thus, its removal incurs high maintenance costs, making application developers and administrators rather reluctant to remove it.

Table schema size and updates. If we observe the middle column of Fig. 4, we can see the relation of a table's schema size at its birth (depicted in the x-axis of each of the scatterplots) with the amount of change the table undergoes (y-axis in the scatterplots). There are two main clusters in the plots: *(a) a large, dense cluster close to the beginning of the axes, denoting small size and small amount of change, and, (b) a sparse set of outliers, broken in two subcategories: (b1) medium schema size tables typically demonstrating medium to large amounts of changes and (b2) "wide" tables with large schema sizes demonstrating small to medium amount of change.* We refer to this distribution as the *comet pattern*.

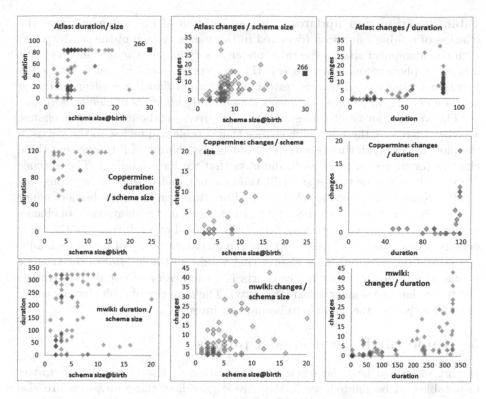

Fig. 4. Correlation of schema size (number of attributes), duration (over versions) and change (total number of changes) of all the tables in three datasets

The first "nucleus" cluster is typically contained within a box of size 10×10 (i.e., no more than 10 attributes typically result in no more than 10 changes). This is attributed to the small pace of change that tables undergo (cf. Sect. 5), resulting in small probabilities of change for small tables. At the same time, in most of the datasets, *the tables with the largest amount of change are not necessarily the largest ones in terms of attributes, but tables whose schema is one standard deviation above the mean.* Typically, medium sized tables demonstrate all kinds of change behaviour as they cover the entire y-axis, whereas the (few) tables with large schema size demonstrate observable change activity (i.e., not zero or small), which is found around the middle of the y-axis of the plot in many cases. An extra observation, not depicted in the figure for lack of space, occurs when we observe the average schema size instead of the schema size at birth. The charts look quite alike with small differences, but in several of the datasets we see tables with medium-sized schema obtaining a large schema via a large number of updates. In other words, *if a table is already wide, there is rarely a tendency for large growth; however, it is more often that a table of medium schema size is augmented a lot to carry more information.*

Table duration and updates. The last column of Fig. 4 demonstrates the relation of a table's duration (depicted in the x-axis of each of the scatterplots) with the amount of change the table undergoes (y-axis in the scatterplots). We observe a phenomenon that we call the *inverse Γ pattern*: *tables with small duration undergo small change, tables with medium duration undergo small or medium change, and, long-lived tables demonstrate all kinds of change behavior.*

The vast majority of tables have "calm" lives, without too much change activity; therefore, there is a high chance that a table will be low in the y-axis of the plot. In detail, all three scatterplots of the last column of Fig. 4 present three clusters (omnipresent in all eight datasets that we have studied). The absence of extremities means that a short lifetime cannot really produce anything but small (typically close to zero) change; medium duration has some higher chance of producing change in the range of 5 to 10 changes; and high amounts of change are only found in tables with long lives. Still, instead of a full triangle, the chart demonstrates mainly an inverse $Γ$: there is a striking scarcity of tables with mid-sized durations and mid-range change that would fill the interior of the triangle (visual clue: color intensity in our charts is an indicator of population density, due to overlapping semi-transparent points). The majority of tables having quiet, calm lives, pushes the triangle to become an inverse $Γ$.

5   Table Birth, Death and Updates

Having examined how the schema size, duration and total amount of update of a table can be related, we further investigate how these measures are also related with table birth and death. One related question that has been puzzling us, concerns the relationship of update activity with the possibility of removal. As it is theoretically possible that a table with intense change activity has a short life, we introduced an extra parameter to the problem that assesses change activity. So, we define the *Average Transitional Update (ATU)* as the fraction of the sum of updates that a table undergoes throughout its life over its duration. Equivalently, one can think of this measure as *the average number of schema updates of a table per transition*. Thus, the average transitional update of a table practically measures how active or rigid its life has been by normalizing the total volume of updates over its duration.

We have assessed the average transitional update of all the tables in all datasets and we have studied its relation to the rest of the characteristics of the tables (schema size, version of birth and death, etc.). To address the *survival vs death* separation task, we will employ the following terminology:

- *survivor* is a table that was present in the last version of the database that we examine, and,
- *non-survivor* refers to a table that was eventually eliminated from the database.

We also discriminate the tables with respect to their update profile as follows:

- *Active tables* or *top-changers* are the ones with (a) a high average transitional update, exceeding the empirically set threshold of 0.1, and, (b) a non-trivial volume of updates, exceeding the also empirically set limit of 5 updates.

Table distribution (pct of tables) wrt their avg transitional update rate

	#tables	DIED				SURVIVED				Aggregate per update type		
		No change	Quiet (0-0.1)	Active (>0.1)	Total	No change	Quiet (0-0.1)	Active (>0.1)	Total	No change	Quiet (0-0.1)	Active (>0.1)
atlas	88	8%	7%	2%	17%	13%	42%	28%	83%	20%	49%	31%
biosql	45	20%	13%	4%	38%	16%	16%	31%	62%	36%	29%	36%
phpbb	70	0%	3%	4%	7%	50%	31%	11%	93%	50%	34%	16%
typo3	32	16%	6%	6%	28%	22%	34%	16%	72%	38%	41%	22%
coppermine	23	4%	0%	0%	4%	30%	57%	9%	96%	35%	57%	9%
ensembl	155	24%	20%	8%	52%	6%	37%	5%	48%	30%	57%	13%
mwiki	71	14%	13%	3%	30%	3%	63%	4%	70%	17%	76%	7%
opencart*	128	9%	2%	0%	11%	42%	44%	3%	89%	51%	46%	3%

Fig. 5. Table classification concerning the average amount of updates per version for all data sets (∗Opencart reports only on the tables born after transition 17)

- *Rigid tables* are the ones with no change at all; we will also refer to the ones that did not survive as *sudden deaths*, as they were removed without any previous change to their schema.
- *Quiet tables* are the rest of the tables, with very few updates or average transitional update less than 0.1. For the tests we made, discriminating further within this category did not provide any further insights.

Observe Fig. 5 that depicts how tables are grouped in categories for all the datasets. We group measurements separately for (a) non-survivors, (b) survivors, and (c) overall. Within each of these groups, apart from the overall statistics, we also provide the breakdown for tables with (a) no changes, (b) quiet change, and (c) active change. Each cell reports on the percentage of tables of a data set that fall within the respective subcategory. We have slightly manipulated a single dataset, for the sake of clarity. OpenCart has an extreme case of tables renamings, involving (a) the entire schema at transitions 17 and 18 and (b) massive renamings in transitions 23 and 35. As these massive renamings involved 108 of the 236 tables of the monitored history, we decided to exclude the tables that were dead before transition 23 from our study, so that their change does not overwhelm the statistics. For example, after the aforementioned short period, only 14 tables were removed in the subsequent 130 revisions.

Clearly, quiet tables dominate the landscape. If we observe the non-survivors, the percentage of tables in each of the categories is typically dropping as the probability of change rises: *sudden deaths are the most frequent, quiet non-survivors come second, and third, a very small percentage of tables in the vicinity of 0–6 % dies after some intense activity.* In the case of the survivors, the percentage of rigid tables is often substantial, but, with the exception of phpbb, the landscape is dominated by quiet tables, typically surpassing 30 % of tables and reaching even to 60 % of the tables for a couple of data sets. Things become quite interesting if one focuses on the more mature data sets (coppermine, ensemble, mediawiki, opencart) whose profile shows some commonalities: in contrast to the rest of the datasets, active survivors are a small minority (approx. 5 %) of the population, and similarly, active over both survivors and non-survivors ranges between 3–13 %. So, *as the database matures, the percentage of active tables drops* (as already mentioned, evolution activity seems to calm down). It is also noteworthy that *despite similarities, and despite the obvious fact that at least two thirds of the tables (in the mature databases: 9 out of 10) lead quiet lives, the internal breakdown of updates obliges us to admit that each database (and developer community) comes with its own update profile.*

In Fig. 6 we present the most interesting insights from the Mediawiki data set. We have chosen Mediawiki as a fairly representative dataset, as it comes with both a non-negligible amount of removed tables, a large duration, and, a fairly large number of tables (72). Our findings are discussed below.

Table birth, duration and updates

– *Surviving top-changers live long.* Observe the top chart of Fig. 6: the vertical line of the *inverse Γ* in long durations, as well as its nearby region are being populated by survivors; the vertical line demonstrates that several of the top-changers are very long lived. Remember that the top-changers are the ones with the highest average transitional update, i.e., their update is normalized over their duration. This is quite different from the total amount of update that is depicted in the y-axis of the chart: in other words, it is theoretically possible (and in fact, there is indeed such a case in the Mediawiki example, found close to the beginning of the axes) that a short lived table with relatively low total updates is a top-changer. In practice, however, longevity and high average transitional update are closely related.
– Concerning the top-changers, there is an interesting correlation of high update activity, overall change, duration and birthdate: *most (although not all) top-changers are born early, live long, have high average transitional update* and, consequently a large amount of total update. Of course, there exist tables born early who survive and have lower update activity and top-changers that violate this rule; however, in all data sets, it is quite uncommon to observe top-changers outside the "box" or early birth and long duration.

Table death, duration and updates

– There is a large concentration of the *deleted tables in a cluster of tables that are newly born, quickly removed, with few or no updates.* Few deleted tables demonstrate either late birth, or long durations, or high average transitional update, or large number of updates (Mediawiki shows just a couple of them in each of the first two categories). In the triangle formed in the bottom figure by the two axes and the line of survivors, the diagonal of survivors is expected: if you are a survivor, the earlier you have been born, the more you live. *The really surprising revelation of the triangle, however, is that (a) the diagonal is almost exclusively made from survivors, and, (b) the triangle is mainly hollow!* Theoretically, it is absolutely legitimate for non-survivors to have long durations and to be born in the middle of the database lifetime: if there was not a pattern of attraction of non-survivors to the beginning of the axes we could legitimately see non-survivors in the area of the diagonal and,

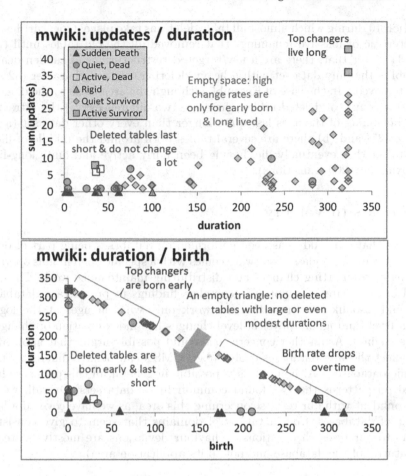

Fig. 6. Update profile with respect to survival-vs-death injected in the study of total change over duration (top) and duration over birthday (bottom)

of course, a uniform spread in the interior of the triangle. In fact, in the case of Mediawiki, we do see a couple (but just a couple) of such cases for both the aforementioned possibilities.

– *It is quite rare to see tables being removed at old age; although each data set comes with a few such cases, typically, the area of high duration is overwhelmingly inhabited by survivors!* See the upper part of Fig. 6 for the Mediawiki data set, showing just a couple of deleted tables with high durations. We believe (although cannot prove) that this has to do with the cost of removing a table after some time: applications have been built around it and the maintenance cost to the surrounding queries shrinks the possibility of removal.

Exceptions to the aforementioned observations do exist. In the case of Typo3, the diagonal is closely followed by a set of medium-change tables. This is mainly attributed to the fact that there is a short period of 5 transitions, around transition 70 during which almost all table deletions take place. Opencart has the characteristic of massive renamings that removed all the old tables until transition 34. After that, there are a few targeted restructurings of sudden deaths. Ensembl is the only data set with a heavy deletion oriented character (52 % of tables removed). In the case of Ensembl, although the area close to the diagonal has just a couple of deleted tables, there are two differences: (a) the attraction to the beginning of the axes last quite longer than every other data set (a box of 135 × 135) and (b) there are several tables born around the middle of life of the data set that eventually die, having been fairly active and quite long-lived (approximately 200 transitions).

6   Threats to Validity

We stress that our study has been performed on databases being part of open-source software. This class of software comes with a specific open modus operandi in terms of committing changes and distributing maintenance work. Thus, we should be very careful to not overgeneralize findings to proprietary databases. We would also like to stress that we work only with changes at the logical schema level (and ignore physical-level changes like index creation or change of storage engine). As another concern, one could possibly argue that application areas affect the way databases evolve (e.g., CMS's can behave differently than scientific databases). Although this is possible in general, in this paper we have focused on patterns that we found common in all data sets. Overall, we are quite confident with our results concerning this area, as we have covered a large number of databases, from a variety of domains that seem to give consistent answers to our research questions (behaviour deviations are mostly related to the maturity of the database and not to its application area).

7   Practical Exploitation and Open Issues

Assume you are the chief architect of a large information system. Evolution is inevitable, maintenance takes up to 70 % of resources that are always limited, time is pressing and you have to keep the system up to date, correct, without failures and satisfactory for the users. In this section, we relate the key findings of our study with *guidelines on designing and building both tables and applications that access them, so as to sustain the evolution of the database part gracefully.*

Life and death of tables: only the thin die young, all the wide ones seem to live for ever. The Γ pattern says: "Large-schema tables typically survive". Yet, more than half of the tables are of short schema sizes and up to 80 % of tables have less than 10 attributes, and in fact, it is the short-sized tables' category being the one with short durations and table deletions. The deletions of these "narrow" tables typically take place early in the lifetime of the project, either due to deletion or due to renaming (which is equivalent from the point of view of the applications: they crash in both cases). To address change, we strongly denounce table denormalization as a solution that reduces the number of potentially evolution-prone, thin tables by introducing few, artificially wide, reference tables. Rather, we suggest the usage of views as an *evolution-buffering, "API-like" mechanism* between applications and databases.

> *Mask your applications from tables prone to change (e.g., tables with few attributes) as much as possible, especially at the early versions of the database where most of the deletions take place. To this end, use views as a buffering mechanism between the database and the applications that masks change in several evolution scenarios like renamings, table splitting or merging, etc.*

Everything in between: quiet growth. Due to the large impact that their update induces, most tables lead quiet lives with few updates (remember: additions and deletions of attributes, data type and key changes). Medium-rated change is scarcely present too: in fact, the *inverse Γ* pattern states that updates are not proportional to longevity, but rather, only few top-changer tables deviate from a quiet, low-change profile. Top-changers tables are long lived, frequently come from the first version of the database, can have large number of updates (both in absolute terms and as a normalized measure over their duration) and, unexpectedly, are not necessarily the wider ones, but frequently their schema is typically medium sized. Although each database comes with its own idiosyncracy, we still can coarsely say that early stages of the database life are more "active" in terms of births, deaths and updates, whereas, later, activity is less "hot", including table additions (which seem to happen throughout the entire timeline), and deletions and updates that become more concentrated and focused. The overall result, also due to the outnumbering of table deletions by table additions is that the schema of the database expands over time.

> *Plan for a quiet expansion of the schema! The database is augmented with new tables all the time; at the same time, although most tables lead quiet lives and do not grow much, some top-changers are "change attractors". To address the database expansion, try to systematically use metadata rich facilities involving the interdependencies of applications and tables to be able to locate where the code should be maintained for added, deleted or modified attributes and tables.*

Open issues. Possibilities for follow-up work are immense. For lack of space, we restrict ourselves to a few points. First, there is always the ambitious scientific goal of having some "weather forecast" for the forthcoming changes of a database (we should, however, warn younger readers on the level of risk this research encompasses). Second, one can deal with the possibility of capturing more types of changes (like renamings, normalization actions, etc.) that have not been part of our fully automated mechanism for transition extraction. Third, one could also work on the arguably feasible engineering goal of having flexible structures for gluing applications to evolving database schemata, thus minimizing application dependency and evolution impact (see [6] for our take on the problem).

Acknowledgments. This work was partially supported from the European Community's FP7/2007-2013 under grant agreement number 257178 (project CHOReOS). We would like to thank the reviewers of the paper for helpful comments and suggestions for solidifying our work.

References

1. Curino, C., Moon, H.J., Tanca, L., Zaniolo, C.: Schema evolution in wikipedia: toward a web information system benchmark. In: Proceedings of ICEIS 2008. Citeseer (2008)
2. Curino, C., Moon, H.J., Deutsch, A., Zaniolo, C.: Automating the database schema evolution process. VLDB J. **22**(1), 73–98 (2013)
3. Dunham, M.H.: Data Mining: Introductory and Advanced Topics. Prentice-Hall (2002)
4. Hartung, M., Terwilliger, J.F., Rahm, E.: Recent Advances in Schema and Ontology Evolution. In: Bellahsene, Z., Bonifati, A., Rahm, E. (eds.) Schema Matching and Mapping, pp. 149–190. Springer, Heidelberg (2011)
5. Lin, D.Y., Neamtiu, I.: Collateral evolution of applications and databases. In: Proceedings of the Joint International and Annual ERCIM Workshops on Principles of Software Evolution (IWPSE) and Software Evolution (Evol) Workshops, IWPSE-Evol 2009, pp. 31–40 (2009)
6. Manousis, P., Vassiliadis, P., Papastefanatos, G.: Automating the adaptation of evolving data-intensive ecosystems. In: Ng, W., Storey, V.C., Trujillo, J.C. (eds.) ER 2013. LNCS, vol. 8217, pp. 182–196. Springer, Heidelberg (2013)
7. Qiu, D., Li, B., Su, Z.: An empirical analysis of the co-evolution of schema and code in database applications. In: Proceedings of the 2013 9th Joint Meeting on Foundations of Software Engineering, pp. 125–135, ESEC/FSE 2013 (2013)
8. Sjøberg, D.: Quantifying schema evolution. Inf. Softw. Technol. **35**(1), 35–44 (1993)
9. Skoulis, I., Vassiliadis, P., Zarras, A.: Open-source databases: within, outside, or beyond Lehman's laws of software evolution? In: Jarke, M., Mylopoulos, J., Quix, C., Rolland, C., Manolopoulos, Y., Mouratidis, H., Horkoff, J. (eds.) CAiSE 2014. LNCS, vol. 8484, pp. 379–393. Springer, Heidelberg (2014)
10. Wu, S., Neamtiu, I.: Schema evolution analysis for embedded databases. In: Proceedings of the 2011 IEEE 27th International Conference on Data Engineering Workshops, ICDEW 2011, pp. 151–156 (2011)

Inferring Versioned Schemas from NoSQL Databases and Its Applications

Diego Sevilla Ruiz[(✉)], Severino Feliciano Morales,
and Jesús García Molina

Faculty of Computer Science, University of Murcia, Campus Espinardo,
Murcia, Spain
{dsevilla,severino.feliciano,jmolina}@um.es

Abstract. While the concept of database schema plays a central role in relational database systems, most NoSQL systems are schemaless: these databases are created without having to formally define its schema. Instead, it is implicit in the stored data. This lack of schema definition offers a greater flexibility; more specifically, the schemaless databases ease both the recording of non-uniform data and data evolution. However, this comes at the cost of losing some of the benefits provided by schemas. In this article, a MDE-based reverse engineering approach for inferring the schema of aggregate-oriented NoSQL databases is presented. We show how the obtained schemas can be used to build database utilities that tackle some of the problems encountered using implicit schemas: a schema diagram viewer and a data validator generator are presented.

Keywords: NoSQL databases · Schemaless databases · Schema inference · Model-driven data reverse engineering · JSON

1 Introduction

Modern applications that have to deal with huge collections of data have evidenced the limitations of relational database management systems. This has motivated the development of a continuously growing number of non-relational systems, with the purpose of tackling the requirements of such applications. Specially, the ability to represent complex data and achieving scalability to manage both large data sets and the increase in data traffic. The NoSQL (*Not SQL/Not only SQL*) term is used to denote this new generation of database systems.

The lack of an explicit data schema (*schemaless*) is probably the most attractive NoSQL feature for database developers. While relational systems require the definition of the database schema in order to determine the data organization, in NoSQL databases data is stored without the need of having previously defined a

---

Work partially supported by the Cátedra SAES of the University of Murcia (http://www.catedrasaes.org), a research lab sponsored by the SAES company (http://www.electronica-submarina.com/).

© Springer International Publishing Switzerland 2015
P. Johannesson et al. (Eds.): ER 2015, LNCS 9381, pp. 467–480, 2015.
DOI: 10.1007/978-3-319-25264-3_35

schema. Being schemaless, a larger flexibility is provided: the database can store data with different structure for the same entity type (non-uniform data), and data evolution is favoured due to the lack of restrictions imposed on the data structure. However, removing the need of declaring explicit schemas does not have to be confused with the absence of a schema, since a schema is implicit into data and database applications. The developers must always keep in mind the schema when they write code that accesses the database. For instance, they have to honor the names and types of the fields when writing insert or query operations. This is an error-prone task, more so when the existence of several versions of each entity is probable. Therefore, the idea is emerging of combining a schemaless approach with mechanisms (e.g. data validations against schemas) that guarantee a correct access to data [6,10]. On the other hand, some NoSQL database tools and utilities need to know the schema to offer functionality such as performing SQL-like queries or automatically migrating data. A growing interest in managing explicit NoSQL schemas is therefore arising [9,10,12,15].

This article presents a reverse engineering strategy to infer the implicit schema in NoSQL databases, which takes into account the different versions of the entities. We call these schemas *Versioned Schemas*. The usefulness of the inferred versioned schemas is illustrated through two possible applications: schema visualization, and automated generation of data validators. The approach has been designed to be applied to NoSQL systems whose data model is aggregate-oriented [13], which is the data model of the three most widely used types of NoSQL stores: *document*, *key-value*, and *column family* stores. Model-Driven Engineering (MDE) techniques, such as metamodeling and model transformations, have been used to implement both the schema inference strategy and the applications, in order to take advantage of the abstraction and automation capabilities that they provide.

There are therefore two main contributions in this work. To our knowledge, this is the first approach that infers conceptual schemas from NoSQL databases discovering all the versions of the inferred entities and relationships. Moreover, we show how the inferred schemas can be used to automatically generate different software artifacts, which help to improve the productivity and code quality. The approach proposed has been validated for the MongoDB, CouchDB, and HBase stores, and the tools implemented may be downloaded from http://www. catedrasaes.org/wiki/NoSQLSchemaVersions.

This article is organized as follows. The following Section explains the notion of aggregate-oriented data model, and presents a running example. Section 3 gives an overview of the approach proposed, and Sect. 4 describes in detail the schema inference strategy. The utilities that have been built are described in Sect. 5, and the related work is discussed in Sect. 6. Finally, conclusions and future work are presented in Sect. 7.

2   Background

This section introduces some key concepts that are used throughout the article and motivates the work. Moreover, a simple NoSQL database is shown, which will be used as a running example.

2.1   Semi-structured Data and the JSON Format

Semi-structured data is mainly characterized by the fact that it has a non-uniform and an implicit structure, which can evolve rapidly [1]. This data is expressed in formats, such as XML and JSON, which allow the representation of information in hierarchical form (i.e. a tree-like structure) by using tags or symbols as separator elements.

JSON (JavaScript Object Notation)[1] is a standard human-readable text format widely used to represent semi-structured data. This notation is taking the place of XML as primary data interchange format because it is more simple and legible. A JSON object or document is formed by a set of key-value pairs (*fields*). The type of a JSON value may be a primitive type (Number, String, or Boolean), an object, or an array of values. null is used to indicate that a key has no value.

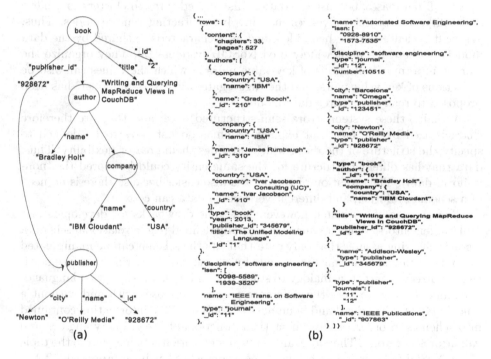

Fig. 1. Example database used and a tree representation of a book.

As indicated in [4], a piece of semi-structured data can be formalized as a tree whose leaf nodes are atomic values of primitive types (e.g. string, integer, float, or boolean) and the root and intermediate nodes are objects (i.e. tuples) or either arrays of objects or values. The edges are labelled with the names of the attributes. A root or intermediate node has a child node by each attribute of the object associated. For instance, Fig. 1(a) shows the tree that corresponds to the JSON object that represent the book with _id=2 in Fig. 1(b). References

----
[1] http://json.org/.

among data may be expressed in a similar way as foreign keys in relational databases, that is, the atomic value of an attribute (e.g. *publisher_id* in *book*) matches a value in another attribute of a different object (e.g. *_id* in *publisher*).

The term *aggregate* is normally used to refer to the object structure that consists of a root object that recursively embeds other objects, so that the tree-like structure of an semi-structured data is aggregate-oriented. In Fig. 1(a), the *book* object aggregates an *author* object, which aggregates, in turn, a *company*.

2.2 Aggregate-Oriented Data Models

While complex data is addressed in relational databases through joins by means of foreign keys (i.e. references between tables), object references and aggregate objects are more appropriate ways to represent such data. Unlike object-oriented databases, aggregate objects are usually preferred to object references in the case of NoSQL databases, because the data is distributed through clusters to achieve scalability, and object references may involve contacting remote nodes. Thus, aggregate-orientation has been identified as a characteristic shared by the data models of the three most widely used NoSQL systems [13]. They organize the storage in form of collections of key-value pairs in which the values can also be collections of key-value pairs, and the "aggregate-oriented data model" has been proposed to refer these three data models.

Actually, these systems store semi-structured data, and they are therefore characterized by the fact that explicit schemas do not have to be defined to specify the structure of the data, which provides them greater flexibility. Thus, data that has different structure for the same entity could be stored (i.e. non-uniform data). The evolution of the data is also easier because there is no need of a schema evolution, and different versions of data can coexist.

The absence of a schema, however, has some drawbacks for developers and tool implementors. When a schema is formally defined (e.g. a relational schema), a static checking assures that only data that fits the schema can be manipulated in application code, and mistakes made by developers in writing code are statically spotted. In fact, the analogy to statically and dynamically typed languages is commonly used to note the difference among databases with and without a schema [4]. On the other hand, a number of tools need the information contained into schemas in order to implement their functionality (e.g. query engines and validator generators). Therefore, an increasing attention is being paid to the topic of the NoSQL schema inference and some approaches has been proposed [12,15].

A schema of an aggregate-oriented data model is basically formed by a set of entities connected through two types of relationships: aggregation and reference. Each entity will have one or more fields that are specified by its name and its data type. Several versions of an entity can exist due to the non-uniformity characteristic and the database evolution.

2.3 A NoSQL Database for the Running Example

For the purposes of this work, we consider a NoSQL database as an arbitrarily large array (i.e. a collection) of JSON objects that include: (a) a field (e.g. *type*) that describes its entity type; and (b) some form of unique identifier for the object (in our case the _id field). This format is non-compromising, and provides system independence. In fact, it is very similar to what it is actually used in most NoSQL database implementations. For example, CouchDB guides recommend the usage of the *type* field. MongoDB creates one collection for each type of object, so that the collection name could provide the *type* field. In HBase, the *type* field of an object could be the name of its *column family*. If the value of the *type* field is not directly obtainable, some heuristics could be used. However, in some cases it may require the user to provide it.

Figure 1(b) shows a simple database that stores objects for the *Book*, *Publisher*, and *Journal* entities. The first *Book* object aggregates an array of authors (*authors* field) and an embedded object for the content (*content* field). In turn, the *author* field aggregates an embedded object that records the company for which he or she works (*company* field). With regards to object references, both *Book* and *Publisher* objects show examples of them. The *Book* objects have a reference to its publisher (*publisher_id* field) and *Publisher* objects hold a reference to the list of journals published (*journal* field).

It is worth noting that aggregates and references are implicit: a parser could identify the embedded objects. However, references require some kind of heuristics if conventions are not used. Some idioms have been therefore proposed to express references in NoSQL databases, some of them considered in Sect. 4.

3   Overview

We shall here outline the general arquitecture of the proposed approach to infer schemas. Moreover, we shall describe the metamodel created to represent NoSQL schemas.

Reverse engineering can take advantage of MDE techniques. Metamodels provide a formalism to represent the knowledge harvested at a high-level of abstraction, and automation is facilitated by using model transformations. Therefore, we have devised an MDE solution to reverse engineer versioned schemas from aggregate-oriented NoSQL databases that we use to create database utilities.

Figure 2 shows the architecture of the solution, which is organized in three stages. Firstly, a Map-Reduce operation is applied in order to extract a collection of JSON objects that contains one object for each version of an entity, i.e. the minimum number of objects that are needed to perform the inference process. Map-Reduce is germane to most NoSQL databases, and gives an advantage in performance, as it is the native processing method when an algorithm has to deal with all the objects in a database. Secondly, that collection is injected into a model that conforms to a JSON metamodel, which is easily obtained by mapping the JSON grammar elements into the metamodel elements. Thirdly, the reverse engineering process is implemented as a model-to-model transformation whose input is the JSON model, and that generates a model that conforms to the NoSQL-Schema metamodel (Fig. 3). The inferred NoSQL-Schema models may be used to build tools that could be classified in two categories: (i) database utilities

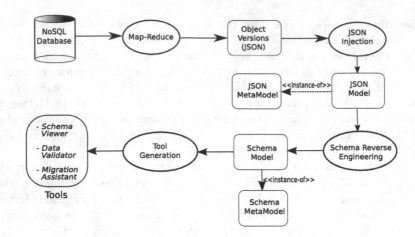

Fig. 2. Overview of the proposed MDE architecture.

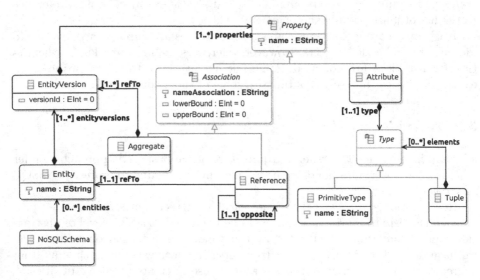

Fig. 3. *NoSQL-Schema* metamodel representing NoSQL schemas.

that require knowledge of the database structure, and (ii) helping developers to deal with problems caused by the absence of an explicit schema, for instance the tools presented below, which are able of generating data validators, migration scripts or schema diagrams.

Figure 3 shows the NoSQL-Schema metamodel that represents schemas of aggregate-oriented NoSQL databases according to the notion of NoSQL schema presented in Sect. 2.2. A schema (metaclass *NoSQLSchema*) is formed by a collection of entities (*Entity*); one or more versions (*EntityVersion*) exist for each entity. A version is defined by a set of properties (*Property*) that may be *Attributes* or *Associations*, depending on whether the property represents a type (either a *PrimitiveType* or a *Tuple*) or a relationship between two entities.

A tuple denotes a collection that may contain primitive types and tuples. An association can be either an *Aggregation* or a *Reference*. The cardinality of an association is captured by the *lowerBound* and *upperBound* attributes, which can take values 0, 1, and −1.

Note that an aggregate is connected to one or more entity versions (*[1..*]refTo*) because an embedded object may aggregate an array with objects of different versions. Instead, a reference is connected to one entity (*[1..1]refTo*), since we need to know that a version holds references to a certain entity, but we decided not to cross object boundaries. The *opposite* self-reference in the *Reference* metaclass is used to make the relationship bidirectional, and specifies the other end.

4 Reverse Engineering Process

Extracting Versioned Schemas from aggregate-oriented NoSQL databases involves discovering entities, versions of each entity, the attributes of each version, and relationships between entities (aggregations and references). The reverse engineering algorithm should traverse all the stored objects (i.e. root entities), and analyze their properties in order to identify all the schema elements.

4.1 Building the *Raw Schema* of an Object

The first step in discovering the versioned schemas is obtaining what we call the *raw schema* of an object, which is a JSON object built honoring two rules: (i) it has the same structure as the described object with respect to fields, nested objects and arrays, and (ii) each primitive value in the described object is substituted in the raw schema by its JSON type (e.g. String or Number).

In our running example (Fig. 1(b)), {*name:String, city:String*} would be the raw schema for the *Publisher* entity with _id=123451, and {*title:String, publisher_id:String, author:{name:String, company:{country:String, name:String}}*} would be the raw schema of the *Book* with _id=2. More visually:

JSON object	Raw Schema
{*name:"Omega", city:"Barcelona"*}	{*name:String, city:String*}

{*title:"Writing and...",*
   publisher_id:"928672",
   author:{name:"Bradley Holt",
     company:{country:"USA",
       name:"IBM Cloudant"} } }
	{*title:String,*
	   publisher_id:String,
	   author:{name:String,
	     company:{country:String,
	       name:String} } }

4.2 Obtaining the Version Collection

To improve the efficiency, we have considered a preliminary stage to the reverse engineering process. In this stage, a Map-Reduce operation is applied to obtain a collection that only contains one object for each entity version, which will be referred to as the *Version Collection*.

For each object, the `map()` operation performs a two-step process. First, it generates the *version identifier*: the string obtained by concatenating the value of the special *type* field with a textual representation of the object's raw schema. Secondly, the *<version identifier, object>* key/value pair is emitted.

Then, the `reduce()` operation is performed once for each version identifier. It receives a set of objects that share the same version identifier and selects one of the objects as the archetype for the group, adding it to the output list. The result is an array of JSON objects following the format explained in Sect. 2.3, and shown in Fig. 1(b), but now containing just one object per object version.

4.3   Obtaining the Schema

The JSON object collection obtained in the previous stage is injected into a JSON model, from which a model-to-model transformation generates the Schema model. The transformation discovers the elements of the schema, and works as follows:

Discovering Entities and Entity Versions. For each JSON object in the model, an *Entity Version* is considered. This usually leads to a new *Entity Version*, but not in all cases, because a similar *Entity Version* may exist already that only differs with the considered one with respect to cardinalities. If this is the case, the cardinalities of the existing *Entity Version* are adjusted to include both specifications, and no new *Entity Version* is created. When the created *Entity Version* is the first one discovered for a particular entity, an *Entity* element is also generated. Each *Entity* holds a list of entity versions, in which each new *Entity Version* is added. Obtaining an entity name differs for root and embedded objects. For *root objects*, the name is given by the *type* field of the object; for *embedded objects*, the name is given by the key of a pair whose value is a JSON object. If the value is an array of objects and the name is plural, then the singular name is used.

An *Entity Version* is named by appending, to the entity name, a suffix with an underscore and a counter of the number of version. For instance, three *Entity Version* would be generated for the *Publisher* root objects of the running example, named *Publisher_1*, *Publisher_2*, and *Publisher_3*, and *Author_1* and *Author_2* would be generated for the *Author* embedded object. Figure 4(b) shows a textual report with all the entity versions.

Discovering Attributes. An *Attribute* is generated for each object's pair whose value is either atomic or an array of either primitive types or nested arrays of primitive types. The attribute name is given by the pair name. With regard to the type, a *PrimitiveType* or a *Tuple* is generated depending on whether the value is atomic or an array. Each created *Attribute* is added to the collection of attributes of the corresponding *Entity Version*. For instance, the pair *"title": "The Unified Modeling Language"* in a version of *Book* would lead to the *Attribute* named "title" and a *PrimitiveType* named "String"; and the pair *"issn": ["0928-8910"*, *"1573-7535"]* in a version of *Journal* would generate an *Attribute* named "issn" and a *Tuple*, as shown in Fig. 4(b).

Discovering Aggregation Relationships. A pair results in an *Aggregate* (i.e. an aggregation relationship) if its value is an object. Each created *Aggregate* must be connected to the *EntityVersion* that corresponds to the object value. Several *EntityVersions* may embed the same aggregated *Entity*. For this, the transformation is organized in two stages. Firstly, *Entities*, *EntityVersions*, *Attributes*, *Types* and *Pairs* are created, and then *Aggregates* and *References* are created in a second stage, once all the *EntityVersions* have been created.

Regarding to the cardinality, the *lowerBound* and *upperBound* attributes take their values depending on the multiplicity of the *Pair*, e.g. it is *one-to-one* (*lowerBound=1* and *upperBound=1*) if the pair value is an object that can be *null*, and the cardinality is *zero-to-many* (*lowerBound=0* and *upperBound=−1*) if the pair value is an array of objects that can take the *null* value.

The *Aggregate* name is the same as the pair name but there are some exceptions. For instance, if the cardinality is *zero-to-many* or *one-to-many*, a singular name is converted into a plural name. In the database example, the *Book_1* entity version would aggregate several *Authors*, so the *Aggregate* name would be *authors*, with cardinality *one-to-many*. The three aggregation relationships discovered for the running example are shown in the diagram of the Fig. 4(a).

Discovering Reference Relationships. A reference implies that a entity's pair identifies to an object of another entity, that is, the pair values of the referencing entity match the values of another pair in the referenced entity. These identifier values are Strings, Integer numbers or arrays of these two primitive types. Two strategies are applied to discover references (i.e. *Reference* elements):

- Some conventions commonly used to express references are checked, such as:
  - If a pair name has the *entityName_id* suffix, then, a entity named *entityName* would be referenced if it exists.
  - MongoDB itself suggests to use a construct like {*$ref: "entityName"*, *$id: "reference_id"*} to express references to objects of the entity named "entityName" [11].
- If a pair name is the name of an existing entity and the pair values match the values of a _id pair of such an entity.

For instance, the *journal* field of a *Publisher* version references to an array of *Journal* objects and the *publisher_id* field of a *Book* version references a *Publisher* object (Fig. 4(a)).

As in the case of aggregations, the references are connected to the corresponding entity in the second stage of the transformation. The cardinality is obtained for references as explained above for the aggregation relationships. Once all the references have been generated, the *opposite* relationship is resolved.

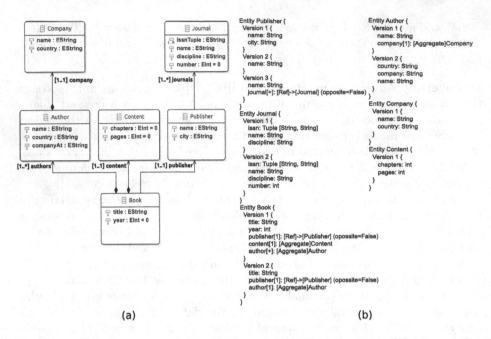

(a)   (b)

Fig. 4. Graphical representation of all the entities with the sum of all fields, and the textual report of versions.

5   Versioned NoSQL-Schema Applications

The inferred schemas are useful to build a number of tools intended to help developers that make use of NoSQL databases. There are tools that require knowledge of the schema in order to provide certain functionality (e.g. SQL query engines). On the other hand, the schema inference may be used to mitigate the problems due to the lack of an explicit schema. For instance, reports, diagrams, validators, and version migration scripts could be automatically generated from the NoSQL Schema models. As a proof of concept, we have created a schema viewer and a validator generator in order to illustrate the possible applications of the inferred schemas. We shall describe these utilities in this Section, and other applications will be outlined in Sect. 7.

Several benefits are gained by representing NoSQL schemas: both reasoning about them and its communication are facilitated, and a documentation separated from the code is obtained. Figure 4(b) shows a textual report of all the entity versions in our running example; attributes (name and type), and aggregate and reference relationships are indicated for each entity version. These reports are automatically generated by a model-to-text transformation that has the schema model as its input. As shown in Fig. 4, the inferred schemas have been also visualized as UML class diagrams. Entities are shown as classes, field as attributes, aggregate as composite associations, and references as associations that are navigable at the end owned by the referenced entity. Note that entity

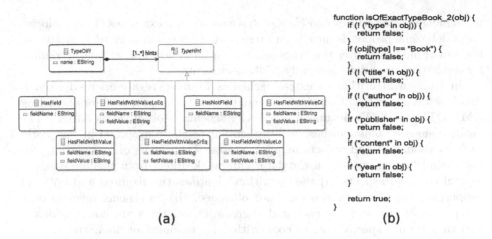

```
function isOfExactTypeBook_2(obj) {
    if (! ("type" in obj)) {
        return false;
    }
    if (obj[type] !== "Book") {
        return false;
    }
    if (! ("title" in obj)) {
        return false;
    }
    if (! ("author" in obj)) {
        return false;
    }
    if ("publisher" in obj) {
        return false;
    }
    if ("content" in obj) {
        return false;
    }
    if ("year" in obj) {
        return false;
    }

    return true;
}
```

(a)   (b)

Fig. 5. TypeRelations metamodel and a simplified code for a specific version.

versions cannot be explicitly represented in class diagrams, but a new kind of
representation is needed. Instead, it is possible to show the elements inferred for
the different versions of a pair, for instance, the *Book* entity contains the *authors*
attribute whose type is a collection of strings along with an aggregate to *Author*.
To generate these class diagrams, we have taken advantage of the tooling provided by EMF/Ecore [14] to represent metamodels as UML class diagrams. This
illustrates the benefits of representing models and metamodels uniformly.

Validation is often needed when dealing with NoSQL databases. For instance,
a developer would want to assure that all the objects retrieved and stored by a
given application conform to a given entity version. When developing a new version of an application, for example, object validators (a.k.a. schema predicates)
could be created so that the programmer can check each object that transfers
to and from the database. Another scenario could be removing a given version
of objects. Validators allow characterizing objects to perform a filter operation
on the database.

Figure 5(a) show the metamodel used to specify relations between versions
of an entity type with respect to the JSON object structure (*TypeRelations*).
These models are obtained via a model-to-model transformation from the NoSQL
Schema, and then a model-to-text transformation generates the validator functions that check the entity version of a given JSON object. Figure 5(b) shows
a simplified code to assert a given entity version. The same approach could be
used to generate specialized queries for specific versions. The metamodel defines
a type discrimination (*TypeDiff*) as a set of hints (*TypeHint*) that a given JSON
object should fulfill to be considered of a given entity version. For example, it has
to contain a given field (*HasField*), or a field with a value (*HasFieldWithValue*)
(e.g. "type" should be "Book").

6   Related Work

The extraction of explicit schemas for JSON-based technologies and applications
is gaining attention as JSON is emerging as a *lingua franca* for information

interchange. Web services and NoSQL systems are two examples of technologies for which some proposals have been presented. This research effort is related to the works published over the years on schema inference and schema versioning for semi-structured data, specially XML documents.

In [10], an algorithm to extract schemas from aggregate-oriented NoSQL databases is presented. This algorithm adapts strategies proposed for extracting XML DTDs to JSON documents. A JSON schema is obtained as output. The authors suggest some database utilities similar to those proposed in this work, such as validators and objects mapper classes. They focus on calculating statistics and finding outliers in the data. Our work differs from this approach in several essential aspects: (i) the algorithm identifies the required and optional properties, but object versions are not obtained; (ii) an schema involves only a type of objects, and reference and aggregation relations are not considered; (iii) they do not specify how to cope with huge amounts of data; and (iv) we obtain a model that conforms to a metamodel, instead of a JSON Schema.

The *JSON Schema* initiative [7] has recently emerged to provide standard specifications for describing JSON schemas. Although its adoption is still very limited, some tools (e.g. validators, schema generators, documentation generators) have evidenced the usefulness of having JSON schemas. The notion of NoSQL schema presented in our work is more expressive than the JSON schemas in the standard, since NoSQL schemas contain aggregate and reference relationships between entities, and also entity versions are represented.

Some tools able of discovering a schema from NoSQL databases have recently emerged. For instance, Spark SQL query engines [15] and Drill [2] are examples of such tools. In Spark SQL, a schema is described as a set of Scala algebraic types and can be inferred for a given set of JSON objects. Spark addresses object versions by means of "sum types", that is, creating types that contain all the keys in all the objects of an entity type, allowing them to be *null* in the objects created or received. As for conflicting types, it generalizes to a String type, that is able to represent any value. This may allow these conflicting types to be addressed without crashing, but it does not offer any guarantee regarding the consistency of the data. Instead, our approach discovers and represents the exact set of versions existing for each object type. Thus, versioned schemas are complete, and allow having a more fine grained control of the objects that enter to and are obtained from a database. Moreover, the reference and aggregation relations between entities are not made explicit in Spark SQL. Drill dynamically discovers the schema during the processing of a query, but it cannot cope with conflicting objects (those that do not comply with the schema). Also, the discovered schema is just used for the purposes of Drill, and cannot be reused by other applications.

MongoDB-Schema [12] is an early prototype of a tool whose purpose is to infer schemas from JSON objects and MongoDB collections. Given a set of objects of the same collection, the inference algorithm obtains an schema that is represented by a JSON document similar to the raw schemas in Sect. 4.1. Moreover, metadata is added to each field in the root and embedded objects in form of a key/value pair. For instance "type" indicates the object type (e.g. Number,

String, or Boolean) and "count" indicates the number of objects that contains a field. Note that this approach has the same limitations of Spark SQL.

A MDE-based approach to infer JSON schemas from REST web services is proposed in [5]. A three-step process is performed to discover the domain model of the services. Firstly, the JSON data for a service is injected into models which conforms to a metamodel similar to that used here. In the second step, a mapping between the JSON metamodel and the Ecore meta-metamodel is established in order to transform the JSON model into a domain model. This JSON-to-Ecore mapping is similar to the one applied here for obtaining a visual representation of NoSQL schemas. In our case, Schema models are represented as Ecore metamodels, which is a more direct mapping. Finally, the domain models obtained for each service are integrated by superposing the common classes. This work is clearly close to our approach but there are however some significant differences between them, namely JSON Discoverer does not tackle the existence of data versions, and the references between objects are not discovered.

In [8] a strategy to infer schemas from heterogeneous XML databases is presented. The schema is provided as a Schema Extended Context-Free Grammar, and the different versions are integrated into a single grammar which is mapped to a relational database schema. In our case, we have used a metamodel to represent the schemas, which has allowed us to apply MDE techniques, and we keep the different versions instead of obtaining a single schema. As in previous related works, our schema is richer, taking into account aggregations and references.

Finally, a design method for aggregate-based NoSQL database is proposed in [3]. This method defines the NoAM model to represent these databases in a system-independent way. NoAM is similar to our Schema metamodel, but it does not consider the possible existence of database object versions. Moreover, the model is simply proposed but it is not implemented in form of a metamodel.

7   Conclusions and Future Work

To bring the well-known benefits of schemas to NoSQL databases, several approaches have been proposed to infer the schema from the data. However, these tools do not cope well with the variability of the schemaless data: they either do not support variation in the structure of the objects of a given type, or they overgeneralize the schema to embrace all the possible variations. In our proposal, the schema takes into account the existing versions of each type: the Versioned Schema has the unique characteristic of completely defining the structure of the data, also showing the high-level relationships, such as aggregation and reference.

The presented approach has proved useful in generating specifications that describe the data, as well as in generating useful applications within the development technical space of the NoSQL databases, such as type validators.

After this initial effort, future directions include generating data visualizations that take into account the type *and* version of the objects in the database, allowing to visually identify the quantities of objects of each type and version.

If a data base has evolved over time, it would be interesting to show which data belong to each version.

Generating object version transformers could also be interesting. A developer can describe, by means of a specialized DSL, the necessary steps to convert one version of an object to another version. These could be used in at least two ways:

- A new application that uses the stored old data may require that all the recovered objects comply with the new version. A version transformer could be generated that removes the unneeded fields, and gives values to new, non-existing fields. This would guarantee that the application would always use object with the correct (new) version, giving all the process more robustness.
- Batch database migration. Map-Reduce jobs could be generated to transform old version objects into new versions. This is possible given the precise version information stored in the schema.

References

1. Abiteboul, S.: Querying semi-structured data. Technical report 1996–19, Stanford InfoLab (1996). http://ilpubs.stanford.edu:8090/144/
2. Apache Foundation: Apache Drill, Visited April 2015. http://drill.apache.org/
3. Bugiotti, F., Cabibbo, L., Atzeni, P., Torlone, R.: Database design for NoSQL systems. In: Yu, E., Dobbie, G., Jarke, M., Purao, S. (eds.) ER 2014. LNCS, vol. 8824, pp. 223–231. Springer, Heidelberg (2014)
4. Buneman, P.: Semistructured data. In: Sixteenth ACM SIGACT-SIGMOD-SIGART Symposium on Principles of Database Systems, pp. 117–121. ACM (1997)
5. Cánovas Izquierdo, J.L., Cabot, J.: Discovering implicit schemas in JSON data. In: Daniel, F., Dolog, P., Li, Q. (eds.) ICWE 2013. LNCS, vol. 7977, pp. 68–83. Springer, Heidelberg (2013)
6. Fowler, M.: Schemaless Data Structures, January 2013. http://martinfowler.com/articles/schemaless/
7. IETF: JSON Schema Specification, Visited April 2015. http://json-schema.org/
8. Janga, P., Davis, K.C.: Mapping heterogeneous XML document collections to relational databases. In: Yu, E., Dobbie, G., Jarke, M., Purao, S. (eds.) ER 2014. LNCS, vol. 8824, pp. 86–99. Springer, Heidelberg (2014)
9. Karpov, V.: Mongoose NPM package, Visited April 2015. https://www.npmjs.com/package/mongoose
10. Klettke, M., Störl, U., Scherzinger, S.: Schema extraction and structural outlier detection for JSON-based NoSQL data stores. In: BTW 2105, pp. 425–444 (2015)
11. Redmond, E., Wilson, J.R.: Seven Databases in Seven Weeks. A Guide to Modern Databases and the NoSQL Movement, Pragmatic Programmers (2013)
12. Rückstieß, T.: mongodb-schema NPM package, Visited April 2015. https://www.npmjs.com/package/mongodb-schema
13. Sadalage, P., Fowler, M.: NoSQL Distilled: A Brief Guide to the Emerging World of Polyglot Persistence. Addison-Wesley, Reading (2012)
14. Steinberg, D., Budinsky, F., Paternostro, M., Merks, E.: Eclipse Modeling Framework. Addison-Wesley, Reading (2008)
15. Zaharia, M., Chowdhury, M., et al.: Resilient distributed datasets: a fault-tolerant abstraction for in-memory cluster computing. In: NSDI, April 2012. http://spark.apache.org

Schema Discovery in RDF Data Sources

Kenza Kellou-Menouer(✉) and Zoubida Kedad(✉)

PRISM - University of Versailles Saint-Quentin-en-Yvelines, Versailles, France
{kenza.menouer,zoubida.kedad}@prism.uvsq.fr

Abstract. The Web has become a huge information space consisting of interlinked datasets, enabling the design of new applications. The meaningful usage of these datasets is a challenge, as it requires some knowledge about their content such as their types and properties. In this paper, we present an automatic approach for schema discovery in RDF(S)/OWL datasets.

We consider a schema as a set of type and link definitions. Our contribution is twofold: (i) generating the types describing a dataset, along with a description for each of them called type profile; (ii) generating the semantic links between types as well as the hierarchical links through the analysis of type profiles. Our approach relies on a density-based clustering algorithm and it does not require any schema-related information in the dataset. We have implemented the proposed algorithms and we present some evaluation results showing the effectiveness of our approach.

Keywords: Schema extraction · Clustering · Semantic Web

1 Introduction

The Web has evolved into a huge global information space, consisting of interlinked datasets, expressed in the Semantic Web standard languages such as RDF, RDFS or OWL. An unprecedented amount of data is made available, enabling the design of new applications in many domains. But if access to these datasets is provided, their meaningful usage is still a challenge, because it requires some knowledge about their content. For example, a user willing to query a dataset needs to know some of the properties and types existing in this dataset. This information could be obtained by randomly browsing the data, but this would be a tedious task. Writing the query would be straightforward with a schematic description of the dataset. This schema would be useful for other purposes, such as creating links between datasets; some interlinking tools have been proposed, such as Silk[1], which was used to link Yago [14] to DBpedia [1], but they require type and property information about the datasets to generate the appropriate *owl:sameAs* links.

An RDF(S)/OWL dataset may provide some information about its schema, such as *rdf:type*, specifying the type of a given entity, but this information is

---

[1] Silk: wifo5-03.informatik.uni-mannheim.de/bizer/silk.

© Springer International Publishing Switzerland 2015
P. Johannesson et al. (Eds.): ER 2015, LNCS 9381, pp. 481–495, 2015.
DOI: 10.1007/978-3-319-25264-3_36

not always complete. Even when datasets are automatically extracted, such as DBpedia which is extracted from Wikipedia, type information can be missing. The experiments presented in [12] show that at most 63.7 % of the data have complete type declarations in DBpedia, and at most 53.3 % in YAGO. We argue that providing a schematic description of a dataset is essential for its meaningful exploitation. In the context of Web data, the notion of schema is understood as a guide, not as a strict representation to which the data must conform. Indeed, languages used to format data on the Web do not impose any constraint on its structure: two entities having the same type may have different properties. In addition, an entity may have several types.

The goal of our work is to extract the schema describing an RDF(S)/OWL dataset and we propose a deterministic and automatic approach for schema discovery. We consider a schema as a set of type and link definitions. Our contribution is twofold: (i) generating the types describing a dataset, along with a description for each of them called type profile, where each property is associated to a probability; (ii) generating the semantic links between types as well as the hierarchical links through the analysis of type profiles. Our approach relies on a density-based clustering algorithm, it does not require any schema related information and it can detect noisy instances. We have implemented the algorithms underlying our approach and we present some evaluation results using different datasets to show the quality of the resulting schema.

The paper is organized as follows. We introduce a formal definition of the problem in Sect. 2 and we give some definitions related to entity description in Sect. 3. In Sect. 4, we present the core algorithms of our schema discovery approach. Section 5 presents the generation of overlapping types and Sect. 6 is devoted to links generation. Our evaluation results are presented in Sect. 7. We discuss the related works in Sect. 8 and finally, Sect. 9 concludes the paper.

2   Problem Statement

Consider the sets R, B, P and L representing resources, blank nodes (anonymous resources), properties and literals respectively. A dataset described in RDF(S)/OWL is defined as a set of triples $D \subseteq (R \cup B) \times P \times (R \cup B \cup L)$. Graphically, a dataset is represented by a labeled directed graph G, where each node is a resource, a blank node or a literal and where each edge from a node e to another node e' labeled with the property p represents the triple (e, p, e') of the dataset D. In such RDF(S)/OWL graph, we define an entity as a node corresponding to either a resource or a blank node, that is, any node apart from the ones corresponding to literals.

Figure 1(a) shows an example of such dataset, related to conferences. We can see that some entities are described by the property *rdf:type*, defining the classes to which they belong, as it is the case for "UVSQ", defined as a university. For other entities, such as "WWW", this information is missing. Two entities having the same type are not necessarily described by the same properties, as we can see for "UVSQ" and "MIT" in our example, which are both associated to the "University" type, but unlike "UVSQ", "MIT" has a "website" property.

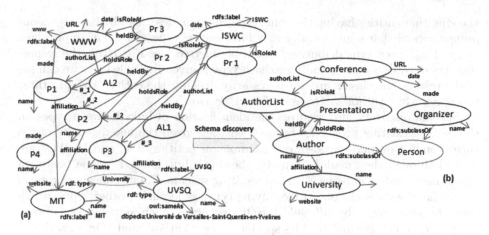

Fig. 1. Example of dataset and generated schema.

Our problem can be stated as follows: given an RDF(S)/OWL dataset with incomplete schema information, how to infer type definitions and links between these types? As a result, we provide an extracted schema defined as follows:

Definition. The extracted schema S of a dataset D is composed of:

- A set of possibly overlapping classes $C = \{C_1, ..., C_n\}$, where each class C_i corresponds to a set of entities in D and defines their type;
- A set of links $\{p_1, ..., p_m\}$, such that each p_i is a property for which both the range and the domain are two types corresponding to two classes in C;
- A set of hierarchical links involving two types corresponding to classes in C, expressing that one is the generic type of the other.

Table 1. Entity types in the conference dataset (Fig. 1(a)).

Types	Entities
"Conference"	"WWW", "ISWC"
"Presentation"	"Pr1", "Pr2"
"AuthorList"	"AL1" and "AL2"
"University"	"UVSQ", "MIT"
"Organizer"	"P1", "P4"
"Author"	"P1", "P2" and "P3"
"Person"	"P1", "P2", "P3" and "P4"

In order to define the schema describing a dataset, we first need to evaluate the similarity between entities and then group the similar ones into types, knowing that entities having the same type could be described by heterogeneous properties and that a given entity may have several types. In the example given in Fig. 1(a), seven type definitions would be inferred (see Fig. 1(b)), the respective sets of entities corresponding to each type is given in Table 1. As we can see in the results, the classes are not necessarily disjoint: the entity "P1" has three types, namely "Author", "Organizer" and "Person".

Schema discovery also requires the identification of links between types on the basis of existing properties between entities, given the heterogeneity of the corresponding property sets. In our example, as entities "P4" and "ISWC" are related through the "made" property, there is a link labeled "made" between "Organizer" and "Conference" in the resulting schema. Finally, another problem we are faced with is the one of identifying types generalizing other types, which could be represented by *rdfs:subClassOf* properties, such as the ones between the generic type "Person" and its specific types "Author" and "Organizer".

3   Entity Description

We consider that an entity is described by different kinds of properties. Some of them are part of the RDF(S)/OWL vocabularies, and we will refer to them as primitive properties, and others are user-defined. We distinguish between these two kinds because all the properties should not be used for the same purpose during type discovery. Some predefined properties may provide information about the schema of the dataset and could be applied to any entity, therefore they should not be considered when evaluating the similarity between entities. Some of these properties can be used to validate the resulting schema: for example, if two entities e and e' are linked with the *owl:sameAs* primitive property, then we can check that the types inferred for e and e' are the same. We define the description of an entity as follows.

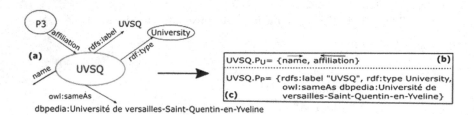

Fig. 2. Example of entity description.

Definition. Given the set of primitive properties P_P and the set of user-defined properties P_U in the dataset D, an entity e is described by:

1. A user-defined property set $e.P_U$ composed of properties p_u from P_U, each one annotated by an arrow indicating its direction, and such that:

- If $\exists (e, p_u, e') \in D$ then $\overrightarrow{p_u} \in e.P_U$;
- If $\exists (e', p_u, e) \in D$ then $\overleftarrow{p_u} \in e.P_U$.

2. A primitive property set $e.P_P$, composed of properties p_p from P_P with their values, such that:

- If $\exists (e, p_p, e') \in D$ then $p_p \in e.P_P$.

Figure 2 shows an example of entity description. In order to infer type definitions, entities are compared according to their structure. Our type discovery algorithm takes as input the set $D_U = \{e_i.P_U : i = 1, ...n\}$, where $e_i.P_U$ represents the set of user-defined properties describing the entity e_i. The set $D_P = \{e_i.P_P : i = 1, ...n\}$, where $e_i.P_P$ represents the set of primitive properties describing the entity e_i, can be used to validate the results of the algorithm and to specify type checking rules. This validation process is out of the scope of this paper.

4   Generating Types

Our requirements for type discovery are as follows: (i) the number of types in the dataset is not known, (ii) an entity can have several types, and (iii) the datasets may contain noise. The most suitable grouping approach is density-based clustering [3] because it is robust to noise, deterministic and it finds classes of arbitrary shape, which is useful for datasets where entities are described with heterogeneous property sets. In addition, unlike the algorithms based on k-means and k-medoid, the number of classes is not required.

Algorithm 1. Density-based Clustering with Type Profile

Require: D_U, ε, $MinPts$
  $setOfTypeProfile \leftarrow \emptyset$
  $class \leftarrow 0$
  while $\exists$ "no marked" $e.P_U \in D_U$ **do**
    mark $e.P_U$
    $setOfNeighbor \leftarrow$ FindNeighbor(D_U, $e.P_U$, ε)
    if $|setOfNeighbor| \geq MinPts$ **then**
      $class + +$;
      $setOfTypeProfile \leftarrow setOfTypeProfile \cup$ ExpandCluster(D_U,
      $e.P_U$, $MinPts$, ε, $class$, $setOfNeighbor$)
    end if
  end while
  return $setOfTypeProfile$, typed D_U

Our density-based algorithm has two parameters: ε, representing the minimum similarity value for two entities to be considered as neighbors, and $MinPts$, representing the minimum number of entities in the neighborhood required for an entity to be a core [3] and to generate a type; $MinPts$ is used to exclude the outliers and the noise. We use Jaccard similarity to compute the closeness between two property sets $e.P_U$ and $e'.P_U$ describing respectively two entities e and e'.

Beside the set of types, we provide a description of each of them called a type profile. The profiles will be used to find links between types and to generate overlapping types, i.e. multiple types for each entity. A profile consists in a property vector where each property is associated to a probability. The profile corresponding to a type T_i is denoted $TP_i = ((p_{i1}, \alpha_{i1}), ..., (p_{in}, \alpha_{in}))$, where each p_{ij} represents a property and where each α_{ij} represents the probability for an entity of T_i to have the property p_{ij}. It is evaluated as the number of entities in T_i having the property p_{ij} over the total number of entities in T_i. In Fig. 1, the profile of the "University" type is: $((\overrightarrow{name}, 1), (\overrightarrow{website}, 0.5), (\overleftarrow{affiliation}, 1))$. The probability of "website" is 0.5 because this property is defined for "MIT" but not for "UVSQ".

In order to group similar entities and build the profile of each class, we have adapted a density-based clustering algorithm (Algorithm 1); it uses the **Find-Neighbors** function which returns for a given entity all the entities having a distance smaller than ε. For each new entity e such that the number of its neighbors is lower than $MinPts$, we expand the class of e using the **ExpandCluster** function, defined in Algorithm 2, which finds all entities belonging to the same class as the current entity e. Each time a new entity e belonging to the current class is found, the type profile of this class is updated using the **UpdateType-Profile** function which adds the properties of e if they are not already in the profile and recomputes the probabilities.

5   Generating Overlapping Types

An important aspect of RDF(S)/OWL datasets is that an entity may have several types [12]. A fuzzy clustering algorithm such as FCM or EM could be used to assign several types to an entity. However, they require the number of classes, as their grouping criterion is the similarity between an entity and the center of each class. This parameter can not be defined in our context, we therefore propose to derive the set of disjoint classes first, and then to form overlapping or fuzzy classes by analyzing the type profiles.

Recall that the importance of a property for a given type is captured by the associated probability. Figure 3(a) shows three classes generated by our algorithm: $C_1 = \{P1\}$, $C_2 = \{P4\}$ and $C_3 = \{P2, P3\}$, described respectively by the following type profiles:

- $TP_1 = ((\overrightarrow{name}, 1), (\overrightarrow{made}, 1), (\overrightarrow{affiliation}, 1), (\overrightarrow{heldBy}, 1), (\overrightarrow{holdsRole}, 1), (\overleftarrow{\#_-}, 1))$
- $TP_2 = ((\overrightarrow{name}, 1), (\overrightarrow{made}, 1))$
- $TP_3 = ((\overrightarrow{affiliation}, 1), (\overrightarrow{heldBy}, 1), (\overrightarrow{holdsRole}, 1), (\overleftarrow{\#_-}, 1))$.

The entity "P1" has all the properties of the two types TP_2 and TP_3 for which the probability is 1, we can therefore assign the corresponding types to "P1".

Algorithm 2. Expand Cluster

Require: D_U, $e.P_U$, $MinPts$, ε, $class$, $setOfNeighbor$
   $nbEntitiesInClass \leftarrow 0$
   Type profile of the class $TP_{class} \leftarrow \emptyset$
   add $class$ to $e.P_U$ types
   UpdateTypeProfile(TP_{class}, $class$, $e.P_U$, $nbEntitiesInClass$)
   $nbEntitiesInClass$++
   while $\exists\, e'.P_U \in setOfNeighbor$ **do**
     if "not marked" $e'.P_U$ **then**
       mark $e'.P_U$
       $setOfNeighbor' \leftarrow$ FindNeighbor(D_U, $e'.P_U$, ε)
       if $|setOfNeighbor'| \geq MinPts$ **then**
         $setOfNeighbor \leftarrow setOfNeighbor \cup setOfNeighbor'$
       end if
     end if
     add $class$ to e' types
     UpdateTypeProfile(TP_{class}, $class$, $e'.P_U$, $nbEntitiesInClass$)
     $nbEntitiesInClass$++
   end while
   return TP_{class}

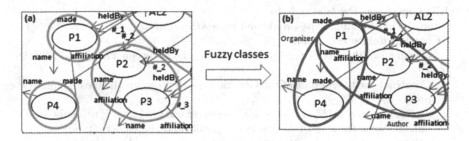

Fig. 3. Generating fuzzy classes

Fuzzy classes are generated using type profiles as follows; consider the type T_i described by the profile TP_i; if all the properties p of TP_i having a probability $\alpha = 1$ belong to another type profile TP_k, then the type T_i falls within the types of entities of the group k. This is expressed by the following rule:

– If $(\forall (p, \alpha)$ in TP_i, $\alpha = 1$: (p, α) in $TP_k)$ then (Add T_i to the types of the entities of class k).

In our example (Fig. 3), the cluster containing "P1" (class k here) is the intersection of the classes C_2 and C_3. Note that instead of comparing only the properties having a probability of 1, we could extend the comparison to the ones having a probability greater that a given threshold.

6   Generating Links

Beside type definitions, links between these types are also important to understand the content of the dataset at hand, as it is shown in Fig. 1(b). We are
interested in two types of links: semantic links, corresponding to user-defined
properties and hierarchical links, corresponding to the *rdfs:subClassOf* property.

Semantic links. Two types T_i, T_j are linked by a property p if $\overrightarrow{p}$ belongs to
the properties of type profile TP_i and $\overleftarrow{p}$ belongs to the properties of type profile
TP_j, as follow:

- If $(\exists\ p: (\overrightarrow{p}, \alpha_i)$ in $TP_i \wedge (\overleftarrow{p}, \alpha_j)$ in $TP_j)$ then (Add $\overrightarrow{p}$ from T_i to T_j).

These generated links are checked by finding two entities $e \in T_i$ and $e' \in T_j$
such that $(e, p, e') \in D$.

Algorithm 3. Hierarchical Links Generation

Require: *setOfTypeProfile*
  setOfHierarchicalLinks ← ∅
  while $|setOfTypeProfile| > 1$ **do**
    Find the most similar type profile and its *bestSimilarity*
    if *bestSimilarity* = 0 **then**
      Group all of the rest of type profile into the Generic type *Thing* and **STOP**
    else
      Construct the type profile of the Generic type that groups the most similar type
      profile
      setOfHierarchicalLinks ← *setOfHierarchicalLinks* ∪ {"rdfs:subClassOf"
      links between these most similar types profile and the Generic type}
      setOfTypeProfile ← *setOfTypeProfile* ∪ {type profile of the Generic type}

      Remove these most similar types profile from the *setOfTypeProfile*
    end if
  end while
  return *setOfHierarchicalLinks*

Hierarchical links. Two types T_i, T_j can be linked by a hierarchical property
(*rdfs:subClassOf*), such as "Organizer" and "Person" in Fig. 1(b). We can use a
hierarchical clustering algorithm on the entire dataset to generate these links but
this would be very costly; to find the best partition, all the generated hierarchy
has to be explored. In addition, the result would consist in many hierarchical links,
not all of them being meaningful. We generate instead the hierarchy from the type
profiles, which is less expensive because the number of profiles is small compared
to the size of the entire dataset. Our procedure is given in Algorithm 3. We have
adapted an ascending hierarchical clustering algorithm; the profile of the generic
type is built at each step of the hierarchy. We define a similarity measure between

two type profiles TP_i, TP_j, inspired from the Jaccard similarity and based on the probability of a property p_k for two classes Ci and Cj. It is defined as follows:

$$ProfileSim(TP_i, TP_j) = \frac{\sum_{\forall p_k \in \{TP_i \cap TP_j\}} Prob_{i,j}(p_k)}{\sum_{\forall p_k \in \{TP_i \cup TP_j\}} Prob_{i,j}(p_k)} \tag{1}$$

where:

$$Prob_{i,j}(p_k) = \frac{\alpha_{ik} \times |C_i| + \alpha_{jk} \times |C_j|}{|C_i| + |C_j|} \tag{2}$$

A generic type is defined from the two most similar type profiles at each level of the hierarchy. The corresponding profile is composed of all the properties of the two types, the probability of a property is calculated as in (2).

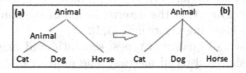

Fig. 4. Eliminating generic type redundancy in a hierarchy.

This approach generates a hierarchy by grouping types in pairs to find their generic type. However, some generic types can be composed of more than two sub-types as in Fig. 4(a). In this case, the user can detect easily intermediate redundant generic types and eliminate them (as in Fig. 4(b)), since the number of types in a dataset is generally not very high.

7   Evaluation

This section presents some experimentation results using our approach. We have evaluated the quality of the generated schema using well-known information retrieval metrics in different real datasets, described in the following section.

7.1   Datasets

For our experiments, we have used three datasets: the Conference[2] dataset, which exposes data for several Semantic Web conferences and workshops with 1430 triples; the BNF[3] dataset which contains data about the French National Library (Bibliothèque Nationale de France) with 381 triples and a dataset extracted from DBpedia[4] with 19696 triples considering the following types: Politician, SoccerPlayer, Museum, Movie, Book and Country.

---

[2] Conference: data.semanticweb.org/dumps/conferences/dc-2010-complete.rdf.
[3] BNF: datahub.io/fr/dataset/data-bnf-fr.
[4] DBpedia: dbpedia.org.

7.2 Metrics and Experimental Methodology

In order to evaluate the quality of the results provided by our algorithms, we have extracted the existing type definitions from our datasets and considered them as a gold standard. We have then run our algorithm on the datasets without the type definitions and evaluated the precision and recall for the inferred types. We have annotated each inferred class C_i with the most frequent type label associated to its entities. For each type label L_i corresponding to type T_i in the dataset and each class C_i inferred by our algorithm, such that L_i is the label of C_i, we have evaluated the precision $P_i(T_i, C_i) = |T_i \cap C_i|/|C_i|$ and the recall $R_i(T_i, C_i) = |T_i \cap C_i|/|T_i|$. We have set $MinPts = 1$ so that an entity is considered as noise if it has no neighbors. For the Conference dataset, we have empirically set $\varepsilon = 0.75$, which leads to very homogeneous classes. We have empirically stated that $\varepsilon = 0.72$ provides a number of classes equal to the number of types initially declared in the BNF dataset; the value is $\varepsilon = 0.5$ for the DBpedia dataset. Note that the determination of the similarity threshold ε is an open issue for clustering algorithms [3,13].

To provide the overall quality of types, semantic and hierarchical links, we have used the precision and recall metrics. The number of generated classes is denoted k, and the number of entities in the dataset, n. To assess the overall type quality, each type is weighted according to its number of entities as follows:

$$P = \sum_{i=1}^{k} \frac{|C_i|}{n} \times P_i(T_i, C_i) \qquad R = \sum_{i=1}^{k} \frac{|C_i|}{n} \times R_i(T_i, C_i)$$

Precision and recall of the generated links are evaluated considering the true/false positives and the false negatives. We have compared the inferred schema to the one of the dataset when it was provided, as for the BNF dataset. If no schema was provided, we have manually designed it based on the information provided in the dataset; to this end, we have built the set D_P as defined in Sect. 3.

7.3 Results

The quality of each discovered type in the Conference dataset is shown in Fig. 5(a). Our approach gives good precision and recall and detects types which were not declared in the dataset: classes 6, 10, 11 and 12 are manually labeled "AuthorList", "PublicationPage", "HomePage" and "City" respectively.

In some cases, types have been inferred relying on incoming properties only. Indeed, for containers, such as "AuthorList", it is necessary to consider these properties as they do not have any outgoing property. Classes 1 and 7 do not have a good precision because they contain entities with different types in the dataset. However, these types have the same structure, it is therefore impossible to distinguish between them. The recall for "Person" is not good because it is split into three classes: class 8 represents persons who have both published and played a role in the conference; class 2 represents persons who have only played a role (e.g. Chair, Committee Member); class 5 represents persons who have only published. Overlapping types are generated based on the analysis of type profiles. This has led to the results shown in Fig. 5(b). Class 8 is associated

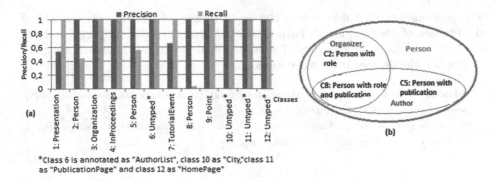

*Class 6 is annotated as "AuthorList", class 10 as "City,"class 11 as "PublicationPage" and class 12 as "HomePage"

Fig. 5. Quality of the generated types (a) and overlapping types (b) in the conference dataset.

to two types: the one of class 2 (manually labeled "Organizer") and the one of class 5 (manually labeled "Author"), which indeed conforms to the entities of the dataset. Note that finding the labels of classes is an open problem that we will address in future works.

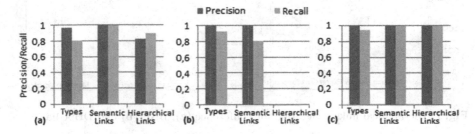

Fig. 6. Evaluation of schema discovery in conference (a) BNF (b) and DBpedia (c).

We can see in Fig. 6 that the approach gives good precision and recall for the generated schema, composed of types, semantic and hierarchical links. For the Conference dataset (see Fig. 6(a)) the precision is not maximum because of classes 1 and 7 as discussed before. The recall is not maximum because the "Person" type is split in three as discussed above. The results for the BNF (see Fig. 6(b)) and DBpedia datasets (see Fig. 6(c)) show that the assignment of types to entities has achieved good precision and recall. The recall is not maximum because noisy instances were detected. For the BNF dataset, some of the semantic links were not declared in the provided schema, which is why the recall is not maximum. However, after checking the entities of the dataset, we found out that these semantic links were valid. For the DBpedia dataset, our algorithm was able to differentiate between entities of the two types "Politician" and "SoccerPlayer" even if they have similar property sets, as it is shown by the corresponding type profiles generated by our algorithm and presented below.

The generated type profiles reflect the heterogeneity of the dataset, e.g. 6 % of the entities of "SoccerPlayer" have a *"deathDate"* outgoing property, and yet the generated grouping was good. The results achieved by our approach are good even when the dataset is heterogeneous.

- Politician: < ($\overrightarrow{name}$, 1), ($\overrightarrow{party}$, 0.73), ($\overrightarrow{children}$, 0.21), ($\overrightarrow{birthDate}$, 0.94), ($\overrightarrow{nationality}$, 0.15), ($\overleftarrow{successor}$, 0.78), ($\overrightarrow{deathDate}$, 0.68), ...>.
- SoccerPlayer: < ($\overrightarrow{name}$, 1), ($\overrightarrow{height}$, 0.46), ($\overrightarrow{surname}$, 0.93), ($\overrightarrow{birthDate}$, 1), ($\overrightarrow{nationalteam}$, 0.86), ($\overleftarrow{currentMember}$, 0.8), ($\overrightarrow{deathDate}$, 0.06), ...>.

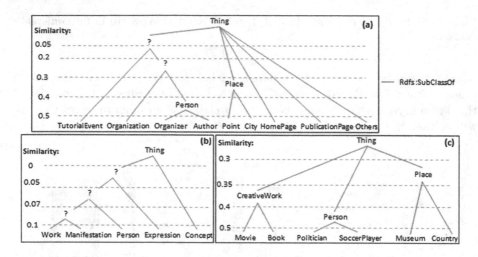

Fig. 7. Hierarchical links generation for conference (a), BNF (b) and DBpedia (c).

The hierarchical links generated for DBpedia are correct as they conform to the existing *rdfs:subClassOf* declarations (see Fig. 7(c)). The BNF dataset has no hierarchical links, the values of both precision and recall are therefore null. When the similarity between two profiles is low, the semantic of the generic type is unclear. This is represented by a question mark in Fig. 7(b). It is the same for some of the hierarchical links generated for the Conference dataset (see Fig. 7(a)): a generic type has been generated for "Person" and "Organization", however, the similarity between their type profiles is low. Our algorithm could not identify a generic type for "HomePage" and "PublicationPage" because their type profiles do not share any property, which explains the recall.

8   Related Works

Schema discovery from semi-structured data has been addressed by some research works. In [16], an approximate DataGuide based on the COBWEB hierarchical clustering algorithm is proposed. The incoming/outgoing edges are considered in the same way, which could be a problem in RDF datasets as it will not differentiate between the domain and the range of properties. The resulting classes are disjoint, and the approach is not deterministic as it is based on COBWEB. In [8], several types are inferred for an entity, but only when a type is more general than another, such as "Employee" and a "Person". The approach distinguishes between incoming and outgoing edges: incoming edges are considered as roles, and potential labels for the inferred types. This is suitable for the OEM [11] model used in the approach, but not to RDF, where incoming edges do not necessarily reflect the type of an entity. The proposed algorithm requires the threshold of the jump which is not easy to define as it depends on the regularity of the data; this parameter is not known in our context. The approach presented in [9] uses bottom-up grouping providing a set of disjoint classes. Unlike in our approach, the number of classes is required. In [2], standard ascending hierarchical clustering is used to build structural summaries of linked data. Each instance is represented by its outgoing properties and the property set of a class is the union of the properties of its entities, while in our approach, the probability of each property is computed for a type. The algorithm provides disjoint classes; the hierarchical clustering tree is explored to assess the best cutoff level, which can be costly. SchemEX [6] finds the relevant data sources for a query by building an RDF triple index. Unlike in our approach, *rdf:type* declarations are required to find classes. The approach in [5] adds structural information to a database from a set of available databases. It searches for a similar database among the set of existing ones in order to make design decisions.

Some works have addressed the problem of enriching an existing schema by adding more structure through RDF(S)/OWL primitives. SDType [12] enriches an entity by several types using inference rules, provided that these types exist in the dataset, and it computes the confidence of each type for an entity. The focus of the approach is therefore on the evaluation of the relevance of the assigned types. In addition, *rdfs:domain*, *rdfs:range* and *rdfs:subClassOf* properties are required. Works in [4,10] infer types for DBpedia only: [10] uses K-NN and [4] finds the most appropriate type for an entity in DBpedia based on descriptions from Wikipedia and links with WordNet and the Dolce ontology. The statistical schema induction approach [15] enriches an RDF dataset with some RDFS/OWL primitives, however type information must be provided. The approach in [17] uses an ascending hierarchical clustering to form classes by exploiting the existing *rdf:type* declarations. The approach finds hierarchical links, but is specific to the bio-medical ontologies DrugBank and Diseasome. In [7], a reverse engineering method dealing with the derivation of inheritance links embedded in a relational database is presented. Decision rules for detecting existence dependencies and translating them into hierarchies among entities are defined.

9   Conclusion

We have proposed an approach for schema discovery in RDF(S)/OWL datasets. In order to generate several types for an entity, we have adapted a density-based clustering algorithm. Each generated type is described by a profile, consisting of a property vector where each property is associated to a probability. These type profiles are used to generate overlapping types as well as semantic and hierarchical links. Our experiments show that our approach achieves good quality results regarding both types and links in the generated schema, even when the entities are very heterogeneous, such as in DBpedia. One important problem that we plan to address is the annotation of the inferred types. Indeed, in addition to identifying a cluster of entities having the same type, it is also useful to find the labels which best capture the semantics of this cluster.

Acknowledgements. This work was partially funded by the French National Research Agency through the CAIR ANR-14-CE23-0006 project.

References

1. Auer, S., Bizer, C., Kobilarov, G., Lehmann, J., Cyganiak, R., Ives, Z.G.: DBpedia: a nucleus for a Web of open data. In: Aberer, K., et al. (eds.) ASWC 2007 and ISWC 2007. LNCS, vol. 4825, pp. 722–735. Springer, Heidelberg (2007)
2. Christodoulou, K., Paton, N.W., Fernandes, A.A.: Structure inference for linked data sources using clustering. In: EDBT/ICDT 2013 Workshops (2013)
3. Ester, M., Kriegel, H.-P., Sander, J., Xu, X.: A density-based algorithm for discovering clusters in large spatial databases with noise. In: Kdd (1996)
4. Gangemi, A., Nuzzolese, A.G., Presutti, V., Draicchio, F., Musetti, A., Ciancarini, P.: Automatic typing of DBpedia entities. In: Cudré-Mauroux, P., et al. (eds.) ISWC 2012, Part I. LNCS, vol. 7649, pp. 65–81. Springer, Heidelberg (2012)
5. Klettke, M.: Reuse of database design decisions. In: Kouloumdjian, J., Roddick, J., Chen, P.P., Embley, D.W., Liddle, S.W. (eds.) ER Workshops 1999. LNCS, vol. 1727, pp. 213–224. Springer, Heidelberg (1999)
6. Konrath, M., Gottron, T., Staab, S., Scherp, A.: Schemex: efficient construction of a data catalogue by stream-based indexing of linked data. WWW **16**, 52–58 (2012)
7. Lammari, N., Comyn-Wattiau, I., Akoka, J.: Extracting generalization hierarchies from relational databases: a reverse engineering approach. Data Knowl. Eng. **63**(2), 568–589 (2007)
8. Nestorov, S., Abiteboul, S., Motwani, R.: Inferring structure in semistructured data. ACM SIGMOD Rec. **26**(4), 39–43 (1997)
9. Nestorov, S., Abiteboul, S., Motwani, R.: Extracting schema from semistructured data. ACM SIGMOD Rec. **27**, 295–306 (1998). ACM
10. Nuzzolese, A.G., Gangemi, A., Presutti, V., Ciancarini, P.: Type inference through the analysis of Wikipedia links. In: LDOW (2012)
11. Papakonstantinou, Y., Garcia-Molina, H., Widom, J.: Object exchange across heterogeneous information sources. In: Proceedings of the Eleventh International Conference on Data Engineering, pp. 251–260. IEEE (1995)
12. Paulheim, H., Bizer, C.: Type inference on noisy RDF data. In: Alani, H., et al. (eds.) ISWC 2013, Part I. LNCS, vol. 8218, pp. 510–525. Springer, Heidelberg (2013)
13. Sánchez-Díaz, G., Martínez-Trinidad, J.F.: Determination of similarity threshold in clustering problems for large data sets. In: Sanfeliu, A., Ruiz-Shulcloper, J. (eds.) CIARP 2003. LNCS, vol. 2905, pp. 611–618. Springer, Heidelberg (2003)

14. Suchanek, F.M., Kasneci, G., Weikum, G.: Yago: a core of semantic knowledge. In: Proceedings of the 16th International Conference on World Wide Web (2007)
15. Völker, J., Niepert, M.: Statistical schema induction. In: Antoniou, G., Grobelnik, M., Simperl, E., Parsia, B., Plexousakis, D., De Leenheer, P., Pan, J. (eds.) ESWC 2011, Part I. LNCS, vol. 6643, pp. 124–138. Springer, Heidelberg (2011)
16. Wang, Q.Y., Yu, J.X., Wong, K.-F.: Approximate graph schema extraction for semi-structured data. In: Zaniolo, C., Grust, T., Scholl, M.H., Lockemann, P.C. (eds.) EDBT 2000. LNCS, vol. 1777, pp. 302–316. Springer, Heidelberg (2000)
17. Zong, N., Im, D.-H., Yang, S., Namgoon, H., Kim, H.-G.: Dynamic generation of concepts hierarchies for knowledge discovering in bio-medical linked data sets. In: ICUIMC. ACM (2012)

Process and Text Mining

Learning Relationships Between the Business Layer and the Application Layer in ArchiMate Models

Ayu Saraswati[1]([✉]), Chee-Fon Chang[2], Aditya Ghose[1], and Hoa Khanh Dam[1]

[1] Decision Systems Laboratory, School of Computer Science
and Software Engineering, University of Wollongong,
Wollongong, NSW 2522, Australia
sa783@uowmail.edu.au, {aditya,hoa}@uow.edu.au
http://www.dsl.uow.edu.au
[2] Centre for Oncology Informatics, University of Wollongong,
Wollongong, NSW 2522, Australia
c03@uow.edu.au

Abstract. Enterprise architecture provides a visualisation tool for stakeholder to manage and improve the current organization strategy to achieve its objectives. However, building an enterprise architecture is a time-consuming and often highly complex task. It involves data collection and analysis in several levels of granularity, from the physical nodes to the business execution. Existing solutions does not provide techniques to learn the relationship between the levels of granularity. In this paper, we proposed a method to correlate the business and application layers in ArchiMate notation.

Keywords: Enterprise architecture · Data-driven model extraction · ArchiMate

1 Introduction

An enterprise architecture (EA) provides a comprehensive map of the assets of an enterprise (both knowledge/conceptual resources as well as physical assets), their inter-relationships, and the manner in which they help realize the goals/strategies of the enterprise [1,2]. A well-defined EA gives an organisation an accurate and deep understanding of the opportunities and priorities for change and innovation, and the means to assess the impact of the proposed changes [1,3]. There is widespread recognition of the value of developing and maintaining enterprise architectures [1–3].

Despite the obvious benefits that accrue from devising and maintaining an EA, organizations often shy away from investing in these exercises. There are two key drivers for this. First, organizations find the investment required for building and maintaining an EA unduly high relative to the benefits that follow (this is often based on an inadequate appreciation of the true nature of

P. Johannesson et al. (Eds.): ER 2015, LNCS 9381, pp. 499–513, 2015.
DOI: 10.1007/978-3-319-25264-3_37

these benefits). It is generally acknolwedged that capturing and analysing all the data that describes the enterprise environment (which can be complex and diverse, consisting, for instance, of in-house services, cloud services, virtual and physical applications and infrastructures in multiple sites) is expensive and time-consuming [3–5]. Second, organizations find it difficult to maintain an EA in the face of rapid change. In the time that the EA model takes to build, it may no longer be an accurate representation of the enterprise environment due to continuously changing enterprise environments [5].

Machinery that extracted an EA from available data (including execution histories) in an automated (or semi-automated) fashion would be a game-changer in this domain, by alleviating the cost and time impediments to EA development and maintenance discussed above. In its ideal form, this machinery would dramatically reduce EA development and maintenance costs. It would also reduce the time involved, thus making EA maintenance in a rapidly changing environment feasible.

The ArchiMate EA framework [6] provides a formal notation to represent the relationships between enterprise entities in a hierarchical fashion with 3 layers: the *business layer*, the *application layer* and the *technology layer* [6] (in the remainder of this paper, we shall use the terms *domain* and *layer* interchangeably). Our work focuses on the Archimate framework due to the ability to leverage the formal notation that it offers (the absence of which is often a critique of other EA frameworks) as well as its widespread adoption and associated tool support. However, the conceptual insights that we generate are equally applicable to other EA frameworks.

The specific contribution of this paper is a technique for data-driven extraction of the correlation between the business and application layers in an ArchiMate model. We take as input a process log [7] and a record of function invocations in applications described in the application layer (we shall refer to this log as an *invocation log*). The proposed method uses a combination of time correlation heuristics [8] and frequent closed sequence pattern mining [9–11] to find the correlations between the business and application domains. Mining these correlations generates a preliminary (first-cut) version of part of an ArchiMate model. This can then be edited by enterprise architects where necessary, but they are able to avoid the effort involved in generating such a model from scratch.

Currently, automated data collection techniques for the purpose of building an EA exist [4, 12, 13] but involve considerable effort to identify the relevant parts of the captured data to map into an EA [4]. The mapping is also not automated. We note that [12, 13] offer techniques that partially extracts a description of the application and technology layers. The data they collect is mainly from network topology discovery tools (network scanners) where the resulting model is mapped into the infrastructural and application services. While process mining [7, 14] can help extract a partial description of the business layer, no proposals have been reported in the literature to correlate the business and application layers.

The rest of this paper is organised as follows. Section 2 provides background on the ArchiMate language and frequent closed sequence pattern mining. Section 3 describes the process and invocation log and how they are mapped into ArchiMate notation. Section 4 describes in detail the correlation environment and technique. Section 5 provides an empirical analysis of this approach. Section 6 discusses the related work of different methods with similar goals before presenting conclusion and directions for future work in Sect. 7.

2 Background

2.1 The ArchiMate Language

ArchiMate describes the enterprise environment through layered view where the higher layers use services that are provided by the lower layers [6,15,16]. The *business layer* represents the front-end services of the enterprise. The *application layer* represents the software implementation that service the *business layer*. The *technology layer* represents the software and hardware component of infrastructural services (e.g., operating systems, processing, storage and communication services) required to run the *application layer*.

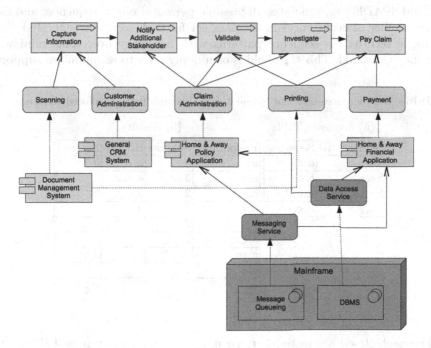

Fig. 1. ArchiMate example, ArchiSurance [17]

Figure 1 shows an example of an ArchiMate model. The business layer consists of *Business Process* that access *Application Service* in the application layer.

In this example, the *Capture Information* uses *Scanning* and *Customer Administration*. These *Application Services* use *Application Components*. In this example, *Scanning* and *Printing* use *Document Management System*. In the technology layer, there are two *System Software*: Message Queueing and DBMS that provide *Infrastructure Services*: Messaging Service and Data access Service respectively.

The business layer in ArchiMate is similar to a workflow design generated by process mining [7]. They both stand for the casually-related activities, which means that process mining can certainly be utilised in building ArchiMate models. However, process mining focused on the lateral activities. Relationships between the business and application layers cannot be discovered from process mining alone.

2.2 Frequent Closed Sequential Pattern

There are many techniques in discovering sequential pattern in a data set [9–11, 18–20]. These techniques are of particular interest because the business layer consists of sequential events. A frequent sequence pattern mining technique, such as Prefix Span [18], Generalized Sequential Pattern algorithm (GSP) [19], and SPADE [20], generates all possible permutations of sequences and calculate number of occurrences in the data set for each sequence generated. The technique returns the sequence patterns with frequency over some minimum frequency threshold. This threshold is commonly refer to as minimum support.

Table 1. Differences between frequent sequence and frequent closed sequence

(a) Sequence Table   (b) Result

Sequence ID	Sequence	Prefix Span	Support	BIDE+	Support
Seq_1	a, b, c	a	3	a, c	3
Seq_2	b, c	a, c	3	a, c, e	2
Seq_3	a, c, d, e	a, c, e	2	b, c	2
Seq_4	a, c, e	a, e	2	c	4
		b	2		
		b, c	2		
		c	4		
		c, e	2		
		e	2		

Frequent closed sequential pattern mining technique, such as BIDE+ [9], ClaSP [10], and CloSP [11], generates all permutation of sequence patterns and calculate number of occurrences in the data set for each sequence generated. The technique returns the sequence patterns with frequency over some minimum threshold that do not contain any subsequence pattern with the same frequency.

Therefore, if the pattern A and pattern A' have the same frequency over some threshold $min_{support}$ and $A' \sqsubseteq A$ (i.e., A' is a subsequence of A) then the frequent sequential pattern mining techniques return both of patterns A and A'. On the other hand, frequent closed sequential pattern mining techniques return only pattern A.

Consider a data set $\{a, b, c, d\}$ with 4 sequences shown in Table 1(a) with the minimum support set to 2 i.e., 50 %. As shown in Table 1(b), frequent sequential pattern mining return sequences $\langle a \rangle$ and $\langle a, c \rangle$. On the other hand, frequent closed sequential pattern mining returns only sequence $\langle a, c \rangle$ but not $\langle a \rangle$ because the pattern $\langle a \rangle$ is a subsequence of $\langle a, c \rangle$ and both has support of 3.

3 Input Data

In this section, we describe the mapping between the data elements to the business and application layers of ArchiMate. Let us assume that the technology layer has been previously documented or discovered using the methods in [12,13]. Each of the application services generates invocation logs where function calls are recorded. An example of invocation log is Application log accessible from Microsoft Windows® event viewer. This information can be used to populate the application layer. We also assume that we have access to a process log in the enterprise to populate the business layer by utilising process mining machinery, such as [7,14].

Table 2. Generalised log format

(a) Invocation Log

Timestamp	Source	Function
t_0^f	App_1	A_1
t_1^f	App_1	A_2
t_3^f	App_2	B_1
t_4^f	App_2	B_2
t_5^f	App_2	B_3
t_6^f	App_3	C_1
t_7^f	App_3	C_2
t_8^f	App_3	C_3

(b) Process Log

Timestamp	ID	Actor	Task
t_0^p	ID_0	$role_1$	P_1
t_2^p	ID_0	$role_1$	P_2
t_6^p	ID_0	$role_1$	P_3

In this paper, we assume the invocation log has the following attributes: *Source*, *Function*, and *Timestamp*. An application services are decomposed into functions. *Source* records the source application services that invoke *Function*. *Timestamp* records the time when the function is invoked. The invocation log may record other information, such as event ID, function category, and thread ID, depending on the application making the function calls. An example of an invocation log is shown in Table 2(a).

The process log has the following attributes: *ID*, *Actor*, *Task* and *Timestamp*. *ID* identifies a process instance. *Actor* typically records the user executing tasks.

In this context, we assume that *Actor* records the user's role. *Task* records the executed task name of a process. *Timestamp* records the time when the processes start. An example of a process log is shown in Table 2(b) that records an execution of process with task sequence $\langle P_1, P_2, P_3 \rangle$ executed by one actor $role_1$. The process log is assumed to conform to these properties: (i) each task refers to a complete instance of processes and (ii) tasks are totally ordered [7].

The relationships between the business layer and application layer can be inferred from the process and invocation log. The mapping from the process log and invocation log to the ArchiMate business and application layers are shown in Fig. 2. Since *Source* is the externally observable behaviour of the application layer, it is mapped to *Application Service* [6]. The corresponding *Application Component* to *Application Service* is recorded as the *Function* in the log. The mapping from process log to the business layer is straightforward. The *Task* in process log is mapped to *Business Process*.

ArchiMate specifies that *Application Service* is accessed by *Business Process* [6]. However, *Business Process* may use different *Application Component* of different *Application Service* so that we propose to correlate the *Business Process* with *Application Component* for better accuracy.

In determining the correlation between invocation and process log, we first have to join the two logs. The joined log consists of a sequence $\langle p_i, \langle F_{seq} \rangle \rangle$ where task p_i followed by function sequences F_{seq}. These sequences are determined by the timing proximities of the tasks and functions [8]. Thus, this joined log is the input to the frequent closed sequence pattern mining [9–11].

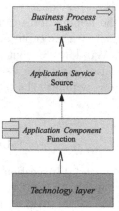

Fig. 2. ArchiMate mapping

The frequent closed sequence pattern mining [9–11] returns the sequence, $\langle p_i, \langle F_{seq} \rangle \rangle$, that satisfy the threshold to represent correlation of task p_i and functions F_{seq}. The threshold, called minimum support, is bounded by the number of distinct cases which a sequence with a task prefix p_i occurs.

The next section describes method to correlate the business and application layers of ArchiMate in details.

4   Learning Task and Function Relationships

In this section we will describe the technique to learn the relationship between the business and application layer in the ArchiMate model as discussed in Sect. 3. Populating the business layer itself can be realised using process mining methods, such as [7,14]. The main goal of this section is to correlate the tasks from process log to the functions in the invocation log. Hence, linking the business and application domain represented by the business layer and the application layer of ArchiMate.

We can conceive the business layer-application layer correlation mining being done in two distinct scenarios: (1) a **complete timestamp scenario** where the start and end times of all process tasks are available in the process log [14] and (2) a **partial timestamp scenario** where only the start time but not the end time of each process task is known.

These two scenarios cover the all the possibilities of the task and function behaviour. A task requests a set of functions to complete. Depending on the protocol between the business layer and the application layer, the application may return to the task an acknowledgement so that the task records its completion, such as TCP communication protocol. However, the application may never return such acknowledgement because the task and function are concurrent or the task requests are stateless, such as UDP communication protocol.

On an orthogonal dimension, we can conceive of two different settings in relation to concurrent task execution: (1) a **unique task setting** where there is a guarantee that only a single task can be executed at any point in time and (2) a **concurrent task setting** where potentially multiple tasks might be executed at the same time. The latter setting may be of interest if the process being executed admits parallel flows, or if multiple instances (of one or more process designs) are executed at the same time.

Under **complete timestamp** scenario with unique task setting, we can create a joined log sorted by the process instance ID and timestamp in ascending order. As mentioned before, the function calls between the task start and end are the functions required for the task to complete. Therefore, the candidate sequence is the task and functions within the time the task starts and end [8].

Let us consider an example of complete timestamp process log with $P = \{p_1, p_2, p_3\}$ and a set of functions $F = \{f_1, f_2, f_3, f_4\}$ with their corresponding timestamp t_m as shown in Table 3(a). In this joined log, we can see there are two instances of the process. This example contains enough information to populate the business layer using the existing algorithm to extract the process model [7]. Since Table 3(a) contains both start and end time for a task p_i, there exists a time window where the functions required by task p_i is executed. Therefore, in this example, one of the sequence candidates is $\langle p_1, \langle f_1, f_2 \rangle \rangle$ because functions $\langle f_1, f_2 \rangle$ are invoked during the time for task p_1 to be executed. Since functions $\langle f_1, f_2 \rangle$ are invoked in both instances of task p_1, task p_1 is correlated with functions $\langle f_1, f_2 \rangle$. The same correlation is made for task p_2 and functions $\langle f_4, f_2 \rangle$ as well as functions $\langle f_2, f_3 \rangle$ are correlated to task p_3.

The same scenario with **concurrent task setting** would return invalid sequences. Hence, the logs are joined by each distinct ID so that it resembles the joined table as shown in Table 3(b). Then the candidate sequences are determined as in previous setting. In this example, functions $\langle f_1, f_2 \rangle$ are invoked for both instances of task p_1 so that $\langle f_1, f_2 \rangle$ is correlated with p_1. The same correlation is made for task p_2 and $\langle f_4, f_2 \rangle$ as well as functions $\langle f_2, f_3 \rangle$ are correlated to task p_3.

In partial timestamp scenario, the complete time assumption is relaxed. The task start time is known but the task end time is unknown while function

Table 3. Complete timestamp scenario

(a) **Unique task setting**

ID	Timestamp	task	Function
ID_1	t_0	$p_1, start$	
	t_1		f_1
	t_2		f_2
ID_1	t_3	p_1, end	
ID_1	t_4	$p_2, start$	
	t_5		f_4
	t_6		f_2
ID_1	t_7	p_2, end	
ID_1	t_8	$p_3, start$	
	t_9		f_2
	t_{10}		f_3
ID_1	t_{11}	p_3, end	
ID_2	t_{12}	$p_1, start$	
	t_{13}		f_1
	t_{14}		f_2
ID_2	t_{15}	p_1, end	
ID_2	t_{16}	$p_2, start$	
	t_{17}		f_4
	t_{18}		f_2
ID_2	t_{19}	p_2, end	
ID_2	t_{20}	$p_3, start$	
	t_{21}		f_2
	t_{22}		f_3
ID_2	t_{23}	p_3, end	

(b) **Concurrent task setting**

ID	Timestamp	task	Function
ID_1	t_0	$p_1, start$	
	t_1		f_1
	t_3		f_2
ID_1	t_5	p_1, end	
ID_2	t_4	$p_1, start$	
	t_7		f_1
	t_{10}		f_2
ID_2	t_{11}	p_1, end	
ID_1	t_8	$p_2, start$	
	t_9		f_4
	t_{12}		f_2
ID_1	t_{13}	p_2, end	
ID_1	t_{15}	$p_3, start$	
	t_{16}		f_2
	t_{17}		f_3
ID_1	t_{19}	p_3, end	
ID_2	t_{18}	$p_2, start$	
	t_{19}		f_4
	t_{21}		f_2
ID_2	t_{22}	p_2, end	
ID_2	t_{20}	$p_3, start$	
	t_{23}		f_2
	t_{24}		f_3
ID_2	t_{25}	p_3, end	

execution start and end time still hold (function execution ends when the next function starts). In this scenario, the sequence candidates are not bounded by a time window [8]. Furthermore, there is not enough information to populate the business layer using existing process mining methods [7, 14] because the task end time is unknown.

Let us observe an example of this scenario with the same task set P and function set F in **complete timestamp scenario** from timestamp t_0 to t_m as shown in Fig. 3 where f_k is the function invoked at timestamp t_m such that $f_k \in F$. We assume that given a task set P, function set F, an ordered sequence set $F' \sqsubseteq F$ and timestamp range that starts at t_0 and ends at t_m such that a joined Log L consists of a set of joined log entry $\{A_u, ..., A_v\}$ where A_x (for $x = u, ..., v$) is a sequence of $(p_i, \langle f'_j, ..., f'_k \rangle)$ where $p_x \in P$ is executed at timestamp t_l and $f'_y \in F'$ (for $y = j, ..., k$) is executed between timestamp t_l and t_m.

What we ended up having is a cascading pattern of the task and function joined table as shown in Table 4(a). Whether it is a unique or concurrent task setting, the sequence of functions would potentially infinitely long for very large

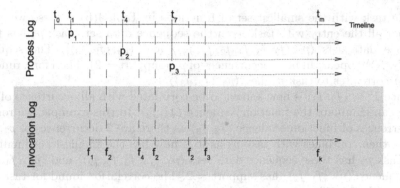

Fig. 3. Task start time known and end time unknown

Table 4. Partial timestamp scenario

(a) Joined Process and Invocation Log

Timeline	task	Function
$t_0 - t_m$	p_1	$\langle f_1, f_2, f_4, f_2, f_2, f_3, ..., f_k \rangle$
$t_4 - t_m$	p_2	$\langle f_4, f_2, f_2, f_3, ..., f_k \rangle$
$t_7 - t_m$	p_3	$\langle f_2, f_3, ..., f_k \rangle$
...	...	....
$t_l - t_m$	p_v	$\langle f_j, ..., f_{k-1}, f_k \rangle$

(b) Translated Sequence Database

Sequence	Function
seq_1	$\langle p_1, \langle f_1, f_2, f_1, f_2, f_4, f_2, f_2, f_3, f_4, f_2, f_2, f_3 \rangle \rangle$
seq_2	$\langle p_1, \langle f_1, f_2, f_4, f_2, f_2, f_3, f_4, f_2, f_2, f_3 \rangle \rangle$
seq_3	$\langle p_2, \langle f_4, f_2, f_2, f_3, f_4, f_2, f_2, f_3 \rangle \rangle$
seq_4	$\langle p_3, \langle f_2, f_3, f_4, f_2, f_2, f_3 \rangle \rangle$
seq_5	$\langle p_2, \langle f_4, f_2, f_2, f_3 \rangle \rangle$
seq_6	$\langle p_3, \langle f_2, f_3 \rangle \rangle$

event logs. We need to use extraneous knowledge about the length of the longest process to decide how long to record sequences consisting of a single task prefix and an arbitrarily long sequence of functions as the suffix.

Consider the log in Table 3(b) under this scenario, the joined log becomes a sequence data set $\{\langle p_i, F'_j \sqsubseteq F \rangle, ..., \langle p_n, F'_k \sqsubseteq F \rangle\}$, shown in Table 4(b). We group the same tasks together and find which set of function is consistent with this task, i.e., when p_i occurs, F'_j also occurs to find the correlation set between task p_i and a set of function F'_j. Since we have this cascading pattern, we can reasonably assume that the smallest cardinality of $|F'|$ would be the best approximate for the task and function correlation. We start with a group of p_i with the smallest cardinality $|F'|$ and calculate the frequent closed sequences [9–11] in a group of p_i.

The task with the smallest set of functions in Table 4(b) is p_3 so we group together all the entry with task p_3 in the sequence data set. Task p_3 has these sequence data sets $\langle p_3, \langle f_2, f_3, f_4, f_2, f_2, f_3 \rangle \rangle$ and $\langle p_3, \langle f_2, f_3 \rangle \rangle$. The sequence $\langle p_3, \langle f_2, f_3 \rangle \rangle$ appears in both occurrences of p_3 (support = 2). Therefore, function $\langle f_2, f_3 \rangle$ correlates to task p_3, i.e., $\langle p_3, \langle f_2, f_3 \rangle \rangle$.

Since $F'_1 = \langle f_2, f_3 \rangle$ is now known to be correlated with all occurrence of task p_3, we can eliminate the function sequence $\langle f_2, f_3 \rangle$. In this example, we remove 2 occurrences of function sequence $\langle f_2, f_3 \rangle$ as there are 2 occurrences of p_3. The process then continues with task p_2 as it is now has the smallest cardinality of $|F'|$. Task p_2 has these sequence data sets $\langle p_2, \langle f_4, f_2, f_4, f_2 \rangle \rangle$ and $\langle p_2, \langle f_4, f_2 \rangle \rangle$. The sequence $\langle p_2, \langle f_4, f_2 \rangle \rangle$ has support = 2. The correlation found for task p_2 is function $\langle f_4, f_2 \rangle$. Then we repeat the process with task p_1. We found that task p_1 is correlated with function $\langle f_1, f_2 \rangle$.

Once all the correlations are made, we can use the last function in the sequence as a proxy for the end time of the task. Then the tasks would have the transition "start" at the timestamp when the task starts and "complete" when the last function in the correlation set ends. The result is joined Table 5 that is similar to Table 3(b) from *complete timestamp scenario*.

Table 5. Partial timestamp scenario result

Timestamp	Task	Function
t_0	$p_1, start$	
t_1		f_1
t_2	p_1, end	f_2
t_4	$p_1, start$	
t_5		f_1
t_6	p_1, end	f_2
t_8	$p_2, start$	
t_9		f_4
t_{10}	p_2, end	f_2
t_{12}	$p_3, start$	
t_{13}		f_3
t_{14}	p_3, end	f_3
t_{16}	$p_2, start$	
t_{17}		f_4
t_{18}	p_2, end	f_2
t_{20}	$p_3, start$	
t_{21}		f_2
t_{22}	p_3, end	f_3

The correlation made under **partial timestamp scenario** can be used as a proxy to the task end time. Given a correlation made between a task of a process $p_i \in P$ and a sequence of functions $F'_j \sqsubseteq F$, the last function $f_k \in F'_j$ is an approximate indicator of the end time of that task (we would use the start time of the function invocation if we did not have access to the end time of that function invocation, by making the assumption that the start and end times of function invocations are typically very close to each other). The task end time is required by some process mining method, such as [14]. Moreover, inferring the duration of the task provides the ability to perform additional sensitivity analysis, e.g., there may be some implementation bottleneck (application layer) such that a task is taking longer than expected.

In the next section, we present an evaluation of this approach. Using existing tools and algorithm, we demonstrate how the relationships of business and application layers in ArchiMate model are learned.

5   Evaluation

We started the evaluation with a simulated telephone repair process log[1] which contains 500 process instances with 7152 entries overall. The process consists of 8 tasks: {Register, Analyze Defect, Repair (Complex), Repair (Simple), Test Repair, Restart Repair, Inform User, Archive Repair}. In this simulated telephone repair process log, the tasks are sorted by instance ID so that the process instances appear in sequence. SPMF pattern mining library [21] is used to run frequent closed sequential pattern mining algorithm BIDE+ [9]. Since the SPMF library has a few limitations, for example it can only process unique items in a set, inaccuracies are expected.

The invocation log was generated from this process log. In generating the invocation log, we assume that the task and function tuples are as follow: {Register,$\langle f_1, f_2 \rangle$}, {Analyze Defect,$\langle f_3, f_4, f_5 \rangle$}, {Repair (Simple),$\langle f_7, f_8 \rangle$}, {Repair (Complex),$\langle f_7, f_8, f_9 \rangle$}, {Test Repair,$\langle f_7, f_9 \rangle$}, {Restart Repair,$\langle f_6 \rangle$}, {Inform User,$\langle f_{10}, f_{11} \rangle$}, and {Archive Repair,$\langle f_{12} \rangle$}. We generated 3 sets of joined log with that conform to these scenarios and settings; Log_1 : Complete timestamp - unique tasks, Log_2 : Complete timestamp - concurrent tasks, and Log_3 : Partial timestamp.

The given telephone repair process log already reflects the complete timestamp scenario (Log_1). The log has unique task entries because it is already sorted by its process instance ID. The corresponding invocation log was generated by assigning the function timestamp in the range between the task start time and end time. This simulates function calls required for a task to complete. The log was then modified to reflect Log_2. We sorted the log in ascending timestamp to focus on the concurrent task setting. We simulated the function call where a timestamp of a function call is after the start time of a task. Log_3 was generated by first eliminating the end time of each tasks in the log then simulate the function call as describe in Log_2. Since the process log is relatively small, we treated the end of file as the end of sequences.

The performance was analysed using recall and precision metric for every task-function correlation made. Perfect recall (recall = 1) for task-function correlation is where a discovered sequence $\langle p_i, \langle f_l, ..., f_k \rangle \rangle$ contains all the function required to realise the task p_i, e.g., discovered sequence $\langle Register, \langle f_1, f_2, f_3 \rangle \rangle$ has recall = 1 for task $\langle Register, \langle f_1, f_2 \rangle \rangle$. Perfect precision (precision = 1) for task-function correlation is when a discovered sequence $\langle p_i, \langle f_l, ..., f_k \rangle \rangle$ contains only the function required to realise the task p_i, e.g., discovered tuple $\langle Register, \langle f_1, f_2, f_3 \rangle \rangle$ has precision = 0.67 for task $\langle Register, \langle f_1, f_2 \rangle \rangle$.

Log_1 resulted in recall = 1 and precision = 1 for all tasks. This is to be expected as complete timestamp was provided. Log_1 guaranteed that only a single task can be executed at any point in time and the functions it required would be called before the task end time. Hence, for every task instance time window, the function sequence required to complete said task always occurred. Log_1 also presents as our baseline in testing our approach. Log_2 resulted in recall = 1 and precision = 1

---

[1] http://www.processmining.org/_media/tutorial/repairexample.zip.

Table 6. Log_3 result

Task	Function sequence	Discovered sequence	Recall	Precision
Register	$\langle f_1, f_2 \rangle$	$\langle f_1, f_2 \rangle$	1	1
Analyze Defect	$\langle f_3, f_4, f_5 \rangle$	$\langle f_3, f_4, f_5 \rangle$	1	1
Repair (Simple)	$\langle f_7, f_8 \rangle$	$\langle f_7, f_8 \rangle$	1	1
Repair (Complex)	$\langle f_7, f_8, f_9 \rangle$	$\langle f_7, f_8 \rangle$	0.67	1
		$\langle f_7, f_9 \rangle$	0.67	1
Test Repair	$\langle f_7, f_9 \rangle$	$\langle f_7 \rangle$	0.5	1
		$\langle f_9 \rangle$	0.5	1
Restart Repair	$\langle f_6 \rangle$	$\langle f_6 \rangle$	1	1
Inform User	$\langle f_{10}, f_{11} \rangle$	$\langle f_{10}, f_{11} \rangle$	1	1
Archive Repair	$\langle f_{12} \rangle$	$\langle f_{12} \rangle$	1	1
Average			0.93	1

for all task. Log_3 result is shown in Table 6 with average recall < 1 as discussed above, the pattern mining library has some limitations. This type of inaccuracies were expected because tasks Repair (Complex), Repair (Simple), Test Repair, and Restart Repair shared functions $\{f_7, f_8, f_9\}$ so that there were some repetition in the sequence that the SPMF pattern mining library [21] was not able to handle properly.

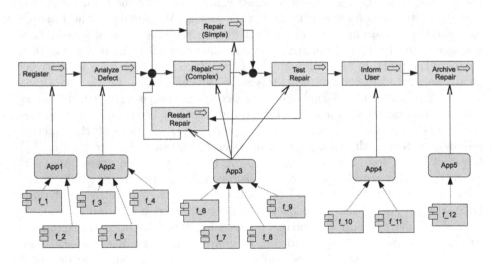

Fig. 4. ArchiMate business and application layer

In constructing the business layer of ArchiMate, we used process mining machinery ProM framework [22]. The model returned from ProM for process log in scenario Log_1 is used as our baseline. The model returned from ProM for scenario Log_2 is identical to Log_1. We added complete timestamp from the correlation result for Log_3. In the case where two sets of sequences returned, we took the last function in the second sequence as the proxy end time. The model returned from ProM for scenario Log_3 is identical to Log_1. Since Log_1 and Log_2 returned perfect recall and precision for the correlations, we only show the resulting business layer and application layer relationships from Log_3 in Fig. 4.

6   Related Work

There are a number of Enterprise Architecture (EA) tools available in the market that provide data collecting feature, for example BizzDesign Architect[2], Troux[3], MooD Business Architect[4], ARIS Business Architect for SAP[5]. The resulting models do not cover all aspects of EA, such as infrastructure services. These tools generate partial EA model and a significant effort of its abstraction is dependent on the EA modeller.

In [4], the paper provides analysis of possible information sources and appropriateness for (EA). They conducted questionnaire to the related EA Professional in different industries. Participants were asked about the existing data quality attributes in the context of EA modelling. They analysed these attributes from low level data collecting tools, such as results from network scanner and Enterprise Service Bus, and from productive tools, such as Configuration Management Database (CMDB) and Change Management tool. The paper found that the data from these tools are too refined to be used in EA context. The data surveyed requires other mechanisms to be translated to EA. However, the paper did not consider data that resides in the systems, e.g., process log.

In [12,13], a method to automatically collect and translate data from network scanners to ArchiMate model is proposed. The model extracted from the data collected is partial (up to the application layer). However, the data did not provide enough information to infer the relationship within application components or the relationship between the application layer and the business layer.

7   Conclusions and Future Work

In this paper, we have provided preliminary evidence to support the hypotheses that the relationship between business and application layer of ArchiMate can be inferred from process and invocation logs within the enterprise. We have performed this evaluation on different scenarios and settings, but much deeper

---

[2] http://www.bizzdesign.com/tools/bizzdesign-architect/.
[3] http://www.troux.com/.
[4] http://www.moodinternational.com/.
[5] https://www.softwareag.com/corporate/products/aris/bpa/products/sap/overview/default.asp.

and careful evaluation remains to be done. We need to evaluate with larger models and refinement of other settings, such as incomplete information and parallel function call.

References

1. Zachman, J.A.: Enterprise architecture: the issue of the century. Database Program. Des. **10**(3), 44–53 (1997)
2. Lankhorst, M. (ed.): Enterprise Architecture at Work: Modelling, Communication, and Analysis. Springer, Heidelberg (2005)
3. Kaisler, S., Armour, F., Valivullah, M.: Enterprise architecting: critical problems. In: Proceedings of the 38th Annual Hawaii International Conference on System Sciences, HICSS 2005, p. 224b, January 2005
4. Farwick, M., Breu, R., Hauder, M., Roth, S., Matthes, F.: Enterprise architecture documentation: empirical analysis of information sources for automation. In: 2013 46th Hawaii International Conference on System Sciences (HICSS), pp. 3868–3877, January 2013
5. Aier, S., Buckl, S., Franke, U., Gleichauf, B., Johnson, P., Nrman, P., Schweda, C.M., Ullberg, J., Gallen, C.-S., Mnchen, T.U.: A survival analysis of application life spans based on enterprise architecture models. In: 3rd International Workshop on Enterprise Modelling and Information Systems Architectures, pp. 141–154 (2009)
6. Lankhorst, M.: ArchiMate language primer, Telematica institute (2004)
7. van der Aalst, W., Weijters, T., Maruster, L.: Workflow mining: discovering process models from event logs. IEEE Trans. Knowl. Data Eng. **16**, 1128–1142 (2004)
8. Fdhila, W., Rinderle-Ma, S., Indiono, C.: Memetic algorithms for mining change logs in process choreographies. In: Franch, X., Ghose, A.K., Lewis, G.A., Bhiri, S. (eds.) ICSOC 2014. LNCS, vol. 8831, pp. 47–62. Springer, Heidelberg (2014)
9. Wang, J., Han, J.: Bide: efficient mining of frequent closed sequences. In: Proceedings of 20th International Conference on Data Engineering, pp. 79–90, March 2004
10. Gomariz, A., Campos, M., Marin, R., Goethals, B.: ClaSP: an efficient algorithm for mining frequent closed sequences. In: Pei, J., Tseng, V.S., Cao, L., Motoda, H., Xu, G. (eds.) PAKDD 2013, Part I. LNCS, vol. 7818, pp. 50–61. Springer, Heidelberg (2013)
11. Yan, X., Han, J., Afshar, R.: Clospan: mining closed sequential patterns in large datasets. In: SDM, pp. 166–177 (2003)
12. Holm, H., Buschle, M., Lagerstrm, R., Ekstedt, M.: Automatic data collection for enterprise architecture models. Softw. Syst. Model. **13**(2), 825–841 (2014)
13. Buschle, M., Holm, H., Sommestad, T., Ekstedt, M., Shahzad, K.: A tool for automatic enterprise architecture modeling. In: Nurcan, S. (ed.) CAiSE Forum 2011. LNBIP, vol. 107, pp. 1–15. Springer, Heidelberg (2012)
14. Wen, L., Wang, J., van der Aalst, W.M.P., Huang, B., Sun, J.: A novel approach for process mining based on event types. J. Intell. Inf. Syst. **32**, 163–190 (2009)
15. Lankhorst, M., Proper, H., Jonkers, H.: The anatomy of the ArchiMate language. Int. J. Inf. Syst. Model. Des. **1**(1), 1–32 (2010)
16. Wierda, G.: ArchiMate 2.0 understanding the basics, White paper, The Open Group, February 2013

17. Jonkers, H., Band, I., Quartel, D.: The archisurance case study, White paper, The Open Group, Spring (2012)
18. Pei, J., Han, J., Mortazavi-Asl, B., Wang, J., Pinto, H., Chen, Q., Dayal, U., Hsu, M.-C.: Mining sequential patterns by pattern-growth: the prefixspan approach. IEEE Trans. Knowl. Data Eng. **16**, 1424–1440 (2004)
19. Zhang, M., Kao, B., Yip, C.-L., Cheung, D.: A GSP-based efficient algorithm for mining frequent sequences. In: Proceedings of IC-AI, pp. 497–503 (2001)
20. Zaki, M.J.: Spade: an efficient algorithm for mining frequent sequences. Mach. Learn. **42**(1–2), 31–60 (2001)
21. Viger, P.F., Gomariz, A., Gueniche, T., Soltani, A., Wu, C.-W., Tseng, V.S.: SPMF: a java open-source pattern mining library. J. Mach. Learn. Res. **15**, 3389–3393 (2014)
22. van Dongen, B.F., de Medeiros, A.K.A., Verbeek, H.M.W.E., Weijters, A.J.M.M.T., van der Aalst, W.M.P.: The ProM framework: a new era in process mining tool support. In: Ciardo, G., Darondeau, P. (eds.) ICATPN 2005. LNCS, vol. 3536, pp. 444–454. Springer, Heidelberg (2005)

Mining Process Task Post-Conditions

Metta Santiputri[✉], Aditya K. Ghose, Hoa Khanh Dam, and Xiong Wen

Decision Systems Lab, School of Computing and Information Technology,
University of Wollongong, Wollongong, NSW 2522, Australia
{ms804,aditya,hoa,xw926}@uowmail.edu.au
http://www.uow.edu.au

Abstract. A large and growing body of work explores the use of semantic annotation of business process designs, but these annotations can be difficult and expensive to acquire. This paper presents a data-driven approach to mining these annotations (and specifically post-conditions) from event logs in process execution histories which describe both task execution events (typically contained in *process logs*) and state update events (which we record in *effect logs*). We present an empirical evaluation, which suggests that the approach provides generally reliable results.

Keywords: Semantic annotation · Data-driven · Process logs · Effect logs

1 Introduction

A large and growing body of work explores the use of semantic annotation of business process designs [4,5,10,16,29,32]. A large body of work also addresses the problem of semantic annotation of web services in a similar fashion [24–26,28]. Common to all of these approaches is the idea that semantic annotation of process tasks or services provides value in ways that the process or service model alone cannot. Our focus in this paper is on *post-conditions* of tasks in the context of process models (pre-conditions are also of interest and we believe that an extension of the machinery presented here can address these, but are outside the scope of the present work). Ideally process designs annotated with post-conditions help answer the following question for any part of a process design: *what changes will have occurred in the process context if the process were to execute upto this point?* Arguably, a sufficiently detailed process model (for instance one that decomposes tasks down to the level of individual read or write operations) will require no additional information to answer this question. However, process models are most valuable when described at higher levels of abstraction, in terms of concepts and activities that stakeholders are familiar with. Processes annotated with post-conditions thus serve a crucial modeling function, providing an effective summary of a substantial body of knowledge regarding the "lower-level" workings of a process. Annotation with post-conditions can also help solve a range of problems such as process compliance management [10], change management [20], enterprise process architectures [13] and the management of the business process life cycle [21].

© Springer International Publishing Switzerland 2015
P. Johannesson et al. (Eds.): ER 2015, LNCS 9381, pp. 514–527, 2015.
DOI: 10.1007/978-3-319-25264-3_38

The modeling and acquisition of these post-conditions poses a particularly difficult challenge. It is generally recognized that process modeling involves significant investment in time and effort, which would be multiplied manyfold if there were an additional obligation to specify semantic annotations. Analysts also tend to find semantic annotation difficult, particularly if the intent is to make these formal (as is required by all of the use cases referred to above). This paper seeks to address this challenge by offering a set of techniques that mine readily available data associated with process execution to generate largely accurate "first-cut" post-conditions for process tasks. Our approach leverages the generally understood notion of *event logging*. The events that occur in a process execution context can be viewed in general terms as being of two types: (1) events that describe the start or end of the execution of process tasks and (2) events that describe state changes in the objects impacted by a process. In many settings, the existing event logging machinery is capable of logging both kinds of events. In other settings, we need to instrument *object state monitors* (for either physical objects or computational objects, or both) to obtain events of the second kind. In line with the literature addressing these, we refer to a time-stamped record of events of the first kind as a *process log*. We shall refer to a time-stamped record of events of the second kind as an *effect log* (in view of the fact that the state transitions being recorded are in fact *effects* of the process in question). We leverage these two types of logs in juxtaposition, and the time-stamped sequences of task execution events and state-change events thus obtained, to generate the *sequence database* taken as input by a sequential rule miner (CMRules [6] in our instance, but others could be used instead). The key idea is to identify commonly occurring patterns of task execution events, followed by sequences of state change events (or *effects*). As we show, the approach is generally quite effective. We also define techniques which leverage a *state update operator* (that defines how a specification of a state of affairs is updated as a consequence of the execution of an action) and the actual history of process execution provided by the juxtaposed process and effect logs to determine whether the mined effects, if accumulated using the state update operator, would indeed generate the available execution histories. This forms a validation step for the mined results.

Our intent is to mine the *context-independent effects* (or *immediate effects*) of each task. These are *contextualized* via iterated applications of the state update operator to obtain the *context-dependent effects* of each task (in the context of a process model)—a complete collection of these for each task or event provides a semantically annotated process model. For instance, the immediate effect of turning a switch on is to complete a circuit. In the context of a light bulb circuit, the context-dependent effect of this task would be to turn the bulb on. In the context of a switching circuit for a chemical reactor, the context-dependent effect of that same task would be to bring the chemical reactor to an operational state. We envisage the machinery we present below being used in the following manner: given as input a process log, an effect log, a process model (or a set of process models in the event that the logs describe the execution of instances of multiple process designs) and a state update operator, the machinery would generate the immediate effects of each task referred to in the process log. These effects could be used directly in annotating process models, or might be viewed as "first-cut" specifications, to be edited and refined by expert analysts.

We provide background on semantic annotation of processes and on process and effect logs in Sect. 2. In Sect. 3, we present our approach to mining task effects and validating these. In Sect. 4, we present results of an experimental evaluation exercise, before presenting concluding remarks.

2   Background

Semantic Annotation: We assume that each task or event in a process is associated with effects written as conjunctive normal form sentences in the underlying formal state description language, which might be propositional or first-order (we do not consider temporal logics in this work, but extensions are possible). We assume that each task or event has context-independent *immediate effects* that can be contextualized via iterated applications of a *state update operator* as in [10] and [16]. We permit the contextualized effects to be non-deterministic— at any given point in a process, the actual effects that might accrue would be one of a set of *effect scenarios*. We need to support this non-determinism for two reasons. First, in any process with XOR-branches, one might arrive at a given task via multiple paths, and the contextualized effects achieved must be contingent on the path taken. Since this analysis is done at design time, we need to admit the possibility of non-deterministic effects since the specific path taken can only be determined at run-time. Second, many state update operators generate non-deterministic outcomes, since inconsistencies (that commonly appear in state update) can be resolved in multiple different ways. Of the two well-known state update operators in the literature—the Possible Models Approach (PMA) and the Possible Worlds Approach (PWA)—our work leverages the PWA [11]. Specifically, we use the operator $\oplus$ defined below. In the following, we assume that all consistency checks implicitly include a background knowledge base ($\mathcal{KB}$) containing rules and axioms. Thus, the statement that $e_i' \cup e_j$ is consistent effectively entails the statement that $e_i' \cup e_j \cup \mathcal{KB}$ is consistent. We omit references to $\mathcal{KB}$ for ease of exposition.

For two effects e_i and e_j, and the knowledge base $\mathcal{KB}$, if $e_i \not\models \bot$ and $e_j \not\models \bot$, then the *pair-wise effect accumulation* (or *state update*) $e_i \oplus e_j$ is defined as:

$$e_i \oplus e_j = \{e_j \cup e_i' \mid e_i' \subseteq e_i \wedge e_i' \cup e_j \cup \mathcal{KB} \not\models \bot \wedge$$
$$\text{there does not exist } e_i'' \text{ such that } e_i' \subset e_i'' \subseteq e_i \wedge$$
$$e_i'' \cup e_j \cup \mathcal{KB} \not\models \bot\}$$

The outcome of the state update operation is not a unique effect specification, but a set of non-deterministic *effect scenarios*. To see why this might be the case, consider a task T with a single associated effect scenario given by $\{p, q\}$ which is followed by task T' whose immediate effect is to make r true. Given a background knowledge base consisting of a single rule $r \rightarrow (\neg p \vee \neg q)$, the $\oplus$ operator would give us two distinct outcomes: $\{p, r\}$ and $\{q, r\}$.

To obtain a complete annotation of a process model, we repeatedly apply the $\oplus$ operator over pairs of contiguous tasks in a process model, with the first argument being an effect scenario associated with the prior task and the second argument being the immediate effect of the later task. Special techniques are provided for dealing with XOR and AND gateways in proposals such as [10, 16] and [32], but these are not directly relevant for our current exposition and we omit details here.

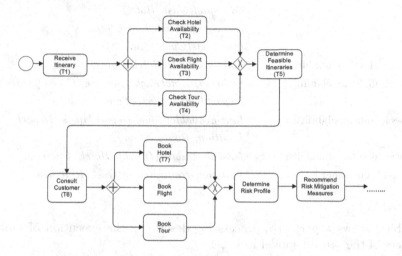

Fig. 1. A semantic effect-annotated BPMN process model for *Holiday Booking* process

Figure 1 illustrates a section of a semantically annotated BPMN process model for a *Holiday Booking* process followed by a travel agent. Table 1 outlines the semantic annotations of some of its tasks. The background knowledge base is given by a set of rules including the following representative examples. We use the standard convention of starting variables with upper-case letters.

(R1) $\forall Hotel, Dates, \exists Cust \; hotel(Hotel, Dates) \rightarrow hotel\text{-}pref(Cust, Hotel)$

(R2) $\forall Flight, Airline, ClassOfTravel, Dates, DepartTime, ArriveTime,$
$\exists Cust \; flight\text{-}available(Flight, Airline, ClassOfTravel, Dates, DepartTime, Arriv\text{-}eTime) \rightarrow travel\text{-}dates(Cust, Dates) \wedge airlines\text{-}preferences(Cust, Airline) \wedge airline\text{-}classoftravel(Cust, ClassOfTravel) \wedge departure\text{-}preferences(Cust, Depart\text{-}Time) \wedge arrival\text{-}prefs(Cust, ArriveTime)$

Process and Effect Logs: We assume that process execution data is available in the form a *process log* and an *effect log*. A process log (or event log) is a set of triples of the form $\langle CaseID, TimeStamp, TaskID \rangle$. The *TimeStamp* value indicates the start time of *TaskID* while *CaseID* identifies the process instance that *TaskID* belongs to. We permit possibly many process instances (of the same process design or distinct process designs) to be executed at the same time.

Table 1. Annotation of holiday booking process in Fig. 1

Obtain customer requirements	*travel-dates(Cust, Dates)* ∧
	airline-preferences(Cust,Airline) ∧
	airline-classoftravel(Cust, ClassOfTravel) ∧
	departure-preferences(Cust,DepartTime) ∧
	arrival-prefs(Cust, ArriveTime) ∧
	meal-constraints(Cust, MealConstraints) ∧
	freq-flyer(Cust, FreqFlyerDetails) ∧
	hotel-pref(Cust, Hotel) ∧
	room-prefs(Cust, RoomPrefs) ∧
	tour-prefs(Cust, TourPrefs)
Check Hotel Availability	*hotel-available(Hotel, Dates)*
Check flight availability	*flight-available(Flight, Airline, ClassOfTravel,*
	DepartTimes, ArriveTimes)
Check tour availability	*tour-available(Tour, DepartTimes, Depart-*
	Location, Route, Stops)
Determine feasibility itinerary	*feasible-itinerary(Flight, Hotel, Tour)*
Consult customer	*customer-confirmation(Cust, Itinerary)*

Table 2 shows a part of a process log describing the execution of multiple instances of the process model in Fig. 1.

An effect log consists of a set of tuples $\langle t_i, e_i \rangle$ where t_i is a timestamp and e_i is an effect assertion in the underlying state description language. The effect log records observed changes in the states of objects impacted by a process. It is important to note that only changes in state are recorded (and not state descriptions that persist). Effect logs can be obtained by instrumenting the process environment with object state monitors (both for physical objects as well as for computational/business objects). The underlying state description language might involve propositional state variables—the changes to describe would then be propositions becoming true or false, or more generally as disjunctions (in case state monitors have limited sensing capabilities). The underlying language might also admit non-Boolean state variables, in which case the effects logged would be the new value assignments to these variables. In many settings, it is convenient to represent effects in terms of first-order sentence schemas. In a travel booking process, the *"Obtain customer requirements"* task would lead to a ground instance of the sentence schema *airline-preference(Cust, Airline)* becoming available. In this setting, the precise grounding of the *Cust* and *Airline* variables are not of particular interest. Indeed, recording the actual values of these variables in the effect log would lead to the effect mining procedure treating different groundings of the variables as distinct effects, when in fact we are only interested in recording the effect that a ground instance of that sentence schema

Table 2. A process log of the BPMN process model in Fig. 1

case ID	task ID	timestamp	case ID	task ID	timestamp
1	T_1	t_1	2	T_6	t_{103}
2	T_1	t_8	1	T_5	t_{125}
2	T_2	t_{25}	3	T_5	t_{145}
3	T_1	t_{27}	3	T_6	t_{156}
1	T_2	t_{28}	1	T_6	t_{172}
3	T_3	t_{32}	1	T_7	t_{174}
2	T_3	t_{33}	4	T_1	t_{377}
1	T_4	t_{36}	4	T_3	t_{379}
3	T_2	t_{71}	4	T_4	t_{438}
2	T_4	t_{80}	4	T_2	t_{440}
3	T_4	t_{88}	4	T_5	t_{461}
1	T_3	t_{95}	4	T_6	t_{471}
2	T_5	t_{98}			

Notes: T_1: *Receive Itinerary*, T_2: *Check Flight Availability*, T_3: *Check Hotel Availability*, T_4: *Check Tour Availability*, T_5: *Determine Feasible Itineraries*, T_6: *Consult Customer*, T_7: *Book Hotel*

becomes available. For effects of this sort, we only record a propositional effect of the form *airline-preference-known*. In other settings, we are interested in the precise instantiations of the variables in a sentence schema of the form $p(X, Y)$, in which case the full ground instance of $p(X, Y)$ is recorded in the effect log.

Table 3 illustrates a part of a effect log describing the effects of the execution of multiple instances of the process in Fig. 1.

3   Mining and Validating Effects

The effect mining technique we describe below takes as input: (1) a semantically annotated process model, (2) a state update operator $\oplus$, (3) a process log, and (4) an effect log (where both logs refer to the same history of process execution) and generates as output the immediate effect e_{T_i} for every task T_i referred to in the process log (these tasks may belong to one or more process designs). Our approach involves first mining a first-cut version of the immediate effects, and then filtering these by validating them against the process execution history as represented in the process log and effect log.

Effect mining procedure: Our approach to effect mining is predicated on the observation that the effects of executing a task occur soon after the execution of the task. Effects that manifest a long period after the execution of a task are typically not effects of that task alone, but of that task plus some others (e.g., one may think of the arrival of a traditional "snailmail" letter 3 days after

Table 3. An effect log of the BPMN process model in Fig. 1

timestamp	effects
t_2	*travel-dates-known*
t_3	*airline-preferences-known*
t_4	*airline-classoftravel-known*
t_4	*departure-preferences-known*
t_5	*arrival-prefs-known*
t_6	*meal-constraints-known*
t_6	*freq-flyer-known*
t_7	*hotel-pref-known*
t_7	*room-prefs-known*
t_7	*tour-prefs-known*
t_9	*travel-dates-known*
t_{10}	*airline-preferences-known*
t_{12}	*airline-classoftravel-known*
t_{13}	*departure-preferences-known*
t_{14}	*arrival-prefs-known*
t_{16}	*meal-constraints-known*
t_{16}	*freq-flyer-known*
t_{17}	*hotel-pref-known*
t_{45}	*room-prefs-known*
t_{22}	*tour-prefs-known*
t_{29}	*flight-available-known*

posting as an effect of the action of letter-posting, when it actually involves several other tasks executed by the postal service). The key pattern we leverage in mining effects is the *sequence* that involves the execution of a task and the manifestation of its effects, using a sequential pattern miner. We are interested in identifying all the effects that occur always (or most of the time) after each task is executed. Since the task executions are recorded in the process log and the effects occurrences are recorded in the effect log, we must first establish the correlations between the two logs to obtain a joined table that serves as the *sequence database* for a sequential rule miner. We use the CMRules algorithm [6] although a number of other candidates exist [3,7,8,15], and the framework is flexible enough to allow the use of any of these.

While our focus is on the sequential patterns that relate tasks to effects, we are not interested in the relative sequencing amongst effects. Indeed, it is undesirable for our purposes to have the sequential rule miner to view the sequences $\langle T, p, q \rangle$ and $\langle T, q, p \rangle$ as being distinct. We therefore enforce the rule that a contiguous sequence of effects in the sequence database must always be represented in lexicographic order (this would require the second sequence above to be re-written as the first sequence).

We consider the problem of effect mining in two settings: (1) Settings characterized by the **unique task assumption** which stipulates that only one task may be performed at any point in time. This permits us to correlate all of the effects observed between the execution of a given task and the start of the next task with the first task. (2) Settings characterized by the **concurrent task assumption** which admits the possibility of multiple tasks executing concurrently (these could be tasks associated with distinct instances of the same process or associated with different processes). The second setting is more general, but the first setting simplifies the effect mining problem, and is worth considering if appropriate. We will apply the CMRules algorithm in both settings.

Both in settings with the *unique task assumption* and in settings with concurrent tasks, we create a joined table from the process log and the effect log. of the form:

$$\langle\langle\langle T_1, \langle\langle e_{11}\rangle, \ldots, \langle e_{1n}\rangle\rangle\rangle, \ldots, \langle T_i, \langle\langle e_{i1}\rangle, \ldots, \langle e_{im}\rangle\rangle\rangle, \ldots \langle T_p, \langle\langle e_{p1}\rangle, \ldots, \langle e_{pk}\rangle\rangle\rangle$$

where each $\langle T_i, T_{i+1}\rangle$ pair represents contiguous tasks and each e_{ij} represents the j-th effect observed after the start of task T_i and before the start of task T_j. This table represents the sequence database provided as input to the sequential rule miner. A special provision is needed for the last task in case it does not have any subsequent task. Instead of using the last record in the process log as the end timestamp, we assume that we have prior information about the maximal time of process execution, ϵ, and use it as the end time of the last task in any case.

We then apply the CMRules algorithm, with the best results obtained when the values of $minSeqSup$ and the $minSeqConf$ are bounded from below by the number of distinct case-ID in which a specific task occurs (as with any association rule mining technique, $minSeqSup$ and $minSeqConf$ represent the support and confidence respectively—higher values of these can give us more reliable results but rule out potentially interesting rules and vice versa). In *unique task* settings with no noise, the sequence of effects following the exection of each task and prior to the execution of the next task in the process instance should be largely identical if the process design is fixed—we apply CMRules mainly to mitigate the effects of noise. In *concurrent task* settings, these could vary significantly since the effects that follow a task might not be its effects but those of a distinct concurrent task. In these settings, the sequential rule miner is essential to identify the commonly occurring patterns of effects following a given task. In general terms, a sequential rule $X \rightarrow Y$ consists of two parts: the antecedent X and the consequent Y, which are both assumed to be sequences of transactions. The rule states that if the elements of X occur in a given sequence in the sequence database being mined, then the elements of Y will follow in the same sequence and in a manner that preserves the sequential relations between the elements of X and between the elements of Y. All sequential rules must also satisfy certain criteria regarding their accuracy (minimum confidence) and the proportion of the data that they actually represent (minimum support).

For example, consider case 1 in Table 2. The first task, task T_1 has timestamp t_1 and the next task in the case, T_2, has timestamp t_{28}; therefore, we group task T_1 with all effects in the effect log with the timestamp t_1 until t_{28}, which gives us the sequence $(T_1)(travel\text{-}dates\text{-}known)(airline\text{-}prefs\text{-}known)(airline\text{-}classoftravel\text{-}known)(departure\text{-}prefs\text{-}known)$ $(arrival\text{-}prefs\text{-}known)$ $(meal\text{-}constraints\text{-}known)$ $(freq\text{-}flyer\text{-}known)$ $(hotel\text{-}prefs\text{-}known)$ $(room\text{-}prefs\text{-}known)$ $(tour\text{-}prefs\text{-}known)$ $(travel\text{-}dates\text{-}known)(airline\text{-}prefs\text{-}known)$ $(airline\text{-}classoftravel\text{-}known)$ $(departure\text{-}prefs\text{-}known)$ $(arrival\text{-}prefs\text{-}known)$ $(meal\text{-}constraints\text{-}known)$ $(freq\text{-}flyer\text{-}known)(hotel\text{-}pref\text{-}known)(room\text{-}prefs\text{-}known)(tour\text{-}prefs\text{-}known)$. Similarly, task T_2 is grouped with all effects with timestamp from t_{28} until t_{36}, and so on. Applying the same process to all the other cases, we obtain the sequences for task T_1, T_2, T_3, until T_7. Next, these sequences are grouped into sequence databases based on their task ID, for instance the sequence for task T_1 from case 1 goes into the same sequence database with the task T_1 sequence from case 2 (along with task T_1 sequences from other cases).

Although the CMRules algorithm is able to generate all sequential rules from the sequence databases, further post-processing is required. Since we are interested only in relations between a task and effects, only rules with a single task ID as antecedent are included in the results and all other rules are discarded.

Validation: We can use the state update operator and the available data to perform additional validation steps on the effects mined in the manner described above. In settings characterized by the *unique task assumption*, an element of the joined process log and effect log can be viewed as *semantic execution trace* of the form:

$$\langle\langle\langle T_1, \langle\langle e_{11}\rangle, \ldots, \langle e_{1n}\rangle\rangle\rangle, \ldots, \langle T_i, \langle\langle e_{i1}\rangle, \ldots, \langle e_{im}\rangle\rangle\rangle, \ldots \langle T_p, \langle\langle e_{p1}\rangle, \ldots, \langle e_{pk}\rangle\rangle\rangle$$

for a process instance (case) with p tasks, with each T_i representing a task ID and each e_{ij} representing the j-th effect associated with task T_i. In other words, the sequence of effects associated with each task in the trace above represents the cumulative effects obtained at that point in the process instance. We shall refer to the sequence of tasks $\langle T_1, \ldots, T_p \rangle$ as the *signature* of the semantic execution trace above, and note that multiple semantic execution traces might be obtained for the same signature (due to the fact that we might find the process in one of many possible non-deterministic effect scenarios after the execution of a sequence of tasks). To validate the immediate effects mined using the procedure described in this section, we must establish for each semantic execution trace in the joined process log and effect log and for each task T_i in that trace that the following condition holds, where es_i is given by $e_{i1} \wedge e_{i2} \wedge \ldots \wedge e_{im}$ with m being the number of effect log entries associated with T_i: $es_i \models e$ for some $e \in e_{T_1} \oplus e_{T_2} \oplus \ldots \oplus e_{T_i}$ where each e_{T_i} denotes the mined immediate effects of task T_i. This represents a *soundness condition*, in the sense that we guarantee that every observed set of accumulated effects includes the accumulation of all mined effects of the tasks executed upto that point. For a sufficiently extensive collection of process and effect logs, we may also require that there must exist, for every $e \in e_{T_1} \oplus e_{T_2} \oplus \ldots \oplus e_{T_i}$, some entry in the joined process-effect log with an es_i associated with T_i such that $es_i \models e$. A completeness condition would reverse the entailment relation (i.e., $e \models es_i$). If the mined effects failed all of these tests, a weaker condition of consistency ($es_i \wedge e \not\models \perp$) would suggest that the mined results were not entirely incorrect. In settings characterized by *concurrent tasks*, we cannot guarantee that the effects observed between the start of a task and the start of the next task in the same process instance are necessarily the effects of the former task (since concurrent tasks from other process instances might have led to these effects). In such settings, we validate by creating modified sequence databases, parameterized by a task sequence length parameter n for use with CMRules. For instance, when the task sequence length parameter is 2, for each contiguous pair of tasks $\langle T_i, T_j \rangle$, we take sequences of the form $\langle T_i, e_{i1}, \ldots, e_{in} \rangle$ and $\langle T_j, e_{j1}, \ldots, e_{jm} \rangle$ where the tasks belong to the same process instance and where the timestamps associated with each e_{ik} is earlier than the start of T_j and create an entry in this modified sequence database of the form $\langle T_i, T_j, e_{i1}, \ldots, e_{in}, e_{j1}, \ldots, e_{jm} \rangle$ by removing any effect e_{ik} associated with the earlier task T_i where $e_{ik} \wedge \mathcal{KB} \models \neg e_{jl}$ for some e_{jl} ($\mathcal{KB}$ is the background knowledge base). Since the effect log records only changes (and not persistent effects or non-changes), we perform the last step to ensure some form of state update is reflected in the combined sequence for $\langle T_i, T_j \rangle$. We use CMRules to obtain rules of the form $\langle T_i, T_j \rangle \rightarrow \langle e_1, \ldots, e_p \rangle$ with the support and confidence being set as earlier to refer only to those process instances where T_i and T_j appear contiguously. The following condition then provides a weak form of validation: $e_1 \wedge \ldots \wedge e_n \models e_{T_i} \oplus e_{T_j}$. The approach generalizes to task sequences of arbitrary length, but we omit details due to space constraints. A general validation strategy is to consider all task sequences of length $i = 1, \ldots, n$ where n is the length of the longest task sequence that conforms to the process design.

4   Evaluation

Evaluation with synthetic process models: Our aim is to establish that our approach generates reasonably reliable results. We ran the first set of experiments with a synthetic semantically annotated process model (i.e., a hand-crafted one with $T_1, T_2, \ldots$ etc., for task names and $p, q, \ldots$ for effects). The model had 8 tasks, with an AND-split nested inside an XOR-split and with each task semantically annotated with 1 or 2 literals (in the 2 literal case, the effects were conjunctions of the 2 literals), and one rule in the $\mathcal{KB}$. We simulated a large number of possible execution traces of this model, and obtained synthetic process and effect logs. These logs involved the execution of multiple concurrent process instances. There were multiple effect scenarios associated with some of the tasks, owing to the fact that XOR gateway contributed to alternative flows that could have led to the same point (none of the effect scenarios were generated by alternative means of resolving inconsistency in the state update operator). We then investigated the effect of scaling up the complexity of the process model, by generating a second synthetic process model with 12 tasks with an XOR-split leading to two alternative flows, one of which included a nested AND-split and the other a nested XOR-split. The semantic annotations were 2 or 3 literals long and involved a mix of conjunctions and disjunctions. The background $\mathcal{KB}$ had 4 rules. There were multiple effect scenarios associated with most of the tasks and these were generated both by alternative flows that could lead to a task (on account of XOR gateways) and by alternative resolutions of inconsistency by the state update operator.

Table 4 below describes the results of 4 experiments with each of these two process models. We used progressively larger numbers of overlapping instances of each process (i.e., T_i in instance 2 would start after the start of T_i in instance 1, but before the start of T_{i+1} in instance 1, and so on). We note that our problem would be no harder if the multiple concurrent process instances were of multiple distinct process models. We obtained progressively larger sizes of the sequence database. We recorded the precision (number of correct effect mined over the total number of effects mined) and recall (the number of correct effects mined over the total number of actual effects. Although not entirely monotonically improving, the results for process 2 confirm the intuition that better results are obtained with larger datasets. The results for process 2 also showed that the effects mined tended to be incorrect for the last task in a process instance (in those settings where precision and recall values were less than 1). This was due to the sequence of effects for the final task not being bounded by the start of the next task, but rather by the end of the log (artificially determined by length of the longest process).

Table 4. The recall and precision measures from the evaluation

	Process model 1				Process model 2			
Number of instances	5	10	100	500	5	10	100	500
Size of sequence DB	48	100	1082	5352	66	133	1297	6512
Recall	1.0	1.0	1.0	1.0	0.953	1.0	0.981	0.989
Precision	1.0	1.0	1.0	1.0	1.0	0.988	1.0	1.0

The synthetic process and effect logs used in these examples considered all possible flows. Real-life data might involve more imperfections (such as certain XOR flows never being executed, certain tasks never being executed and so on). We have also considered cases where noise is artificially added to the logs - we do not report these results here due to space constraints, but, as expected, precision and recall suffer as noise increases. We do not present an evaluation of the technique we propose for settings where we can make a *unique task assumption*, both because of space constraints and because the technique is simple, and, in our experience, almost always accurate. We do not present in detail an evaluation of the validation techniques from the previous section, but our experience suggests that it is effective in filtering out inaccurate effects.

User-mediated evaluation: To evaluate our approach in a more real-life setting, we took a real-life semantically annotated process model (Fig. 1) and obtained a set of process and effect logs from an expert process modeler. We obtained a process log describing 10 execution instances (many of them with temporal overlaps) with a total of 110 entries, and an effect log with 154 entries.

We found that about 1 in 9 tasks, the effects mined were incorrect. The best explanation of this appears to be the fact that in the user-generated process log, there were other tasks that were exactly concurrent with the task for which the wrong effects were mined.

5   Related Work

A few examples of the benefits that can be exploited from semantically annotated business process models including compliance checking [4, 10, 14, 17], management level strategic alignment of business processes [22], and exception handling [9]. Two groups of studies can be defined based on the purpose of adding semantics to business process model: (1) to specify the dynamic behaviour of the business process [32, 34], and (2) to specify the meaning of the entities in the business process model itself [4, 22, 30]. In order to assist business analysts and process designers in the process of business process modelling, many studies have proposed frameworks to provide more user-friendly tools to add semantics into business process models [2, 16, 18].

In regards of assisting designers and analysts, many studies have emerged in harnessing historical data, specifically on software repositories, to discover useful information, with data such as programmer interaction history [23], software revision history [35, 36], email history [1], visited web pages [27], and bug reports [19]. Particularly in business process modeling, there are extensive studies on process mining that exploit the historical data of process model executions, i.e. event logs. Process mining algorithms—such as alpha algorithm [31], heuristic miner [33], and fuzzy miner [12]—extract the structure of the process model.

Our research integrates these two approaches of (1) mining historical data to discover useful information and (2) adding semantics to business process models, in order to support the business process modeling. We use data sources similar to those used in process mining, which is the business process execution history in the form of event log or process log, but our research does not generate the structure of the process model, instead it discovers the semantics of the business processes.

6   Conclusions and Future Work

This paper offers an approach to mining business process task post-conditions (or effects) from process logs and logs of state changes in the process context. The empirical evaluation suggests that the results are generally reliable, pointing to prospects for further development of techniques that leverage these post-conditions in semantic analysis.

References

1. Bacchelli, A., Lanza, M., Humpa, V.: RTFM (Read The Factual Mails)–augmenting program comprehension with REmail. In: Proceedings of 15th IEEE European Conference on Software Maintenance and Reengineering (CSMR), pp. 15–24. IEEE (2011)
2. Born, M., Dörr, F., Weber, I.: User-friendly semantic annotation in business process modeling. In: Weske, M., Hacid, M.-S., Godart, C. (eds.) WISE Workshops 2007. LNCS, vol. 4832, pp. 260–271. Springer, Heidelberg (2007)
3. Chan, K.C., Au, W.H.: An effective algorithm for mining interesting quantitative association rules. In: Proceedings of the 1997 ACM Symposium on Applied Computing, pp. 88–90. ACM (1997)
4. Di Francescomarino, C., Ghidini, C., Rospocher, M., Serafini, L., Tonella, P.: Reasoning on semantically annotated processes. In: Bouguettaya, A., Krueger, I., Margaria, T. (eds.) ICSOC 2008. LNCS, vol. 5364, pp. 132–146. Springer, Heidelberg (2008)
5. Fensel, D., Facca, F., Simperl, E., Toma, I.: Web service modeling ontology. Semantic Web Services, pp. 107–129. Springer, Heidelberg (2011)
6. Fournier-Viger, P., Faghihi, U., Nkambou, R., Nguifo, E.M.: CMRules: mining sequential rules common to several sequences. Knowl. Based Syst. 25(1), 63–76 (2012)
7. Fournier-Viger, P., Nkambou, R., Tseng, V.S.M.: RuleGrowth: mining sequential rules common to several sequences by pattern-growth. In: Proceedings of the 2011 ACM Symposium on Applied Computing, pp. 956–961. ACM (2011)
8. Garofalakis, M.N., Rastogi, R., Shim, K.: SPIRIT: sequential pattern mining with regular expression constraints. Proc. VLDB 99, 7–10 (1999)
9. Ghidini, C., Rospocher, M., Serafini, L.: A formalisation of BPMN in description logics. FBK-irst, Technical Report TR, 06–004 (2008)
10. Ghose, Aditya K., Koliadis, George: Auditing business process compliance. In: Krämer, Bernd J., Lin, Kwei-Jay, Narasimhan, Priya (eds.) ICSOC 2007. LNCS, vol. 4749, pp. 169–180. Springer, Heidelberg (2007)
11. Ginsberg, M.L., Smith, D.E.: Reasoning about action I: a possible world approach. Artif. Intell. 35(2), 165–195 (1988)
12. Günther, C.W., van der Aalst, W.M.P.: Fuzzy mining – adaptive process simplification based on multi-perspective metrics. In: Alonso, G., Dadam, P., Rosemann, M. (eds.) BPM 2007. LNCS, vol. 4714, pp. 328–343. Springer, Heidelberg (2007)
13. Hall, C., Harmon, P.: The 2005 Enterprise Architecture. Process Modeling and Simulation Tools Report, Technical report (2005)
14. Happel, H.J., Stojanovic, L.: Ontoprocess–a prototype for semantic business process verification using SWRL rules. In: Proceedings of the 3rd European Semantic Web Conference (ESWC), June 2006
15. Harms, S.K., Deogun, J.S.: Sequential association rule mining with time lags. J. Intell. Inf. Syst. 22(1), 7–22 (2004)
16. Hinge, K., Ghose, A., Koliadis, G.: Process SEER: a tool for semantic effect annotation of business process models. In: Proceedings of the IEEE International Enterprise Distributed Object Computing Conference (EDOC 2009), pp. 54–63, September 2009
17. Hoffmann, J., Weber, I., Governatori, G.: On compliance checking for clausal constraints in annotated process models. Inf. Syst. Front. 14(2), 155–177 (2012)
18. Hornung, T., Koschmider, A., Oberweis, A.: A recommender system for business process models. In: 17th Annual Workshop on Information Technologies and Systems (WITS) (2009)

19. Kim, M., Notkin, D.: Discovering and representing systematic code changes. In: Proceedings of the 31st International Conference on Software Engineering, pp. 309–319. IEEE Computer Society (2009)
20. Koliadis, G., Ghose, A.: Correlating business process and organizational models to manage change. In: Proceedings of the Australasian Conference on Information Systems, December 2006
21. Koliadis, G., Vranesevic, A., Bhuiyan, M., Krishna, A., Ghose, A.: A combined approach for supporting the business process model lifecycle. In: Proceedings of the 10th Pacific Asia Conference on Information Systems (PACIS 2006) (2006)
22. Koliadis, G., Ghose, A., Bhuiyan, M.: Correlating business process and organizational models to manage change. In: Proceedings of the Australasian Conference on Information Systems, pp. 1–10 (2006)
23. Lee, S., Kang, S., Kim, S., Staats, M.: The impact of view histories on edit recommendations. IEEE Trans. Softw. Eng. **41**(3), 314–330 (2015)
24. Martin, D., Paolucci, M., McIlraith, S.A., Burstein, M., McDermott, D., McGuinness, D.L., Parsia, B., Payne, T.R., Sabou, M., Solanki, M., Srinivasan, N., Sycara, K.: Bringing semantics to web services: the OWL-S approach. In: Cardoso, J., Sheth, A.P. (eds.) SWSWPC 2004. LNCS, vol. 3387, pp. 26–42. Springer, Heidelberg (2005)
25. Meyer, H.: On the semantics of service compositions. In: Marchiori, M., Pan, J.Z., Marie, C.S. (eds.) RR 2007. LNCS, vol. 4524, pp. 31–42. Springer, Heidelberg (2007)
26. Montali, M., Pesic, M., van der Aalst, W.M.P., Chesani, F., Mello, P., Storari, S.: Declarative specification and verification of service choreographiess. ACM Trans. Web **4**, 1–62 (2010)
27. Sawadsky, N., Murphy, G.C., Jiresal, R.: Reverb: recommending code-related web pages. In: Proceedings of the 2013 International Conference on Software Engineering, pp. 812–821. IEEE Press (2013)
28. Smith, F., Missikoff, M., Proietti, M.: Ontology-based querying of composite services. In: Ardagna, C.A., Damiani, E., Maciaszek, L.A., Missikoff, M., Parkin, M. (eds.) BSME 2010. LNCS, vol. 7350, pp. 159–180. Springer, Heidelberg (2012)
29. Smith, F., Proietti, M.: Rule-based behavioral reasoning on semantic business processes. In: ICAART, pp. 130–143. SciTePress (2013)
30. Thomas, O., Fellmann, M.: Semantic EPC: enhancing process modeling using ontology languages. In: SBPM, vol. 251 (2007)
31. van der Aalst, W., Weijters, T., Maruster, L.: Workflow mining: discovering process models from event logs. IEEE Trans. Knowl. Data Eng. **16**(9), 1128–1142 (2004)
32. Weber, I., Hoffmann, J., Mendling, J.: Semantic business process validation. In: Proceedings of the 3rd International Workshop on Semantic Business Process Management (SBPM08), CEUR-WS Proceedings, vol. 472 (2008)
33. Weijters, A., van Der Aalst, W.M., De Medeiros, A.A.: Process mining with the heuristics miner-algorithm. Technische Universiteit Eindhoven, Technical Report WP, vol. 166, pp. 1–34 (2006)
34. Wong, P.Y., Gibbons, J.: A relative timed semantics for BPMN. In: Proceedings of 7th International Workshop on the Foundations of Coordination Languages and Software Architectures, vol. 229 of ENTCS, July 2008
35. Ying, A.T., Murphy, G.C., Ng, R., Chu-Carroll, M.C.: Predicting source code changes by mining change history. IEEE Trans. Softw. Eng. **30**(9), 574–586 (2004)
36. Zimmermann, T., Zeller, A., Weissgerber, P., Diehl, S.: Mining version histories to guide software changes. IEEE Trans. Softw. Eng. **31**(6), 429–445 (2005)

Conceptual Modeling for Financial Investment with Text Mining

Yang Gu[1], Veda C. Storey[2], and Carson C. Woo[1(✉)]

[1] Sauder School of Business, University of British Columbia,
Vancouver, BC, Canada
yangguamy@gmail.com, carson.woo@ubc.ca
[2] University Plaza, Georgia State University, Atlanta 30302, USA
vstorey@gsu.edu

Abstract. Although text-mining, sentiment analysis, and other forms of analysis have been carried out on financial investment applications, a significant amount of associated research is ad hoc searching for meaningful patterns. Other research in finance develops theory using limited data sets. These efforts are at two extremes. To bridge the gap between financial data analytics and finance domain theory, this research analyzes a specific conceptual model, the Business Intelligence Model (BIM), to identify constructs and concepts that could be beneficial for matching data analytics to domain theory. Doing so, provides a first step towards understanding how to effectively generate and validate domain theories that significantly benefit from data analytics.

Keywords: Text-mining · Big data · Small data · Conceptual modeling · Finance domain theory · Domain model · Business intelligence model (BIM)

1 Introduction

Much research on text mining for sentiment analysis, pattern-identification, etc. are increasingly valued due to the explosive amount of data available, known as big data. One domain in which there is an abundance of data generated at an increasingly fast pace is that of finance (e.g., investors reactions to media news). Research in finance, however, has focused primarily on studying investment models (e.g., Zhang [12]), usually relying on relatively small amounts of data without utilizing the wealth of unstructured data found in social media, news, discussion forums, and other such sources.

Consider the following. Loughran and McDonald [8] argue that a generic lexicon for sentiments, the Harvard Psychosocial Dictionary, may not be appropriate for assessing sentiment in finance because certain words, such as "liability" and "tax," do not carry sentiment polarity in finance, although they are generally considered negative. Kitchens et al. [7] argue that this conclusion is incorrect because the Harvard Dictionary works with a program, General Inquirer, which can disambiguate between a word's many meanings and classify words appropriately, given its context. Loughran and McDonald [8], however, did not use General Inquirer; whereas other studies that did (e.g., [11]) obtained very good results. This leads one to conclude that some kind of

© Springer International Publishing Switzerland 2015
P. Johannesson et al. (Eds.): ER 2015, LNCS 9381, pp. 528–535, 2015.
DOI: 10.1007/978-3-319-25264-3_39

domain-specific context must be taken into account when using text-mining. A systematic approach to aiding domain experts in utilizing data mining techniques, facilitated by conceptual modeling, should avoid this problem.

The objective of this research, then, is to analyze the potential use of conceptual models with text mining for investment within the finance domain. To do so, we apply one specific model, the Business Intelligence Model (BIM) [6] to text mining in finance. The contribution of this research is to bridge a perceived gap between data analytics and domain theory, and to generate similar interests in further developing this bridge in other domains.

2 Related Research

Separate research efforts have been carried out that focus on either data analytics or theory. For data analytics, Dhar [3] focuses on building complex models with quantitative data for understanding financial markets. This, and similar, research is primarily focused on data mining while somewhat ignoring potential theories to both drive and interpret results. Nuij et al. [9], for example, identified a set of events that impact the stock market based on previous studies, and developed a trading strategy using text-mining. However, the authors have not revisited why their strategy or set of events performed the way they did. Similarly, Schumaker and Chen [10] studied mined news articles but did not consider which kinds of news articles were included.

Much research in finance, on the other hand, focuses on statistical calculations. Highly-regarded finance journals (e.g., the *Journal of Finance*) reveal that many research efforts focus on fundamental measures such as the number of specified words that appear in the *Wall Street Journal* and how they might impact stock prices. Fang and Peress [5] and Engelberg and Parsons [4], for example, examine the effect of media coverage in newspapers on the stock market, based on keyword searches on company names from an article database. Zhang [12] investigates how new information can affect investor behavior, relying exclusively on quantitative data from investment analysts' reports and earnings announcements. Zhang [12] identifies factors such as "uncertainty," "good news," and "bad news," defined solely using quantitative information, even though these factors also imply a qualitative dimension. These studies neglect the textual information already present in their datasets and seem hesitant to connect advanced big data techniques to financial theory.

We apply a conceptual model to the work of Tetlock [11] who used negative words from a newspaper column to predict the stock market. The number of negative words in the "Abreast of the Market" column from the *Wall Street Journal* was counted and a measure of "media pessimism" created to predict the movement of the stock market. Multiple financial theories that could explain the relationship between news and the stock market were analyzed. For the purposes of the research reported in this paper, we define financial theory as a framework or model that explains how and why the stock market behaves or makes decisions [2].

3 Application of the Business Intelligence Model to Finance

The Business Intelligence Model (BIM) [6] aims to help business users make sense of the "vast amounts of data about the enterprise and its external environment" and was motivated, in part, by the trend towards big data. The model uses constructs commonly associated with business decision-making, in contrast to most systems that are closely linked to technical aspects of data processing. BIM's high-level business constructs could be very useful for connecting text-mining techniques to investment theories. It comprises the basic constructs described below.

- *Goal*: A goal is "an objective of the business." Goals can be refined into sub-goals using *refinement*-links. In an AND-refinement, all sub-goals must be satisfied in order for the parent goal to be satisfied. In an OR-refinement, satisfaction of any of the sub-goals will satisfy the parent goal. A user can reason and deduce the probabilities or conditions needed to satisfy a parent goal by examining the satisfaction of related sub-goals. Parent goals can be refined into domain assumptions, which are situations that must be true to satisfy the parent goal. The *achieves*-link between a process and a goal is used to indicate "how" a process can achieve a goal.
- *Situation*: A situation describes a partial state of the world "in terms of things that exist in that state, their properties, and interrelation". In business modeling and analysis, organizations must consider situations such as strengths (S) and weaknesses (W) of the organization, as the well as the opportunities (O) and threats (T) in the environment (abbreviated to SWOT). BIM specifically identifies situations because their occurrences can influence the satisfaction of goals.
- *Influences*: Influences explain the contribution relationships between goals, situations, and domain assumptions. For example, if the domain assumption of "High Demand" is true, then "Increased Competition" will be less likely to occur. A positive influence captures this relationship. The strengths of influences can be described quantitatively or qualitatively; influences can be logical or probabilistic.
- *Indicators*: These metrics evaluate performance with respects to some objective. They are associated with goals using an *evaluates*-link.

3.1 Application to Financial Theories

BIM is applied to the finance domain to act as a case study for applying the model. Figure 1 shows work in the finance domain, based upon Tetlock's study [11] of the relationship between the stock market and negative words in a newspaper column. It uses BIM to capture the content of the entire study at a high level. The top-level goal is "to understand the role of media in the stock market" and is decomposed into sub-goals until we can identify specific processes to achieve these sub-goals.

In Fig. 1, "Negative words in news predict increase in trading volume" is portrayed as one situation. There is no graphical way to observe the inverse relationship between volume and price when negative words are present. However, it is possible to use one situation to capture this relationship between two entities. The advantage of modeling this relationship is that we can link it to other goals and situations that specifically

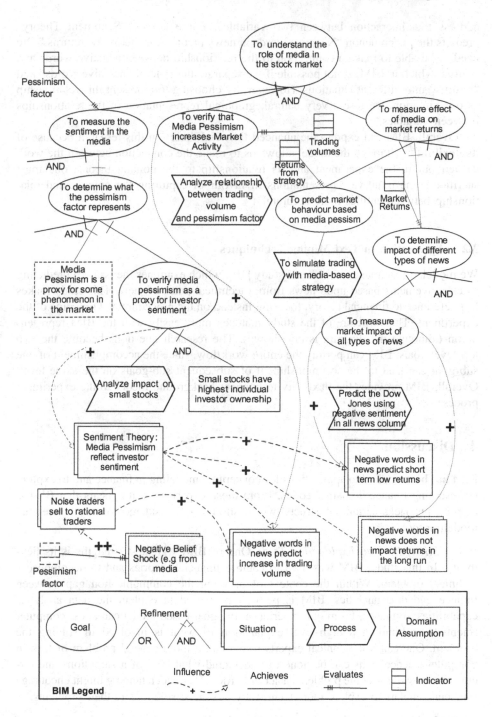

Fig. 1. Portion of BIM diagram [11].

address this interaction between two variables. For example, "Sentiment Theory" predicts the phenomenon "Negative words in news predict short term low returns." We need to be able to make a connection to the relationship between negative words and returns, which in BIM, is not possible if we separate the state of "negative words" and "returns" into different situations. However, by choosing to represent this relationship as one situation, we lose a very natural, graphical representation of the relationships between variables.

Overall, BIM can express the nuanced financial theory in this research because of its flexibility. Although the model allows us to make the connection between the work carried out in the experiment and its relationship to the domain theory, one must sacrifice some intuitive representations in favour of capturing the complicated relationship between entities more accurately.

3.2   Application to Text Mining Techniques

We modeled Schumaker and Chen's study [10], which examines the issue of predicting stock movements based upon news from a technical point of view. This study makes less reference to financial theory, focusing instead on the procedure and outcome of the experiment. This structure of the study matches more easily with the BIM representation (although the diagram is not shown). The research questions become the two top-level goals. BIM can portray the entire workflow, since the accomplishment of one sub-goal can lead to the accomplishment of subsequent sub-goals on the same level. Overall, BIM provides the flexibility to connect background theory to the experiment process.

4   Discussion

BIM has been applied to applications for conceptual modeling in finance and to explore representing finance domain theories. From these experiences, it appears that BIM is helpful, but lacks certain constructs, which should be considered for any conceptual modeling tool.

BIM's Strength as an Explanation Tool: One of BIM's greatest strengths is its flexibility. In this study, BIM was used to capture financial theories and to decompose a technical problem. Within the context of bridging the communication gap between finance and data analytics, BIM is perhaps best used to explain the data analytics capabilities to finance. From our experience, the goal-based model can assist computer scientists in building a high-level explanation of what is involved in solving the problem. One can ask technical experts to explain a more general problem or task in computer science. This can be done to understand what kind of assumptions and circumstances will make the problem easier or harder. This understanding might encourage financial scholars to explore more technically advanced solutions to their problems.

BIM's Inability to Link Explicitly to Business Intelligence: A common theme found in our conceptual models is that none of them make a clear connection to Business

Intelligence, despite the fact that this is the intent of the model. The Indicator is intended to make a connection from data to the business goals. In this case study, the goals are already defined very narrowly around a machine learning or data-mining problem, but, still, it is difficult to pinpoint the data-mining component. Instead, the indicators employed are very generic, such as Mean Squared Errors (MSE) or stock returns.

A possible reason is because indicators are meant to evaluate a goal, whereas the bulk of business intelligence operations are concentrated in realizing that goal. Text-mining and machine learning do not occur in the goals with indicators attached, but in some process at the bottom of the diagram. In Fig. 1, a Pessimism Factor is directly linked to the business intelligence process. However, it is a very high-level indicator constructed by a complicated process, and very far removed from the raw data from which it is composed. It is arguable that, unless we designed the Factor ahead of time, using the model alone would not assist in doing so.

A potential remedy for this situation is to include in BIM a representation of the raw data underlying the data-mining process. BIM can be constructed using a "top-down" or "bottom-up" approach [6]. We can make use of data in strategic decision making in a similar way. It should be possible to indicate in the model some type of data that will be relevant to satisfy a goal. This top-down approach can help identify resources that we need to collect or acquire. Likewise, if BIM includes raw data already possessed by the organization, their presence in the model may help users derive a strategy that can exploit them. Instead of only focusing on the goals we are trying to accomplish, and the indicator by which we are being evaluated, we also have the building blocks for the tools we intend to use. This representation may be able to provide a more organic link between the BIM model and data in the organization.

Limitations of BIM: BIM does not have a mechanism to address the passage of time, which is common in many business models. Popular models, such as Business Process Modeling Notation (BPMN) and the Unified Modeling Language (UML) sequence diagram, can explicitly communicate the passage of time or relative order of events. Clearly, as one follows the arrows from one process to the next, time has elapsed. In BIM, a sequence of events can only be shown indirectly through goal refinement and influence links. We assume that the sub-goal must be satisfied before the parent goal, and that series of influences can imply a sequential relationship. However, there is no time dimension attached to these concepts. If one model component influences two other components, it is difficult to distinguish the relative order in which the two components were influenced. Since many issues in finance are time-sensitive, being able to address the passage of time is an important aspect of financial research. For example, Tetlock [11] measures the effect of media pessimism at two different points in time: the day after the article is released, and five days after release. He found that high media pessimism could lead to low returns in the short run and stable returns in the long run. Although both effects can be captured using influence links, there is no natural or intuitive way to determine that high returns will appear quickly, whereas stables returns will take more time.

BIM also stakes a centralized view of the situation. The model is centered around the goals of a particular organization, although the situations and domain assumptions outside of the organization are not classified. It may be helpful to organize them using something similar to an entity as defined in the Entity-relationship (ER) model [1], or a Class in an UML model. For example, currently we need two separate BIM situations for stocks returns: "High long term returns" and "Low short term returns," because media pessimism influences them differently. However, intuitively, we know that these situations reflect one construct, "stock returns," at different points in time. The level of returns changed during this time by a process that is not yet known. Having a mechanism to show that these two different situations belong to the same entity may result in a better understanding of how the construct changes over time. Furthermore, BIM cannot easily identify that there are two agents involved and that this interaction occurs between the agents.

5 Conclusion

This research has identified the need for conceptual models to be applied to data analytics for big data within the finance domain. The Business Intelligence Model (BIM) was applied and assessed for its potential use for this application domain. The research illustrates that BIM is flexible and potentially useful in explaining big data analytics capabilities to finance. Additionally, the research illustrates an instantiation of interdisciplinary research in finance and conceptual modeling, reflecting the challenge of using a business intelligence model for capturing finance concepts. This leads to several insights for future research in incorporating these concepts (e.g., passage of time and agents playing different roles) into business intelligence modeling. Perhaps the most important implication of this research is the identification of the need and usefulness of developing a conceptual model to capture concepts related to big data and the theory of a domain through the matching of data analytics to domain theory. Specifically, applying the notion of matching may be most useful for other areas of research on data analytics and theory, which will require further research in multiple domains.

References

1. Chen, P.P.: The entity-relationship model – toward a unified view of data. ACM Trans. Database Syst. 1(1), 9–36 (1976)
2. Copeland, T.E., Weston, J.F., Shastri, K.: Financial Theory and Corporate Policy, vol. 3. Addison-Wesley, Massachusetts (1983)
3. Dhar, V.: Prediction in financial markets: the case for small disjuncts. ACM Trans. Intell. Syst. Technol. (TIST) 2(3), Article 19 (2011)
4. Engelberg, J.E., Parsons, C.A.: The casual impact of media in financial markets. J. Finance 66, 67–97 (2011)
5. Fang, L., Peress, J.: Media coverage and the cross-section of stock returns. J. Finance 64, 2023–2052 (2009)
6. Horkoff, J., Barone, D., Jiang, L., Yu, E., Amyot, D., Borgida, A., Mylopoulos, J.: Strategic business modeling: representation and reasoning. Softw. Syst. Model. 13(3), 1015–1041 (2014). Springer

7. Kitchens, B.M., Mitra, D., Johnson, J., Pathak, P.: 'When Is a Liability Not a Liability' Revisited: Should Financial Text Analysis Abandon General Lexicons? SSRN: http://ssrn.com/abstract=2554656 or http://dx.doi.org/10.2139/ssrn.2554656. Accessed 1 July 2014
8. Loughran, T., McDonald, B.: When is a liability not a liability? Textual analysis, dictionaries, and 10-ks. J. Finance **66**(1), 35–65 (2011)
9. Nuij, W., Milea, V., Hogenboom, F., Frasincar, F., Kaymak, U.: An automated framework for incorporating news into stock trading strategies. IEEE Trans. Knowl. Data Eng. **26**(4), 823–835 (2014). doi:10.1109/TKDE.2013.133
10. Schumaker, R.P., Chen, H.: Textual analysis of stock market prediction using breaking financial news: the AZFin text system. ACM Trans. Inf. Syst. (TOIS) **27**(2), Article 12 (2009)
11. Tetlock, P.C.: Giving content to investor sentiment: the role of media in the stock market. J. Finance **62**, 1139–1168 (2007)
12. Zhang, X.F.: Information uncertainty and stock returns. J. Finance **61**(1), 105–137 (2006)

Applications and Domain-based Modeling

Breaking the Recursivity: Towards a Model to Analyse Expert Finders

Matthieu Vergne[1,2](✉) and Angelo Susi[1]

[1] Center for Information and Communication Technology, FBK-ICT,
Via Sommarive, 18, 38123 Povo, TN, Italy
{vergne,susi}@fbk.eu
[2] Doctoral School in Information and Communication Technology,
Via Sommarive, 5, 38123 Povo, TN, Italy
matthieu.vergne@unitn.it

Abstract. Expert Finding (EF) techniques help in discovering people having relevant knowledge and skills. But for their validation, EF techniques usually rely on experts, meaning using another EF technique, generally not properly validated, and exploit them mainly for output validations, meaning only at late stages. We propose a model, which builds on literature in Psychology and practice, to identify generic concepts and relations in order to support the analysis and design of EF techniques, thus inferring potential improvements during early stages in an expert-free manner. Our contribution lies in the identification and review of relevant literature, building the conceptual model, and illustrating its use through an analysis of existing EF techniques. Although the model can be improved, we can already identify strengths and limitations in recent EF techniques, thus supporting the usefulness of a model-based analysis and design for EF techniques.

Keywords: Expert finding · Concept formalization · Model-driven analysis · Design support

1 Introduction

Expert finding (EF), also called expertise location or expert recommendation [9], aims at recommending *experts*, or at least the most knowledgeable or skilled people we find within a community of people, on a given domain. EF is broadly useful, because it allows to acquire knowledge and skills through hiring [8], to support decision making and solve problems [8,9], help in requirements elicitation [11] or even to validate models and approaches in research (e.g. soundness, practicality). One performs EF by evaluating the expertise of performers within the available community, before to rank them or to select the ones to recommend. More precisely, you are an expert when "*having* or *showing* special *skill* or *knowledge* because of what you have been *taught* or what you have *experienced*", as defined by the Merriam-

© Springer International Publishing Switzerland 2015
P. Johannesson et al. (Eds.): ER 2015, LNCS 9381, pp. 539–547, 2015.
DOI: 10.1007/978-3-319-25264-3_40

Webster[1] dictionaries. One can notice that it implies to look at the intrinsic properties of the performer (i.e. having skill or knowledge) as well as the perception of some evaluators (i.e. showing skill or knowledge). Research literature in Psychology also identifies properties for *expertise*, which builds on long experience and high performance [13], as well as *expert*, who is identified through such expertise as well as social recognition [3].

While such a literature exists, designing EF techniques in Computer Science remains rather intuition-based [13], and recent works still validate their approaches by evaluating the output of their technique through domain experts [7,14,17]. Usually, they design their own technique based on indicators they think are of relevance, and validate it through experts identified based on social recognition [14,17], self-evaluations [7], or other resources they did not use in their own technique [7]. This kind of validation brings significant threats: (i) output validations occur only at late stages, delaying the identification of inadequate techniques, and (ii) we need an already valid EF technique to find these domain experts. We faced this situation for our own EF approach [15] and it hurts the reliability of the validation process, thus we need to find a way to validate EF techniques without relying on domain experts, or at least not only on them.

After clarifying the problem and issues we want to tackle in Sect. 2, this paper contributes to the research community by (i) identifying some *relevant literature* in Psychology and EF techniques in Sect. 3, (ii) starting the building of a generic, grounded *conceptual model* for expertise evaluation in Sect. 4.1, and (iii) performing a *model-driven analysis* of the described EF techniques to illustrate its use in Sect. 4.2. While our research exploits both the perspectives of the *performer* and the *evaluator* provided by the literature, we restrict here to the latter and add a perspective on the *evaluation* (perception of the evaluator) to focus on EF techniques. Our model, which relies on scientific evidences in expertise evaluation in general, focuses on the foundational basis for the early stages (design and implementation) of an EF technique, thus offering a good complement to output validations through domain-specific experts. We are convinced that other interpretations from the modelling community and other references could help in building a more complete and reliable expertise evaluation model. In the long term, having such a complete model with proper guidelines could help evaluating existing EF techniques through a model-driven analysis by identifying strengths and limitations, and to fasten the design of new EF techniques by suggesting expertise indicators.

2   Expert Finding for Expert Finding: A Recursive Problem

EF is an important task, especially in research where we exploit the knowledge of domain experts to validate conceptual models and the outputs built based on them. A significant problem is that it also applies to EF techniques themselves, which aim at recommending domain experts from a given community, and thus

---

[1] http://www.merriam-webster.com/.

to validate their recommendations through domain experts [7,14,17]. They find their "validation experts" through social recognition [14,17], self-evaluations [7], or other resources not used in their own approach [7].

This "recursive" problem makes EF techniques hard to validate, because the domain experts could be biased [1] and have limited knowledge on the actual expertise of other people in the community [9], leading to a poor validation. We could think for instance about Open Source forums or international companies, where hundreds of people can be involved, thus making it hard to know everyone and in particular who are the most experts. One could consider different cases of application of the EF technique to mitigate this issue, but trying to find and involve the relevant experts could require a significant amount of time and effort. Moreover, this kind of validation focuses on the output of the EF techniques, meaning that we could assess the effectiveness of the technique only in late implementation stages. One could rely on EF techniques already employed in the community, assuming they are empirically validated, but it could lead to techniques which are hard to generalize to other contexts [9].

In this paper, we build an initial, generic conceptual model of expertise evaluation, to support early analysis without relying on domain experts. In particular, we would like to know (i) which concepts and relations are generic enough to appear in this model, implying the review of some relevant literature, (ii) which strengths and limitations can already be found in existing EF techniques, thus analysing them in the light of our model, (iii) and which parts of the model should be completed or refined, thus discussing the current state of the model in the light of the previous analysis. Consequently, this paper provides a model which can already support such analysis, but which could be further improved and validated.

3 Expert Finding Literature Review

3.1 Recent Expert Finding Techniques in Computer Science

Some EF techniques rely on *direct* contributions of performers to evaluate their expertise. For example, Mockus and Herbsleb [10] analyse the amount of code written in a piece of a software to identify knowledgeable programmers. They rank them relatively to the number of changes they made on the source code, possibly restricting the counting to a given period of time. Similarly, Serdyukov and Hiemstra [12] analyse the content of many documents to identify the contributions of their different authors, which helps in identifying their potential knowledge (i.e. terms used). They compute the probability that a given document or a given term relates to a given author and, when looking for experts related to a specific term, sum up the corresponding probabilities to rank the authors.

Other EF techniques rely on *indirect* indicators, especially how much people are recognized as experts into a given community. Zhang et al. [17] look at question/answers forums in an online community to identify people seeking and providing knowledge. In their work, they compare several algorithms to rank people, starting from the simple counting of answers, assuming it is positively correlated with the level of expertise. Another algorithm combines it with the number of questions written, which should be negatively correlated to the level of expertise. A third algorithm propagates these values over the community (PageRank-like), so that people answering questions from experts are themselves considered as more experts.

Finally, some works combine both indicators, direct as well as indirect, for evaluating expertise. Karimzadehgan et al. [7] exploit the content of the e-mails of employees to retrieve their potential knowledge (i.e. terms and topics), but also exploit hierarchical similarities among employees. They compute probabilities similarly to Serdyukov and Hiemstra [12], but smooth the results between hierarchically-related employees to mitigate the potential lack of data for some of them. We also proposed our own approach [15] which explicitly intended to exploit direct evidences of knowledge (i.e. terms and topics) added to social aspects (i.e. roles). We counted co-occurrences, such as how many times someone used a term, a term is used in a topic, or a role is assigned to someone, to build a weighted graph and compute and propagate probabilities all over it, allowing us to rank people.

3.2   Expertise in Psychology

Behind the fact that some EF techniques use specific indicators, we are also interested in how, generally, people build their own expertise, in order to find what are the relevant indicators to consider. Ericsson [2] summarizes a broad literature on this purpose. In particular, an *acceptable level of proficiency* requires some months of experience during which the performer will focus on the actions to perform while avoiding gross mistakes, like in school or any other training course. An *average, independent professional proficiency*, which means performing in an autonomous way, requires often several years, what we call a *lengthy domain-related experience*, to become fluent in the domain-relevant activities. However, what differentiates the average professional, who maintains his level by executing routine work, from the domain *expert* (or *master*) is the continuation of *deliberate practice* to fix weaknesses [5].

Focusing more on the perspective of someone looking for experts, the main perspective for EF, we can consider the review of Chi [1] who presents the two main approaches used to study expertise. The *absolute approach*, on one hand, studies exceptional people to understand what distinguishes them from the masses, in order to identify the properties which allow to reach the top (potentially some innate capacities). The *relative approach*, on the other hand, focuses on distinguishing people within a common, domain-related group, in order to identify what can be provided to the less experts to reach the level of the more experts. Chi [1] also summarizes the *properties* which seem to characterize experts, who excel for example by generating better solutions faster, perceiving deep features, identifying lacks and errors, and managing better their resources (e.g. skill, knowledge, sources of information). However, she also highlights that experts fail in showing similar excellence in different domains and in judging non-expert abilities, as well as they can be over-confident in their abilities, overlook details, and show more biases when their expertise does not apply.

While the literature provide us useful indicators to consider, Ericsson [2] notices that people evaluating the expertise of a performer often rely on simple experience-based indicators, which do not help in finding the highest experts. In these "good but not best" indicators, we can find the length of experience in the domain, the accumulated accessible knowledge, the completed education and the social reputation. In order to identify the highest experts, one need to look at *reproducibly superior performance* on representative, authentic tasks which require domain-specific experience, like a chess master should find the best move on a chess board already set up. However, when such direct evidences are lacking, we think that evidences of deliberate practice could help identify expert-like behaviours, complementing the simple experience-based indicators criticized by Ericsson [2].

4   Conceptual Model of Expertise Evaluation

4.1   Conceptual Model

In order to differentiate common terms from the concepts introduced by the model, shown in Fig. 1, we use This Font for the concepts of the model.

We start by modelling the the *domain* relating the performer and evaluator, which starts from the Domain root concept (e.g. databases, or DB) and assume that it relates to a Domain Community. Within this Domain Community, we will find the Performers (e.g. DB programmers) who produces the Outcomes (i.e. products, services or ideas) relevant to the Domain. We will also find the Evaluators (e.g. recruiters) evaluating these Performers based on their Outcomes, and who will influence/be influenced by some Social Recognition. In order to perform well and to produce creative ideas, a Performer hopefully consider already existing Outcomes which have been recognized as Domain Prior Achievements, meaning Outcomes which have received a significant amount of Social Recognition from previous Evaluators (Ericsson [4]).

From the perspective of the Evaluator, one considers the Outcomes of a Performer in order to build the Perceived Expertise. More precisely, the Evaluator should identify evidences of Lengthy Domain-Related Experience (e.g. 10 years experience in DB) to assess a reasonable level of expertise, and evidences of Reproducibly Superior Performance (e.g. several projects with complex data) for the highest levels of expertise (Sonnentag et al. [13], Ericsson [2]). Additionally to these direct evidences inferred from the Outcomes, some other Evaluators could have provided their own Perceived Expertise (e.g. LinkedIn endorsements), leading to the building of some Social Recognition, which can be reused by the current Evaluator to refine or complete his or her judgement (Ericsson [3]). Going further in detail, we decompose the Perceived Expertise into both the Perceived Domain Knowledge and the Perceived Domain Skills. They correspond to the specific items supporting the identification of Lengthy Domain-Related

Experience and Reproducibly Superior Performance. These items, however, should be known by the Evaluator in order for him to identify them, thus he should have some Owned Domain Knowledge.

Because we focus on EF techniques, we go even further in the granularity of the *evaluator* by modelling the *evaluation* he produces. A Performance Evaluation represents the Perceived Expertise of an Evaluator, and can be an Absolute Performance Value or a Relative Performance Value (Chi [1]). Typically, we use Performance Levels to express Absolute Performance Values, while we compare Performers through Performers Orderings to express Relative Performance Values. Going more in detail for the Performance Level, a concrete scale can be used, such as the Novice-Master scale described by Chi [1] (Table 2.1, p. 22). If the Evaluator cannot use a concrete scale, he can rely on evidences of Lengthy Domain-Related Experience to assess an Average Performance Level, while additional evidences of Reproducibly Superior Performance would help identifying the Highest Performance Levels (Sonnentag et al. [13], Ericsson [2]).

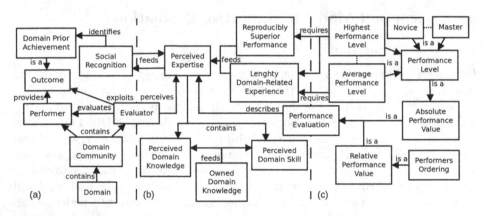

Fig. 1. Conceptual model of the Domain (a), Evaluator (b), and Performance Evaluation (c).

4.2 Preliminary Analysis of Expert Finding Techniques

By analysing the works presented in Sect. 3.1, we can see that Serdyukov and Hiemstra [12] focus mainly on Perceived Domain Knowledge items by identifying the terms used. In particular, by evaluating how much a person contributes compared to all the others (via normalization), these approaches infer Absolute Performance Values (i.e. probabilities) and recommend the people having the highest ones. While we could imagine that the values computed could help to infer Performance Levels, this approach would need to be completed with correlations between their values and proper levels. Moreover, while such approach is probably efficient to build the Perceived Domain Knowledge, it lacks the Perceived Domain Skill dimension. Going further, these approaches probably identify evidences of domain-related experience but not necessarily of Lengthy

Domain-Related Experience, making it difficult to assess even an average level, unless the assumption of a lower bound expertise can be supported by the specific type of documents considered (e.g. peer-reviewed papers accepted for publication). Such assumptions, however, would probably not help in discriminating good from exceptional Performers, meaning finding evidences for Reproducibly Superior Performance.

Summarizing on the other works presented in Sect. 3.1, social approaches like Zhang et al. [17] consider Social Recognition indicators to provide Performers Orderings (i.e. one is less expert than another) or Performance Levels (e.g. *Newbie* or *Top Java expert*). Once again, this approach lacks the identification of Perceived Domain Skills, and they also suffer the same difficulties than Serdyukov and Hiemstra [12] to identify clear evidences for Lengthy Domain-Related Experience as well as Reproducibly Superior Performance. We retrieve these difficulties in approaches combining documents and social analysis, like Karimzadehgan et al. [7] and our own approach [15]. Although they combine Social Recognition indicators (hierarchy for the former, roles for the latter) with Perceived Domain Knowledge indicators (terms and topics), they ignore the Perceived Domain Skills.

Only Mockus and Herbsleb [10] provide a rather complete approach by considering the commits (changes on a software) made by programmers. Commits are at the same time good indicators of Perceived Domain Skills (coding skills are major skills in software) as well as Perceived Domain Knowledge (module modified, names of the variables added/removed/changed, etc.). The number of commits made over time can also show a Lengthy Domain-Related Experience, while frequencies of commits per month could show reproducible performances, although it does not necessarily support the high quality required by Reproducibly Superior Performances. Thus, while they already provide supports and results, our model highlights why they are able to do so and identifies potential improvements (i.e. identifying the highest levels of expertise). Though, these good results should be contrasted with the fact that this approach targets a specific Domain (software implementation) while the other approaches try to be more generic, making the task more difficult.

5   Discussion and Conclusion

Through this paper, we have seen that EF techniques for finding experts in a given domain involves generally a "recursive" problem, by relying on domain experts to validate it, while other validation methods could be used. In particular, we showed that we can build a generic, grounded model to analyse EF techniques during early stages (design and implementation) by relying on literature in other domains, like Psychology.

However, we should notice a significant incompleteness in our model, like the inability to relate the specific formulae of the existing techniques to specific concepts in the model (i.e. modelling the *evaluation processes*). We also miss notions like time, which is critical to identify Lengthy Domain-Related Experience, and relations between Perceived Domain Knowledge/Skill and Lengthy Domain-Related Experience. Going back to the literature already cited, we did not consider the expert properties provided by Chi [1] (i.e. generate better solutions faster, fail in judging non-expert abilities, etc.), while it could provide relevant indicators to exploit. Similarly, we rely exclusively on literature in Psychology to identify the main concepts (top-down), while it could be complemented with systematic literature reviews of existing EF techniques to identify relevant lower level concepts (bottom-up, like [16]). Other perspectives could also be considered, like creativity [4] (i.e. producing something new and useful), which seems to be a way to identify some of the highest experts.

Based on this conceptual model and its limitations, we think that discussions within the research community about EF design and validation could be of relevant interest, and we encourage people to exchange interpretations and further formalization. From these exchanges, relevant future works could be to have a better formalization of this model, not only more complete but also more rigorous, for instance by using ontologies like in [6]. We also think that a systematic literature review of the existing EF techniques could be useful, not only to identify concrete indicators, but also to see how the existing techniques could be classified with such a model. For example, categories of EF techniques focusing on knowledge indicators could be particularly suited for contexts lacking skills indicators, leading to recommend the right EF techniques depending on the context at hand.

References

1. Chi, M.T.H.: Two approaches to the study of experts' characteristics. In: Ericsson, K.A., Charness, N., Feltovich, P.J., Hoffman, R.R. (eds.) The Cambridge Handbook of Expertise and Expert Performance, pp. 21–30. Cambridge University Press, New York (2006)
2. Ericsson, K.A.: The influence of experience and deliberate practice on the development of superior expert performance. In: Ericsson, K.A., Charness, N., Feltovich, P.J., Hoffman, R.R. (eds.) The Cambridge Handbook of Expertise and Expert Performance, pp. 683–703. Cambridge University Press, New York (2006)
3. Ericsson, K.A.: An Introduction to cambridge handbook of expertise and expert performance: its development, organization, and content. In: Ericsson, K.A., Charness, N., Feltovich, P.J., Hoffman, R.R. (eds.) The Cambridge Handbook of Expertise and Expert Performance. Cambridge University Press, New York (2006)
4. Ericsson, K.A.: Creative expertise as superior reproducible performance: innovative and flexible aspects of expert performance. Psychol. Inquiry 10(4), 329–333 (1999)
5. Ericsson, K.A., Krampe, R.T., Tesch-romer, C.: The role of deliberate practice in the acquisition of expert performance. Psychol. Rev. 100(3), 363–406 (1993)
6. Fazel-Zarandi, M., Fox, M.S., Yu, E.: Ontologies in expertise finding systems: modeling, analysis, and design. In: Ahmad, M.N., Colomb, R.M., Abdullah, M.S. (eds.) Ontology-Based Applications for Enterprise Systems and Knowledge Management, pp. 158–177. IGI Global, Hershey (2013)
7. Karimzadehgan, M., White, R.W., Richardson, M.: Enhancing expert finding using organizational hierarchies. In: Boughanem, M., Berrut, C., Mothe, J., Soule-Dupuy, C. (eds.) ECIR 2009. LNCS, vol. 5478. Springer, Heidelberg (2009)

8. Maybury, M.T.: Expert Finding Systems. MITRE Center for Integrated Intelligence Systems, Bedford (2006)
9. McDonald, D.W., Ackerman, M.S.: Just talk to me: a field study of expertise location. In: Proceedings of the Conference on CSCW, pp. 315–324. ACM, New York (1998)
10. Mockus, A., Herbsleb, J.D.: Expertise browser: a quantitative approach to identifying expertise. In: Proceedings of the 24th ICSE, pp. 503–512. ACM, New York (2002)
11. Mohebzada, J., Ruhe, G., Eberlein, A.: Systematic mapping of recommendation systems for requirements engineering. In: 2012 ICSSP, pp. 200–209 (2012)
12. Serdyukov, P., Hiemstra, D.: Modeling documents as mixtures of persons for expert finding. In: Macdonald, C., Ounis, I., Plachouras, V., Ruthven, I., White, R.W. (eds.) ECIR 2008. LNCS, vol. 4956, pp. 309–320. Springer, Heidelberg (2008)
13. Sonnentag, S., Niessen, C., Volmer, J.: Expertise in software design. In: Ericsson, K.A., Charness, N., Feltovich, P.J., Hoffman, R.R. (eds.) The Cambridge Handbook of Expertise and Expert Performance. Cambridge University Press, New York (2006)
14. Tang, J., Zhang, D., Yao, L.: Social network extraction of academic researchers. In: 7th IEEE ICDM, pp. 292–301 (2007)
15. Vergne, M., Susi, A.: Expert finding using Markov networks in open source communities. In: Jarke, M., Mylopoulos, J., Quix, C., Rolland, C., Manolopoulos, Y., Mouratidis, H., Horkoff, J. (eds.) CAiSE 2014. LNCS, vol. 8484, pp. 196–210. Springer, Heidelberg (2014)
16. Yimam-Seid, D., Kobsa, A.: Expert-finding systems for organizations: problem and domain analysis and the DEMOIR approach. J. Organ. Comput. Electron. Commer. **13**(1), 1–24 (2003)
17. Zhang, J., Ackerman, M.S., Adamic, L.: Expertise networks in online communities: structure and algorithms. In: Proceedings of the 16th International Conference on WWW, pp. 221–230. ACM, New York (2007)

Static Weaving in Aspect Oriented Business Process Management

Amin Jalali[✉]

Stockholm University, Stockholm, Sweden
aj@dsv.su.se

Abstract. Separation of concerns is an important topic in business process modelling that aims to reduce complexity, increase the re-usability and enhance the maintainability of business process models. Some concerns cross over several business processes (known as cross-cutting concerns), and they hinder current modularization techniques to encapsulate them efficiently. Aspect Oriented Business Process Modelling aims to encapsulate these concerns from business process models. Although many researchers proposed different aspect-oriented business process modelling approaches, there is no analysis technique to check these models in terms of soundness. Thus, this paper proposes a formal definitions and semantics for aspect-oriented business process models, and it enables the analysis of these models in terms of soundness at design time through defining a static weaving algorithm. The algorithm is implemented as an artefact that support weaving aspect-oriented business process models. The artefact is used to analyse different scenarios, and the result of analysis reveals the situations that can introduce different problems like deadlock. In addition, an example of such scenario is given that shows how the artefact can detect the problems at design time. Such analysis enables process modellers to discover the problems at design time, so the problems will not be left to be discovered at runtime - which apply a lot of costs to correct them.

Keywords: Business process modelling · Aspect orientation · Weaving

1 Introduction

Separation of concerns has long been used as an effective method by people to deal with a complex phenomenon by reducing the dimensions of the complexity through increasing the level of abstraction. Using these techniques, a system is decomposed into different other modules that specify the functionality of the system in overall. Modularizing a system not only increases the level of abstraction but also supports re-usability of modules, so designing a system can become simpler with the wise choice of modularization techniques. Some concerns like security in information systems are not limited to one module, and they cross over different modules - which are known as cross-cutting concerns. It is challenging to encapsulate these concerns in a module since they are scattered through different modules, and those modules are tangled to these concerns.

P. Johannesson et al. (Eds.): ER 2015, LNCS 9381, pp. 548–557, 2015.
DOI: 10.1007/978-3-319-25264-3_41

Aspect Orientation is a paradigm in information systems that aims to encapsulate the cross-cutting concerns. This paradigm is well researched in the programming area, where languages like AspectJ are developed to support encapsulation of cross-cutting concerns in software codes [11]. It is also investigated in Requirement Engineering where Aspect Oriented Requirements Engineering (AORE) aims to support encapsulation of cross-cutting concerns [13]. In the Business Process Management (BPM) area, although it has been investigated how these models should be enacted at runtime (dynamic weaving) [9,10], there is still a gap with regard to static weaving (merging the aspect oriented business process models at design time) and its semantics. Thus, it is not possible to investigate if an aspect oriented business process model is sound. This gap applies a lot of cost in enacting such models, because it is not possible to identify problems in these models at design time. Therefore, the problems are left to be discovered at runtime, which can introduce a lot of difficulties to address them correctly.

Therefore, this paper defines the formal definition and semantics for Aspect Oriented Business Process Modelling (AO-BPM) using Petri-nets. The semantics is implemented as an artefact using "Business Process Technologies 4 Java" (jBPT) [12] to weave and analyse the aspect oriented business process models. The implemented artefact is used for analysing different scenarios.

It should be mentioned that we have defined the operational semantics of dynamic weaving for aspect oriented business processes models in [9,10]. These works specify how the Workflow Management Systems should enact these process models at runtime. In contrast, this work focus on weaving AO-BPM at design time. There are also different modeling approaches for AO-BPM which are compared and explained in [8].

The remainder of this paper is organized as follows. Section 2 introduces basic terminologies in aspect oriented business process management. Section 3 gives a brief definition of Petri nets and Workflow nets. Section 4 defines aspect oriented nets. It also defines an algorithm to weave these models to translate them to workflow nets. Section 5 explains the implemented artefact and experience of using it for analysing different scenarios. Finally, Sect. 6 concludes the paper and introduces future works.

2   Basic Terminologies

This section introduces basic terminologies in AO-BPM through a fictitious scenario. Figure 1 shows the scenario, where several cross-cutting concerns should be considered by different business processes.

The left side of the figure shows the relation between these concerns and business processes, where concerns like security, logging, auditing and traceability should be considered by "Transfer Money", "Deal for speculation", "Change asset deal", and "Issue a bank draft" processes. A concern should be considered by a process model in this example if it crosses over the process. For example, the "Transfer Money Process" should consider security and logging concerns.

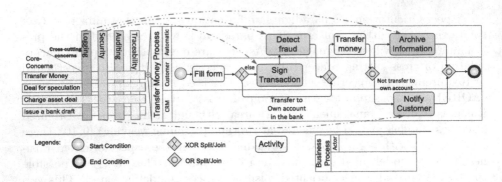

Fig. 1. Cross-cutting in business processes

The right side of the figure shows the "Transfer Money Process" which is modelled using standard BPMN. The process starts when the customer fills a form. As a security concern, the customer should sign the transaction if (s)he aims to transfer money to someone else account. In such a case, the system investigates potential fraud automatically. Thereafter, the money can be transferred. If the money is transferred to someone else account, the customer should be notified. The information about money transformation should be archived as a logging concern.

AO-BPM suggests encapsulating cross-cutting concerns into separated modules called *advices* (marked by a red and green pentagons in Fig. 2). `Save` and `confirm` are two advices that encapsulate logging and security cross-cutting concerns in our example. The encapsulation of similar cross-cutting concerns are called *aspects*, i.e. `Logging Aspect` and `Security Aspect` in the figure. After separating these concerns from process models, the models only contains the *core-concerns* (marked by a white pentagon in the figure), i.e. `Transfer Money Process`. Each activity in the core concern is called a *join point*, e.g. `Fill Information` activity. To relate core and cross-cutting concerns, AO-BPM introduces a set of rules, called *pointcuts*. Each pointcut relates a join point to a set of advices. The join point that is related to an advice is called an *advised join point*, e.g. `Transfer Money` activity.

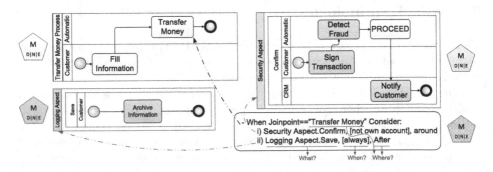

Fig. 2. Separation of cross-cutting concerns (Color figure online)

Pointcuts contain information about what advice should be considered when and where. For example, for considering the logging cross-cutting concern, the `Save` advice (what) should be considered `always` (when) `After Money Transfer` advised join point (where). Each advice can be considered before, after or around of an advised join point. Around advices has a special activity called *proceed* that defines the place of the advised join point around which the advice should be considered, e.g. `PROCEED` activity in the `Confirm` advice.

3   Preliminaries

This section gives a summary of definitions of Petri nets and workflow nets that are used to define aspect oriented nets later. The definitions are borrowed from [2,4].

Definition 1 (P/T-nets [2,4]). A Place/Transition net (P/T-net) is a tuple $N = (P^N, T^N, F^N)$,[1] where:

- P^N is a finite set of *places*;
- T^N is a finite set of *transitions*, such that $(P^N \cap T^N = \emptyset)$; and
- $F^N \subseteq (P^N \times T^N) \cup (T^N \times P^N)$ is a set of *directed arcs*, called the flow relation.

A place is represented by a circle, and a transition is represented by a rectangle. A flow is represented by a directed arrow connecting a place to a transition, or vice versa. The *preset* of an element $x \in P^N \cup T^N$ in the net N is defined as $\overset{N}{\bullet}x = \{\forall y \in (P^N \cup T^N) | (y, x) \in F^N\}$. The *postset* of an element $x \in P^N \cup T^N$ in the net N is defined as $x\overset{N}{\bullet} = \{\forall y \in (P^N \cup T^N) | (x, y) \in F^N\}$. P, T and F represent the universe of all places, transitions and flows, respectively.

The state of a system is determined by distribution of tokens in the places of the net. The distribution of tokens in the places of a net is called the *marking* of the net. The transitions of the net can change the markings of the net. A transition is *enabled* iff there is at least one token in every input places. The result of execution of the transition is a new marking, where a token is removed from every input place of the transition and a token is added to every output place of the transition. The initial state of the system is defined by an initial marking. A marking is called *reachable* iff there exists a sequence of enabled transitions whose firing leads from the initial marking to that marking. The sequence of all executed transition from a marking to another is called the *firing sequence*.

A net is *weakly connected*, or simply connected iff for every two nodes in the net, there are number of flows and nodes that connect one of the node to the other. The net is *strongly connected* iff for every two nodes in the net, there are number of flows and nodes that connect each of them to the other. A marked net is *bounded* iff the set of reachable markings is finite. It is *safe* iff, for any reachable marking, the number of tokens in every place is less or equal than one.

---

[1] The superscript N for a set specifies the net that contains the elements of the set.

A transition in the net is called *dead* iff there is no reachable markings such that the transition be enabled. The net is called *live* iff for every reachable marking, there is also at least another reachable marking from that marking.

Definition 2 (Workflow Nets [2,4]). Let $N = (P^N, T^N, F^N)$ be a P/T-net and $\bar{t}$ an element not in $P^N \cup T^N$. N is a workflow net (WF-net) *iff*: P^N contains an input place i such that $\overset{N}{\bullet}i = \emptyset$; P^N contains an output place o such that $o\overset{N}{\bullet} = \emptyset$; $\bar{N} = (P^N, T^N \cup \{\bar{t}\}, F^N \cup \{(o, \bar{t}), (\bar{t}, i)\}$ is strongly connected.

A workflow net is sound iff these four properties are hole, i.e. (i) Safeness: the net with the initial given marking is safe; (ii) Proper completion: for any marking that is reachable from the initial marking and has a token in the output place, there is no other token in other places; (iii) Option to complete: for any marking that is reachable from the initial marking, there is reachable marking that has a token in the output place (I call this marking as end marking); and (iv) Absence of dead tasks: the net contains no dead transitions.

4   Aspect Oriented Nets

I provide formal definitions and semantics for aspect oriented nets in this section. The formal definitions are derived from requirements that are defined in [5,7,10]. As explained earlier, an aspect oriented business process model consists of at least three models for (i) core-concern, (ii) cross-cutting concerns (advices) and (iii) pointcuts. These elements are defined as follows. Then, the static weaving is defined, so these models can be merged to a WF-net. Thus, these models can be analysed through existing analysis techniques for WF-nets.

Definition 3 (Core Net). A core net is a workflow net that encapsulates the core functionality of a business process. The set of all Core-nets are denoted by $\mathcal{C}$. Every tasks in a core net is called a join point.

Definition 4 (Advice). An advice $W = (P^W, T^W \cup \mathbb{P}(\{t^W_{proceed}\}), F^W)$, is a workflow net that encapsulates a cross-cutting functionality, where:

- $W \notin \mathcal{C}$. The advice net is not a part of core nets, and
- $t^W_{proceed}$ is a special task in the advice net which is a place-holder for a join point in a Core net. $\mathbb{P}$ denotes the power set, so an advice can have zero or one place-holder.

An advice W has zero or one $t^W_{proceed}$ place-holder. The advice with one place-holder is called an around advice. The set of all around advices are denoted by $\mathcal{A}_{\bullet}$. The advice with zero place-holder is called a free advice. The set of all free advices are denoted by $\mathcal{A}_{\circ}$. The set of all advices are denoted by $\mathcal{A} = \mathcal{A}_{\bullet} \cup \mathcal{A}_{\circ}$, where $\mathcal{A}_{\bullet} \cap \mathcal{A}_{\circ} = \emptyset$.

Definition 5 (Pointcut). A pointcut is a function that relates a join point to an advice. There are three sort of pointcuts, i.e. before, after and around - denoted by $\mathcal{P}_{\circ\square}$, $\mathcal{P}_{\square\circ}$ and $\mathcal{P}_{\substack{\bullet\\\square}}$ respectively. The set of all pointcuts are denoted by $\mathcal{P} = \mathcal{P}_{\circ\square} \cup \mathcal{P}_{\square\circ} \cup \mathcal{P}_{\substack{\bullet\\\square}}$.

- A before pointcut is a function $\mathcal{P}_{\circ\square} : T \rightarrow 2^{\mathcal{A}_{\circ}}_{\square}$.
- An after pointcut is a function $\mathcal{P}_{\square\circ} : T \rightarrow 2^{\mathcal{A}_{\circ}}_{\square}$.
- An around pointcut is a function $\mathcal{P}_{\substack{\bullet\\\square}} : T \rightarrow 2^{\mathcal{A}_{\bullet}}_{\square}$.

This means that before and after advices can relate a set of free advices to a join point, and around advices can relate a set of around advices to a join point. It is also possible that a pointcut relates an empty set of advices to a join point. The join points that are related to at least one advice are called advised join points.

Definition 6 (Aspect Oriented Nets). An Aspect Oriented net (AO-net) is a tuple $AO = (C, \mathcal{A}, \mathcal{P})$, where: $C \in \mathcal{C}$ is a core net.

Definition 7 (AO-net Soundness). An AO-net $AO = (C, \mathcal{A}, \mathcal{P})$ is sound iff $W = staticWeaving(AO)$ is sound. Algorithm 1 defines the $staticWeaving$.

This algorithm gets an AO-net and converts it to a Workflow net, so the nets can be analysed through techniques available for Workflow nets. It weaves related advices to the advised join points in the core net. To discover advised join points, the algorithm investigates if every task in the core net is an advised join point. If a task t is an advised join point, the algorithm should weave (i) "before advices" before the task; (ii) "around advices" before and after the task; and (iii) "after advices" after the task. Therefore, it adds a helper split and join tasks for weaving advices (line 3–13). Thereafter, it weaves before (line 14–16), after (line 17–19) and around advices (line 20–30) to the net. To weave around advices, the algorithm should also replace the proceed to the advised join point (line 22–28). Finally, the algorithm return the woven net. The function $addAdvice$ relates an advice to an advised join point, which is used in weaving before, after and around advices.

5   Experiments

I have implemented a prototype tool of the algorithm to investigate the soundness of different scenarios of aspect oriented business process models. I implemented the static weaving algorithm in Java using the jBPT library [12]. This library makes it easier to extend the implementation to support the weaving of other process model notations like EPC and BPMN. The algorithm gets a core-net and set of advices as Petri nets in the Petri Net Markup Language (PNML) format. It also gets the pointcuts in the XML format which is defined in [10]. The result is a woven net as Petri nets in the PNML format. This format is supported by many Petri nets editors. To analyse the result, I used WoPeD [6].

Algorithm 1. Algorithm for weaving an AO-net

---

1 **Algorithm** staticWeaving$(AO = (C = (T^C, P^C, F^C), \mathcal{A}, \mathcal{P}))$
2     $W \leftarrow C$;
3   **foreach** *task t in* T^W **do**
4   **if** $\mathcal{P}_{\bullet}^{\square}(t) \neq \{\}$ **then**
5           $T^W \leftarrow T^W \cup \{t_{t.split}^W\} \cup \{t_{t.join}^W\}$;
6   **else**
7   **if** $\mathcal{P}_{\circ\square}(t) \neq \{\}$ **then**
8             $T^W \leftarrow T^W \cup \{t_{t.split}^W\}$;
9   **if** $\mathcal{P}_{\square\circ}(t) \neq \{\}$ **then**
10             $T^W \leftarrow T^W \cup \{t_{t.join}^W\}$;
11   **foreach** *Advice* A' *in* $\mathcal{P}_{\circ\square}(t)$ **do**
12           $addAdvice(W, A', t_{t.split}^W, t)$;
13   **foreach** *Advice* A' *in* $\mathcal{P}_{\square\circ}(t)$ **do**
14           $addAdvice(W, A', t, t_{t.join}^W)$;
15   **foreach** *Advice* A' *in* $\mathcal{P}_{\bullet}^{\square}(t)$ **do**
16           $addAdvice(W, A', t_{t.split}^W, t_{t.join}^W)$;
17   **foreach** *Place* p *in* $\bullet t_{proceed}^A$ **do**
18             $F^W \leftarrow F^W \cup \{(t, p)\}/\{(t_{proceed}^A, p)\}$;
19   **foreach** *Place* p *in* $t_{proceed}^A \bullet$ **do**
20             $F^W \leftarrow F^W \cup \{(p, t)\}/\{(p, t_{proceed}^A)\}$;
21           $T^W \leftarrow T^W / t_{proceed}^A$;
22   **return** W;

    Function addAdvice$(W \in \mathcal{C}, B \in \mathcal{A}, initiator \in T^W, terminator \in T^W)$
1     $A \leftarrow B$;
2     $T^W \leftarrow T^W \cup T^A$;
3     $P^W \leftarrow P^W \cup P^A$;
4     $F^W \leftarrow F^W \cup F^A \cup \{(initiator, i^A)\} \cup \{(o^A, terminator)\}$;
5   **return** W;

WoPeD supports analysis of Petri nets using Woflan - an advanced Petri-net-based workflow analyser [3].

    The result of analysing different scenarios shows that the woven result is always sound only if (i) there is no advised join point in the model for which more than one advice is defined, or (ii) no around advice is defined for an advised join point for which more than one advice is defined. Otherwise, the soundness must be checked by the tool.

    Figure 3 shows three nets for a scenario that can have deadlock after weaving. It contains a core process containing three tasks, i.e. A, B and C. A pointcut relates two advices to task B in the main process model. One of the advices is "BEFORE" advice (named before), and the other one is "AROUND" (named around_loop). As it can be seen in the around advice, it is possible to fire the

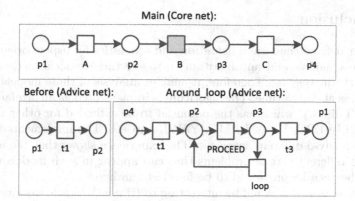

Fig. 3. Aspect oriented nets defined for our sample scenario

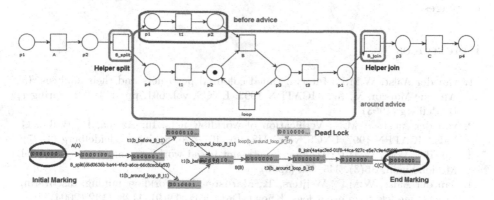

Fig. 4. Weaving the sample scenario

PROCEED by firing loop transition. Figure 4 shows the result of weaving of the nets for this scenario.

The graph bellow the figure shows the reachability graph, which is generated by WoPeD. The reachability graph shows all possible reachable states in the process model. The analysis based on the reachability graph is a very powerful method because it can be used to prove or disprove all kinds of properties [1].

The reachability graph shows that there is a possible deadlock in this process model. The deadlock can occur if the loop transition is fired. The problem is that the advised join point (task B) requires a token in both input places. However, there will be no token in the output place of the "before advice" net because it is consumed by the advised join point previously. This introduces a deadlock in the woven net.

6 Conclusion

This paper defined the formal definition and semantics for aspect oriented business process models and an algorithm to weave these models at design time, known as static weaving. Therefore, it enables analysis of these models in terms of soundness at design time. The algorithm is implemented as an artefact in Java using jBPT library, which has the potential to be extended for other modelling languages like EPC, UML, BPMN, YAWL and etc. The implemented artefact is used to analyse different scenarios. The experience shows that the algorithm can detect different sort of problems that can appear in such models at design time, so they can be prevented to be faced at runtime.

For future works, it would be interesting to (i) use the static weaving through real case studies; (ii) combine static and dynamic weaving to investigate hybrid weaving; and (iii) extend the artefact to support other business process modelling languages.

References

1. van der Aalst, W.M.P.: Interval timed coloured petri nets and their analysis. In: Ajmone Marsan, M. (ed.) ICATPN 1993. LNCS, vol. 691, pp. 453–472. Springer, Heidelberg (1993)
2. van der Aalst, W.M.P.: Verification of workflow nets. In: Azéma, P., Balbo, G. (eds.) ICATPN 1997. LNCS, vol. 1248, pp. 407–426. Springer, Heidelberg (1997)
3. van der Aalst, W.M.P.: Woflan: a petri-net-based workflow analyzer. Syst. Anal. Model. Simul. **35**(3), 345–358 (1999)
4. van der Aalst, W.M.P., Weijters, T., Maruster, L.: Workflow mining: discovering process models from event logs. Knowl. Data Eng. **16**(9), 1128–1142 (2004)
5. Charfi, A., Müller, H., Mezini, M.: Aspect-oriented business process modeling with AO4BPMN. In: Kühne, T., Selic, B., Gervais, M.-P., Terrier, F. (eds.) ECMFA 2010. LNCS, vol. 6138, pp. 48–61. Springer, Heidelberg (2010)
6. Eckleder, A., Freytag, T.: Woped a tool for teaching, analyzing and visualizing workflow nets. Petri Net Newslett. **75**, 3–8 (2008)
7. Cappelli, C., et al.: Reflections on the modularity of business process models: the case for introducing the aspect-oriented paradigm. BPM J. **16**, 662–687 (2010)
8. Jalali, A.: Assessing aspect oriented approaches in business process management. In: Johansson, B., Andersson, B., Holmberg, N. (eds.) BIR 2014. LNBIP, vol. 194, pp. 231–245. Springer, Heidelberg (2014)
9. Jalali, A., Wohed, P., Ouyang, C.: Operational semantics of aspects in business process management. In: Herrero, P., Panetto, H., Meersman, R., Dillon, T. (eds.) OTM-WS 2012. LNCS, vol. 7567, pp. 649–653. Springer, Heidelberg (2012)
10. Jalali, A., Wohed, P., Ouyang, C., Johannesson, P.: Dynamic weaving in aspect oriented business process management. In: Meersman, R., Panetto, H., Dillon, T., Eder, J., Bellahsene, Z., Ritter, N., De Leenheer, P., Dou, D. (eds.) ODBASE 2013. LNCS, vol. 8185, pp. 2–20. Springer, Heidelberg (2013)
11. Kiczales, G., Hilsdale, E., Hugunin, J., Kersten, M., Palm, J., Griswold, W.G.: An overview of AspectJ. In: Lindskov Knudsen, J. (ed.) ECOOP 2001. LNCS, vol. 2072, pp. 327–354. Springer, Heidelberg (2001)

12. Polyvyanyy, A., Weidlich, M.: Towards a compendium of process technologies: The jbpt library for process model analysis. In: CAiSE 2013 Forum, pp. 106–113 (2013)
13. Rashid, A., Sawyer, P., Moreira, A., Araújo, J.: Early aspects: a model for aspect-oriented requirements engineering. In: Requirements Engineering, pp. 199–202. IEEE (2002)

Tangible Modelling to Elicit Domain Knowledge: An Experiment and Focus Group

Dan Ionita[✉], Roel Wieringa, Jan-Willem Bullee, and Alexandr Vasenev

University of Twente, Services, Cybersecurity and Safety Group,
Drienerlolaan 5, 7522 NB Enschede, The Netherlands
{d.ionita,r.j.wieringa,j.h.bullee,a.vasenev}@utwente.nl
http://scs.ewi.utwente.nl/

Abstract. Conceptual models represent social and technical aspects of the world relevant to a variety of technical and non-technical stakeholders. To build these models, knowledge might have to be collected from domain experts who are rarely modelling experts and don't usually have the time or desire to learn a modelling language. We investigate an approach to overcome this challenge by using physical tokens to represent the conceptual model. We call the resulting models **tangible** models. We illustrate this idea by creating a tangible representation of a socio-technical modelling language and provide initial evidence of the relative usability and utility of tangible versus abstract modelling. We discuss psychological and social theories that could explain these observations and discuss generalizability and scalability of the approach.

Keywords: Participatory modelling · Tangible modelling · Socio-technical · Enterprise models · Usability experiment

1 Introduction

A conceptual model consists of concepts and relations among the concepts, by which people understand a part of the world. They are usually represented in (software) modelling tools using abstract graph-like structures containing boxes, arrows, and other symbols. The problem with these abstract representations is that domain experts whose input or feedback is needed to construct an adequate model may be unfamiliar with the notation, and may not be willing or able to learn it.

To mitigate this problem, we propose communicating with domain experts using a tangible representation of a conceptual model, that we call a **tangible model.** This facilitates participation of and collaboration among domain experts and modelling experts, and the evidence presented in this paper suggests that the resulting models are more accurate than abstract representations of conceptual models, and cost less effort to build.

We start by discussing related work (Sect. 2) and use established theories to formulate a hypothesis (Sect. 3). We illustrate the approach for a given language (Sect. 4) and use it validate our hypothesis empirically by means of a small-scale usability experiment (Sect. 5) and a focus group (Sect. 6).

© Springer International Publishing Switzerland 2015
P. Johannesson et al. (Eds.): ER 2015, LNCS 9381, pp. 558–565, 2015.
DOI: 10.1007/978-3-319-25264-3_42

2   Related Work

Barijs [1] claims that in enterprise modelling, *participation* of key stakeholders is crucial for creating shared understanding of how an enterprise operates in practice, and that *collaboration* of modellers, analysts and domains experts is needed to construct accurate models in a timely manner. Modelling tools should be *interactive* to allow easy manipulation of the constructed models and see the effects of changes. We will see that tangible modelling satisfies these three criteria.

Grosskopf [6] provides empirical evidence that by tangible process modelling, practitioners achieved a better understanding, higher consensus and a higher rate of adoption of the results. Fleischmann et al. [4] claim that tangible modelling of organisational processes stimulate stakeholder engagement and ensure coherence of representations on the organisational level, thus paving the way towards acceptance of the results. Garde [5] reports that domain experts with no design skills or process modelling skills successfully designed and specified a complex new procedure within a few hours by means of a participatory and tangible board game, while reporting high levels of commitment and even enjoyment.

In the related field of Tangible User Interfaces (TUIs) digital concepts are given physical form to leverage people's lifelong experience with the physical world [3]. However, a TUI is aimed at providing a *human-computer interface* [2], whereas our research is focused on extracting domain knowledge.

3   Theoretical Background

We hypothesize that a tangible collaborative modelling approach can speed up the modelling process and improve the quality of the resulting models when these models need to integrate knowledge from various fields and various stakeholders. We decompose our hypothesis as follows (Fig. 1):

H1 Physical representations of a conceptual model are easier to understand and manipulate than abstract representations;
H2 The participatory aspect encourages engagement, collaboration and agreement between stakeholders;
H3 Physical and participatory modelling, similar to board games, increases engagement while reducing repetitiveness.

The **theory of cognitive load** says that human performance is increased when a task matches human cognitive architecture, leading to less effort (i.e. cognitive load) to perform the task [13]. Reduction of effort can be achieved by exploiting dimensional stimuli such as color, sound, material and space to support memory [11]. This motivates hypothesis H1.

The **theory of cognitive fit** says that cognitive load is reduced when a problem representation matches the problem [15]. For example, using physical tokens to represent physical elements of a problem, as we do in tangible modelling, should improve cognitive fit. This motivates H1 too.

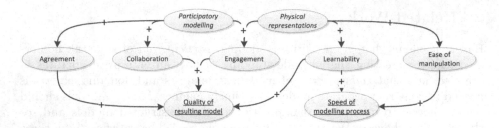

Fig. 1. Causal graph describing our hypotheses. The nodes in *italics* are the variables we hope to influence. The underlined nodes are the target variables

Gamification is the "the application of gaming metaphors to real life tasks to influence behaviour, improve motivation and enhance engagement" [10]. By building the models around a table instead of in front of a screen, and selecting colourful game-like tokens such as Lego characters and cards we hope to enhance the board-game metaphor. As confirmed by the findings of Garde [5], participants have more fun doing this than they have when manipulating abstract representations, thus creating more participation and engagement (H2, H3).

Constructionism says that learning about a product improves when people are engaged in constructing it [8]. Thus, learnability of both the modelling concepts and the model itself can be enhanced by engaging a group of people in the construction of the model (H1, H2).

4   Case Study: The TREsPASS Language

To illustrate the idea of tangible modelling, we create a tangible representation of a socio-technical modelling language being developed within the TREsPASS project[1] to model the socio-technical infrastructure of an organization. This model is then used in an information security risk assessment, but this is out of scope of this paper.

The TREsPASS modelling language contains concepts such as: actors, (physical and virtual) locations, (physical and virtual) assets and access policies. The language allows four types of relationships to be defined between these concepts: position, possession, connection and containment [14]. Table 1 shows a mapping of these concepts to tangible tokens. This is not the only possible mapping, and we us it here for illustration only.

5   Usability Experiment

We set up an experiment in order to obtain an initial indication as to whether tangible modelling is more usable than abstract modelling and why. We describe the object of study, the treatment, and the measurement procedures, and then analyse the results [17].

---

[1] http://www.trespass-project.eu/.

Table 1. Mapping of concepts to representations

Concept	Software representation	Tangible representation
Actor	Stickman	LEGO®character
Asset (physical)	Solid circle	LEGO®item[a]
Asset (digital)	Dotted circle	LEGO®mini-brick[b]
Location (digital)	Box (green)	Card
Location (physical)	Box (yellow)	Box
Access policy	Text-box	Sticky-note
Relationship (position)	Solid line	Physical overlap
Relationship (possession)	Dotted line	Physical attachment
Relationship (containment)	Directional arrow	Physical overlap
Relationship (connection)	Bi-directional arrow	Line

[a]A LEGO®item resembles a real-world object and can be placed in a LEGO® characters' hand
[b]A LEGO®mini-brick is the smallest LEGO®brick available, usually of circular or cylindrical shape and can be placed on a LEGO®characters' head

5.1 Object of Study

We are interested in comparing the relative usability of tangible and abstract representations of conceptual models for domain experts. For the experiment, we invited Business Administration and Computer Science students at the University of Twente. We divided a total of eight volunteers into two independent groups, each containing a mix of students from various tracks.

The sample and treatment allocation were not random, and statistical inference is not possible. Moreover, students are not necessarily representative of experienced domain experts. However, there is still some basis for generalizing from our experiment to domain experts: The cognitive theories identified in Sect. 3 are equally applicable to both experts and students. If the experimental outcome can be explained by these general theories, then this provides some support for the claim that similar outcomes will occur for domain experts. Such a claim would of course have to be substantiated by further research.

While we tried to balance the groups as much as possible, there is still the possibility that some of the participants had more modelling experience, layouting expertise or were simply more skilled. This is a threat to internal validity, as it is a possible cause of differences in group outcomes, unrelated to the difference in modelling approaches. To measure this threat to validity, any variations in group behaviour and dynamics were noted throughout the experiment.

The modelling target should ideally be a socio-technical system the participants are familiar with, such as their own organization. Thus, we asked the student volunteers to create a model representing the physical layout, network infrastructure (both servers and clients), important roles and associated access policies of their own Student Associations. Students from Computer Science might be less familiar wth the student association of Business Administration, and vice-versa. This simulates the disjunct knowledge individual domain experts might have with regard to the model and therefore improves external validity of the choice of modelling target. To limit variation due to lack of knowledge, as well as the effect of pre-existing knowledge, each group was provided with a half-page description listing the core components of each association.

Since we are measuring how well the tools fit the task, not how familiar each participant is with the modelling target, the participants were allowed to ask questions with regard to the modelling target at any time.

5.2   Treatment Design

Each group was shown a brief description of the system and an outline of the task they have to perform. These were identical for both groups. The groups are given as much time as they need to understand this description.

Each group was then given a specification of the modelling concepts of the TREsPASS language. One group received the mapping of concepts to conceptual representations available in the software tool (the first two columns of Table 1), the other a mapping to tangible tokens (the first and last column of Table 1).

Each group was allowed to ask questions pertaining to the concepts before starting. When ready, each group was given an unlimited amount of time to create a model of the given system – to the best of their abilities – using the only the concepts provided. Once the modelling started, the moderator only intervened when the participants had questions about the modelling target or when the group agreed that they were done. At the end of the experiment, each student was rewarded with a 50 Euro gift-card, with bonus movie tickets awarded by means of a raffle.

Experimenter expectancy is the phenomenon that subjects try to satisfy what they think are the expectations of the experimenter, which could lead to favorable results for tangible modeling in our case. To avoid this phenomenon, we told both groups that their goal is to finish as fast as possible while maintaining consistency with the system description.

A threat to validity we were unable to mitigate was the quality of the software tool. The ability of the TREsPASS tool to manipulate diagrams, as well as the developer's choice on how to represent the concepts directly impacts its usability and thus may have biased the results in favour of the tangible approach. To find out if this threat has materialized, we need to repeat the experiment with another tool.

5.3   Measurement Design

We are interested in the usability of tangible modelling versus computer-based modelling. We operationalize usability in terms the four indicators [12]: *Learnability* is operationalized by the time needed to understand the modelling language, and the number of question asked with respect to it. *Efficiency* is operationalized by the inverse of the totel time needed to construct model. *Correctness* is the number of errors at the end of the modelling process. We distinguish three types of errors: (1) Placing an element where none was expected, (2) Missing element where one was expected and (3) Using a wrong concept to represent an element. *Satisfaction* is measured via an exit-questionnaire, containing questions on ease of task, time on task, tool satisfaction and group agreement, to be answered on a 5-point semantic differential scale with labelled end-points [9].

In order to make sure that we are indeed measuring the effects of the method, and not something else, we need to control variation due to other causes. Such causes might be internal (for example, due to improper description of the concepts), or contextual [16]. We controlled for internal causes by providing the same definitions to all groups, irrespective of the method used. Contextual causes include variation across the subjects applying the method, the environment during application or other aspects related to the context in which the method was applied. We tried to minimise these by using the same room, layout and instructions for both groups. Variations might still appear due to different skill levels within each group, which are harder to control for.

5.4   Summary of Results

We observed several improvements when using the tangible method. While the time needed to learn the concepts was similar, the number of questions when applying them was almost double for the abstract-representation group. The tangible group finished 52 % faster and with half as many errors. The tangible group also indicated 54 % higher satisfaction with the tool provided and 12.5 % more agreement with the resulting model while perceiving the task, on average, as being 25 % easier and 24 % faster. Furthermore, the tangible group did not divide tasks, suggesting increased collaboration. Detailed measurements and observations are available in an internal report [7].

6   A Focus Group to Assess Utility

In order to gain insight with regard to the utility of the method in practice, we conducted a demo session at BiZZdesign[2], a company providing consultancy on the enterprise architecture modelling. Eight consultants participated in a 2-h demo and workshop aimed to generating feedback and discussions on the topic of tangible modelling for enterprise architectures. After a presentation describing the overall approach, introducing the TREsPASS language and its tangible mapping, they were asked to collaboratively create a tangible TREsPASS model and later to envision the possible usefulness of a similar approach to ArchiMate.

They indicated several application scenarios where tangible modelling of enterprise architectures might prove useful:

---

[2] www.bizzdesign.nl.

- Architecture modelling sessions with domain experts (not modelling or architecture experts). Tangible modelling will allow less technical people tend to have a stronger impact on the model, as they can now manipulate the concepts themselves, and do not rely on a "modeller" to parse their input.
- Early stages of design where different types of stakeholders have to come up with an architecture; the participative aspect increases collaboration and encourages imagination.
- Models built with the goal of increasing awareness and feeling of involvement of employees with regard to the internal structure of the company. Non-technical people can more easily understand the tangible model, which could be displayed somewhere in the company. Potentially, employees could be allowed to tweak it, thus taking enterprise architecture out of the architecture department.

7   Conclusions and Future Work

In our experiment, both measured and self-reported usability was higher when using tangible tokens versus software. This provides initial evidence that such an approach can be more intuitive and understandable by domain experts and encourages collaboration and engagement while fostering discussion and ultimately agreement. While we were unable to control for all other contextual causes (such as participant skill or limitations of the software tool), both participants in the experiment and practitioners recognized the value of a tangible modelling approach over a tool-supported approach with abstract notations when the modellers are not familiar with the modelling language. Because these positive effects can be explained by general theories of human cognition, and are similar to the results reported by other researchers [4–6], we expect similar benefits in similar situations.

However, there are limits to generalizability. We have evaluated our approach only on a small scale. Due to practical reasons such as the limited availability of physical tokens or the space to place them on, we do not expect our approach to be scalable to large systems or organizations. We further restrict the scope of our generalization to situations where formal analyses and a strict syntax adherence are secondary. Finally, our focus group indicates that a tangible modelling approach may be especially useful at the start of enterprise architecture processes, when awareness and commitment of domain experts is required. We intend to explore this further by replicating this experiment with business analysts and an enterprise modelling language.

Acknowledgments. The research leading to these results has received funding from the European Union Seventh Framework Programme (FP7/2007–2013) under grant agreement ICT-318003 (TREsPASS) as well as from the Joint Program Initiative (JPI) Urban Europe via the IRENE project.

References

1. Barjis, J.: Collaborative, participative and interactive enterprise modeling. In: Filipe, J., Cordeiro, J. (eds.) Enterprise Information Systems. LNBIP, vol. 24, pp. 651–662. Springer, Heidelberg (2009)
2. Fishkin, K.P.: A taxonomy for and analysis of tangible interfaces. Pers. Ubiquitous Comput. **8**(5), 347–358 (2004)
3. Fitzmaurice, G.W., Ishii, H., Buxton, W.A.S.: Bricks: laying the foundations for graspable user interfaces. In: Proceedings of the SIGCHI Conference on Human Factors in Computing Systems. CHI 1995, pp. 442–449. ACM Press/Addison-Wesley Publishing Co, New York (1995)
4. Fleischmann, A., Schmidt, W., Stary, C.: Tangible or not tangible – a comparative study of interaction types for process modeling support. In: Kurosu, M. (ed.) HCI 2014, Part II. LNCS, vol. 8511, pp. 544–555. Springer, Heidelberg (2014)
5. Garde, J.A., van der Voort, M.C.: The procedure usability game: a participatory game for development of complex medical procedures and products. In: Proceedings of the CIRP IPS2 Conference 2009 (2009)
6. Grosskopf, A., Edelman, J., Weske, M.: Tangible business process modeling – methodology and experiment design. In: Rinderle-Ma, S., Sadiq, S., Leymann, F. (eds.) BPM 2009. LNBIP, vol. 43, pp. 489–500. Springer, Heidelberg (2010)
7. Ionita, D., Wieringa, R.J., Bullee, J.H., Vasenev, A.: Investigating the usability and utility of tangible modelling of socio-technical architectures. Technical report TR-CTIT-15-03, University of Twente
8. Kafai, Y., Harel, I.: Learning through design and teaching: exploring social and collaborative aspects of constructionism. In: Harel, I., Papert, S. (eds.) Constructionism, pp. 85–110. Ablex, Norwood (1991)
9. Lewis, J.R.: Psychometric evaluation of an after-scenario questionnaire for computer usability studies: the ASQ. SIGCHI Bull. **23**(1), 78–81 (1991)
10. Marczewski, A.: Gamification: A Simple Introduction. Andrzej Marczewski (2013)
11. Miller, G.: The magical number seven, plus or minus two: some limits on our capacity for processing information. Psychol. Rev. **63**, 81–97 (1956)
12. Nielsen, J.: Usability 101: Introduction to usability. Jakob Nielsen Alertbox (2003)
13. Sweller, J.: Cognitive load during problem solving: effects on learning. Cogn. Sci. **12**(2), 257–285 (1988)
14. The TREsPASS Project: D1.3.1. Initial prototype of the socio-technical security model, Deliverable D1.3.1 (2013)
15. Vessey, I., Galletta, D.: Cognitive fit: an empirical study of information acquisition. Inf. Syst. Res. **2**(1), 63–84 (1991)
16. Vriezekolk, E., Etalle, S., Wieringa, R.: Experimental validation of a risk assessment method. In: 21st International Working Conference on Requirements Engineering: Foundations for Software Quality. REFSQ 2015. Springer (2015)
17. Wieringa, R.J.: Design Science Methodology for Information Systems and Software Engineering. Springer (2014)

The REA Accounting Model: Enhancing Understandability and Applicability

Walter S.A. Schwaiger[✉]

Institute of Management Science, Vienna University of Technology, Vienna, Austria
schwaiger@imw.tuwien.ac.at

Abstract. The REA accounting model developed by McCarthy conceptualizes the economic logic of the double-entry bookkeeping without referring to debits, credits and accounts. The conceptual core elements of the model are the economic resources, economic events and economic agents as well as the relationships that link the underlying stock flows according to the duality principle. In this paper the debit and credit notations are included as a meta concept to promote the model's understanding within the traditional accounting logic. By specifying additional economic resource types in form of liabilities and equity the model is completed with respect to the essential balance sheet positions, so that the REA accounting model is ready for accounting applications.

Keywords: REA accounting model · Debit and credit · Liability and equity resources · ALE accounting model · Financial derivatives

1 Introduction

For developing the REA accounting model, which is based on the constituting elements in form of resources (R), events (E) and agents (A), McCarthy started from the following considerations [11, pp. 559–560]: "It is a primary contention of this paper that the semantic modeling of accounting object systems should not include elements of double-entry bookkeeping such as debits, credits, and accounts. As noted previously …, these elements are artifacts associated with journals and ledgers (that is, they are simply mechanisms for manually storing and transmitting data). As such, they are not essential aspects of an accounting system. It is possible to capture the essence of what accountants do and what things they account for by modeling economic phenomena directly in the conceptual schema. Any double-entry manipulations desired by particular users can then be effected only in the external schemata presented to those users".

The REA accounting model was extended by Geerts/McCarthy [4, 5] to the REA business ontology. In this ontology the REA accounting model provides the accounting infrastructure, which is extended by the policy infrastructure that contains the future related elements needed for the formulation of business policies. Due to its origin in the information systems research it is obvious that the REA business ontology (including the REA accounting model) is used in the enterprise information system (EIS) literature (see e.g. Dunn/Cherrington/Hollander [3]) and the accounting information system (AIS)

© Springer International Publishing Switzerland 2015
P. Johannesson et al. (Eds.): ER 2015, LNCS 9381, pp. 566–573, 2015.
DOI: 10.1007/978-3-319-25264-3_43

literature (see e.g. Steinbart/Romney [12]). But amazingly it does not appear in the accounting literature (see e.g. Harrison/Horngren/Thomas/Suwardy [6] and Horngren/ Harrison/Oliver [7]).

This non-existence of the REA business ontology in the accounting literature leads to the two primary research questions of this paper:

- Why is the REA business ontology despite its conceptual merits not present in the traditional accounting literature?
- What has to be done in order to promote the understanding of the REA business ontology within the accounting community and to assure its applicability in the accounting domain?

To answer these questions the working and the requirements of the traditional accounting logic is analyzed. Then the identified requirements are compared with the conceptual elements of the REA business ontology to detect the shortcomings. Finally the REA business ontology will be modified to the "REA-based ALE accounting ontology", which conceptualizes the traditional accounting logic.

The paper is organized as follows. In the subsequent section the extension from the REA accounting model to the REA business ontology is shown. Next to that the traditional accounting logic is investigated. Due to its starting point in terms of the accounting equation, where the equality of assets (A) with liabilities (L) and equity (E) is postulated, the derived logic is called the "ALE accounting model". In the following section the REA-based ALE accounting ontology is derived by suitably adjusting the REA business ontology. In the final section the paper is concluded.

2   REA Business Ontology: Inclusion of Current and Future Events

In Fig. 1 the REA accounting ontology is modeled in form of a class diagram, which is used in the ISO/IEC 15944-4:2006 standard related to the Accounting and Economic Ontology (AEO) to model business transactions [10, p. 33] and by Abmayer/Schwaiger [1] to model REA-related ontologies. The accounting infrastructure of the REA business ontology consists of the REA accounting model. The focus of this infrastructure lies on the resource flows that occur in economic transactions between the involved agents. The duality relationship expresses the economic principle that scarce resources have a positive price that has to be paid in an exchange transaction. The linkage to the resources is termed as "resource flow", which in the REA business ontology can be an increment or a decrement event. Hruby [8] provides different examples of business patterns, where the increment and decrement event structure is applied.

The policy infrastructure allows the modeling of future related business policies and it is set on top of the accounting infrastructure. For simplicity reasons only the elements of the policy infrastructure, which are important for financial instruments accounting, are shown in Fig. 1. Economic contracts are defined as economic bundles of economic commitments, which fulfill the reciprocity principle. The reciprocity principle is the conceptual analogue to the duality principle and it relates to future events in the form

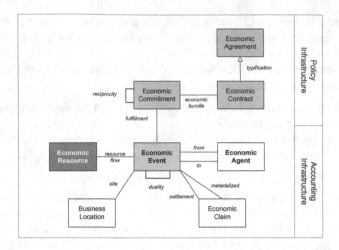

Fig. 1. REA business ontology (accounting relevant elements)

of economic commitments. The economic contracts are themselves specializations of economic agreements.

In this article future related elements of the policy infrastructure are used for extending the traditional accounting perspective. Traditionally current events are recorded over time and financial statements are generated thereof. With the additional elements the future commitments of financial assets, debt and equity instruments can be included as well.

3   Traditional Accounting Logic: ALE Accounting Model

The traditional accounting logic starts with the accounting equation, which specifies the resources of the enterprise as assets and the claims to those resources as liabilities and equity. The assets are owned and the liabilities are owed by the enterprise. The equity is the owner's claim to the net value of the enterprise, which is defined as the difference of the assets and the liabilities.

"The account category (asset, liability, equity) governs how we record increases and decreases. For any given account, increases are recorded on one side, and decreases are recorded on the opposite side. The following T-accounts provide a summary:

Assets	Liabilities and Owner's Equity
Increase = Debit | Decrease = Credit	Decrease = Debit | Increase = Credit

These are the *rules of debit and credit*. Whether an account is increased or decreased by a debit or a credit depends on the type of account. Debits are not "good" or "bad". Neither are credits. Debits are not always increases or always decreases - neither are credits" [7, p. 92].

In Fig. 2 the ALE categorization of resources is used to define the matrix, which contains all nine possible combinations of ALE resource changes. The nine combinations constitute the nine elementary accounting transaction types that have to be recorded in an enterprise. Their meanings are as follows:

1. accounting exchange on the assets side – e.g. acquisition transaction
2. balance sheet extension – e.g. debt financing transaction
3. accounting exchange on the liabilities side – e.g. refinancing transaction
4. balance sheet contraction – e.g. loan redemption transaction
5. expenses (balance sheet contraction) – e.g. depreciation expenses
6. expenses (balance sheet "neutral") – e.g. provisioning
7. revenues (balance sheet extension) – e.g. revaluation profits
8. revenues (balance sheet "neutral") – e.g. unused provision release
9. accounting exchange on the equity side – e.g. creation of equity reserves

Transaction Types		Credit		
		A-	L+	E+
Debit	A+	1	2	7
	L-	4	3	8
	E-	5	6	9

Fig. 2. Categorization of accounting transactions – 9 transaction types

In each accounting transaction two ALE resources are involved. The increment and decrement event structure of the REA business ontology works well for physical resources and cash. But the intuitiveness gets lost once resources related to liabilities and equity are considered. Let's take e.g. a debt financing transaction. The financial instruments involved in debt financing are distinguished from physical resources and cash by having contractually defined future commitments. In the debt financing transaction the loan stock is increased. At the same time the cash inflow increases the cash stock as well. As there are two increment events in this transaction the increment/decrement duality of the REA business ontology is violated. Consequently the increment/decrement notation is not useful in ALE accounting and it has to be replaced by the debit/credit notation. In this notation the debt financing transaction is recorded by debiting the cash resource for the cash inflow and crediting the debt resource for the future obligation.

In the traditional ALE accounting the future payment structures behind the obligations of financial assets, debt and equity instruments are not directly modeled. For this purpose the future related elements of the REA business ontology are suitable. The debt financing transaction constitutes an economic contract, which includes the future payments specified in the contract as commitments. As these future cash payments are outgoing payments from the perspective of the enterprise the commitments are credited.

The inclusion of the economic contracts and commitments is useful for modeling financial derivatives (see Hull [9]) as well. These instruments normally cause problems in ALE accounting, because they are difficult to handle within the traditional accounting logic. The usage of the future related elements of the REA business ontology solves these difficulties by allowing the modeling of commitments on the asset side as well as on the liability side. For understanding the valueless property of some derivative instruments the present value restriction from the no-arbitrage theory (see e.g. Black/Scholes [2]) is important. Applied to swap contracts this principle says that if the present value of the debit commitments is equal to the present value of the credit commitments, then the value of the swap is zero. If this would not be the case, then arbitrage profits can be earned by engaging in the swap contract and performing the adequate swap duplication strategy.

4   REA-Based ALE Accounting Ontology: Integrating Finance into Accounting

In the previous section the requirements of the ALE accounting were identified in the context of recording ALE resource changes in the nine accounting transaction types according to the debit/credit notation. Furthermore the future related elements of the REA business ontology showed beneficial especially for explicitly modeling the payment structure of derivative and non-derivative financial instruments.

The REA-based ALE accounting ontology is a modification of the REA business ontology in order to include the requirements from ALE accounting. This ontology is presented in Fig. 3. The distinguishing features of the ALE accounting ontology in the accounting infrastructure are as follows:

- It includes the accounting transaction object, which is missing in the REA business ontology, as a composition of debited and credited ALE resource changes. The business transaction object is the conceptual starting point of accounting professionals and academics. Its inclusion anchors the ontology in the accounting domain.
- It uses the value restriction requirement related to debit events and credit events, which is also missing in the REA business ontology. This is the central attribute of the accounting transaction object.
- It uses the value flow relationship related to debit and credit events instead of the stock flow relationship related to increment and decrement events in the REA business ontology. The reason for this modification lies in the fact that not only resource flows have to be accounted for in the REA-based ALE accounting ontology. In accrual accounting the periodic income also includes profits and losses that result e.g. from changing resource prices. Such value changes occur without resource flows. On the other hand all resource flows are related to value flows, so that in the value flow relationship all accounting transactions can be recorded.
- It covers all resources related to the assets, liabilities and equity instead of the primary focus on physical assets and cash in the REA business ontology. The inclusion of all ALE resources is needed to cover all nine elementary accounting transaction types defined in Fig. 2.

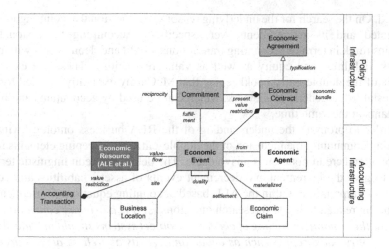

Fig. 3. REA-based ALE accounting ontology

The distinguishing features of the REA-based ALE accounting ontology in the policy infrastructure are as follows:

- It includes the economic contract object as a composition of commitments and economic events. The link to the economic events is important for modeling e.g. debt financing instruments, where cash is paid to the loan taker at the contract instantiation date against the commitment to pay it back in the future. If only economic events are included in the business transaction, then this is a spot market contract without future commitments. The swap contract is an example of an economic contract that contains only commitments, which are credited and debited.
- It allows an appropriate modeling of economic claims in terms of economic contracts. In the REA business ontology claims are seen as resulting from imbalances in related event sets. "However, in the outline of the generalized framework, resources were materialized as base objects while claims were not. In actual practice, this disparity in treatment may not always be warranted, especially when the processing require-ments and decision usefulness of some claims are projected" [11, p. 571]. Economic contract models developed in the finance domain are nowadays the adequate repre-sentation of temporal imbalances in financial instruments. In this case the economic claim disappears and appears as an economic contract.

5 Conclusions

At the center of this article is the REA business ontology, which was developed by McCarthy [11] and Geerts/McCarthy [4, 5]. This seminal contribution inspires the enterprise and accounting information systems research, where the ontology is used to design enterprise and accounting databases. Amazingly, in the traditional accounting literature, which is the actual home base of the ontology, the ontology is mainly neglected. On the search for the underlying reasons the traditional accounting logic was investigated and its requirements were specified. Accounting professionals and academics think in term of accounting transactions, credit and debit events with respect to assets, liabilities and equity as well as value restrictions. These are exactly the elements of the double-entry bookkeeping that McCarthy explicitly avoided for being able to establish a generic framework, which can be used by accountants and by non-accountants at the same time.

In order to promote the understanding of the REA business ontology within the accounting community the inclusion of the double-entry bookkeeping elements in form of the debit and credit notation is unavoidable. The debit and credit linguistic terms are needed to give the increment and decrement events of assets, liabilities and equity a consistent interpretation within the ALE-based accounting equation. Consequently the main finding related to the two research questions is: *It is a primary contention of this paper that the semantic modeling of accounting object systems should include elements of double-entry bookkeeping such as debits and credits as well as all resource types related to assets, liabilities and equity.* Otherwise the understanding of the REA business ontology will not be promoted among accounting professionals and academics and its applicability will remain restricted to a subset of predominately asset related business transactions.

The extensions and modification of the REA business ontology needed to integrate the ALE accounting model resulted in the REA-based ALE accounting ontology. This ontology specifies the traditional accounting logic in terms of accounting transaction objects that contain the debit and credit events related to the changes of assets, liabilities and equity and that satisfy the value restriction property.

Beyond that the REA-based ALE accounting ontology is by its economic contract foundation also able to adequately integrate all kind of financial resources. Consequently this ontology constitutes a fusion of the accounting and the finance domains. Such a fusion is surly needed in current times, where many derivative instruments are considered off-balance. Modeling these products as economic contracts would bring them on-balance, so that its magnitudes and risk are explicitly shown to the benefit of the investors and other stakeholders. This special feature of the REA-based ALE accounting ontology might be a promising starting point for future legislation as well as for future enterprise and accounting systems research.

References

1. Abmayer, M., Schwaiger, W.: Accounting and management information systems – a semantic integration. In: Weippl, E., Indrawan-Santiago, M., Steinbauer, M., Kotsis, G., Khalil, I. (eds.) 15th International Conference on Information Integration and Web-based Application and Services (iiWAS 2013), pp. 345–351. ACM 2013, Vienna (2013). ISBN 978-1-4503-2113-6
2. Black, F., Scholes, M.: The pricing of options and corporate liabilities. J. Polit. Econ. **81**(3), 637–654 (1973)
3. Dunn, Ch., Cherrington, J.O., Hollander, A.: Enterprise Information Systems: a Pattern-Based Approach, 3rd edn. McGraw-Hill, Boston (2006)

4. Geerts, G., McCarthy, W.E.: Policy level specification in REA enterprise information systems. J. Inf. Syst. **20**(2), 37–63 (2006)
5. Geerts, G., McCarthy, W.E.: An ontological analysis of the economic primitive of the extended REA enterprise information architecture. Int. J. Acc. Inf. Syst. **3**, 1–16 (2002)
6. Harrison, W., Horngren, Ch., Thomas, W., Suwardy, Th.: Financial Accounting – International Financial Reporting Standards, 9th edn. Pearson, Boston (2014)
7. Horngren, Ch., Harrison, W., Oliver, S.: Accounting, 9th edn. Pearson, Boston (2012)
8. Hruby, P.: Model-Driven Design Using Business Patterns. Springer, Heidelberg (2006)
9. Hull, J.: Options, Futures, and Other Derivatives, 9th edn. Prentice Hall, New Jersey (2014)
10. ISO/IEC-Accounting and Economic Ontology Standard. Information Technology – Business Operational View – Part 4: Business Transaction Scenarios – Accounting and Economic Ontology. ISO/IEC 15944-4:2006 (2006)
11. McCarthy, W.: The REA accounting model – a generalized framework for accounting systems in a shared data environment. Acc. Rev. **LVII**(3), 554–578 (1982)
12. Steinbart, P., Romney, M.: Accounting Information Systems, 12th edn. Pearson, Boston (2012)

Data Models and Semantics

A Schema-Less Data Model for the Web

Liu Chen$^{(\boxtimes)}$, Mengchi Liu, and Ting Yu

State Key Lab of Software Engineering, School of Computer, Wuhan University,
Wuhan, China
{dollychan,yuting}@whu.edu.cn, mengchi@scs.carleton.ca

Abstract. To extract and represent domain-independent web scale data,
we introduce a schema-less and self-describing data model called Object-
oriented Web Model (OWM), which is rich in semantics and flexible in
structure. It represents web pages as objects with hierarchical structures
and links in a web page as relationships to other objects, so that objects
form a network. Taking use of web segmentation techniques, data from
data-intensive web pages can be extracted, represented and integrated
as OWM objects.

Keywords: Web data model · Schema-less · Web information extraction

1 Introduction

The Web has become the world's largest information resource. Researchers have
been working on methods to extract and query distributed web data. Web information extraction systems translate unstructured, semi-structured and structured web data to uniform structured data for post-processing such as data
mining and search. Intermediate systems use schema mapping and global views
to map distributed webs [1–3].

Some web data extraction approaches are designed to solve specific problems
and operate in ad-hoc domains [4,5], or to extract data from one or similar web
sites with some templates or layouts [6]. Some domain-independent methods aim
at structured data in web pages, like tables [7–9]. A common issue of domain-
specific and domain-independent methods is developing a consensus schema to
capture the diverse semantics of data from different sources, and additionally
keeping up with new concepts as data sources are increasing and new data needs
to be added to the integrated system from time to time [10,11].

To extract domain-independent web scale data, and to bypass the development of consensus and well-defined schemas for the integrated systems, schema-
free data representation is a good solution because it captures the semantics of
web data and helps to get rid of the definition of an uniform schema. Hence,
we introduce such a schema-less data model called Object-oriented Web Model

This work is supported by National Natural Science Funds of China under grant No.
61202100.

P. Johannesson et al. (Eds.): ER 2015, LNCS 9381, pp. 577–584, 2015.
DOI: 10.1007/978-3-319-25264-3_44

(OWM) to represent and search data distributed throughout the web in a more feasible and flexible way.

Firstly, a web page is represented as an object and hyperlinks to other pages inside the web page are relationships between objects. The properties of an object are subject to same variations, such as missing attribute, multivalued, multiordering and etc. [2]. Hence, OWM provides flexible and recursive data types to represent object properties. In addition, to represent the classification of an object as this kind of knowledge can be obtained, OWM supports some object-oriented features, that are object classification and class hierarchy.

Secondly, for the flexibility of OWM, there is no limitation on the structural extent of data extracted from the Web, and data can be a plain text or a well-structured value. A naive and extreme case when modeling web data in OWM is directly converting a web page into an OWM object and taking the whole web page content as a plain text value of the object, while it is oversimplified and not satisfying and usually more semantics should be extracted if possible. Instead of pre-specifying the target data during web data extraction, via OWM all available web data can be extracted and generated in a flexible semi-structured format, and letting querying filter and return the expected results. One reason is that identifying distinct information during web data extraction is exhausted and susceptible to failure for the versatile of web layouts and complexity of natural languages. The other reason is that in most of extraction approaches only records in ad-hoc domain are extracted while meaningful unexpected data are usually neglected. Thus, instead of extracting fine-grained targeted attribute records, taking advantage of segmentation techniques [12,13] for data-intensive web pages and vision-based web data extraction methods [14,15], a web page can be translated to an OWM object without prior knowledge of the domain and web templates or layouts.

Thirdly, the interior of an OWM object has a hierarchical structure and objects form a network. Tree traversal methods and graph queries can be used to query the integrated web data in OWM.

The rest of this paper is organized as follows: OWM data model in Sect. 2, web data extraction and representation in Sect. 3, objects and object networks in Sect. 4, and finally the conclusion.

2   Web Data Model

2.1   Dynamic Structure with Object-Oriented Features

An important feature of web data is that contents and structures are flexible and unpredictable. Manually reading the data and summarizing a uniform schema is impossible.

In OWM, a web page is modeled as an *object*; and where there is a link, there is a *reference* to an object. The schema of the data needs not defined in advance. An object can has more than one names that are hyperlink labels other pages use to link to this object. The value of an object can be any one of the four types: plain text, object reference, attribute-value pair and list. Attribute-value pair is the usual attribute type. Value can be a list, and OWM supports multivalued attribute and the order of the property is kept. Wherever there is a link, there is an object, and the hyperlinks in a web page are represented as relationships of OWM object. For these relationships, there may or may not be attribute names specifying the labels of the relationships. Hence, both the attribute-object reference pairs and object references represent the relationships. Furthermore, different types of data can be aggregated and nested just as how the web displays them, and in this way, the semantics of the web data can be captured and retained as much as possible.

In addition, the dynamic structure of objects and relationships is extended with some object-oriented features, that are object classification and class hierarchy. Class information will contribute to data query and analysis. If the knowledge is given, like the domain information is known by relevant web page crawler program or the content of some web pages may indicate class information, OWM can represent this extra semantics. Hence, an OWM object is identified by its URI as in Semantic Web and it can have multiple classifications. Classification information can be added with web data extraction or later after getting the knowledge.

Data from various sources have different time lines and quality, and there is also record linkage problem that is identifying whether the objects from different sources is the same one or not [16]. The URIs are taken as identifiers, hence an object displayed in different sources is not combined and the conflicts are left as they are. Data fusion is not considered at this stage with data model.

2.2 Syntax

Before we define the notions formally, we assume the existence of the following sets:

1. a set C of class names;
2. a set N of object names;
3. a set I of Uniform Resource Identifiers of webs;
4. a set A of attribute names;
5. a set S of text strings.

Definition 1. An *object* is a quadruple $o = (C, N, i, v)$ where

1. $C = [c_1, \ldots, c_l]$ with $l \geq 0$ is an optional set of class names that the object belongs to, where $c_i \in C$ with $1 \leq i \leq l$ is a class name;
2. $N = n_1, \ldots, n_m$ with $m \geq 0$ is an optional set of object names, where $n_i \in N$ with $1 \leq i \leq m$ is an object name;
3. $i \in I$ is the object identifier;
4. v is the value associated with the object that is defined recursively as follows:
   (a) v is a text string s where $s \in S$;
   (b) v is a reference to an object of the form $\langle n, i \rangle$ where $n \in N$ is the object reference name and $i \in I$ is the identifier of the referenced object.

(c) v is an attribute value of the form $a : v'$ where $a \in \mathcal{A}$ is the attribute name and v' is a value.

(d) v is a list of values $\{v_1, \ldots, v_n\}$ with $n \geq 1$ and each v_i with $1 \leq i \leq n$ is a value.

2.3   Class Definition

Class information classifies objects and contributes to data analysis. Domain information can be obtained by web crawler program, and there are some websites providing the class hierarchy information and object classification information.

Definition 2. *The class definition is an expression of the form:*

$$class\ c\ isa\ \{c_1, \ldots, c_n\}\ subsume\ \{c'_1, \ldots, c'_m\}$$

where $c_1, \ldots, c_n, c, c'_1, \ldots, c'_m \in C$ *with* $n \geq 0$ *and* $m \geq 0$. *Keywords isa and subsume respectively specifies that* $c_1, \ldots, c_n$ *are superclasses of c and* $c'_1, \ldots, c'_m$ *subclasses of c.*

Example 1. The following definitions connect three classes Conference, Academic Conference and DB Conference, and represents the class hierarchy.
   class *"Academic Conference"* isa *Conference*;
   class *"DB Conference"* isa *"Academic Conference"*;

3   Web Data Extraction and Representation in OWM

A data-intensive web page is the description of an entity, which is what most web applications developed for, and the contents of a web page are properties of the entity, where some values refer to other web pages, that are relationships with other entities. The steps of extracting and modeling a web page using OWM briefly are: dividing a web into several segments and if the number of segments is more than one, the type of object value is list; for each segment, there are four categories: (1) if the segment is plain text, it is a text type value; (2) if the segment is a hyperlink, it is an object reference value; (3) if the segment has secondary titles, it is represented as attribute name and the attribute value is modeled in the same way recursively; (4) if the segment is a mix of above three types, like repetitive structures, a list is generated.

HTML is noisy and losing semantics for switching between semantic and visual markup (e.g. <EM> VS. <I>), misusing <TABLE> structures for positioning non-tabular elements as well as completely neglecting HTML semantics for the layout. Hence, the DOM tree is no longer suitable to take as web content structure.

Except the information in the natural language, visual structure of web is what human users receive. Opposed to DOM structure, visual structure is more concise and significant. By virtue of the flexible data model, OWM has no restrains of the data structural extent, and using OWM data extraction process can ignore some meaning and semantics of the text and natural language. Hence, we combine the analysis of DOM structure and visual features, and extract data capturing the semantics delivered by web page visual layout.

Using aforementioned approach, we have extracted and integrated data from more than 20 thousands of web pages, mainly information of Chinese universities, faculty and academic conferences.

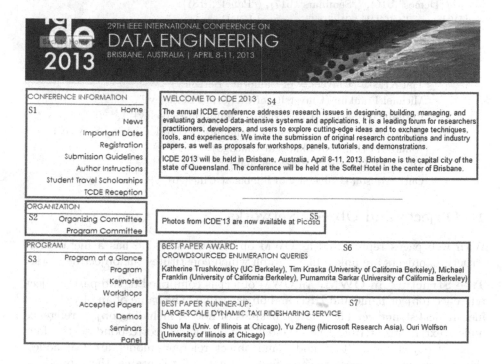

Fig. 1. The home page of conference "ICDE 2013" (Color figure online).

Example 2. Figure 1 is the home page of Conference "ICDE 2013". It shows how to represent the whole content of the web page in OWM as follows.

In OWM, the conference "ICDE 2013" is an object of the class DB Conference and as it shows, there are 7 attributes respectively according to the 7 segments in the web page. The coloured and bold titles are taken as attribute names, such as "Conference Information", "Organization", and etc. For the segments which have no explicit attribute names, such as the property "Photos from ICDE'13 are now available at Picasa (with a hyperlink)", we just use the type of list to represent the aggregated pairs of a text item and a hyperlink (object reference) one. Besides titles which are taken as attribute name, attribute values can be partitioned to different items according to their styles and layouts on web pages, like attribute "Best Paper Award" and "Best paper Runner-up", which are nested attributes. Also, variables $c1$, $b1$, $b2$, ... are short for object URIs.

[**DB Conference**] "ICDE 2013" ⟨c1⟩{
"Conference Information":{ ⟨"Home",b1⟩, ⟨"News",b2⟩, ⟨"Important Dates",b3⟩,
    ⟨"Registration",b4⟩, ⟨"Submission Guidelines",b5⟩, ⟨"Author Instructions",b6⟩,
    ⟨"Student Travel Scholarships",b7⟩, ⟨"TCDE Reception",b8⟩},
"Organization":{ ⟨"Organizing Committee",b9⟩, ⟨"Program Committee",b10⟩},
"Program":{ ⟨"Program at a Glance",b11⟩, ⟨"Program",b12⟩, ⟨"Keynotes",b13⟩,
    ⟨"Workshops",b14⟩, ⟨"Accepted Papers",b15⟩,
    ⟨"Demos",b16⟩, ⟨"Seminars",b17⟩, ⟨"Panel",b18⟩},
"Welcome To ICDE 2013":"The annual ICDE conf. ....",
{"Photos from ICDE'13 are now available at", ⟨ "Picasa",b19⟩},
"Best Paper Award": "Crowdsourced Enumeration Queries":
        {"Katherine Trushkowsky (UC Berkeley)",
        "Tim Kraska (University of California Berkeley)",
        "Michael Franklin (University of California Berkeley)",
        "Purnamrita Sarkar (University of California Berkeley)"},
"Best Paper Runner-Up": "Large-scale Dynamic Taxi Ride Sharing Service":
        {"Shuo Ma (Univ. of Illinois at Chicago)",
        "Yu Zheng (Microsoft Research Asia)",
        "Ouri Wolfson (University of Illinois at Chicago)"}};

4   Objects and Object Networks

With web pages represented as OWM objects whose value has a hierarchical structure, objects are linked like a graph which constitutes the object networks.

Tree Structure in OWM. An OWM object is comprised of two parts: object reference (object name and URI) and object value. The value of an object has a hierarchical structure. There are four kinds of nodes, including object reference node, text node, attribute node and list node, which correspond to the four types of object value. Text nodes and object reference nodes are leaf nodes; the children of list nodes are value nodes that the list contains; the subnode of attribute node is the attribute value node.

Figure 2 displays a part of the hierarchical structure of object "ICDE 2013" in Sect. 3 and every rectangle represents a node where the first line explains the kind of the node and the second line the name of the node. As it shows, the value of "ICDE 2013" is a list, the *List Node 1*, which is also the root node. The name of an attribute node is the attribute name, that of a text node is what the plain text is, and that of a list node is null. For an object reference node, it has two properties: one is the object name, the other one is the URI.

Graph Structure in OWM. OWM does not have a classical graph structure, but we can abstract nodes and edges from OWM objects and relationships and make use of graph querying to provide better analyzing interfaces. An OWM graph G is also a pair of (V, E), where V is a finite set of objects and E is a finite set of links in objects connecting pairs of objects.

Figure 3 shows the graph structure abstracted from the objects and relationships. Firstly, every edge is directed so it is a directed graph. Secondly, graph

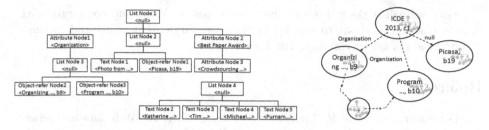

Fig. 2. Hierarchical structure of object "ICDE 2013".

Fig. 3. A part of object networks

vertices are objects. Thirdly, edges are represented by dash lines because they may not be actual direct connections because the relationships in OWM can be hierarchical. As in Fig. 3, *"ICDE 2013"* connects to *"Organization..."* by direct link "Organization".

5   Conclusion and Future Work

In this paper, we have proposed a schema-less and object-oriented data model OWM, which can represent a web page as an object with a hierarchical structure and links in a web page as connections to other objects. Because of the flexibility of model and the full web page conversion mode, web data can be directly converted into semi-structured OWM objects with complex interior hierarchical structures, and the linking networks across objects are formed.

The reusability of data depends on the extent to which it is structured. The more regular and well-defined the structure of the data is, the more easily people can create tools to reliably process it for reuse. As OWM has no limitation on structural extent of data, the structure of OWM objects can be any extent between unstructured to well-structured and we roughly divide the extent into three stages. The first and most naive stage of modeling data in OWM is that taking a web page as a plain text string that is the value of an OWM object. OWM objects at this first stage lack semantics and only a few simple queries can make sense and return meaningful results. In the second stage, the visual and superficial structures are extracted and constructed, which capture partial structures and semantics of web content. Now, the extraction method based on OWM is at this second stage. The third stage is that texts are analyzed, natural languages processed and well-structured attribute value pairs generated. In this stage, web data are well structured with high precision and detailed semantics.

Upon the data model, we will design an inclusive and comprehensive schema-less query language to traverse the dynamic data structure that is a hybrid of tree structure and graph structure. Tree expressions traverse the interior hierarchical structures and graph expressions excavate the connections across objects. Additionally, reasoning concepts can be attached to the query processing and more intelligent and powerful analyzing query can be conducted.

Also, we will make full use of the advantages of OWM and cooperate with data extraction methods to capture as much semantic as possible and increase the precision of query results in the future.

References

1. Ferrara, E., De Meo, P., Fiumara, G., Baumgartner, R.: Web data extraction, applications and techniques: a survey. CoRR (2012)
2. Chang, C.H., Kayed, M., Girgis, M.R., Shaalan, K.F.: A survey of web information extraction systems. IEEE Trans. Knowl. Data Eng. **18**(10), 1411–1428 (2006)
3. Sarawagi, S.: Automation in information extraction and integration. In: Tutorial of The 28th International Conference on Very Large Data Bases (VLDB) (2002)
4. Su, W., Wang, J., Lochovsky, F.H.: Ode: Ontology-assisted data extraction. ACM Trans. Database Syst. (TODS) **34**(2), 12 (2009)
5. Embley, D.W.: Toward semantic understanding: an approach based on information extraction ontologies. In: Proceedings of the 15th Australasian database conference, vol. 27, pp. 3–12. Australian Computer Society Inc (2004)
6. Crescenzi, V., Mecca, G., Merialdo, P., et al.: Roadrunner: towards automatic data extraction from large web sites. VLDB **1**, 109–118 (2001)
7. Mulwad, V., Finin, T., Joshi, A.: A domain independent framework for extracting linked semantic data from tables. In: Ceri, S., Brambilla, M. (eds.) Search Computing. LNCS, vol. 7538, pp. 16–33. Springer, Heidelberg (2012)
8. Michael, J., Cafarella, A.H., Wang, D.Z., Wang, E., Zhang, Y.: Webtables: exploring the power of tables on the web. Proc. VLDB Endowment **1**(1), 538–549 (2008)
9. Bohannon, P., Dalvi, N., Filmus, Y., Jacoby, n., Keerthi, S., Kirpal, A.: Automatic web-scale information extraction. In: Proceedings of the 2012 ACM SIGMOD International Conference on Management of Data, pp. 609–612. ACM (2012)
10. Madhavan, J., Halevy, A.Y., Cohen, S., Dong, X.L., Jeffery, S.R., Ko, D., Yu, C.: Structured data meets the web: a few observations. IEEE Data Eng. Bull. **29**(4), 19–26 (2006)
11. Talukdar, P.P., Ives, Z.V., Pereira, F.: Automatically incorporating new sources in keyword search-based data integration. In: Proceedings of the 2010 ACM SIGMOD International Conference on Management of data, pp. 387–398. ACM (2010)
12. Zeng, J., Flanagan, B., Xiong, Q., Wen, J., Hirokawa, S.: A web page segmentation approach using seam degree and content similarity. In: Lee, R.Y. (ed.) Applied Computing and Information Technology, pp. 91–103. Springer, Berlin (2014)
13. Kohlschtter, C., Nejdl, W.: A densitometric approach to web page segmentation. In: CIKM 2008, pp. 1173–1182 (2008)
14. Cai, D., Yu, S., Wen, J., Ma, W.-Y.: Extracting content structure for web pages based on visual representation. In: Zhou, X., Zhang, Y., Orlowska, M.E. (eds.) APWeb 2003. LNCS, vol. 2642, pp. 406–417. Springer, Heidelberg (2003)
15. Liu, W., Meng, X., Meng, W.: Vide: a vision-based approach for deep web data extraction. IEEE Trans. Knowl. Data Eng. **22**(3), 447–460 (2010)
16. Dong, X.L., Srivastava, D.: Big data integration. In: 2013 IEEE 29th International Conference on Data Engineering (ICDE), pp. 1245–1248. IEEE (2013)

An Analysis and Characterisation of Publicly Available Conceptual Models

C. Maria Keet[1]([⊠]) and Pablo Rubén Fillottrani[2,3]

[1] Department of Computer Science, University of Cape Town,
Cape Town, South Africa
mkeet@cs.uct.ac.za

[2] Departamento de Ciencias e Ingeniería de la Computación,
Universidad Nacional Del Sur, Bahía Blanca, Argentina

[3] Comisión de Investigaciones Científicas de la Provincia de Buenos Aires,
Bahía Blanca, Argentina
prf@cs.uns.edu.ar

Abstract. Multiple conceptual data modelling languages exist, with newer version typically having more features to model the universe of discourse more precisely. The question arises, however, to what extent those features are actually used in extant models, and whether characteristic profiles can be discerned. We quantitatively evaluated this with a set of 105 UML Class Diagrams, ER and EER models, and ORM and ORM2 diagrams. When more features are available, they are used, but few times. Only 64 % of the entities are the kind of entities that appear in all three language families. Different profiles are identified that characterise how a typical UML, (E)ER and ORM diagram looks like.

Keywords: UML class diagram · EER · ORM · Quantitative analysis · Language feature

1 Introduction

Many conceptual data modelling languages (CDMLs) exist, due to, among others, having originated from different communities (e.g., relational databases vs. OO software), addressing a specific modelling issue (e.g., spatial, temporal), and design decisions for leanness or expressiveness. There is a general trend toward more modelling features in CDMLs over time; compare, e.g., the early UML with the latest v2.4.1 [9] (a.o., with identifiers), ORM vs ORM 2 (a.o., more ring constraints, role values) [5], and ER vs EER (a.o., subsumption, disjointness) [12,13]. There are also many opinions on what is the 'best' approach to this feature richness, its relation to model quality [8] and fidelity of representing the customer's needs. But: *which features are actually used in conceptual data models 'out there' in the field?* An answer to this question on actual usage will be useful in general for language development, affordances and usability of modelling tools, modelling methodologies, and can feed into the teaching of

© Springer International Publishing Switzerland 2015
P. Johannesson et al. (Eds.): ER 2015, LNCS 9381, pp. 585–593, 2015.
DOI: 10.1007/978-3-319-25264-3_45

conceptual modelling. We could find only one *quantitative* analysis of feature usage in conceptual models [11], which used 168 ORM models developed by one modeller with a proprietary tool. The GenMyModel UML diagram repository [https://repository.genmymodel.com] does show some counts of model elements in its web interface, so it will have the statistics of its models, but aggregate data are not available. There are many works on the use of conceptual models as a whole [2], however, to the best of our knowledge, a quantitative study into actual use of CDML features and across CDML families has not occurred.

Besides the general usefulness of such insights into the actual usage of CDML features, we also seek to practically apply this in the development of inter-model assertions and model transformations [3], as, in theory, very many rules and patterns would be needed to cover all the features across CDM languages. This also has the advantage of having a unifying metamodel [7] of the three main conceptual data modelling language families (UML Class Diagrams, EER, and ORM), so that it allows a comparison of the data across models represented in the different languages. This brings us to the following hypotheses to test:

A. When more features are available in a language, they are used in the models.
B. Following the "80–20 principle", about 80 % of the entities present in the models are from the set of entities that appear in all three language families.
C. Given the different (initial) purposes of UML class diagrams, (E)ER, and ORM, models in each language still have a different characteristic 'profile'.

To falsify the hypotheses, we collected 105 public available UML Class Diagrams (henceforth UML CD), ER and EER (abbreviated as (E)ER), and ORM and ORM2 (abbreviated as ORM/2) models and categorised all entities in terms of their unifying metamodel [7] for data analysis and comparisons. Although only 64 % of the entities are within the exact intersection, this is almost 90 % with a suitable transformation between attributes and value types. Most modelling features are indeed used, though some very sparingly. Each CDML family has a profile how a typical model in such language looks like, using characteristics such as entity type to relationship ratios and subsumption.

We first describe the materials and methods (Sect. 2), which is followed by the results (Sect. 3), discussion (Sect. 4), and conclusions (Sect. 5).

2   Materials and Methods

The experimental design is summarised as follows:

1. Collect UML CDs, (E)ER, and ORM/2 diagram, 35 each, sourced from: GenMyModel, scientific articles (e.g., ER'13), textbooks, and online diagrams.
2. For each static, structural entity (i.e., not behavioural, implementation, or organisational element)
   (a) Classify it in terms of the metamodel entities of [4], for a named or unnamed element, including constraints;
   (b) Add any comments in the comment field of the spreadsheet;
   (c) Note any violations (syntax mistakes, not semantic ones);

3. Data analysis: Compute mean and median of element/model, percentage the entity is present in the model at all (presence) and as percentage of the total amount of entities (prevalence), for each family and aggregated; examine other characteristics, such as attribute:class ratio and binaries:n-aries.

The dataset of 105 models and spreadsheet (xls) with analyses are available from http://www.meteck.org/SAAR.html.

The unifying metamodel was introduced in [7] and formalised in [4]. It has Entity as top-type with four main subclasses: Relationship with 11 subclasses, Role, Entity type with 9 subclasses, and Constraint with 49 subclasses.

3   Results and Analysis of the Classification of Entities

Four models had to be discarded; 101 models were classified manually and used in the analysis. Their average 'model size' (more precisely: vocabulary), calculated as (Object Types + Relationships + Subsumption + [Attributive property or Value Type]), is similar for each family: UML CD 51.1, ORM/2 46.6, and (E)ER 50.8.

3.1   Common Features in the Language Families

There was a total of 8037 entities, of which 5191 (64.6 %) appear in all three families (UML CD, (E)ER, ORM/2), 1737 in two of the three (21.6 %), and 1109 (13.8 %) in one. **Thus, Hypothesis B is falsified.** Only if one relaxes CDML feature overlap to also include the obvious transformation rules between UML and (E)ER's attributes and ORM/2's value types (as described in [3]), then it reaches 87.6 %. These values remain similar (± 2 %) regardless whether 20, 25, or 35 models of each family were classified.

The common entities across the model families were one or more Object type, Relationship, Object type cardinality, Subsumption (object type), Single identification, Disjoint and Complete object types. Single identification, however, was very rare in UML (7 occurrences in two models). The metamodel's Object type cardinality does not distinguish between computationally important differences between a 1..* participation and other number restrictions (e.g., ≥ 2, 3..5). We enumerated the presence/absence, whose aggregate data is shown in Table 1. Other number constraints appear in at most 20 % of the models, and while optional and mandatory constraints occur in roughly the same amount of models, 0..1 and 1 appear in notably fewer (E)ER models.

3.2   Usage of Features in the Language Families

Prevalence—as percent of total for that family—is included in Table 2 for both the top-5 and bottom-5 for each family. In the top-5 list, note that for UML and (E)ER, there are substantially more attributes than object types, whereas this is roughly the same for ORM's object types and value types. Dimensional value types were observed ($n = 16$ scattered in 29 % of the models), of which 9 are

Table 1. Presence/absence of cardinality constraints in the models, aggregated.

	No. of models	* or 0..* (optional)	0..1 (functional)	1..* (mandatory)	1 (exactly 1)	Other nr constraint
UML CD	34	22	21	17	25	7
ORM/2	33	20	28	14	29	6
(E)ER	34	19	16	19	11	2

about date and time and the others with measurements, such as Blood pressure (Pa) in `InfoModelerDiagram2` and Area (sq.m) in `campbellAbs`. Several types of entities did not appear at all. For UML, they were: Nested Object Type, Qualified relationship, Attribute value constraint, and Disjoint roles; for (E)ER, they were: Subsumption on roles or relationships, Inclusive mandatory, and Disjoint roles; for ORM/2: Join-disjointness and Value comparison constraint.

Overall, while some features are used little, the vast majority are, and **thus, hypothesis A is validated**.

Analysing the uncommon entities, one can look at the percent present in the model, and the median; see online material for details. What is immediately clear in the general case, is that many ORM constraints are hardly used, but most of them are used at least in some models. From a computational complexity viewpoint, there are several noteworthy observations, of which we touch upon disjointness and completeness constraints and ORM's ring constraints.

Disjointness and completeness constraints for class hierarchies feature prominently in ontologies, and they are useful for automated reasoning. Object types in CDMs are assumed to be disjoint, but it has to be declared explicitly for subsumption. There were only 13 disjointness and 21 completeness constraints in 11.8 % of the full set of models, each family has some, and they were observed only in models for teaching conceptual data modelling. (E)ER had the most disjointness and completeness constraint (8 and 9, respectively). An additional 13 disjoint roles and 2 disjoint relationships were present in the ORM models.

Ring constraints are not computationally well-behaved, yet ORM has 6 different ones and ORM2 has 11. Of the 33 ORM/2 models, there were 23 relationship constraints in total (median: 0), and 32 % of the ORM models had at least one of them. Of the 23 relationship constraints, 11 were irreflexive, 4 acyclic, 3 symmetric, 3 intransitive, 1 asymmetric, and 1 probably purely reflexive (ambiguous icon). 16 models had at least one recursive relationship; e.g., `ERmodelROMULUS`'s hasModule between Ontology clearly could have been declared transitive and acyclic (and thus also asymmetric, antisymmetric, and irreflexive) if it had been available in EER. The reason for a low incidence of relationship constraints is unclear. Perhaps it is not perceived to be needed either semantically or in the implementation, or both, or is unknown or too hard for an 'average' modeller.

Table 2. Prevalence of particular entity in the models, as percent of total number of entities for that family, aggregated by model family and rounded off to one decimal. OT: Object type; VT: Value type; Rel.: Relationship; Int. Unique.: Internal uniqueness constraint; ID: Identifier.

Top-5		
UML CD	ORM/2	(E)ER
Attribute (31.2 %)	OT cardinality (29.0 %)	Attribute (39.5 %)
OT (21.2 %)	OT (14.5 %)	OT cardinality (22.1 %)
OT cardinality (17.5 %)	2-ary Rel. (14.4 %)	2-ary Rel. (11.6 %)
2-ary Rel. (12.4 %)	Int. unique. (13.1 %)	OT (11.5 %)
OT subsumption (9.6 %)	VT (10.4 %)	single ID (7.7 %)

Bottom-5		
UML CD	ORM/2	(E)ER
3-ary Rel., Subsumption (Rel.), Disjoint OT (0.1 %)	Compound cardinality, Equality (Rel.), Join equality (0.0 %)	Attribute cardinality (0.0 %)
SingleID (0.3 %)	Disjoint Rel., Join subset, Role value constraint, Disjoint OT, Completeness, Subsumption (Rel.), 4-ary Rel. (0.1 %)	4-ary Rel., Nested OT (0.1 %)
Completeness (0.4 %)	Role equality, 5-ary Rel. (0.2 %)	Multivalued attribute, Disjoint OT (0.3 %)
Attribute value constraint (0.7 %)	Disjoint roles, External ID, Subsumption (role), Disjunctive mandatory (0.4 %)	Completeness (0.4 %)
Attribute card. (0.9 %)	Dimensional VT (in ref mode) (0.5 %)	3-ary Rel. (0.5 %)

3.3   Salient Aggregate Characteristics for Each Family

We first consider a set of ratios that contribute to formulating a 'characteristic profile' of a family; they are included in Table 3 and elaborated on in the remainder of this section. While the average model sizes are fairly similar, in ratio to the overall amount of entities, UML has relatively few constraints compared to ORM. This also matches the notion that ORM has more constraint types: if you have them, they are used some time somewhere.

There are large differences among UML CDs, ORM/2 and (E)ER in their use of attributes or value types per object type. This may be due in part to the sharability of an ORM value types among object types, whereas attributes are

Table 3. A selection of ratios of entities aggregated by family and combined.

Ratio	UML	ORM/2	(E)ER	Combined
model size:total entities	0.8	0.5	0.7	0.6
Attribute or Value type:Object type	1.5	0.7	3.5	1.7
binaries:n-aries	180.5	12.4	20.9	20.4
Subsumption(class):Object type	0.5	0.3	0.2	0.3
Relationship (non isa):Object type	0.8	1.1	1.1	1.0
Object type cardinality:other constraint	7.4	1.2	2.2	1.8
Single identification:other ID	–	17.3	5.4	8.4
role:relationship naming	4.3	(readings, mostly)	0.1	N/A

exclusive to the class in UML and (E)ER so that multiple attributes have to be modelled for what is semantically the same attribute (e.g., **Age**).

There are relatively few class subsumption hierarchies in general, though UML CD's ratio is twice as high as that of (E)ER and a third higher than in ORM/2. The ratio Relationship (non isa):Object type are similar for ORM/2 an (E)ER, and they have about 35% more relationships than the UML models. Together with the Subsumption (of object types):Object type ratio, it shows UML is much more object type-oriented. This may be expected, as it comes from OO history, and is thus also recognisable from our dataset.

Among the relationships, noteworthy are UML CD's aggregation and binaries vs. n-aries ($n > 2$). The ratio of 'plain' association:aggregate is 2.6. Whether they are modelled correctly and whether they are implicitly present in ORM/2 and (E)ER through the name of the role or relationship, and whether they would be used in the latter if there were an icon for it, is an interesting avenue for further work. The ratio of binaries to n-aries (also in Table 3) differs greatly between UML CDs vs ORM and (E)ER. The data does not explain why UML models have mostly just binaries (261); it may be an avoidance strategy or lack of affordance in the tool, and it has been shown that modelling n-aries in UML is problematic due to notation, whereas ER does not have this problem [10].

The ratio of object type cardinality constraints to other constraints are as one may expect, with UML 3 to 6 times higher than (E)ER and ORM/2, respectively, as there are not many other constraints to add in UML CDs. The much lower values for ORM/2 and (E)ER is largely due to the manifold identification constraints in ORM and EER, which UML does not require.

3.4   Characteristic Features of a Family

Using the data and analysis, we construct the following feature-based characteristic profiles of the three families, therewith **validating Hypothesis C.**

The analysed UML diagrams are characterized by expressing more 'object-oriented' features: mostly classes and binary associations, naming of association ends, relatively high use of class subsumption, cardinalities, attribute value constraints, and aggregative composition of object types. More than 99 % of all the elements in studied UML class diagrams are one of these features.

(E)ER diagrams are characterised by those features that describe relationships: extended use of binary and ternary relationships that are named, complex identification schemes (multiattribute and weak entity types), composite and multivalued attributes, and associative object types. These features together entity types and attributes, also obtains more than 99 % of all the elements in the studied (E)ER diagrams.

Despite that ORM/2 is the most expressive language family, we can also obtain a characteristic profile based on relatively few features, although with a less high coverage than for UML and (E)ER (more than 98 % of all the elements in the analysed ORM/2 models). In general, distinguished features of ORM diagrams are: fact type-orientedness, constraints over arbitrary n-ary fact types, subsumption constraints between object types and between roles, nested object types, disjointness between roles, internal and external uniqueness constraints, value constraints, and internal and external identification constraints.

3.5   Data Analysis of Potential Interfering Factors

Other observations regarding the current scope concern mainly syntax and tooling. Especially the (E)ER models had a mix of notational styles, but the real syntax errors from a modelling viewpoint were the identifiers; e.g., giving subtypes new identifiers (e.g., ABM), object types without one (e.g., BritellER14-er), a wrong Weak identification with underlined attribute instead of dashed (Bill in ER_no_2big), and adding identifiers to a relationship's attribute (malli1). This has been noted as early as 1993 (Batra in [2]), yet easily could be checked by a metamodel or logic-based syntax checker, which NORMA [1] already does for the first two types of mistake. More problematic to detect are oddities of the wide use of the graphical notations beyond its original scope, notably as graphical notation of a relational model (fewer constraints) and ontology visualisation (no identifiers). Further, the language version and, with that, the set of available language features, is difficult to discern, hampering differentiation between what is 'available, but not needed' and what is 'needed, but not available'. In most cases, it is unclear which CASE or drawing tool has been used to develop the model. For instance, ORMmulti...all is drawn in DOGMA Modeler [6] that permits only binary relations and no value types. The free GenMyModel (v0.32.2) allows only binary associations and 0..1, 1, 1..*, or * multiplicity, therewith pushing down their respective counts, whereas the (for payment) Edraw (v7.8) allows for n-aries and arbitrary multiplicities with easy drag-'n'-drop icons.

The size of the dataset does influence the obtained aggregates and ratios, but not much and many features remained at very similar or the same averages and median regardless the number of models classified (see dataset for analysis).

4   Discussion

The results reveal some interesting patterns, and help prioritising rules for model transformations and inter-model assertions. There is considerable overlap in which features are being used and how often, but this holds for only two-thirds on the conceptual level, negatively impacting feasibility and success of transformation rules and algorithms and validation of inter-model assertions to effectively link components of a complex software system. It may induce further investigation into various issues, such as why some features are used so little and prevalence of part-whole relations.

There are two main limitations to the data collection and results. First, arguably, 'appropriate' conceptual data models are safely guarded in IT departments in companies and our dataset may neither be the best nor representative. Realistically, there is no way to ascertain that. Notwithstanding, when computer science students and graduates search online for examples and reuse, these are the kind of models they will find and emulate, for better or worse.

Second, great care has been taken in the manual classification, and a selection of models was classified several times (data not included), but automation may enhance precision. However, conceptual models are typically made available as figures, not their XML-serialised version, and even if they were available as XML file, the problem of multiple XSD format arises, which has yet to be resolved.

That said, comparing our data with the only other dataset, consisting of ORM models [11], there are only minor differences in averages and ratios for their set of models. For example, their 0.2 for subsumption:object type (0.3 in our data), and 3.2 for relationship:object type, which is higher (1.1 in our data) but anyway confirming the importance of relations cf. UML CDs, and also supporting the low value type:object type ratio (0.5 cf. our 0.7).

5   Conclusions

The quantitative evaluation of features in a set of 105 publicly available UML Class Diagrams, ER and EER models, and ORM and ORM2 diagrams is, to the best of our knowledge, the first of its kind, and constitutes a public dataset that can be used for further in-depth analyses. The quantitative evaluation showed that 64 % of the entities were classified in those features shared by all three model families, and also that when more features are available, they are mostly used in at least one model. Although graphical notations may not always be strictly according to syntax and purpose, each family still yielded different characteristic profiles that typify how a typical diagram in that family looks like.

The outcomes inform a prioritisation of mapping rules for automated validation of inter-model assertions [3], and raised multiple questions, from UML's aggregation to the affordances and features of modelling tools.

Acknowledgements. This work is based upon research supported by the National Research Foundation of South Africa (Project UID: 90041) and the Argentinian Ministry of Science and Technology.

References

1. Curland, M., Halpin, T.: Model driven development with NORMA. In: Proceeding of the 40th International Conference on System Sciences (HICSS-40), pp. 286a. IEEE Computer Society, Los Alamitos, Hawaii (2007)
2. Davies, I., Green, P., Rosemann, M., Indulska, M., Gallo, S.: How do practitioners use conceptual modeling in practice? DKE **58**, 358–380 (2006)
3. Fillottrani, P.R., Keet, C.M.: Conceptual model interoperability: a metamodel-driven approach. In: Bikakis, A., Fodor, P., Roman, D. (eds.) RuleML 2014. LNCS, vol. 8620, pp. 52–66. Springer, Heidelberg (2014)
4. Fillottrani, P., Keet, C.M.: KF metamodel formalisation. Technical report 1412.6545v1, December 2014. arxiv.org
5. Halpin, T., Morgan, T.: Information Modeling and Relational Databases, 2nd edn. Morgan Kaufmann, San Francisco (2008)
6. Jarrar, M., Demy, J., Meersman, R.: On using conceptual data modeling for ontology engineering. J. Data Seman. **1**(1), 185–207 (2003)
7. Keet, C.M., Fillottrani, P.R.: Toward an ontology-driven unifying metamodel for UML class diagrams, EER, and ORM2. In: Ng, W., Storey, V.C., Trujillo, J.C. (eds.) ER 2013. LNCS, vol. 8217, pp. 313–326. Springer, Heidelberg (2013)
8. Moody, D.L.: Theoretical and practical issues in evaluating the quality of conceptual models: current state and future directions. DKE **55**, 243–276 (2005)
9. Object Management Group: Superstructure specification. Standard 2.4.1, Object Management Group (2012). http://www.omg.org/spec/UML/2.4.1/
10. Shoval, P., Shiran, S.: Entity-relationship and object-oriented data modeling–an experimental comparison of design quality. DKE **21**, 297–315 (1997)
11. Smaragdakis, Y., Csallner, C., Subramanian, R.: Scalable satisfiability checking and test data generation from modeling diagrams. ASE **16**, 73–99 (2009)
12. Song, I.Y., Chen, P.P.: Entity relationship model. In: Liu, L., Özsu, M.T. (eds.) Encyclopedia of Database Systems, vol. 1, pp. 1003–1009. Springer, Heidelberg (2009)
13. Thalheim, B.: Extended entity relationship model. In: Liu, L., Özsu, M.T. (eds.) Encyclopedia of Database Systems, vol. 1, pp. 1083–1091. Springer, Heidelberg (2009)

An Extended ER Algebra to Support Semantically Richer Queries in ERDBMS

Moritz Wilfer and Shamkant B. Navathe[(✉)]

Georgia Institute of Technology, Atlanta, USA
{moritz.wilfer,shamkant.navathe}@cc.gatech.edu

Abstract. In this paper we present the foundations for a semantically rich main-memory DBMS based on the entity-relationship data model. The DBMS is fully operational and performs all queries that are illustrated in the paper. So far, the ER model is mainly used as a conceptual model and mapped into the relational model. Semantics like the relationships among entities or the cardinality ratio constraints are not explicit in the relational model. This paper treats the ER model as a logical model for the user and we use the relational as the physical model in our ER model based DBMS - ERDBMS. We use CISC (complex instruction set computing) operators but implement them efficiently in main-memory data storage. This paper concentrates on the extended ER algebra. Our high-level query language ERSQL and the main memory implementation are elaborated in [14].

Keywords: ER model · Semantics · Algebra · Main-memory · Logical model · CISC operators

1 Introduction

Chen's ER model [2] introduces mapping (derivation) techniques to the relational, network, and entity-set model as a set of logical data models. We use the term *logical data model* to represent a model used to form queries and write transactions using a DBMS that supports that model. There were many approaches to provide an algebraic [1,9,10] or a calculus [5,12] foundation for the ER model but none of them were widely accepted and implemented commercially. Further, multiple high-level query languages have been proposed to transform the ER model into a logical data model [4,7,8,15]. But they either lack a semantic foundation or have navigational characteristics which makes them rather complex. Hohenstein and Engels propose an SQL/EER language [6] and map it into an EER calculus [5] providing a sound mathematical foundation. But the results of their calculus expressions are no longer covered by concepts of their targeted EER model. Parent et al. [11–13] proposed an algebra and an equivalent calculus that returns an entity set as a result of an algebraic or calculus expression. All the mentioned proposals shared one major issue: due to the complexity of the ER model, the proposed functional primitives are also more complex than

© Springer International Publishing Switzerland 2015
P. Johannesson et al. (Eds.): ER 2015, LNCS 9381, pp. 594–602, 2015.
DOI: 10.1007/978-3-319-25264-3_46

their relational counterparts. It was hard or even impossible to implement them in an efficient fashion with acceptable response times for the user. Nowadays, modern implementation techniques and increased computational power make CISC (complex instruction set computing) operators efficiently implementable. Our proposal in this paper is to stay with the core concepts of the ER model but to use complex functions to operate on it and implement them efficiently in a main-memory DBMS called ERDBMS. We have also designed a high level query language ERSQL to express complex queries that are much more difficult to express in SQL. The details of our ERDBMS system and ERSQL language are deferred to another paper [14].

The remainder of this paper is organized as follows. Section 2 provides motivational aspects for the development of a DBMS based on the ER model. Section 3 briefly introduces the targeted EER model. Section 4 introduces the extended ER algebra which constitutes functional primitives of an ERDBMS. Finally, Sect. 5 presents ongoing work and conclusions.

2   Motivation for an ER Model Based DBMS

Although ER model has been around for 39 years, we feel it deserves to be implemented as a DBMS considering the current wave of NOSQL implementations.

- **Semantic perception gap:** Entities and their interrelationships are modelling constructs that are close to the natural perception of human beings. Treating the ER model as a logical data model bridges the semantic gap in the views of a database designer and a database user, simplifying query formulation.
- **Satisfying richer constraints:** The relational model offers key constraints, entity integrity constraints, and referential integrity constraints. The ER model offers more advanced constraints to control the database state and affords the user ease of query formulation. Relationship cardinality ratio constraints allow the user to exercise control over relationships with respect to their participating entity types. Automatic enforcement of cardinality ratio and value set constraints lends richness to update operations.
- **Redundancy in the flat relational model:** Normalization yields designs with a consistent set of non-redundant relations. But results of queries with joins can have a built-in redundancy due to the flatness of the relational model. We avoid this by using the multi-valued attribute construct efficiently in our operators.

3   The Enhanced ER Model

The *enhanced entity relationship (EER)* model, the target of our algebra, is based on the original ER model [2] and follows the convention and notation of Elmasri and Navathe [3]. Figure 1 shows the example schema used in the rest of

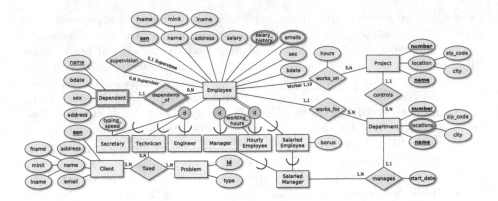

Fig. 1. EER schema for examples in this paper (in graphical notation from [3])

this paper. Our data model allows composite and multivalued attributes (even in nested form).

Aggregation with multivalued attributes: Gogolla and Hohenstein [5] claimed that aggregation should be supported by the calculus. It is easier to incorporate aggregation into our EER model directly with multivalued attributes. We support aggregation functions like *count, average, min,* or *max* for all multi-valued attributes. In case of a multivalued composite attribute, the aggregation functions are also defined for each component and recursively for their components in case of multivalued nested attributes.

4   An Extended ER Algebra for ERDBMS

The functional primitives for our ERDBMS are a slightly adapted version of the algebra proposed by Parent and Spaccapietra [11,13]. This section has three goals: First, we briefly introduce the algebra operators adopted from Parent et al. [11,13]. Next, we present the extensions we made to their operators. Finally, we present new operators we have added. We strongly recommend the reader consult the original papers [11,13] first. The algebra is closed. *The result of a query* from an expression in our algebra is an appropriate subschema with entity, relationship, and attribute instances that qualify for the result.

4.1   Operators Adapted from Parent and Spaccapietra

We have adapted the following operations from Parent and Spaccapietra's work and have provided for their main-memory implementation in ERDBMS.

- **Nested Cartesian Product** $E_1 \vec{\times} E_2$: The entity type E_1 is extended by a multivalued composite attribute E_2. The components of E_2 are the attributes of the operand entity type E_2. The population of the multivalued E_2 attribute

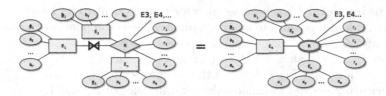

Fig. 2. Relationship join - ER transform schema

for $e \in E_1$ consists of all entities in the operand entity type E_2. Basically, every entity $e \in E_1$ *gets the whole population of* E_2 *nested in a new multivalued attribute* E_2. (*Name in original paper: Cartesian Product*)

- **Relationship Join** $E_1 \bowtie_R (E_2, ..., E_n)$: The relationship type R defined among n entity types is used to join the entity type E_1 with the entity types $E_2, ..., E_n$. The result of the join is to extend the entity set E_1 *by a multivalued composite attribute* R. The components of R are the attributes of the relationship type R and the composite attributes $E_2, ..., E_n$. The components of E_i are the attributes of the operand entitiy type E_i, $\forall i \in [2, ..., n]$. The population of the multivalued attribute R for $e \in E_1$ can be computed by following all relationships in R, that the entity e paritcipates in, and storing the corresponding values of the relationship attributes and all participating entities for the relationship instance. Figure 2 illustrates the resulting entity type from the relationship join $E_1 \bowtie_R E_2, ..., E_n$. *Employee* $\bowtie_{works\_on}$ *Project*: This query returns the entity set employees with each employee's projects nested in a multivalued attribute *works_on*.

- **Selection** $\sigma_p(E)$: A selection filters entities $e \in E$ based on the predicate p. The predicates for monovalued attributes are the same which are used in the relational selection operator. Predicates on multivalued attributes can be expressed using $\exists$ and $\forall$ quantifiers.

- **Reduction** $\chi_p(E)$: While the selection operator only filters entities $e \in E$ based on the predicate p, a reduction can be used to remove values from a multivalued attribute that fulfill the predicate p. The predicate p can be written in the following form to reduce a multivalued attribute: $x \in mv\_attr(p_i)$. $\chi_{x \in salary\_history(x>100000 \lor x<10000)}(Employee)$ removes all salaries that are bigger than $100,000$ or smaller than $10,000$ from the salary history of an employee. If a reduction is expressed on an attribute path with multiple levels of multivalued attributes, the reduction is performed for every single instance of the multivalued attribute that should be reduced.

- **Projection** $\Pi_A(E)$: This operator projects an entity type onto a subset A of its attributes. It can also project a composite attribute onto a subset of its components. $\Pi_{name(fname,lname)}(Employee)$ projects the *name* attribute onto its first and last component. $\Pi_{name.fname}(Employee)$ gets the firstname component of *name*. This works also with multivalued composite attributes.

- **Renaming** $\rho_R(E)$: The rename operator is equivalent to its relational coun-
  terpart. It is also possible to rename a (multivalued) component of a composite
  attribute: $\rho_{name.fname \to firstname}(Employee)$.
- **Union** $E_1 \cup E_2$: Calculates the set union of E_1 and E_2.
- **Intersection** $E_1 \cap E_2$: Calculates the set intersection of E_1 and E_2.
- **Difference** $E_1 - E_2$: Calculates the set difference of E_1 and E_2.

4.2   Extensions to the Adopted Operators

Quantifiers for Relationship Join $E_1 \bowtie_R (\exists E_2, ..., \forall E_n)$. Parent et al. [11,13]
implicitly used an $\exists$ quantifier for all entity types on the right side of a relationship join. That means a join of an entity $e \in E_1$ with $E_2, ..., E_n$ is considered
as successful, if there is a join partner for each of the right side entity types or
in other words, if there is a relationship in R that connects $e \in E_1$ with participating entities from the right side entity types. We extended the relationship
join by a $\forall$ quantifier. If a right side entity type E_i with $i \in [2, ..., n]$ is qualified with $\forall$, the relationship join of $e \in E_1$ with E_i is considered as successful,
if e is joinable with all $e' \in E_i$. The right side entity types of a relationship
join can be qualified with a mix of $\exists$ and $\forall$ quantifiers. A join of $e \in E_1$ with
such mixed qualified entity types $E_2, .., E_n$ is considered as successful, if the
join of e with each entity type E_i, $\forall i \in [2, ..., n]$, is successful with the given
quantifiers. If the standard join mode is used, entities $e \in E_1$, that are not successfully joinable, are omitted. If the outer join mode is used, such entities are
kept with no values for the multivalued composite attribute R. The notation
for an outer join is: $\stackrel{\supset}{\bowtie}_R$. The following examples should clarify any confusion:
$Technican \bowtie_{fixed} (\forall Client, \exists Problem)$ returns all technicans (with their clients
and fixed problems) who have fixed at least one problem and who have fixed
some problem for all clients. $Technican \bowtie_{fixed} (\forall Client, \forall Problem)$ returns all
technicans, who have fixed all instances of problems and who have fixed some
problem instance for every client.

Multivalued Selection Predicates. A multivalued selection predicate has
either the form $\exists_i x \in mv\_attr (p_i)$ or $\forall x \in mv\_attr (p_i)$ where mv_attr is
a path to a multivalued attribute and p_i is a selection predicate itself. The
bind variable is used to iterate over all values of the multivalued attribute
during evaluation. The inner predicate p_i can be expressed on all attributes
of the source entity type and on the bind attribute x which contains a single value of the multivalued attribute at a time. The multivalued predicates
can be applied recursively if x consists of another multivalued attribute. $\exists_i x \in$
$mv\_attr (p_i)$ evaluates to true if there are i values in mv_attr for a given entity
e that make p_i true. $\forall x \in mv\_attr (p_i)$ is true if all values for e make p_i true.
$\sigma_{\forall x \in salary\_history(x >= 50000 \land x <= 100000)}(Employee)$ returns employees that have a
multivalued set for salary_history all of whose elements are between 50,000 and
100,000. It is also allowed to use an attribute path to a component of a multivalued composite attribute. The bind variable binds itself to the value of the

component: $\sigma_{\exists_1 x \in locations.city(x=' Atlanta' \lor x=' Houston')}(Department)$ returns all departments that have a locations multivalued attribute that contains values Atlanta or Houston (or both).

4.3 Additional Operators

- **Cartesian Product** $E_1 \times E_2$: The cartesian product is used to combine entities from two entity sets. It is basically the same as its counterpart in the relational algebra. The attributes of E_1 are combined in a composite attribute E_1. The same happens to the attributes of E_2. We call this technique *entity packing* and it is used to avoid attribute naming ambiguities. If an entity set is already packed, it is not packed again.
- **Merge Join** $E_1 \bowtie_x^\succ E_2$: The merge join is used to join two operand entity types based on a compatible attribute that is part of a multivalued attribute in both entity types. Compatible means identical name and value-set (also for components in case of a composite attribute x). Usually, such compatible attributes are the results of joining the same entity type on the right side in a prior relationship join. *Entity packing* is used again to avoid naming ambiguities. To calculate the population of the result entity type after a merge join, we first introduce the concept of a *multivalued attribute join*.

Multivalued attribute join based on matching x from participating multivalued attributes: Let e_1 be an entity from E_1 and e_2 be an entity from E_2. The population of the multivalued attribute join based on the compatible attribute x from multivalued attributes $m_1(e_1)$ and $m_2(e_2)$ is defined as:
$$m_1(e_1) \bowtie_x m_2(e_2) = \{v \mid v \in m_1(e_1), s.t. \exists v' \in m_2(e_2), x(v) = x(v')\}$$

Entities $e_1 \in E_1$ and $e_2 \in E_2$ are joinable if the multivalued attribute join $m_1(e_1) \bowtie_x m_2(e_2)$ is not empty. The resulting entity is generated by keeping the values of e_1 for all attributes of E_1 with m_1 reduced to $m_1(e_1) \bowtie_x m_2(e_2)$ and by keeping the values of e_2 for all attributes of E_2 with m_2 reduced to $m_2(e_2) \bowtie_x m_1(e_1)$. For *ERSQL*, we have extended our merge join operator to join based on multiple compatible attributes $x_1, ..., x_n$ which are all part of distinct multivalued attributes. A join of e_1 and e_2 is successful, if all multivalued attribute joins based on all $x_i, \forall i \in [1, ..., n]$, are not empty. The operator is mainly used to answer queries like: Give me all pairs of employees and departments such that the department controls at least one project the employee works on: $(Employee \bowtie_{works\_on} Project) \bowtie_{Project}^\succ (Department \bowtie_{controls} Project)$. The examples at the end of the section show some usecases.
- **Collapse** $\xi_c(E)$: The collapse operator collapses the entity set of an entity type E into a single entity by introducing a new multivalued composite attribute c which has all attributes of E as its components. An entity in E becomes a value for c. $\xi_e(\sigma_{sex=' f'}(Employee))$, for example, returns a single entity with a multivalued composite attribute e which contains all female employees. The entity type name of the source entity type is preserved.
- **Casting** $\Phi_C(E)$: The casting operator composes multiple attributes into a composite attribute or transforms a monovalued attribute into a multivalued

one. It can be used to make two entity types compatible for set operations like union, intersection, or difference. $\Phi_{(bdate,name)\to info,\{salary\}}(Employee)$ compounds the attributes *bdate* and *name* in the composite attribute *info* and transforms the monovalued attribute *salary* into a multivalued attribute with at most a single value of salary for each employee.

Note that our algebra is as powerful as the relational algebra as it supports the minimal set required to make it relationally complete, namely, cartesian product, selection, projection, renaming, union, and set difference operator [3]. Given a set of relations, any relational algebra query can be expressed in our algebra using each relation as a distinct entity type. The following queries should give a good idea about how the presented operators can be used to answer queries (based on schema in Fig. 1):

– Get engineers, who work on all projects the *Research* department controls:

$$Engineer \bowtie_{works\_on} \forall(Project \bowtie_{controls} \sigma_{name='Research'}(Department))$$

Note that the $\forall$ quantifier is applied here to the list of projects that are returned from the join that are controlled by the research department.
– Get the name, number, and locations of all departments and projects:

$$Department \cup \Phi_{\{location\}}(Project)$$

Note the use of the cast operator that converts location from monovalued to a multivalued attribute and makes *Department* and *Project* union compatible.
– Get all department names, the maximum salary among its technicans, and the technican names:

$$\Pi_{name,works\_for.Technican.salary.max(),works\_for.Technican.name}(Department \bowtie_{works\_for} Technican)$$

Note that the Technician.name is a multivalued attribute generated from the join that contains names of all technicians.
– Get all pairs of employees such that the employees work for the same department and on at least one project together:

$$\sigma_{E_1.ssn<E_2.ssn}(((\rho_{Employee\to E_1}(Employee) \bowtie_{works\_for} Department) \bowtie_{works\_on} Project)$$
$$\bowtie_{Department,Project} ((\rho_{Employee\to E_2}(Employee) \bowtie_{works\_for} Department) \bowtie_{works\_on} Project))$$

Note here how we conduct a merge join among two *Employee* entity-sets on the nested multivalued attributes *Department* and *Project*.
– Get all employee-department pairs such that the employee works on at least one project for at least 10 hours, that the department controls:

$$\chi_{x\in works\_on.hours(x<10)}(Employee \bowtie_{works\_on} Project) \bowtie^{\succ}_{Project} (Department \bowtie_{controls} Project)$$

Note how we used the reduction operator by eliminating projects on which an employee works for less than 10 hours.

5 Ongoing Work and Conclusions

In this paper we presented the functional primitives for an ER model based DBMS that allow specification of much more complex queries compared to standard relational algebra. Our versions of joins go beyond the traditional relational join and are extended to include multivalued simple and multivalued composite attributes. The result of our join creates an entity set where instead of repeating matched values from the operands (or concatenation of tuples as in relational join), we keep the left argument entity set and create a multivalued attribute for the joined values. This enables a powerful join operation using nested multivalued attributes. We are currently working on a transaction model for the ERDBMS that would guarantee eventual consistency. We feel that the overall idea of a main-memory ER model based DBMS and efficient implementation techniques (e.g. data-centric code generation) of CISC operators is timely and worth pursuing in keeping with the current popularity of NOSQL systems. We plan to develop our ERSQL language [14] and the ERDBMS prototype system further in these directions.

References

1. Chen, P.P.: An algebra for a directional binary entity-relationship model. In: Proceedings of the First International Conference on Data Engineering, pp. 37–40. IEEE Computer Society, Washington, DC (1984)
2. Chen, P.P.S.: The entity-relationship model - toward a unified view of data. ACM Trans. Database Syst. 1(1), 9–36 (1976)
3. Elmasri, R., Navathe, S.: Fundamentals of Database Systems, 7th edn. Addison-Wesley Publishing Company, USA (2015)
4. Elmasri, R., Wiederhold, G.: GORDAS: a formal high-level query language for the entity-relationship model. In: Proceedings of the Second International Conference on the Entity-Relationship Approach to Information Modeling and Analysis, pp. 49–72, ER 1981, North-Holland Publishing Co., Amsterdam, The Netherlands (1983)
5. Gogolla, M., Hohenstein, U.: Towards a semantic view of an extended entity-relationship model. ACM Trans. Database Syst. 16(3), 369–416 (1991)
6. Hohenstein, U., Engels, G.: SQL/EER - syntax and semantics of an Entity-Relationship-based query language. Inf. Syst. 17(3), 209–242 (1992)
7. Lawley, M., Topor, R.W.: A Query Language for EER Schemas. In: Australasian Database Conference, pp. 292–304 (1994)
8. Markowitz, V.M., Raz, Y.: ERROL: an entity-relationship, role oriented, query language. In: ER, pp. 329–345 (1983)
9. Markowitz, V.M., Raz, Y.: An entity-relationship algebra and its semantic description capabilities. J. Syst. Softw. 4(23), 147–162 (1984). Entity-Relationship approach to databases and related software

10. Omodeo, E.G., Doberkat, E.E.: Algebraic semantics of ER-models in the context of the calculus of relations: I: static view. Electron. Notes Theoret. Comput. Sci. **44**(3), 136–152 (2003). relMiS 2001, Relational Methods in Software (a Satellite Event of ETAPS 2001)
11. Parent, C., Spaccapietra, S.: An algebra for a general entity-relationship model. IEEE Trans. Softw. Eng. **11**(7), 634–643 (1985)
12. Parent, C., Rolin, H., Yetongnon, K., Spaccapietra, S.: An ER calculus for the entity-relationship complex model. In: ER, pp. 361–384 (1989)
13. Parent, C., Spaccapietra, S.: A model and an algebra for entity-relation type database. Technol. Sci. Inform. **6**(8), 623–642 (1987)
14. Wilfer, M., Navathe, S.: ERSQL - A User-friendly Query Language for the Extended ER Model (2015). Georgia Tech Technical report, Submitted for publication
15. Wuu, G.: SERQL: an ER query language supporting temporal data retrieval. In: 1991 Tenth Annual International Phoenix Conference on Computers and Communications, Conference Proceedings, pp. 272–279, March 1991

Enhancing Entity-Relationship Schemata for Conceptual Database Structure Models

Bernhard Thalheim and Marina Tropmann-Frick[✉]

Department of Computer Science, Christian-Albrechts-University Kiel,
24098 Kiel, Germany
mtr@is.informatik.uni-kiel.de

Abstract. The paper aims at development of well-founded notions of database structure models that are specified in entity-relationship modelling languages. These notions reflect the functions a model fulfills in utilisation scenarios.

1 Utilisation Scenarios of Conceptual Models

Conceptual models are used as an artifact in many utilisation scenarios. Design science research [4] and ER schema development methodologies (e.g. [3,8,11]) developed so far a good number of such scenarios

Communication and negotiation scenario: The conceptual model is used for exchange of meanings through a common understanding of notations, signs and symbols within an application area. It can also be used in a back-and-forth process in which interested parties with different interests find a way to reconcile or compromise to come up with an agreement. The schema provides negotiable and debatable propositions about the understanding of the part of the reality but does not have well-developed justificatory explanations.

Conceptualisation scenario: The main application area for extended entity-relationship models is the conceptualisation of database applications. Conceptualisation is typically shuffled with discovery of phenomena of interest, analysis of main constructs and focus on relevant aspects within the application area. The specification incorporates concepts injected from the application domain.

Description scenario: In a description scenario, the model provides a specification how the part of the reality that is of interest is perceived and in which way augmentations of current reality are targeted. The model says what the structure of an envisioned database is and what it will be.

Prescription scenario: The conceptual model is used as a blueprint for or prescription of a database application, especially for prescribing the structures and constraints in such applications. The schema proposes what the structure of a database is on the one hand and how and where to construct the realisation on the other hand. ER schemata can be translated to relational, XML or other schemata based on transformation profiles [11] that incorporate properties of the target systems.

© Springer International Publishing Switzerland 2015
P. Johannesson et al. (Eds.): ER 2015, LNCS 9381, pp. 603–611, 2015.
DOI: 10.1007/978-3-319-25264-3_47

These scenarios are typically bundled into *use spectra*. For instance, design science uses three cycles: the relevance cycle based on the design cycle based on description and communication and negotiation scenario, and the rigor cycle based on a knowledge discovery and experience propagation scenario. Database development is mainly based on description, conceptualisation, and construction scenarios. The re-engineering and system maintenance use spectrum is based on combination of documentation scenarios with an explanation and discovery scenario from one side and communication and negotiation scenario from the other side. Models are also used for documentation scenarios, explanation and discovery scenarios for applications or systems, and for knowledge experience scenario. We concentrate here on the four scenarios.

Contribution of the Paper

The first main contribution of this paper is an analysis whether an entity-relationship schema suffices as a model for database structures. We realise that the four scenarios require additional elements for the ER schema in order to become a model. The second main contribution of this paper is a proposal for an enhancement of ER schemata which allows to consider the artifact as a model within the given four scenarios. The paper partially presupposes our research (esp. [15], see also other papers in [16]).

2   The General Notion of a Model

Science and technology widely use models in a variety of utilisation scenarios. Models function as an artifact in some utilization scenario. Their function in these scenarios is a combination of functions such as explanation, optimization-variation, validation-verification-testing, reflection-optimization, exploration, hypothetical investigation, documentation-visualization, and description-prescription as a mediator between a reality and an abstract reality that developers of a system intend to build. The model functions determine the *purposes* of the deployment of the model.

The following notion of the model has been developed [17] after an intensive discussion in workshops with researchers from disciplines such as Archeology, Arts, Biology, Chemistry, Computer Science, Economics, Electrotechnics, Environmental Sciences, Farming and Agriculture, Geosciences, Historical Sciences, Humanities, Languages and Semiotics, Mathematics, Medicine, Ocean Sciences, Pedagogical Science, Philosophy, Physics, Political Sciences, Sociology, and Sport Science.

Definition 1. *A model is a well-formed, adequate, and dependable artifact that represents origins. Its criteria of well-formedness, adequacy, and dependability must be commonly accepted by its community of practice within some context and correspond to the functions that a model fulfills in utilisation scenarios. As an artifact, a model is grounded in its community's sub-discipline and is based on elements chosen from the sub-discipline.*

This notion has been tested against the notions of a model that are typically used in these disciplines. We could state that these notions are covered by our notion. Origins of a model [7] are artifacts the model reflects. Adequacy of models has often been discussed, e.g. [6,9,10]. Dependability is only partially covered in research, e.g. [5].

Models have several *essential properties* that qualify an artifact as a model [15,16]:

– An artifact is *well-formed* if it satisfies a well-formedness criterion.
– A well-formed artifact is *adequate* for a collection of origins if (i) it is analogous to the origins to be represented according to some analogy criterion, (ii) it is more focused (e.g. simpler, truncated, more abstract or reduced) than the origins being modelled, and (iii) it sufficient satisfies its purpose.
– Well-formedness enables an artifact to be *justified*: (i) by an empirical corroboration according to its objectives, supported by some argument calculus, (ii) by rational coherence and conformity explicitly stated through formulas, (iii) by falsifiability that can be given by an abductive or inductive logic, and (iv) by stability and plasticity explicitly given through formulas.
– The artifact is *sufficient* by its *quality* characterisation for internal quality, external quality and quality in use or through quality characteristics [13] such as correctness, generality, usefulness, comprehensibility, parsimony, robustness, novelty etc. Sufficiency is typically combined with some assurance evaluation (tolerance, modality, confidence, and restrictions).
– A well-formed artifact is called *dependable* if it is sufficient and is justified for some of the justification properties and some of the sufficiency characteristics.
– An artifact is called **model** if it is *adequate* and *dependable*. The adequacy and dependability of an artifact is based on a *judgement* made by the community of practice.
– An artifact has a *background* consisting of an undisputable grounding from one side (paradigms, postulates, restrictions, theories, culture, foundations, conventions, authorities) and of a disputable and adjustable basis from other side (assumptions, concepts, practices, language as carrier, thought community and thought style, methodology, pattern, routines, commonsense).
– A model is used in a *context* such as discipline, a time, an infrastructure, and an application.

The Taxonomy of Conceptual Models

The starting point in our investigation was the observation that there is no unique and commonly agreeable notion of the conceptual database structure model as such. The model supports different purposes and has different functions in utilisation scenarios. Therefore, we must have different notion of the conceptual database structure model.

The conceptual model functions within the utilisation scenarios in different roles with different rigidity, modality and confidence. Models are used as *perception* models (reflection of one partys current understanding of world; for understanding the application domain), *situation* models (reflection of a given state of affairs), *conceptual* models (based on formal concepts and conceptions) *experimentation* models (as a guideline and basis for experimentation), *formal* models (based some formalism within a well-based formal language), *mathematical* models (in the language of mathematics), *computational* models (based on some (semi-)algorithm), *physical* models (as physical artifact), *visualisation* models (for representation using some visualisation), *representation* models (for representation of some other notion), *diagrammatic* models (using a specific language, e.g. UML), *exploration* models (for property discovery), and *heuristic* models (based on some Fuzzyness, probability, plausibility, correlation), etc. The large variety of known and used model notions (e.g. see [14,15,17] and the collection [16]) mainly reflects these different kinds. The situation model might use a rigorous structured English (OMG proposal) and represents the nature of the business within the language of the business. In this case it is also called *reality* model.

Due to space limitations we concentrate here on the four utilisation scenarios described in Sect. 1. The other scenarios, such as documentation scenario, explanation and discovery scenario for applications, explanation and discovery scenario for systems and knowledge discovery and experience propagation scenario, are supported by specific conceptual models in a similar form.

3   Conceptual Database Structure Models for Communication and Negotiation

Communication aims at exchange of meanings among interested parties. The model is used as a means for communication. It truly represents some aspects of the real world. It enables clearer communication and negotiation about those aspects of the real world. It has therefore potentially several meanings in dependence on the parties. Communication acts essentially follow rhetoric frames[1], i.e. they are characterised through "who says what, when, where, why, in what way, by what means" (Quis, quid, quando, ubi, cur, quem ad modum, quibus adminiculis). In our case, the model ("what") incorporates the meaning of parties (semantical space; "who") during a discourse ('when') within some application with some purpose ("why") based on some modelling language.

Typically, artifacts used for communication and negotiation follow additional principles: Viewpoints and specific semantics of users are explicitly given. The artifact is completely logically independent from the platform for realisation. The name space is rather flexible. The model is functioning and effective if methods for reasoning, understanding, presentation, exploration, explanation, validation, appraisal and experimenting are attached.

Conceptual model for communication: *The conceptual database structure model comprises the database schema, reflects viewpoints and perspectives of different*

---

[1] It relates back to Hermagoras of Temnos or Cicero more than 2000 years ago.

involved parties $\mathcal{U}$ and their perception models, and implicitly links to (namespaces or) concept fields of parties. Adequacy and dependability are based on the association of the perception models to viewpoints and of the viewpoints with the schema..

A partial communication model does not use a schema and does not associate viewpoints to schema elements.

Therefore, the model can be formally defined as a quintuple

$$(\mathcal{S}, \{(\mathcal{V}_i, \Phi_i) \mid i \in \mathcal{U}\}, \{(\mathcal{P}_i, \Psi_i) \mid i \in \mathcal{U}\}, \mathcal{A}, \mathcal{D})$$

that relates elements of the conceptual schema $\mathcal{S}$ to the perception model $\mathcal{P}_i$ of the given party i. The perception model is reflected in the schema via viewpoints $\mathcal{V}_i$. It implicitly uses concept fields $\mathfrak{C}_i$ of parties i. The mapping $\Psi_i : \mathcal{P}_i \rightarrow \mathcal{V}_i$ associates the perception model of a given party i to the agreed viewpoint. In the global-as-design approach, the viewpoint $\mathcal{V}_i$ is definable by some constructor Φ_i defined on $\mathcal{S}$. The adequacy $\mathcal{A}$ is directly given by the second and third parts of the model. The justification $\mathcal{J}$ and the dependability $\mathcal{D}$ are extracted from the properties of Φ_i and Ψ_i.

The negotiation scenario can thus be understood as stepwise construction of the mappings, stepwise revision of the schema and the viewpoints, and analysis whether the schema represents the corresponding perception model.

4   Conceptual Database Structure Models for Conceptualisation

Conceptualisation is based on one or more concept or conception spaces of business users. Given a business user community $\mathcal{U}$ with their specific concept fields $\{\mathfrak{C}_i \mid i \in \mathcal{U}\}$. Let us assume that the concept fields can be harmonised or at least partially integrated into a common concept field of users $\mathfrak{C}^{\mathcal{U}}$ similar to construction approaches used for ontologies

Conceptual model for conceptualisation: *The conceptual database structure model comprises the database schema, a collection of views for support of business users, and a mapping for schema elements that associates these elements to the common concept field..*

Therefore, the model can be formally defined as a quintuple

$$(\mathcal{S}, \mathfrak{V}, \mathcal{M}, \mathcal{A}, \mathcal{D})$$

consisting of the conceptual schema $\mathcal{S}$ and a mapping $\mathcal{M} : \mathcal{S} \rightarrow \mathfrak{C}^{\mathcal{U}}$. The adequacy $\mathcal{A}$ is based on the mapping. The justification $\mathcal{J}$ and the dependability $\mathcal{D}$ are derived from the concept fields.

5 Conceptual Database Structure Models for Description and Prescription

An artifact that is used as a conceptual model for database system description can be either understood as a representation, refinement and amplification [13,16] of situation or reality models or as a refinement and extension of the communication model. The main approach to conceptual modelling for system construction follows the first option. The second option would however be more effective but requires a harmonisation of the perception models. The first option may start with reality models that reflect the nature of the business in terms and in the language of the business. It includes also the top management view, a corporate overview, and a sketch of the environment. The reality models are reasonable complete, are described in terms of the business and use general categories that are convergent.

Conceptual model for description: *The conceptual database structure model comprises the database schema, a collection of views for support of business users, a collection of a commonly accepted reality models that reflects perception or situation models with explicit association to views, and the declaration of model adequacy and dependability..*

Therefore, the model can be formally defined as a quintuple

$$(\mathcal{S}, \mathfrak{V}, (\mathfrak{R}, \Psi), \mathcal{A}, \mathcal{D}) .$$

The conceptual model may be enhanced by an association Φ of views to the schema. This enhancement is however optional.

Descriptive models adequately explicate main concepts [12] from the reality models and combine them into views. The descriptive model reflects the origins and abstracts from reality by scoping the model to the ideal state of affairs.

Prescriptive models that are used for system construction are filled with anticipation of the envisioned system. They deliberately diverge from reality in order to simplify salient properties of interest, transforming them into artifacts that are easier to work with.They may follow also additional paradigms and assumption beyond the classical background of conceptual database structure models: Salami slicing of the schema by rigid separation of concern for all types; conformance to methods for simple (homomorphic) transformation; adequateness for direct incorporation;hierarchical architecture within the schema, e.g. for specialisation and generalisation of types; partial separation of syntax and semantics; tools with well-defined semantics; viewpoint derivation; componentisation and modularisation; integrity constraint formulation support; conformance to methods for integration; variations for the same schema for more flexible realisation etc.

Directives (or pragmas) [1] prescriptively specify properties for the realisation. Transformation parameters [11] for database realisation are, for instance,

Table 1. Conceptual database structure models that extend the conceptual database schema in dependence on utilisation scenarios

Scenario	Model origin	Add-ons to the conceptual database schema
Communication and negotiation	Perception (and situation) models	Views representing the viewpoint variety and associated with the perception models
Conceptualisation	Perception and reality models	Associations to concepts and conceptions, semantics and meanings, namespaces
Description	Reality model	View collection, associations to origins
Prescription	Reality (and situation) models	View collection, realisation template

treatment of hierarchies, controlled redundancy, NULL marker support, constraint treatment, naming conventions, abbreviation rules, set or pointer semantics, handling of weak types, and translation options for complex attributes. Based on [2] we give an explicit specification of directives for the realisation. The prescription model also consists of a general description of a realisation style and tactics, of configuration parameters (coding, services, policies, handlers), of generic operations, of hints for realisation of the database, of performance expectations, of constraint enforcement policies, and of support features for the system realisation. These parameters are combined to the realisation template T. The realisation template can be extended by quantity matrix for database classes $\mathfrak{Q}$ and other performance constraints $\mathfrak{C}$ and by business tasks and their reflections through business data units $\mathfrak{B}$. Directives can be bound to one kind of platform and represent in this case a technological twist, e.g. by stating how data is layered out. They are typically however bound to several platforms in order to avoid evolution-proneness of models.

Conceptual model for prescription: *The conceptual database structure model comprises the database schema, a collection of views for both support of business users and system operating, a realisation template, and the declaration of model adequacy and dependability..*

Therefore, the model can be formally defined as a quintuple

$$(\mathcal{S}, \mathfrak{B}, \mathcal{T}, \mathcal{A}, \mathcal{D}).$$

6   Conclusion

This paper shows that the ER schema is a central unit in a conceptual database structure model. The conceptual database structure model contains also other elements in dependence on its function in utilisation scenarios. As long as we use a global-as-design approach, the ER schema is essential and the kernel of such database structure models.

We may combine the conceptual models to description/prescription models

$$(\mathcal{S}, \mathfrak{V}, (\mathfrak{R}, \Psi), \mathcal{T}, \mathcal{A}, \mathcal{D}) \ .$$

and to description/prescription models with conceptualisation

$$(\mathcal{S}, \mathfrak{V}, (\mathfrak{R}, \Psi), \mathcal{M}, \mathcal{T}, \mathcal{A}, \mathcal{D}) \ .$$

The combination with communication/negotiation is more problematic since the corresponding models are based on divergent perception models that might represent the very personal viewpoint of business users in different context and work organisation (Table 1).

The notion of the model for conceptual database structure models can be summarised in dependence on their utilisation scenario.

References

1. ISO/IEC JTC 1/SC 22: Information technology - Programming languages - C, ISO/IEC 9899:2011
2. AlBdaiwi, B., Noack, R., Thalheim, B.: Pattern-based conceptual data modelling. In: 24th International Conference on Information Modelling and Knowledge Bases (EJC 2014). Information Modelling and Knowledge Bases, vol. XXVI, Kiel, Germany, 3–6 June 2014, pp. 1–20 (2014)
3. Batini, C., Ceri, S., Navathe, S.B.: Conceptual Database Design: An Entity-Relationship Approach. Benjamin-Cummings Publishing Co. Inc., Redwood City (1992)
4. Gregor, S.: The nature of theory in information systems. MIS Q. **30**(3), 611–642 (2006)
5. Halloun, I.A.: Modeling Theory in Science Education. Springer, Berlin (2006)
6. Kaschek, R.: Konzeptionelle Modellierung, habilitation, University Klagenfurt (2003)
7. Mahr, B.: Information science and the logic of models. Softw. Syst. Model. **8**(3), 365–383 (2009)
8. Mannila, H.A., Räihä, K.J.: The Design of Relational Databases. Addison-Wesley, Wokingham (1992)
9. Stachowiak, H.: Modell. In: Seiffert, H., Radnitzky, G. (eds.) Handlexikon zur Wissenschaftstheorie, pp. 219–222. Deutscher Taschenbuch Verlag GmbH & Co. KG, München (1992)
10. Steinmüller, W.: Informationstechnologie und Gesellschaft: Einführung in die angewandte Informatik. Wissenschaftliche Buchgesellschaft, Darmstadt (1993)
11. Thalheim, B.: Entity-Relationship Modeling: Foundations of Database Technology. Springer, Heidelberg (2000)
12. Thalheim, B.: The conceptual framework to user-oriented content management. In: Duz, M., Jaakkola, H., Kiyoki, Y., Kangassalo, H. (eds.) Information Modelling and Knowledge Bases. Frontiers in Artificial Intelligence and Applications, vol. XVIII, pp. 30–49. IOS Press, Amsterdam (2007)

13. Thalheim, B.: Towards a theory of conceptual modelling. J. Univ. Comput. Sci. **16**(20), 3102–3137 (2010)
14. Thalheim, B.: Handbook of Conceptual Modeling: Theory, Practice, and Research Challenges. Springer, Heidelberg (2011)
15. Thalheim, B.: The conceptual model ≡ an adequate and faithful artifact enhanced by concepts. In: 23rd European-Japanese Conference on Information Modelling and Knowledge Bases (EJC 2013). Information Modelling and Knowledge Bases, vol. XXV, Nara, Japan, 3–7 June 2013, pp. 241–254 (2013)
16. Thalheim, B.: Models, to model, and modelling - Towards a theory of conceptual models and modelling - Towards a notion of the model, collection of recent papers (2014). http://www.is.informatik.uni-kiel.de/~thalheim/indexkollektionen.htm
17. Thalheim, B., Nissen, I. (eds.): Wissenschaft und Kunst der Modellierung. De Gruyter, Berlin (2015)

Author Index

Printed in the United States
By Bookmasters